LONDON MATHEMATICAL SOCIETY LECTURE NOTE SERIES

The titles below are available from booksellers, or, in case of difficulty, from Cambridge University Press.

46 *p*-adic Analysis: a short course on recent work, N. KOBLITZ
59 Applicable differential geometry, M. CRAMPIN & F.A.E. PIRANI
66 Several complex variables and complex manifolds II, M.J. FIELD
86 Topological topics, I.M. JAMES (ed)
87 Surveys in set theory, A.R.D. MATHIAS (ed)
88 FPF ring theory, C. FAITH & S. PAGE
89 An F-space sampler, N.J. KALTON, N.T. PECK & J.W. ROBERTS
90 Polytopes and symmetry, S.A. ROBERTSON
92 Representations of rings over skew fields, A.H. SCHOFIELD
93 Aspects of topology, I.M. JAMES & E.H. KRONHEIMER (eds)
96 Diophantine equations over function fields, R.C. MASON
97 Varieties of constructive mathematics, D.S. BRIDGES & F. RICHMAN
98 Localization in Noetherian rings, A.V. JATEGAONKAR
99 Methods of differential geometry in algebraic topology, M. KAROUBI & C. LERUSTE
100 Stopping time techniques for analysts and probabilists, L. EGGHE
104 Elliptic structures on 3-manifolds, C.B. THOMAS
105 A local spectral theory for closed operators, I. ERDELYI & WANG SHENGWANG
107 Compactification of Siegel moduli schemes, C.-L. CHAI
109 Diophantine analysis, J. LOXTON & A. VAN DER POORTEN (eds)
113 Lectures on the asymptotic theory of ideals, D. REES
114 Lectures on Bochner-Riesz means, K.M. DAVIS & Y.-C. CHANG
116 Representations of algebras, P.J. WEBB (ed)
119 Triangulated categories in the representation theory of finite-dimensional algebras, D. HAPPEL
121 Proceedings of *Groups - St Andrews 1985*, E. ROBERTSON & C. CAMPBELL (eds)
128 Descriptive set theory and the structure of sets of uniqueness, A.S. KECHRIS & A. LOUVEAU
130 Model theory and modules, M. PREST
131 Algebraic, extremal & metric combinatorics, M.-M. DEZA, P. FRANKL & I.G. ROSENBERG (eds)
132 Whitehead groups of finite groups, ROBERT OLIVER
133 Linear algebraic monoids, MOHAN S. PUTCHA
134 Number theory and dynamical systems, M. DODSON & J. VICKERS (eds)
137 Analysis at Urbana, I, E. BERKSON, T. PECK, & J. UHL (eds)
138 Analysis at Urbana, II, E. BERKSON, T. PECK, & J. UHL (eds)
139 Advances in homotopy theory, S. SALAMON, B. STEER & W. SUTHERLAND (eds)
140 Geometric aspects of Banach spaces, E.M. PEINADOR & A. RODES (eds)
141 Surveys in combinatorics 1989, J. SIEMONS (ed)
144 Introduction to uniform spaces, I.M. JAMES
146 Cohen-Macaulay modules over Cohen-Macaulay rings, Y. YOSHINO
148 Helices and vector bundles, A.N. RUDAKOV *et al*
149 Solitons, nonlinear evolution equations and inverse scattering, M. ABLOWITZ & P. CLARKSON
150 Geometry of low-dimensional manifolds 1, S. DONALDSON & C.B. THOMAS (eds)
151 Geometry of low-dimensional manifolds 2, S. DONALDSON & C.B. THOMAS (eds)
152 Oligomorphic permutation groups, P. CAMERON
153 L-functions and arithmetic, J. COATES & M.J. TAYLOR (eds)
155 Classification theories of polarized varieties, TAKAO FUJITA
156 Twistors in mathematics and physics, T.N. BAILEY & R.J. BASTON (eds)
158 Geometry of Banach spaces, P.F.X. MÜLLER & W. SCHACHERMAYER (eds)
159 Groups St Andrews 1989 volume 1, C.M. CAMPBELL & E.F. ROBERTSON (eds)
160 Groups St Andrews 1989 volume 2, C.M. CAMPBELL & E.F. ROBERTSON (eds)
161 Lectures on block theory, BURKHARD KÜLSHAMMER
162 Harmonic analysis and representation theory, A. FIGA-TALAMANCA & C. NEBBIA
163 Topics in varieties of group representations, S.M. VOVSI
164 Quasi-symmetric designs, M.S. SHRIKANDE & S.S. SANE
166 Surveys in combinatorics, 1991, A.D. KEEDWELL (ed)
168 Representations of algebras, H. TACHIKAWA & S. BRENNER (eds)
169 Boolean function complexity, M.S. PATERSON (ed)
170 Manifolds with singularities and the Adams-Novikov spectral sequence, B. BOTVINNIK
171 Squares, A.R. RAJWADE
172 Algebraic varieties, GEORGE R. KEMPF
173 Discrete groups and geometry, W.J. HARVEY & C. MACLACHLAN (eds)
174 Lectures on mechanics, J.E. MARSDEN
175 Adams memorial symposium on algebraic topology 1, N. RAY & G. WALKER (eds)
176 Adams memorial symposium on algebraic topology 2, N. RAY & G. WALKER (eds)
177 Applications of categories in computer science, M. FOURMAN, P. JOHNSTONE & A. PITTS (eds)
178 Lower K- and L-theory, A. RANICKI
179 Complex projective geometry, G. ELLINGSRUD *et al*
180 Lectures on ergodic theory and Pesin theory on compact manifolds, M. POLLICOTT
181 Geometric group theory I, G.A. NIBLO & M.A. ROLLER (eds)
182 Geometric group theory II, G.A. NIBLO & M.A. ROLLER (eds)

183 Shintani zeta functions, A. YUKIE
184 Arithmetical functions, W. SCHWARZ & J. SPILKER
185 Representations of solvable groups, O. MANZ & T.R. WOLF
186 Complexity: knots, colourings and counting, D.J.A. WELSH
187 Surveys in combinatorics, 1993, K. WALKER (ed)
188 Local analysis for the odd order theorem, H. BENDER & G. GLAUBERMAN
189 Locally presentable and accessible categories, J. ADAMEK & J. ROSICKY
190 Polynomial invariants of finite groups, D.J. BENSON
191 Finite geometry and combinatorics, F. DE CLERCK *et al*
192 Symplectic geometry, D. SALAMON (ed)
194 Independent random variables and rearrangement invariant spaces, M. BRAVERMAN
195 Arithmetic of blowup algebras, WOLMER VASCONCELOS
196 Microlocal analysis for differential operators, A. GRIGIS & J. SJÖSTRAND
197 Two-dimensional homotopy and combinatorial group theory, C. HOG-ANGELONI *et al*
198 The algebraic characterization of geometric 4-manifolds, J.A. HILLMAN
199 Invariant potential theory in the unit ball of $\mathbf{C}^n$, MANFRED STOLL
200 The Grothendieck theory of dessins d'enfant, L. SCHNEPS (ed)
201 Singularities, JEAN-PAUL BRASSELET (ed)
202 The technique of pseudodifferential operators, H.O. CORDES
203 Hochschild cohomology of von Neumann algebras, A. SINCLAIR & R. SMITH
204 Combinatorial and geometric group theory, A.J. DUNCAN, N.D. GILBERT & J. HOWIE (eds)
205 Ergodic theory and its connections with harmonic analysis, K. PETERSEN & I. SALAMA (eds)
207 Groups of Lie type and their geometries, W.M. KANTOR & L. DI MARTINO (eds)
208 Vector bundles in algebraic geometry, N.J. HITCHIN, P. NEWSTEAD & W.M. OXBURY (eds)
209 Arithmetic of diagonal hypersurfaces over finite fields, F.Q. GOUVÊA & N. YUI
210 Hilbert C*-modules, E.C. LANCE
211 Groups 93 Galway / St Andrews I, C.M. CAMPBELL *et al* (eds)
212 Groups 93 Galway / St Andrews II, C.M. CAMPBELL *et al* (eds)
214 Generalised Euler-Jacobi inversion formula and asymptotics beyond all orders, V. KOWALENKO *et al*
215 Number theory 1992–93, S. DAVID (ed)
216 Stochastic partial differential equations, A. ETHERIDGE (ed)
217 Quadratic forms with applications to algebraic geometry and topology, A. PFISTER
218 Surveys in combinatorics, 1995, PETER ROWLINSON (ed)
220 Algebraic set theory, A. JOYAL & I. MOERDIJK
221 Harmonic approximation, S.J. GARDINER
222 Advances in linear logic, J.-Y. GIRARD, Y. LAFONT & L. REGNIER (eds)
223 Analytic semigroups and semilinear initial boundary value problems, KAZUAKI TAIRA
224 Computability, enumerability, unsolvability, S.B. COOPER, T.A. SLAMAN & S.S. WAINER (eds)
225 A mathematical introduction to string theory, S. ALBEVERIO, J. JOST, S. PAYCHA, S. SCARLATTI
226 Novikov conjectures, index theorems and rigidity I, S. FERRY, A. RANICKI & J. ROSENBERG (eds)
227 Novikov conjectures, index theorems and rigidity II, S. FERRY, A. RANICKI & J. ROSENBERG (eds)
228 Ergodic theory of $\mathbf{Z}^d$ actions, M. POLLICOTT & K. SCHMIDT (eds)
229 Ergodicity for infinite dimensional systems, G. DA PRATO & J. ZABCZYK
230 Prolegomena to a middlebrow arithmetic of curves of genus 2, J.W.S. CASSELS & E.V. FLYNN
231 Semigroup theory and its applications, K.H. HOFMANN & M.W. MISLOVE (eds)
232 The descriptive set theory of Polish group actions, H. BECKER & A.S. KECHRIS
233 Finite fields and applications, S. COHEN & H. NIEDERREITER (eds)
234 Introduction to subfactors, V. JONES & V.S. SUNDER
235 Number theory 1993–94, S. DAVID (ed)
236 The James forest, H. FETTER & B. GAMBOA DE BUEN
237 Sieve methods, exponential sums, and their applications in number theory, G.R.H. GREAVES *et al*
238 Representation theory and algebraic geometry, A. MARTSINKOVSKY & G. TODOROV (eds)
239 Clifford algebras and spinors, P. LOUNESTO
240 Stable groups, FRANK O. WAGNER
241 Surveys in combinatorics, 1997, R.A. BAILEY (ed)
242 Geometric Galois actions I, L. SCHNEPS & P. LOCHAK (eds)
243 Geometric Galois actions II, L. SCHNEPS & P. LOCHAK (eds)
244 Model theory of groups and automorphism groups, D. EVANS (ed)
245 Geometry, combinatorial designs and related structures, J.W.P. HIRSCHFELD *et al*
246 p-Automorphisms of finite p-groups, E.I. KHUKHRO
247 Analytic number theory, Y. MOTOHASHI (ed)
248 Tame topology and o-minimal structures, LOU VAN DEN DRIES
249 The atlas of finite groups: ten years on, ROBERT CURTIS & ROBERT WILSON (eds)
250 Characters and blocks of finite groups, G. NAVARRO
251 Gröbner bases and applications, B. BUCHBERGER & F. WINKLER (eds)
252 Geometry and cohomology in group theory, P. KROPHOLLER, G. NIBLO, R. STÖHR (eds)
253 The q-Schur algebra, S. DONKIN
254 Galois representations in arithmetic algebraic geometry, A.J. SCHOLL & R.L. TAYLOR (eds)
255 Symmetries and integrability of difference equations, P.A. CLARKSON & F.W. NIJHOFF (eds)
257 An introduction to noncommutative differential geometry and its physical applications, 2ed, J. MADORE
260 Groups St Andrews 1997 in Bath, I, C.M. CAMPBELL *et al* (eds)
261 Groups St Andrews 1997 in Bath, II, C.M. CAMPBELL *et al* (eds)
263 Singularity theory, BILL BRUCE & DAVID MOND (eds)
264 New trends in algebraic geometry, K. HULEK, F. CATANESE, C. PETERS & M. REID (eds)

London Mathematical Society Lecture Note Series. 263

Singularity Theory

Proceedings of the European Singularities Conference, Liverpool, August 1996. Dedicated to C.T.C. Wall on the occasion of his 60th birthday.

Edited by

Bill Bruce
University of Liverpool

David Mond
University of Warwick

CAMBRIDGE
UNIVERSITY PRESS

CAMBRIDGE UNIVERSITY PRESS
Cambridge, New York, Melbourne, Madrid, Cape Town, Singapore,
São Paulo, Delhi, Dubai, Tokyo

Cambridge University Press
The Edinburgh Building, Cambridge CB2 8RU, UK

Published in the United States of America by Cambridge University Press, New York

www.cambridge.org
Information on this title: www.cambridge.org/9780521658881

First published 1999

A catalogue record for this publication is available from the British Library

ISBN 978-0-521-65888-1 Paperback

Transferred to digital printing 2010

Contents

Global Singularity Theory

Singularities of Mappings

Applications of Singularity Theory

Preface

Singularity Theory is a broad subject, involving a substantial number of mathematicians in most European countries. It was natural that a group should put together an application to the EU to set up a European Singularities Network (ESN) in 1993. It was equally natural that Terry Wall, of the University of Liverpool, should be selected to head up the bid and, subsequently, the organising committee.

One of the activities funded by the ESN was an international review meeting. Given that Terry Wall was born in 1936, a number of his friends, colleagues, and ex-students (some in all three categories) decided that, to honour him on his 60th birthday, the meeting should bear his name. It was held from the 18th to the 24th of August 1996, at the University of Liverpool. It is a sign of Terry's continued vigour in the subject that we felt it best to give a Web Page address (see the end of this preface) for a list of his current publications, rather than print that list at the time of going to press.

The meeting was attended by 88 mathematicians, 74 of them from 14 different countries outside the UK, and there were 61 talks. The festivities included a party hosted by Terry and his wife Sandra at their home, a multinational football (soccer) match, and a trip to the Lake District to experience a traditional English downpour.

The papers presented here are a selection of those submitted to the Editors for inclusion in the proceedings of the meeting. Pressure of space has meant that a substantial number of other high quality submissions could not be included. In the introduction we have given a brief review of the subject and have attempted to set the scene for the various contributions in this collection.

We would like to thank the EU for sponsoring the ESN, and consequently much of the expense associated to the meeting. We would also like to thank the London Mathematical Society for its financial help, which made it possible to include 10 mathematicians from the Moscow School of singularity theory in our invitations. We are grateful to Andrew Ranicki for coming and telling us of Terry Wall's exploits in his earlier life as a topologist, and to our tea-ladies Emily and Joan, for serving up the refreshments at the conference with such style. Finally many thanks to Wendy Orr for doing so much of the organisational work in her usual calm, friendly and efficient manner, and to Neil Kirk for his enormous help with a great deal of the technical editing.

Web Address

The first URL below refers to the Web pages for The Department of Mathematical Sciences at The University of Liverpool, England. Follow the link to the Pure Mathematics Research Division (this can be reached directly via the second URL if you prefer) and then follow the link to staff and postgraduates. (We decided not to give a direct link as these have the annoying habit of changing over time. The addresses below should be fairly stable!)

`http://www.liv.ac.uk/Maths/`
`http://www.liv.ac.uk/PureMaths/`

October 1998

Bill Bruce
University of Liverpool

David Mond
University of Warwick

Introduction

We start with a rapid survey, which we hope may be of some use to non-experts.

Singularity theory is a broad subject with vague boundaries. It is concerned with the geometry and topology of spaces (and maps) defined by C^∞, polynomial or analytic equations, which for one reason or another fail to be smooth, (or submersions/immersions in the case of maps). It draws on many (most?) other areas of mathematics, and in turn has contributed to many areas both within and outside mathematics, in particular differential and algebraic geometry, knot theory, differential equations, bifurcation theory, Hamiltonian mechanics, optics, robotics and computer vision.

It can be seen as a crossroads where a number of different subjects and projects meet. In order to classify its current diverse productions, we centre our discussion around the contribution of five major figures: Whitney, Thom, Milnor, Mather and Arnold.

The first of these to work on singularities was Whitney, who was led to study singularities in the process of proving his immersion theorem. An n-manifold M can be immersed in $(2n-1)$-space, even though immersions are not dense in the space of all maps $M \to \mathbb{R}^{2n-1}$: singularities persist under small deformations. To remove them one needs a large deformation, and a good understanding of the singularities themselves. Whitney wanted a short list of the singularities that persist; besides finding the list for maps $M^n \to \mathbb{R}^{2n-1}$ ([18]) he also did so for maps $M^2 \to \mathbb{R}^2$ ([19]). The persistence of certain singularities under deformation leads to the idea of a stable map: one which is essentially unchanged if we deform it a little. Two maps $M \to N$ are *right-left equivalent* if one can be converted into the other by composing with diffeomorphisms of source and target; a smooth map is *stable* if its equivalence class is open, in some reasonable topology, in $C^\infty(M, N)$. The weaker notion of topological stability replaces diffeomorphisms by homeomorphisms in source and target. Whitney characterised the stable mappings $M^n \to \mathbb{R}^{2n-1}$ and $M^2 \to \mathbb{R}^2$, and in doing so introduced many of the central ideas and techniques in the subject.

Thom was led to study singularities while considering the question of whether it is possible to represent homology classes in smooth manifolds by embedded submanifolds. With his transversality theorem ([16]) he gave the subject a push towards a kind of modern Platonism. Given smooth manifolds M and N, the jet bundle $J^k(M,N)$ is the space of Taylor polynomials of degree k of germs of smooth maps from M to N. It has a natural structure as a smooth bundle over $M \times N$, using the coefficients in the Taylor polynomials as fibre coordinates. If $f : M \to N$ is a smooth map, then by taking its k-th degree Taylor polynomial at each point, we obtain a map $j^k f : M \to J^k(M,N)$, the jet extension map. Thom saw the jet bundle as a version of the Platonic world of disembodied ideas, partitioned into attributes (the orbits of the various groups which act naturally on jets) as yet unattached to the objects (functions and mappings) which embody them. A map $f : M \to N$ embodies an attribute W faithfully if its jet extension map $j^k f : M \to J^k(M,N)$ is transverse to W. Thom showed that for all M, N and W, most maps $M \to N$ (actually a residual set of maps) have this property. This theorem put the "generic" into every subsequent theorem about generic behaviour of maps, in particular Mather's theorems that stable maps $M \to N$ are dense in $C^\infty(M,N)$ if $(\dim M, \dim N)$ are in his range of nice dimensions (Mather [8]–[13]), and that topologically stable maps are dense, in all dimensions [14]. It does the same for the applications of singularity theory to differential geometry, a field which is represented in this collection by the papers of Ishikawa and Donelan-Gibson.

Thom also contributed the idea of a versal unfolding. Closely related to the idea of a slice to a group orbit, a versal unfolding of a map or map-germ f is a finite-dimensional family $f_\lambda, \lambda \in \Lambda$ which explores every nearby possibility of deformation, up to some specified notion of equivalence. Versal unfoldings of minimal dimension are unique up to isomorphism, and the structure of its miniversal unfolding provides one of the best ways of understanding a singularity. A germ $f : (M,x) \to (N,y)$ has a versal unfolding with respect to a given equivalence relation if it has "finite codimension"; that is, if its orbit in the jet bundle $J^k(M,N)$, with respect to this equivalence relation, has codimension which is eventually independent of k. The term "versal" is the intersection of "universal" and "transversal", and one of Thom's insights was that the singularities of members of families of functions or mappings are versally unfolded if the corresponding family of jet extension maps is transverse to their orbits (equivalence classes) in jet space.

This insight, and Thom's Platonist leanings, led him to Catastrophe Theory. He identified and described the seven orbits of function singularities which can be met transversely in families of four or fewer parameters: these were his seven elementary catastrophes, which were meant to underly all abrupt changes (bifurcations) in generic four-parameter families of gradient dynamical systems. With the eye of faith some of his followers were able to see the elementary catastrophes in every field of science; without it, their critics objected to the invocation that this entailed of such invisibles as real variables and smooth potential functions in fields like politics and prison riots, and to the triteness of some of its conclusions. Nevertheless many of Thom's ideas in bifurcation theory and gradient dynamical systems have provided the basis for later development, and the controversy surrounding

Catastrophe Theory should not mask the importance of his contribution to the subject.

Besides proving Thom's conjectures about the genericity of stability, Mather made possible the systematic classification of germs of functions and maps. There are several standard equivalence relations, induced by the action of the various groups of diffeomorphisms on the space of map-germs. Mather showed that if a germ $f : (M, x) \to (N, y)$ has finite codimension with respect to a given equivalence relation, then it is *finitely determined* with respect to that relation: there is an integer k such that any other germ having the same k-jet as f, is equivalent to it. Subsequently other authors, beginning with Jean Martinet, Terry Gaffney and Andrew du Plessis, improved on Mather's estimates for the determinacy degree (the smallest such k); see the excellent survey article of Wall [17]. Their work led to a number of papers providing lists of polynomial normal forms for equivalence classes of low codimension. Terry Wall is also a prodigious list-maker — see for example [105],[110],[129] in his list of publications. Indeed, the presence of a large taxonomy is one of the features of contemporary singularity theory, and significant theoretical advances have often come from the desire to understand features empirically observed in the lists. The paper of Houston and Kirk in this collection uses recently developed techniques from classification theory to obtain a new list, namely a list of right-left equivalence classes of singularities of maps from 3-space to 4-space.

One of the most influential mathematicians working in singularity theory is V.I. Arnold, also a prodigious list-maker. In the early seventies, in connection with a project to extend the stationary phase method for the estimation of oscillatory integrals in quantum mechanics, Arnold produced extensive lists of singularities of functions ([1]–[4]) making spectacular advances in the techniques of classification, and in understanding the structure of the objects and the lists he found. He also introduced his notions of Lagrangian and Legendrian singularities, which systematise many of the mathematical ideas behind Catastrophe Theory. The reader is recommended Arnold's little book [5], both for a lively, personal, and probably overly harsh view of Catastrophe Theory and, more interestingly, for its exposition of the important contributions made by his school to a number of interesting geometrical problems, using techniques from singularity theory.

One of Arnold's contributions was the crucial concept of *simple singularity* — one whose classification does not involve continuous invariants — which has proved extremely fruitful. One of the oddities of a subject with many lists is that the same lists may occur with different headings, and none more so than the short list of simple singularities determined by Arnold. The singularities in Arnold's list had already appeared as Kleinian singularities, and as the "singularities which do not affect adjunction" in a paper of Du Val in 1934. In [6], Durfee gives 15 different characterisations. Several of them, due to Brieskorn, come from the theory of algebraic groups. Arnold also noticed a mysterious duality between some of his lists, which was subsequently interpreted by a number of mathematicians and extended by Terry Wall and Wolfgang Ebeling. Ebeling's paper in this collection surveys this and further extensions of Arnold's strange duality, and relates it to mirror symmetry.

Singularity theory typically focuses on local behaviour, on germs of spaces and maps. Isolated singular points of algebraic and analytic varieties are studied by considering only what goes on inside a conical neighbourhood, small enough to exclude the global topology and geometry of the variety. One then looks not only at the singular space, or map, but also at what goes on within this neighbourhood as the singularity is perturbed. Building on earlier work of Burau, Zariski, Mumford and Brieskorn, Milnor developed this approach in his 1968 book [15] (see also [7]), showing that if the zero set of a polynomial $f : \mathbb{C}^n \to \mathbb{C}$ has an isolated singular point, then the intersection of a nearby non-singular level set of f with a small ball B_ε around the singular point is homotopically equivalent to the wedge of a finite number of n-dimensional spheres. The number of these spheres reflects the complexity of the singularity; it is known as the Milnor number, μ, and the intersection itself is the Milnor fibre. In fact the restriction of f to B_ε defines a fibration, the Milnor fibration, over a small punctured disc around the critical value. A loop around zero in this disc lifts to a diffeomorphism of the Milnor fibre, inducing an automorphism of its homology. The determination of the monodromy automorphism of the singularities in Arnold's lists was one of the great achievements of his school. The study of the topology and algebraic geometry of isolated singular points of complex hypersurfaces remains the focus of a great deal of activity. It has developed a formidable technical armoury, crowned by Deligne's mixed Hodge theory, which places canonical filtrations on the cohomology of the Milnor fibre. The application of mixed Hodge theory to singularity theory was developed independently in the mid seventies by Joseph Steenbrink and Alexander Varchenko. Steenbrink's paper in these proceedings is concerned with his extension of this work to the construction of a mixed Hodge structure on the cohomology of the Milnor fibre of an isolated complete intersection singularity.

The idea of looking at what goes on in the neighbourhood of a singular point generalises naturally to other contexts in singularity theory: one should look for the nearby stable object. For an unstable map-germ f, the analogue of the Milnor fibre is therefore a *stable perturbation*; for a non-generic hyperplane arrangement, one sees what happens when it is moved into general position. Both cases are treated in papers in this volume: Marar-Montaldi-Ruas and Fukui compute the number of stable singularities of given type appearing in stable perturbations of certain unstable map-germs, using commutative algebra techniques centred on the properties of Cohen-Macaulay rings, and Damon proves a conjecture of Varchenko on multi-valued functions on the complements of affine hyperplane arrangements, by viewing the arrangements as "singular Milnor fibres" (nearby objects, stable in an appropriate class of spaces) of certain singularities.

Singularity theory is also concerned with the global behaviour of singular spaces. Singular algebraic and analytic spaces can be *stratified*, that is, given a locally finite partition into manifolds. One of Whitney's most fruitful contributions was to give geometric conditions (which now bear his name) for the local triviality (product-like structure) of a stratification. The study of stratified spaces led to the development, by Goresky and MacPherson, of Intersection Cohomology, in which a version of Poincaré duality for singular spaces is recovered, and to the introduction of the category of perverse sheaves. Whitney stratifications were also the key to Mather's proof of the topological stability theorem, and to work of many authors on conditions for topological triviality of families of functions and maps. Recent developments are surveyed here in the paper of Terry Gaffney and David Massey. Two papers on stratifications, by Andrew du Plessis, provide the first correct proofs of key results in the area of real stratification theory, and the paper of Brasselet and Legrand looks at differential forms on stratified spaces and describes a version of Connes's celebrated theorem in which the de Rham cohomology of a smooth manifold is computed from the cyclic homology of its algebra of smooth functions.

Singular sets arise naturally as the images and discriminants of smooth mappings. The paper of Kevin Houston presented here describes a spectral sequence for computing the homology of such sets from the alternating homology of the multiple point sets of the map.

One of the areas of greatest interest over the last few years has concerned the jacobian conjecture: that a polynomial map $\mathbb{C}^n \to \mathbb{C}^n$ with nowhere-vanishing (and therefore constant) jacobian determinant, is an isomorphism. Attempts by Lê and Weber to prove this conjecture by looking at the behaviour of polynomials at infinity have led to a number of papers on this topic, represented here by the contribution of Mihai Tibar.

We hope we have convinced you that singularity theory is indeed a broad subject. This brief introduction can only hope to cover a part of the large area of mathematics it encompasses. Indeed, the topics represented by the articles in this volume, some of which are mentioned above, give a good indication of the diversity of the subject. The following section provides short summaries of each paper, which we hope the reader will find useful.

References

[1] V.I. Arnold, Remarks on the stationary phase method and Coxeter numbers, *Russian Math. Surveys*, 23:5 (1968), 19-48.

[2] V.I. Arnold, Normal forms of functions in neighbourhoods of degenerate critical points, *Russian Math. Surveys*, 29:2 (1974), 19-48.

[3] V.I. Arnold, Critical points of smooth functions and their normal forms, *Russian Math. Surveys*, 30:5 (1975), 1-75.

[4] V.I. Arnold, Critical points of functions on a manifold with boundary, the simple Lie groups B_k, C_k and F_4 and singularities of evolutes, *Russian Math. Surveys*, 33:5 (1978), 99-116.

[5] V.I. Arnold, *Catastrophe theory*, Springer-Verlag, New York, Heidelberg, and Berlin, 1986.

[6] A. Durfee, Fifteen characterisations of rational double points and simple critical points, *Enseign. Math.*, 25 (1979), 131-163.

[7] E.J.N. Looijenga, *Isolated singular points of complete intersections*, Cambridge Univ. Press, London and New York, 1984.

[8] J.N. Mather, Stability of C^∞-mappings. I: The division theorem, *Annals of Math.*, 87, No. 1 (1968), 89-104.

[9] J.N. Mather, Stability of C^∞-mappings. II: Infinitesimal stability implies stability, *Annals of Math.*, 89, No. 2 (1969), 254-291.

[10] J.N. Mather, Stability of C^∞-mappings. III: Finitely determined map germs, *Publ. Math. I.H.E.S.*, 35 (1968), 127-156.

[11] J.N. Mather, Stability of C^∞-mappings. IV: Classification of stable germs by their $\mathbb{R}$-algebras, *Publ. Math. I.H.E.S.*, 37 (1969), 223-248.

[12] J.N. Mather, Stability of C^∞-mappings. V: Transversality, *Advances in Mathematics*, 4, No. 3 (1970), 301-336.

[13] J.N. Mather, Stability of C^∞-mappings. VI: The nice dimensions, *Proceedings of the Liverpool Singularities Symposium I*, Lecture Notes in Mathematics 192, Springer-Verlag, New York, 1971.

[14] J.N. Mather, *Notes on topological stability*, Preprint Harvard University, 1970.

[15] J.W. Milnor, *Singular points of complex hypersurfaces*, Ann. of Math. Stud., vol 61, Princeton Univ. Press, Princeton, NJ, 1968.

[16] R. Thom, Une lemme sur les applications differentiables, *Bol. Soc. Mat. Mexicana*, (1956), 59-71.

[17] C.T.C. Wall, Finite determinacy of smooth map-germs, *Bull. London Math. Soc*, 13 (1981), 481-539.

[18] H. Whitney, The singularities of a smooth n-manifold in $(2n-1)$-space, *Ann. of Math.*, (2) 45 (1944), 247-293.

[19] H. Whitney, On singularities of mappings of euclidean spaces, I. Mappings of the plane into the plane, *Ann. of Math.*, (2) 62 (1955), 374-410.

Summaries of the Papers

Complex Singularities

Singularities Arising from Lattice Polytopes
K. ALTMANN

Toric geometry begins with a procedure for assigning to each integral polyhedral cone σ in $\mathbb{R}^n$ an algebraic variety V_σ, and with each inclusion of one cone in another a morphism of the corresponding varieties. The varieties obtained in this way are called *toric varieties*, and include affine space, projective space, and many other important algebraic varieties. Since its introduction twenty years ago, toric geometry has become an important tool for algebraic geometers; the combinatorics of lattice polytopes encodes important information about the algebraic geometry of the toric varieties they give rise to. Klaus Altmann's paper begins with a brief introduction to toric geometry and goes on to survey some of the links between the geometry of polytopes and the singularities of toric varieties, including their deformation theory.

Critical Points of Affine Multiforms on the Complements of Arrangments
J.N. DAMON

A hypersurface $(D, 0)$ in a complex manifold $(X, 0)$ is a *free divisor* if the sheaf $\mathrm{Der}(\log D)$ of vector fields on X tangent to D is free as an $\mathcal{O}_X$-module. Free divisors occur throughout singularity theory, as discriminants of stable maps and of versal deformations of singularities. If $\phi : (Y, y) \to (X, 0)$ is transverse to the free divisor D outside y then $E = \phi^{-1}(D)$ is an *almost free divisor*, and if ϕ_t is a deformation of ϕ which is everywhere transverse to D, then $E_t = \phi_t^{-1}(D)$ is a *singular Milnor fibre* of E. The freeness of D allows one to calculate the rank of the vanishing homology of E_t as the length of a certain module defined in terms of ϕ and $\mathrm{Der}(\log D)$, and Damon has made spectacular use of this fact in a number of papers. Here he further develops these ideas to prove a strengthened version of a conjecture of Varchenko on critical points of multivalued functions defined on the complement of a hyperplane arrangement. Varchenko's conjecture originated in his work on the KZ equation described in this volume in the paper of Looijenga.

Strange Duality, Mirror Symmetry and the Leech Lattice
W. EBELING

This paper surveys a series of curious dualities, originally concerning Arnold's list of 14 exceptional unimodal (or "triangle") singularities, and later extended by the author and Terry Wall to cover a wider class. The author describes interpretations of this strange duality, due to Pinkham, and, independently, to Dolgachev and Nikulin, and shows that these too can be extended to cover the Ebeling-Wall version. He also describes the connection, observed by M. Kobayashi, between Arnold's strange duality and mirror symmetry, and the relation found by K. Saito with the Leech lattice, famous as a hunting ground for sporadic simple groups.

Geometry of Equisingular Families of Curves
G.-M. Greuel and E. Shustin
The simplest, and for many the most familiar singularities, are those presented by plane curves. Their local topology was first investigated in the late 1920's, and they have provided an excellent proving ground for theories and theorems ever since. This paper addresses three central problems concerning plane curve singularities. Suppose we are given a natural number d and a finite list of types of plane curve singularities. We can then ask:
(1) Is there an (irreducible) curve of degree d in the complex projective plane $\mathbb{CP}^2$ whose singularities are precisely those in the given list?
(2) Is the family of such curves smooth, and of the expected dimension?
(3) Is this set connected?

These questions have a long history, appearing in the works of Plücker, Severi, Segre and Zariski. They are part of a more general problem: that of moving from local information to global results. The authors have made spectacular recent progress, which is surveyed in this paper. They extend their discussion to curves on more general surfaces. Combined with recent ideas of Viro their results also have application in real algebraic geometry.

Arrangements, KZ Systems and Lie Algebra Homology
E.J.N. Looijenga
The Knizhnik-Zamalodchikov differential equation first appeared in theoretical physics in connection with models for quantum field theory. A rather difficult paper of Varchenko and Schechtman established connections between this equation and the cohomology of local systems on the complements of hyperplane arrangements: under certain circumstances a complete set of solutions to the KZ equation is provided by generalised hypergeometric integrals associated to the hyperplane arangement. Looijenga's paper gives a largely self-contained account of the work of Varchenko and Schechtman and other related work, in the process providing considerable clarification of some of the main ideas. In particular it contains a clear and concise sheaf-theoretic treatment of the cohomology of the complement of a hyperplane arrangement.

The Signature of $f(x,y)+z^N$

A. NEMETHI

If $f:(\mathbb{C}^m,0)\to(\mathbb{C},0)$ and $g:(\mathbb{C}^n,0)\to(\mathbb{C},0)$ have isolated singularity then so does the sum $f+g:(\mathbb{C}^{m+n},0)\to(\mathbb{C},0)$. Classical theorems of Thom and Sebastiani describe the Milnor fibre F_{f+g} of $f+g$ and its monodromy automorphism in terms of the Milnor fibres and monodromy automorphisms of f and g. In case $m=2$ and $n=1$, F_{f+g} is a 4-manifold, so that the intersection pairing on its middle dimensional homology is symmetric, and thus a quadratic form. One of the most important invariants of a simply connected 4-manifold is the signature of this quadratic form; it is often referred to simply as the signature of the manifold. In this paper Andras Nemethi collects together results, many of them his own, which compute the signature of F_{f+g} in terms of invariants of f and g.

The Spectra of $\mathcal{K}$-Unimodal Isolated Singularities of Complete Intersections

J.M. STEENBRINK

Steenbrink and W. Ebeling recently extended Steenbrink's earlier construction of a mixed Hodge structure (MHS) on the Milnor fibre of a hypersurface singularity: an MHS can now be defined on the Milnor fibre of any isolated complete intersection singularity (ICIS) (X,x). The construction involves embedding (X,x) as a hypersurface in an ICIS (X',x) of minimal Milnor number. In fact, as in the case of hypersurface singularities, the MHS itself is very hard to calculate explicitly, and the closest one can get to it is the Hodge numbers (the dimension of the graded pieces of the associated bigraded structure) and the *spectrum* of the singularity, which is defined in terms of the monodromy action on the MHS. The spectrum is a collection of rational numbers, symmetric about its midpoint. In a certain precise sense it is semi-continuous under deformation, and indeed constant under μ-constant deformation. This semi-continuity is, together with its calculability, the key to its importance: it can be used to show that certain adjacencies of singularities do not occur.

In this paper Steenbrink calculates the spectrum for the unimodal ICIS's of embedding codimension at least 2, the calculation for hypersurface singularities having been made by Goryunov some years ago. The modality of a singularity is the number of continuous parameters involved in its classification; thus, "simple" = 0-modal. The unimodal hypersurface singularities were classified by Arnold in 1973, and the remainder of the list of unimodal ICIS's was determined in 1983 by Terry Wall (paper 105 in the list of his publications).

Dynkin Graphs, Gabriélov Graphs and Triangle Singularities

T. URABE

It is of interest to describe not only the topology of the Milnor fibre of a singularity (X,x), but also the constellations of singularities which can appear on singular fibres in a deformation of (X,x). In this paper the author gives a simple algorithm for determining which constellations of simple singularities can appear on the fibres of a deformation of a triangle singularity. The 14 triangle singularities are a class of unimodal hypersurface singularity; under the name "exceptional unimodal singularities" they are the subject of the strange duality reported on by Ebeling in his paper in this volume. The name "triangle singularities" arose because they can be realised as quotients of $\mathbb{C}^2$ by certain finite groups of automorphism, constructed from triangles in the hyperbolic plane. The algorithm is in terms of the combinatorics of Dynkin diagrams (weighted graphs which encode information about the intersection form on the Milnor fibre).

Stratifications and Equisingularity Theory

Differential Forms on Singular Varieties and Cyclic Homology

J.-P. BRASSELET AND Y. LEGRAND

Brasselet has worked for a number of years on the global topology of stratified sets. In collaboration with Goresky and MacPherson, the creators of intersection homology, he has proved a version of the de Rham theorem for simplicial spaces. Another special feature of manifolds is provided by the de Rham theorem which (roughly) asserts that the cohomology of the complex of differential forms and exterior derivatives yields (over the reals) a ring isomorphic to the singular cohomology ring. The introduction of intersection homology theory led naturally to a search for a result analagous to de Rham's for a Whitney stratified set. Clearly what is required is some extension of the notion of smooth differential forms to stratified spaces.

This paper is inspired by the celebrated theorem of Alain Connes, which recovers the de Rham cohomology of a smooth manifold by means of the cyclic homology of the Fréchet algebra of smooth functions on the manifold. Brasselet and Legrand prove a version of this result for Whitney stratified spaces, in the process introducing a new definition of differential form on a stratified space. The key is the definition of a Fréchet algebra of functions on the regular part of the stratification. In this paper the authors associate to these objects a complex of differential forms and a Hochschild complex of forms on the regular part, with poles along the singular part. They then relate the de Rham cohomology of the first part, and the cyclic cohomology of the second, with the intersection homology of the stratified set.

The next two papers are motivated by questions emerging from the proof of the topological stability theorem.

Continuous Controlled Vector fields

A.A. DU PLESSIS

Frequently smooth equivalence of map-germs is too fine; it preserves too much detail, distinguishing phenomena one wished it did not. One then has to replace diffeomorphism by homeomorphism. Whereas diffeomorphisms are produced by integrating smooth vector fields, to produce the homeomorphisms needed to prove topological stability, for example, one has to relax the smoothness condition on the vector fields while retaining enough control to ensure that one can integrate and obtain continuous flows. This is done by stratifying the space and constructing the vector fields on the individual strata. To ensure that the flows on the strata fit together to yield a family of homeomorphisms, the stratification has to satisfy Whitney's regularity conditions, and the vector fields themselves must satisfy certain technical conditions (be "controlled").

In this paper the author looks at the following important special case. Suppose we are given a stratified manifold and a projection to a space which is a submersion on strata. This is the situation which arises when trying to prove that a given stratification is a topological product. If the stratification is Whitney regular (or indeed satisfies a weaker condition known as C- regularity, due to K. Bekka) then one can lift a smooth vector field in the target to one in the source which is controlled, hence integrate it and obtain the required trivialisation. Here du Plessis shows that one can also ensure that the lifted field is actually continuous.

Finiteness of Mather's Canonical Stratification

A.A. du Plessis

One of the triumphs of singularity theory is Mather's theorem that topologically stable maps are dense. In a series of papers Mather showed how to construct a "canonical" Whitney-regular stratification of each jet space, with the property that a proper map whose jet-extension is transverse to the stratification is topologically stable. In particular such mappings are open and dense in the space of all proper mappings.

Mather also claimed that this stratification has only finitely many connected strata. This is an important fact, since it implies that there are only finitely many topological types of topologically stable map-germs for any given source and target dimension. This paper gives the first complete proof of this fact. It also gives a new simpler construction of the canonical stratification, which is of substantial independent interest.

Trends in Equisingularity Theory

T.J. Gaffney and D. Massey

In many situations classification up to smooth equivalence yields continuous families of inequivalent types (or moduli). Finding coarser relations which yield finite classifications is the main aim of equisingularity theory. There are a number of different approaches: the purely topological, developed first by Lê and Ramanujam, using Smale's h-cobordism theorem; the stratification-theoretic, which looks for invariants whose constancy should guarantee Whitney regularity, developed initially by Teissier; and a third approach using filtrations on the spaces of map-germs, due mainly to Damon. In this paper the authors survey work on the second approach, discussing extensions (for which they are responsible) of the early work of Teissier, to cover the case of families of complete intersection singularities, and of non-isolated hypersurface singularities.

Regularity at Infinity of Real and Complex Polynomial Functions
M. TIBAR
Let $f : K^n \to K$ be a polynomial mapping, with K the real or complex numbers. Classical singularity theory gives us a good understanding of the local topology of the fibres of f, but the global topology still presents a challenge. A value $a \in K$ is called typical if the map f is a local fibration near a. There are only finitely many atypical values. Some are critical values, where the mapping fails to be a fibration for obvious reasons, and the local results alluded to above explain how the fibres of f change near a. The other atypical points arise, roughly speaking, because of problems "at infinity". To obtain control of the mapping at infinity the author uses two regularity conditions: t- and ρ-regularity. The first hinges on a compactification of the map f, the second seeks a control function in the affine space which defines a codimension one foliation which can be used to control lifts of a constant (non-zero) vector field in K. The paper investigates the relationships between these triviality conditions and affine polar curves, over both $\mathbb{R}$ and $\mathbb{C}$.

Global Singularity Theory

A Bennequin Number Estimate for Transverse Knots
V.V. GORYUNOV AND J.W. HILL
Vassiliev's approach to knot invariants, motivated by ideas from singularity theory, has proved astonishingly successful. The paper presented uses this approach to consider Legendrian knots in the standard contact 3-space $\mathbb{R}^3$. It is well known that any topological knot type in a contact 3-manifold has a Legendrian representative, that is, there is a curve of the same knot type whose tangent lines are contained at each point in the associated contact plane. When the contact structure is coorientable any Legendrian knot inherits a natural framing. A result of Bennequin shows that the self-linking (or Bennequin) numbers of all canonically framed Legendrian representatives of a fixed knot are bounded from one side. A new proof is presented here of the fact that these Bennequin numbers are bounded from below by the negative of the lowest degree of the framing variable in its HOMFLY polynomial. For knots with at most 8 double points in their diagrams these two numbers coincide.

Elimination of Singularities: Thom Polynomials and Beyond
O. SAEKI AND K. SAKUMA
The theory of elimination of singularities is concerned with the following global problem: given a smooth map $f : M \to N$, does there exist a smooth map g homotopic to f which has no singularities of a prescribed type Σ? The

first, and most basic, obstruction to the existence of such a map g comes from the Thom polynomial, a polynomial in the Stiefel-Whitney classes of M and N which is dual to the homology class represented by the closure of the set of points of type Σ, since this only depends on the homotopy type of f. This article discusses circumstances in which this is the only obstruction.

Abelian Branched Covers of the Projective Plane
A. LIBGOBER
It is a classical problem, first studied by Zariski and van Kampen, to understand how the type and relative position of the singularities of a plane projective curve are reflected in the fundamental group of its complement. In this paper the author outlines a relationship between the fundamental group of the complement of a reducible plane curve, and certain geometric invariants depending on the local type and configuration of the singularities on the curve.

Singularities of Mappings

Multiplicities of Zero-Schemes in Quasihomogeneous Corank-1 Singularities
W.L. MARAR, J.A. MONTALDI AND M.A.S. RUAS

Butterflies and Umbilics of Stable Perturbations of Analytic Map-Germs
T. FUKUI
In the nice dimensions, every map-germ f of finite codimension can be perturbed so that it becomes stable. Since the bifurcation set is a proper analytic subset of the (smooth) base of a versal unfolding, any two stable perturbations are smoothly equivalent, and thus there is a well-defined stable perturbation f_t of f. Understanding it is an important problem, both for theoretical reasons, and for applications to other subjects. Besides the homology of its image, there are other integer invariants of importance, such as the number of nodes on a stable perturbation of a germ of a parametrised plane curve, or the number of critical points of a stable perturbation of a function-germ. These two papers give formulae for related 0-dimensional invariants of weighted homogenenous map-germs, in terms of their weights and degrees. Both use the principal of conservation of multiplicity for Cohen-Macaulay modules.

Fukui's paper considers map-germs $f : (\mathbb{C}^5, 0) \to (\mathbb{C}^4, 0)$. Here there are two local singularity types occuring at isolated points in a stable perturbation, namely A_4 and D_4. Marar, Montaldi and Ruas look at corank-1 map-germs $(\mathbb{C}^n, 0) \to (\mathbb{C}^n, 0)$ and compute all 0-dimensional invariants, including those involving multi-germs.

An Introduction to the Image-Computing Spectral Sequence
K.A. HOUSTON
The images of analytic maps of complex manifolds are generally highly singular, even when the maps are stable: parametrised plane curves have nodes,

and surfaces mapped stably into 3-space have normal crossings and Whitney umbrellas. On the other hand, Sard's theorem assures us that the generic fibres of a map are smooth. Consequently, in order to understand the topology of a space which is the image of a finite map between smooth manifolds, one has to use methods which take account of this fact. A useful handle on such questions is provided by the *multiple point spaces* of the map: the closure of the set of k-tuples of pairwise distinct points with the same image. The symmetric group S_k acts naturally on the k-th multiple point space, permuting the points. This paper surveys work on a spectral sequence which relates the homology of the image to the alternating part of the homology of the multiple point spaces, and discusses applications to the computation of the homotopy type of generic images.

On the Classification and Geometry of Corank-1 Map-Germs from Three-Space to Four-Space

K.A. Houston and N.P. Kirk

Classifying map-germs up to right-left equivalence is far from easy, and rather few lists of equivalence classes have been obtained. The tables in this paper summarise a new classification and also show geometrical information about the germs listed. The key geometrical invariant of a germ $f : (\mathbb{C}^n, 0) \to (\mathbb{C}^{n+1}, 0)$ is its *image Milnor number* $\mu_I(f)$, the rank of the middle homology of the image of a stable perturbation. It is conjectured that the image Milnor number and the $\mathcal{A}_e$-codimension of a map-germ satisfy the Milnor-Tjurina relation: $\mu_I(f) \geq \mathcal{A}_e\text{-codim}(f)$ with equality if f is weighted homogeneous, in analogy with the relation between Milnor number and Tjurina number of an isolated complete intersection singularity. This has been proved only when $n = 1$ and $n = 2$. The examples here provide the first support for the conjecture outside the range of dimensions where it has been proved. The paper also provides a brief survey of current classification techniques, and of results on the topology of map-germs $f : (\mathbb{C}^n, 0) \to (\mathbb{C}^p, 0)$ for $n < p$.

Applications of Singularity Theory

Singularities of Developable Surfaces

G. Ishikawa

Differential geometry, in particular the differential geometry of curves and surfaces in 3-space, has provided an exciting area of application for singularity theory. Initially it seems slightly surprising that singularity theory should have anything new to say about such a well-studied area. In fact its success here has been quite dramatic.

It has taken place on three fronts: the understanding of higher order phenomena (classical differential geometry concerns itself largely with quadratic forms); the understanding of generic phenomena, via transversality and classification theorems (classical geometry illustrates itself with a surprisingly small collection of often very special surfaces); and the understanding of the singular objects which arise naturally in connection with smooth embeddings (such as the developable surface of a curve, and the dual and focal sets of a surface).

In this paper the author surveys results on developable surfaces, those whose Gaussian curvature vanishes identically. It is a classical result that there are three types of analytic developable surface: cones, cylinders, and tangent developables (the union of the lines tangent to a space curve). In a natural sense the most general and interesting of these are the tangent developables. They are highly singular along the curve itself, and classical differential geometers found these singularities a source of embarrassment. In singularity theory they are, however, an object of great interest. After a brief discussion of some of the classical ideas using projective duality, Ishikawa describes the smooth and topological classification of the singularities of these surfaces.

Singular Phenomena in Kinematics

P.S. DONELAN AND C.G. GIBSON

The study of the kinematics of rigid bodies received substantial impetus during the industrial revolution, and has more recently received new stimulus from robotics. Classical results such as the solution by Watt and Peaucellier of the problem of the conversion of circular to rectilinear motion, and Kemp's theorem that any plane algebraic curve can be traced by a mechanism, lie on a fascinating interface between geometry and engineering. The advances of kinematics in the 19th century can, in part, be traced to developments in the algebraic geometry of plane curves.

The thesis set out here is that the generic phenomena of rigid body kinematics can be understood using singularity theory. The basic object under review can be described as follows. Suppose given a mechanism of some form, for simplicity in the plane, with r degrees of freedom. If we attach another copy of the plane to the mechanism we will obtain an r-parameter family of motions of the plane, and in particular for each point of the plane a map from an r-dimensional space to the plane. If $r = 1$ this gives a 2 parameter family of curves. Most are immersed with transverse intersections, but for certain special tracing points we must expect, for example, cusps. Such points will be of some engineering interest. Understanding, in a general context, the types of singularity that can occur, and the nature of the corresponding exceptional tracing points is part of the task set here. The paper gives a detailed survey of the results obtained to date in this area.

Singularities of Solutions of First Order Partial Differential Equations

S. IZUMIYA

Singularity theory has applications to any area whose fundamental tools involve the calculus, and there is now a substantial body of work on its applications to bifurcation problems in ordinary differential equations. This part of the subject now has an independent existence, rather separate from singularity theory itself.

Izumiya's paper surveys some applications of singularity theory to partial differential equations. He confines himself to Hamilton-Jacobi equations and single conservation laws. Both equation types arise in a number of applications (eg. to the calculus of variations, differential games, and gas dynamics). Classically, solutions are obtained by the method of characteristics; initially such characteristics are smooth, but beyond a certain critical time they become multivalued. The classification of the associated singularities is discussed, as is the evolution of shock waves, in the two variable case.

Singularities Arising from Lattice Polytopes

Klaus Altmann

1 Introduction

(1.1) Assume that an affine variety $Y \subseteq \mathbb{C}^w$ is defined by certain binomials $\underline{z}^a - \underline{z}^b$ $(a, b \in \mathbb{N}^w)$; for example take $Y := [z^n - xy] \subseteq \mathbb{C}^3$. Then, the ring of regular functions on Y equals the semigroup algebra $\mathbb{C}[S]$ with S obtained from $\mathbb{N}^w$ via identifying a and b. If, moreover, the semigroup S is easy to handle (for instance, if S is the set of lattice points in a rational, polyhedral, convex cone in some $\mathbb{R}^k$), then one might hope that important features of the algebraic variety Y can be encoded with combinatorial data.
This is, more or less, the main idea of the concept of (affine) toric varieties. A lot more has been done: if a bunch of polyhedral cones comes in a so-called fan, then the affine toric varieties associated to these cones glue together; if the fan arises from the inner normals of a lattice polytope, then we obtain a projective variety. These ideas have been developed over the last 20 years, and many textbooks are now available. For a detailed treatment we refer to [Da], [Fu], [Ke], or [Od]. For a short introduction to the subject without proofs see §2.

(1.2) As just mentioned, lattice polyhedra are related to projective toric varieties. Hence it is no surprise that (affine) cones over projective varieties, in the algebro-geometric sense, arise from cones over lattice polytopes in the sense of convex geometry.
Doing toric geometry, one has to deal with cones and their duals as well (cf. (2.1)). This implies that lattice polytopes have a second chance to induce a certain class of affine toric varieties; it was first observed by Ishida in [Ish], that this class consists exactly of the affine toric Gorenstein varieties. A more detailed explanation of both methods to construct singularities from lattice polytopes is given in (2.6) and (2.7).

(1.3) The main purpose of the present paper is to give a survey of the deformation theory of toric singularities (or equivalently, of affine toric varieties) known so far. In the very beginning, deformation theory appeared as the investigation of how complex structures may vary on a fixed compact, smooth manifold (cf. [Kod]). In a similar manner, we may regard deformations of germs of analytic spaces. If $Y = (Y, 0)$ is such a germ (often called "singularity", since smooth germs are not the interesting ones), we define the following functor:

$$\mathrm{Def}_Y((T,0)) := \{\text{isomorphism classes of flat } g : Z \to T\,, \text{ together with } g^{-1}(0) \xrightarrow{\sim} Y\}\,.$$

Good references for facts about deformation theory of germs are Artin's Lecture notes [Art], the long introduction to Palamodov's paper [Pa], or Stevens' Habilitationsschrift [St 2]. In many cases, for instance for isolated singularities, there exists the so-called mini-versal deformation. By definition, it yields every possible deformation via specialization of parameters, i.e. via base change. The mini-versal deformation is, up to non-canonical isomorphism, uniquely determined and may be considered a source of numerical invariants of the original singularity. If Y is a complete intersection, then every perturbation of the defining equations yields a (flat) deformation of Y; in particular, its mini-versal deformation is a family over a smooth base space with well-known dimension.
However, as soon as we leave this class of singularities, the structure of the versal family or even the base space will be more complicated. The first example showing that the situation is not always as boring as in the complete intersection case was given by Pinkham, cf. [Pi]. He studied the cone over the rational normal curve of degree four; here the versal base space consists of two irreducible components of dimension one and three, respectively. In the following we will see similar examples in higher dimensions; even non-reduced points will occur as versal base spaces.

(1.4) What are the major issues we are concerned with? In the sequel we will discuss the following three:

(1) Study the vector spaces T^1_Y of infinitesimal deformations and T^2_Y containing the obstructions for lifting deformations onto larger base spaces. There are different ways of defining them (cf. (3.1)-(3.3)); but in any case, they are as multigraded as the semigroup algebra $\mathbb{C}[S]$ is. Hence, the problem is to spot the multidegrees R contributing to these vector spaces and to determine the dimensions of $T^p_Y(-R)$. (The minus sign just makes some of the formulas easier.)

(2) Let us return to the trivial example of the A_{n-1}-singularity mentioned in the very beginning. The algebra $A := \mathbb{C}[x,y,z]/(z^n - xy)$ is $\mathbb{Z}^2$-graded via $\deg x := [n,-1]$, $\deg y := [0,1]$, and $\deg z := [1,0]$. The infinitesimal deformations equal $T^1_Y = \mathbb{C}[z]/(z^{n-1})$, and the one-parameter deformation assigned to the element $z^{n-k} \in T^1_Y$ equals $(z^n - xy) + t^{(k)} z^{n-k}$. Substituting $T := z^k + t^{(k)}$, the total space of this deformation is defined by a binomial again, by $Tz^{n-k} - xy$.

The general goal is to look for so-called genuine deformations, i.e. for those which are no longer infinitesimal, but defined over parameter spaces which are reduced or even smooth. To be able to use the language of polyhedral cones, we would like to remain somehow in the category of toric varieties. It turns out to be a good idea to look for deformations having toric total spaces as just seen for A_{n-1}.

(3) The best possible result is the description of the whole versal deformation. This might be done by listing equations, by providing information about its irreducible components, or by different methods.

After a short introduction into the subject of toric varieties we will discuss the progress in each of these questions in a separate section. Emphasis will be put on both cones over projective varieties and toric Gorenstein singularities, i.e. on those toric varieties induced by lattice polytopes.
Our survey does not contain any proofs; the claims are either standard, or easy exercises, or we refer to the original papers.

2 Convex geometry and toric varieties

(2.1) We are going to introduce briefly the notions we need from convex geometry. It should be considered a good opportunity to fix notation, on the one hand, and to get readers from algebraic geometry in the mood for cones and polytopes, on the other hand. References for the details can be found in the new book [Zi] or the appendix in [Od].

Convex cones: Throughout the paper we use the word *cone* for rational, convex, polyhedral cones. If N, M are two mutually dual, free Abelian groups of finite rank, then a cone $\sigma \subseteq N_{\mathbb{R}} := N \otimes_{\mathbb{Z}} \mathbb{R}$ can be given either by its fundamental generators

$$\sigma = \langle a^1, \ldots, a^m \rangle := \sum_{i=1}^{m} \mathbb{R}_{\geq 0} \cdot a^i \qquad (a^1, \ldots, a^m \in N)$$

or by finitely many inequalities

$$\sigma = \{a \in N_{\mathbb{R}} \,|\, \langle a, r^j \rangle \geq 0;\, j = 1, \ldots, K\} \qquad (r^1, \ldots, r^K \in M).$$

The elements $a^i \in N$ and $r^j \in M$ can be normalized by asking for primitive ones, i.e. which are not proper multiples. (I hope the reader will not be confused by abuse of notation: We use the symbol $\langle \ldots \rangle$ for both the pairing between the mutually dual lattices N, M as well as for indicating the generators of cones; "$<$" also denotes the face relation.)

The concept of duality interchanges both representations: the cone dual to σ is defined as

$$\sigma^\vee := \{r \in M_{I\!R} \mid \langle a, r\rangle \geq 0 \text{ for all } a \in \sigma\}.$$

It has $r^1, \dots, r^K \in M$ as fundamental generators, or it can be given by the inequalities provided by $a^1, \dots, a^m \in N$.

(2.2) *Polytopes and polyhedra:* Let $(L_{I\!R}, L)$ be a finite-dimensional real vector (or maybe affine) space with a lattice. Rational polyhedra in $(L_{I\!R}, L)$ are given as intersections of finitely many rational half spaces. If additionally compact, they will be called polytopes. A polyhedron is said to be a lattice polyhedron if its vertices are contained in L.

Definition: *For two polyhedra $Q', Q'' \subseteq L_{I\!R}$ we define their Minkowski sum as the polyhedron $Q' + Q'' := \{p' + p'' \mid p' \in Q', p'' \in Q''\}$. Obviously, this notion also makes sense for translation classes of polyhedra, hence for affine instead of vector spaces $L_{I\!R}$.*

Example: The plane hexagon

$$Q := \text{Conv}\,\{(0,0), (1,0), (2,1), (2,2), (1,2), (0,1)\} \subseteq I\!R^2$$

(1, 2) (2, 2)

(0, 1) (2, 1) hexagon Q

(0, 0) (1, 0)

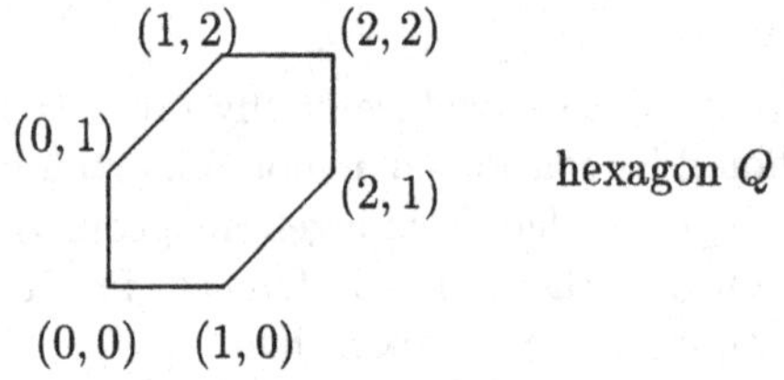

splits into

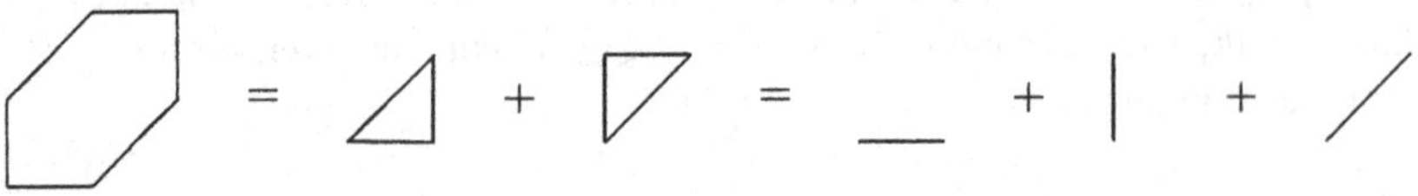

= + = + +

Every polyhedron $Q \subseteq L_{I\!R}$ is decomposable into the Minkowski sum $Q = Q^c + Q^\infty$ of a (compact) polytope $Q^c \in L_{I\!R}$ and the so-called cone of unbounded directions Q^∞; the latter one is contained in the *vector* space associated to $L_{I\!R}$ which, however, will be identified with $L_{I\!R}$. The cone Q^∞ is uniquely determined by Q, the compact summand is not. However, we can take for Q^c the minimal one – given as the convex hull of the vertices of Q itself. If Q was already compact, then $Q^c = Q$ and $Q^\infty = 0$.

Example: Let $\sigma \subseteq N_{I\!R}$ be a cone, and fix some primitive element $R \in M$. Then $L_{I\!R} := [R = 1] := \{a \in N_{I\!R} \mid \langle a, R\rangle = 1\} \subseteq N_{I\!R}$ is an affine space

with lattice $L := [R = 1] \cap N$. We define the crosscut of σ in degree R as the polyhedron $Q := \sigma \cap [R = 1] \subseteq L_{\mathbb{R}}$. It has the cone of unbounded directions $Q^{\infty} = \sigma \cap [R = 0] \subseteq N_{\mathbb{R}}$. The compact part Q^{c} of Q is obtained by describing its vertices: obviously, they correspond exactly to those fundamental generators a^i of σ meeting $\langle a^i, R\rangle \geq 1$ – the actual vertices equal $\bar{a}^i = a^i/\langle a^i, R\rangle$.
Fundamental generators contained in $R^{\perp}$ can still be "seen" as edges in Q^{∞}, but those with $\langle \bullet, R\rangle < 0$ are "invisible" in Q. In particular, we can recover the cone σ from Q if and only if $R \in \sigma^{\vee}$.

(2.3) One of the most frequently used notions will be that of *Minkowski summands* of a given polyhedron $Q \subseteq L_{\mathbb{R}}$. Of course, a Minkowski summand Q' of Q should be at least a summand in the usual sense, i.e. there has to be a Q'' such that $Q = Q' + Q''$. However, since $Q = Q' + Q^{\infty}$ is true for every $Q^{c} \subseteq Q' \subseteq Q$, this might not be enough; we would like to avoid additional face structure of Q' (not "coming" from Q). We take the following definition from [Sm]:

Definition: *A polyhedron Q' is called a Minkowski summand of Q if there is a Q'' such that $Q = Q' + Q''$ and if $(Q')^{\infty} = Q^{\infty}$.*

It is not difficult to see that the faces of $Q' + Q''$ equal the Minkowski sums of the corresponding faces (defined by the same hyperplane in $L_{\mathbb{R}}$) of Q' and Q''. In particular, up to dilatation, the set of edges of $Q' + Q''$ equals the union of the corresponding sets for Q' and Q'', respectively. That means, a Minkowski summand has not only the same cone of unbounded directions, but, up to dilatation with a factor ≥ 0, also the same compact edges as the original polyhedron.

This is the moment to describe the "moduli space" of all Minkowski summands of Q, following [Al 3]. After choosing orientations, denote the compact edges of Q by $d^1, \ldots, d^N \in L_{\mathbb{R}}$:

Definition: *For every two-dimensional compact face ("two-face") $\varepsilon < Q$ we define its sign vector $\underline{\varepsilon} = (\varepsilon_1, \ldots, \varepsilon_N) \in \{0, \pm 1\}^N$ by*

$$\varepsilon_i := \begin{cases} \pm 1 & \textit{if } d^i \textit{ is an edge of } \varepsilon \\ 0 & \textit{otherwise} \end{cases}$$

such that the oriented edges $\varepsilon_i \cdot d^i$ fit to a cycle along the boundary of ε. This determines $\underline{\varepsilon}$ up to sign, and any choice will do. In particular, $\sum_i \varepsilon_i d^i = 0$. Then define $V(Q)$ to be the vector space

$$\{(t_1, \ldots, t_N) \in \mathbb{R}^N \mid \sum_i t_i\, \varepsilon_i\, d^i = 0 \textit{ for every compact two-face } \varepsilon < Q\}.$$

It is obvious that the points of the cone $C(Q) := V(Q) \cap \mathbb{R}^N_{\geq 0}$ parametrize the set of Minkowski summands of positive multiples of Q via measuring the dilatation factors of the compact edges. In particular, $\underline{1} \in C(Q)$ corresponds to Q itself.

(2.4) *Affine toric varieties:* Let N, M be two mutually dual, free Abelian groups of finite rank. From now on, the lattice structure of cones and polyhedra becomes important. In particular, isomorphisms between those objects are always assumed to be induced from isomorphisms of the lattices. We are going to describe how to get affine algebraic varieties from convex cones:

Definition: *If $\sigma \subseteq N_{\mathbb{R}}$ is a cone with apex, then we define by $Y_\sigma := \operatorname{Spec} \mathbb{C}[\sigma^\vee \cap M]$ the associated, affine toric variety. ($\mathbb{C}[\sigma^\vee \cap M]$ denotes the semigroup ring - obtained by regarding elements $r \in \sigma^\vee \cap M$ as exponents of some "abstract symbol" x.)*
Let $\sigma_1 \subseteq N^1_{\mathbb{R}}$, $\sigma_2 \subseteq N^2_{\mathbb{R}}$ be two cones. Then, a $\mathbb{Z}$-linear map $f : N^1 \to N^2$ such that $f(\sigma_1) \subseteq \sigma_2$ induces an algebraic morphism $f : Y_1 \to Y_2$ in an obvious way. Those maps will be regarded as the morphisms in the category of affine, toric varieties.

The semigroup $S := \sigma^\vee \cap M$ is generated by the finite set E of its irreducible elements. E is often called the Hilbert basis of that semigroup. Assigning to each element $r \in E$ a variable z_r, our affine toric variety Y_σ can be embedded into $\mathbb{C}^E$. It is defined by the binomial equations obtained from "raising" linear dependencies between the r's into the exponents of the z_r's. Just to give an example, the relation $r + 2s = 3t + u$ turns into $z_r\, z_s^2 = z_t^3\, z_u$.

Examples:

(1) The cone $\sigma := \mathbb{R}^k_{\geq 0} \subseteq \mathbb{R}^k$ with $N := \mathbb{Z}^k$ yields $\sigma^\vee \cap M = \mathbb{N}^k$, hence $Y_\sigma = \mathbb{C}^k$.

(2) Let $E \subseteq \sigma^\vee \cap M$ be the Hilbert basis for an arbitrary cone $\sigma \subseteq N_{\mathbb{R}}$. Then, assigning to each $a \in N$ the E-tuple $(\langle a, r\rangle)_{r \in E} \in \mathbb{Z}^E$, defines a $\mathbb{Z}$-linear map $N \to \mathbb{Z}^E$ sending σ into $\mathbb{R}^E_{\geq 0}$. At the level of toric varieties, this yields exactly the embedding $Y_\sigma \hookrightarrow \mathbb{C}^E$ described above.

(3) Let $n, q \in \mathbb{Z}$ be relatively prime numbers. With $(\alpha, \beta) \mapsto (-q\,\alpha + \beta,\, n\,\alpha)$, we obtain a $\mathbb{Z}$-linear map $f : \mathbb{Z}^2 \to \mathbb{Z}^2$ sending $\mathbb{R}^2_{\geq 0}$ onto $\sigma := \langle (1, 0), (-q, n)\rangle \subseteq \mathbb{R}^2$. (By the way, any two-dimensional cone can be written in this way.) At the toric level this means we have a morphism $\pi : \mathbb{C}^2 \to Y_\sigma$.
The dual cone equals $\sigma^\vee = \langle [0, 1],\, [n, q]\rangle$, and we obtain for $f^*(\sigma^\vee \cap \mathbb{Z}^2)$ the semigroup

$$f^*(\sigma^\vee \cap \mathbb{Z}^2) = \mathbb{N}^2 \cap \{(r_1, r_2) \in \mathbb{Z}^2 \mid r_1 + q\, r_2 \equiv 0 \bmod n\}\,.$$

Hence, the affine coordinate ring $\mathbb{C}[\sigma^\vee \cap \mathbb{Z}^2]$ of Y_σ equals the subring of $\mathbb{C}[z_1, z_2]$ consisting of polynomials invariant under $\begin{pmatrix} \xi & 0 \\ 0 & \xi^q \end{pmatrix}$ with ξ

being a primitive n^{th} root of unity. In particular, Y_σ is a cyclic quotient singularity, and π is the quotient map.
For $q = -1$ we obtain the A_{n-1}–singularity mentioned in the very beginning; $n = 4$, $q = 1$ yields Pinkham's example for singularities with reducible versal base space.

We have seen that almost all two-dimensional cones yield singular toric varieties. This reflects the general situation - smooth, affine toric varieties are boring: If σ is a top-dimensional cone, then Y_σ *is smooth if and only if* σ *is a simplex generated by a* $\mathbb{Z}$*-basis of* N, i.e. the determinant of its fundamental generators has to be ± 1. Then, Y_σ is isomorphic to the affine space.
Toric varieties got their name because they always contain the torus $T = \operatorname{Spec} \mathbb{C}[M]$. This algebraic group acts on them and causes a (finite) stratification into T-orbits. The unique closed orbit in an affine toric variety (if σ is top-dimensional, then it is a point) is the most singular one. In higher dimensions, most cones are no longer simplicial. This means that the singularities get worse than quotient singularities.

(2.5) *General toric varieties:* As already mentioned, morphisms between affine toric varieties arise from $\mathbb{Z}$-linear maps $f : N^1 \to N^2$ such that $f(\sigma_1) \subseteq \sigma_2$. A very important special case is where $f : N \to N$ is the identity map and σ_1 is a face of σ_2. If $r \in \sigma_2^\vee \cap M$ actually cuts out this face (i.e. $\sigma_1 = \sigma_2 \cap r^\perp$), then $\mathbb{C}[\sigma_2^\vee \cap M]$ equals the localization of $\mathbb{C}[\sigma_1^\vee \cap M]$ by the element x^r. In particular, the induced map $Y_{\sigma_1} \to Y_{\sigma_2}$ is an open embedding identifying the first variety with the open subset $[x^r \neq 0] \subseteq Y_{\sigma_1}$. Moreover, every open embedding in our category arises that way.

Definition: *If* Σ *is a fan in* $N_{\mathbb{R}}$ *(i.e. a finite collection of cones such that* $\sigma, \tau \in \Sigma$ *always implies* $\tau \cap \sigma \leq \tau, \sigma$*, and such that* Σ *contains with every cone all of its faces), then the toric variety* Y_Σ *is obtained by gluing together the affine pieces* Y_σ $(\sigma \in \Sigma)$ *along common faces of* Σ*-cones.*
A map between toric varieties $Y_{\Sigma^1}, Y_{\Sigma^2}$ *is given by a* $\mathbb{Z}$*-linear* $f : N^1 \to N^2$ *such that for each* $\sigma_1 \in \Sigma^1$ *there is some* $\sigma_2 \in \Sigma^2$ *meeting* $f(\sigma_1) \subseteq \sigma_2$.

It is well known that complete toric varieties arise from fans with $|\Sigma| := \bigcup_{\sigma \in \Sigma} \sigma = N_{\mathbb{R}}$. If, moreover, Σ is the inner normal fan of some lattice polytope P in M, then $Y_P := Y_\Sigma$ is projective. The polytope itself reflects the projective embedding; the lattice points of P correspond in an easy way to a natural basis of the global sections of an ample line bundle $\mathcal{L}_P$.
More generally, there is a one-to-one correspondence between certain lattice polytopes, on the one hand, and globally generated invertible sheaves on Y_Σ, on the other; Minkowski addition translates into the tensor product. Hence, it seems quite natural that, for projective toric varieties, the vector space $V(P)$ of (2.3) appears as $V(P) = \operatorname{Pic} Y_P \otimes_{\mathbb{Z}} \mathbb{R}$.

Examples:

(1) The k-dimensional fan spanned by the canonical basis vectors $e^1, \ldots, e^k$ of $\mathbb{Z}^k$ and $-e$, with $e := e^1 + \cdots + e^k$, defines the projective space $\mathbb{P}^k$.

(2) The subdivision of the smooth cone $\sigma = \mathbb{R}^k_{\geq 0} = \langle e^1, \ldots, e^k \rangle$ into the union of $\sigma_i := \langle e^1, \ldots, \hat{e}^i, \ldots, e^k, e \rangle$ $(i = 1, \ldots, k)$ describes the blowing up of the origin in the affine k-space.

(3) If $\sigma = \langle a^1, \ldots, a^m \rangle$ is an arbitrary cone (with apex), then the normalized blowing up of Y_σ in the closed orbit is given by the subdivision of σ into the union of $\sigma_r := \{a \in \sigma \mid \langle a, r \rangle \leq \langle a, E \rangle\}$ with r running through the Hilbert basis E of $\sigma^\vee \cap M$.

In general, every fan can be subdivided into a "smooth" fan. That means, every toric variety admits a toric desingularization.

(2.6) Let $P \subseteq (L_{\mathbb{R}}, L)$ be a lattice polytope. Embedding P via $P \subseteq L_{\mathbb{R}} \cong L_{\mathbb{R}} \times \{1\} \hookrightarrow L_{\mathbb{R}} \times \mathbb{R} =: M_{\mathbb{R}}$ into the next dimension, we obtain rational, polyhedral cones $\sigma^\vee := \operatorname{cone}(P) := \mathbb{R}_{\geq 0} \cdot P \subseteq M_{\mathbb{R}}$ and $\sigma := \sigma^{\vee\vee} \subseteq N_{\mathbb{R}}$. Since

$$\mathbb{C}[S] := \mathbb{C}[\sigma^\vee \cap M] = \oplus_{d \geq 0} \mathbb{C}[d\,P \cap M] = \oplus_{d \geq 0} H^0(Y_P, \mathcal{L}_P^{\otimes d}),$$

the toric variety Y_σ equals the *affine cone over the projective variety* $(Y_P, \mathcal{L}_P)$. There is a distinguished point $a^* := (\underline{0}, 1) \in \sigma \subseteq N_{\mathbb{R}}$; the equation $[a^* = 1]$ recovers P from $\sigma^\vee$. Moreover, a^* may be used to make the ring $\mathbb{C}[\sigma^\vee \cap M]$ $\mathbb{Z}$-graded ($\deg x^r := \langle a^*, r \rangle$), and we obtain $Y_P = \operatorname{Proj} \mathbb{C}[\sigma^\vee \cap M]$ while $Y_\sigma = \operatorname{Spec} \mathbb{C}[\sigma^\vee \cap M]$.
We may elucidate the relation between Y_P and its affine cone Y_σ also from another point of view: the open subset $Y_\sigma \setminus \{0\} \subseteq Y_\sigma$ is given by the fan $\partial\sigma$ contained in $N_{\mathbb{R}}$, and the projection $\pi : Y_\sigma \setminus \{0\} \to Y_P$ is provided by the $\mathbb{Z}$-linear map $\pi : N \twoheadrightarrow N/\mathbb{Z} \cdot a^* = L_{\mathbb{R}}{}^*$. It sends proper faces of σ isomorphically onto Σ-cones, meaning that $Y_\sigma \setminus \{0\}$ is a $\mathbb{C}^*$-bundle over Y_P.

(2.7) Finally, we would like to explain Ishida's relation (cf. [Ish], Theorem 7.7.) between lattice polytopes and Gorenstein singularities: up to sign, the differential form $\omega_0 := \frac{dx_1}{x_1} \wedge \cdots \wedge \frac{dx_k}{x_k}$ on the torus $(\mathbb{C}^*)^k \cong \operatorname{Spec}\mathbb{C}[M] = Y_0 \subseteq Y_\sigma$ does *not* depend on the special choice of coordinates. Multiples $x^r \cdot \omega_0$ are holomorphic on Y_σ if and only if r belongs to the interior of $\sigma^\vee$. Since Y_σ is normal, this means that we can describe the canonical module as $\omega_Y = \big(\oplus_{r \in (\operatorname{int} \sigma^\vee) \cap M} \mathbb{C} \cdot x^r \big) \cdot \omega_0$. In particular, Y_σ is Gorenstein, i.e. ω_Y is invertible if and only if there is an $R^* \in M$ such that $(\operatorname{int} \sigma^\vee) \cap M = R^* + (\sigma^\vee \cap M)$. Replacing ω_Y by its g-th tensor power, we have obtained the following criterion:

Let $Y_\sigma = \mathrm{Spec}\,\mathbb{C}[\sigma^\vee \cap M]$ be an affine toric variety given by a cone $\sigma = \langle a^1, \ldots, a^m\rangle$. Then, Y is $\mathbb{Q}$-Gorenstein if and only if there is a primitive element $R^* \in M$ and a natural number $g \in \mathbb{N}$ such that

$$\langle a^i, R^*\rangle = g \quad \textit{for each } i = 1, \ldots, m.$$

Y is Gorenstein if and only if, additionally, $g = 1$.

In particular, toric Gorenstein singularities are obtained by putting a lattice polytope $Q \subseteq (L_{\mathbb{R}}, L)$ into the affine hyperplane $L_{\mathbb{R}} \times \{1\} \subseteq L_{\mathbb{R}} \times \mathbb{R} =: N_{\mathbb{R}}$ and defining $\sigma := \mathrm{cone}(Q) = \mathbb{R}_{\geq 0} \cdot Q$. The canonical degree R^* equals $[\underline{0}, 1]$ in this setting. As an example, lattice intervals of length n provide the two-dimensional A_{n-1}-singularities; see (2.4)(3).

3 Infinitesimal deformations

(3.1) If $Y \subseteq \mathbb{C}^w$ is defined by an ideal $I = (f_1, \ldots, f_s) \subseteq \mathbb{C}[z_1, \ldots, z_w]$, then we denote by $A := \mathbb{C}[\underline{z}]/I$ the algebra of regular functions, by $\mathcal{R} \subseteq \mathbb{C}[\underline{z}]^s$ the $\mathbb{C}[\underline{z}]$-module of linear relations between $f_1, \ldots, f_s$, and by $\mathcal{R}_0 \subseteq \mathcal{R}$ the so-called Koszul relations generated by all $f_j e^k - f_k e^j \in \mathbb{C}[\underline{z}]^s$. In particular, we have the exact sequences

$$0 \to \mathcal{R} \to \mathbb{C}[\underline{z}]^s \to I \to 0 \quad \text{and} \quad 0 \to I \to \mathbb{C}[\underline{z}] \to A \to 0\,.$$

Then, the easiest way to define the vector spaces T^1_Y and T^2_Y is

$$T^1_Y := \mathrm{Hom}_A({}^{I}\!/_{I^2}, A) \Big/ \mathrm{Hom}_A(A^w, A)$$

via

$$d : {}^{I}\!/_{I^2} \to A^w \quad \text{with} \quad d(f_j) := \Big(\frac{\partial f_j}{\partial z_1}, \ldots \frac{\partial f_j}{\partial z_w}\Big)$$

and

$$T^2_Y := \mathrm{Hom}_{\mathbb{C}[z]}({}^{\mathcal{R}}\!/_{\mathcal{R}_0}, A) \Big/ \mathrm{Hom}_{\mathbb{C}[z]}(\mathbb{C}[\underline{z}]^s, A)\,.$$

These definitions seem to depend on the embedding of Y, but they do not. In the above straightforward formulas, the relation of the vector spaces T^p_Y to deformation theory becomes apparent. For instance, if $\xi \in \mathrm{Hom}_A({}^{I}\!/_{I^2}, A)$, then the infinitesimal deformation represented by ξ may be obtained by replacing f_j with the perturbed equation $f_j + \varepsilon\,\xi(f_j)$.

(3.2) A more fancy way to obtain T^1_Y and T^2_Y is to consider them as the first and second André-Quillen cohomology groups, respectively. This cohomology

theory is obtained from the cotangent complex which is defined for any $\mathbb{C}$-algebras; it is closely related to Hochschild and Harrison cohomology. A nice introduction into this subject may be found in [Lo], §3.5 - §4.5.
To calculate the vector spaces T^p_Y for affine toric varieties Y_σ, the Harrison cohomology approach was most successful. It yields the following results (cf. [AlSl]):

The T^p_Y admit an M-(multi) grading as does any other natural module over $A = \mathbb{C}[\sigma^\vee \cap M]$; let us fix an element $R \in M$. For any face $\tau \leq \sigma$ we define

$$K^R_\tau := \left[\sigma^\vee \cap (R - \operatorname{int} \tau^\vee) \cap M\right] \setminus \{0\}.$$

Definition: Let $K \subseteq M$ be an arbitrary subset of the lattice M. A function $f : K \to \mathbb{C}$ is called *quasilinear* if $f(r) + f(s) = f(r+s)$ for any r and s with $r, s, r+s \in K$. The vector space of quasilinear functions is denoted by $\overline{\operatorname{Hom}}(K, \mathbb{C})$.

The sets K^R_τ admit the following properties:

(i) $K^R_0 = (\sigma^\vee \cap M) \setminus \{0\}$, and $K^R_i := K^R_{a^i} = \{r \in K_0 \mid \langle a^i, r\rangle < \langle a^i, R\rangle\}$ with $i = 1, \ldots, m$ are "thick strips" along the facets of $\sigma^\vee$.

(ii) For $\tau \neq 0$ the equality $K^R_\tau = \bigcap_{a^i \in \tau} K^R_i$ holds. Moreover, if σ is a top-dimensional cone, $K^R_\sigma = K_0 \cap (R - \operatorname{int} \sigma^\vee)$ is a (diamond shaped) finite set.

(iii) The dependence of the sets K^R_τ on τ is a contravariant functor. This gives rise to the complex $\overline{\operatorname{Hom}}(K^R_\bullet, \mathbb{C})$ with

$$\overline{\operatorname{Hom}}(K^R_p, \mathbb{C}) := \oplus_{\tau \leq \sigma, \dim \tau = p} \overline{\operatorname{Hom}}(K^R_\tau, \mathbb{C}) \qquad (0 \leq p \leq \dim \sigma)$$

and the usual differentials.

(iv) If $\tau \leq \sigma$ is a smooth face, then the injections $\operatorname{Hom}(\operatorname{span}_{\mathbb{C}} K^R_\tau, \mathbb{C}) \hookrightarrow \overline{\operatorname{Hom}}(K^R_\tau, \mathbb{C})$ are also isomorphisms. Moreover,

$$\operatorname{span}_{\mathbb{C}} K^R_\tau = \bigcap_{a^i \in \tau} \operatorname{span}_{\mathbb{C}} K^R_i,$$

and the latter vector spaces equal $\operatorname{span}_{\mathbb{C}} K^R_i = M_{\mathbb{C}}$, $(a^i)^\perp$, or 0 if $\langle a^i, R\rangle \geq 2$, $= 1$, or ≤ 0, respectively.

Now we can express the André-Quillen cohomology groups T^p_Y in terms of the sets K^R_τ:

Theorem: ([AlSl])
1) Let σ be an arbitrary rational, polyhedral cone with apex in 0. *Then, for every $R \in M$,*

$$T^p_Y(-R) = H^p(\overline{\operatorname{Hom}}(K^R_\bullet, \mathbb{C})) \quad \textit{for } p = 0, 1, 2.$$

Moreover, for $p = 0, 1$, this vector space equals $H^p((\operatorname{span}_{\mathbb{C}} K^R_\bullet)^*)$.
2) If Y_σ is Gorenstein in codimension two, then $T^2_Y(-R) = H^2((\operatorname{span}_{\mathbb{C}} K^R_\bullet)^*)$.

3) Let Y_σ be an isolated singularity. Then, the André-Quillen cohomology in degree $-R$ equals

$$T^p_Y(-R) = \begin{cases} H^p(\overline{\operatorname{Hom}}(K^R_\bullet, \mathbb{C})) = H^p((\operatorname{span}_{\mathbb{C}} K^R_\bullet)^*) & \textit{for } 0 \le p \le \dim\sigma - 1 \\ H^{\dim\sigma}(\overline{\operatorname{Hom}}(K^R_\bullet, \mathbb{C})) & \textit{for } p = \dim\sigma \\ HA^{p-\dim\sigma+1}(K^R_\sigma; \mathbb{C}) & \textit{for } p \ge \dim\sigma + 1\,. \end{cases}$$

Here, the vector spaces $HA^\bullet(K; \mathbb{C})$ are defined as the cohomology of the complex

$$C^q(K; \mathbb{C}) := \left\{\varphi : \{(r_1, \ldots, r_q) \in K^q \mid \textstyle\sum_v r_v \in K\} \to \mathbb{C} \,\middle|\, \varphi \text{ is shuffle invariant}\right\}$$

with differential $\delta^q : C^{q-1}(K; \mathbb{C}) \to C^q(K; \mathbb{C})$

$$\begin{aligned}(\delta^q\varphi)(r_1, \ldots, r_q) \;&:=\; \varphi(r_2, \ldots, r_q) + \sum_{v=1}^{q-1}(-1)^v \varphi(r_1, \ldots, r_v + r_{v+1}, \ldots, r_q) \\ &\quad + (-1)^q \varphi(r_1, \ldots, r_{q-1})\,.\end{aligned}$$

Remark: The general framework for the above theorem is the existence of a spectral sequence $E_1^{p,q} = \oplus_{\dim\tau=p} HA^q(K^R_\tau; \mathbb{C}) \Longrightarrow T^{p+q-1}_Y(-R)$ together with the vanishing result $HA^q(K^R_\tau; \mathbb{C}) = 0$ for smooth faces $\tau \le \sigma$ and $q \ge 2$.

(3.3) The results for T^1_Y and T^2_Y were previously obtained, under some additional assumptions and by different methods, in [Al 2]. The paper uses the formula $T^p_Y = \operatorname{Ext}^p(\Omega_{A|\mathbb{C}}, A)$, which is true for normal rings A and $p = 1, 2$. Moreover, [Al 2] contains the relation between the above combinatorial formulas for T^p_Y and those mentioned in (3.1). For T^1_Y it looks like the following:

Recall from (2.4) that $E \subseteq \sigma^\vee \cap M$ denotes the Hilbert basis. It provides a natural map $\pi : \mathbb{Z}^E \twoheadrightarrow M$ which is dual to $N \to \mathbb{Z}^E$ mentioned in Example (2.4)(2). Then, the ideal $I \subseteq \mathbb{C}[\underline{z}]$ defining Y_σ is generated, as a $\mathbb{C}$-vector space, by the binomials $\underline{z}^a - \underline{z}^b$ with $a, b \in \mathbb{N}^E$ and $\pi(a) = \pi(b)$. With $E^R_\tau := E \cap K^R_\tau$ we denote by $L_{\mathbb{C}}(E^R_\tau)$ the $\mathbb{C}$-vector space consisting of the linear relations among E^R_τ-elements, i.e. $L_{\mathbb{C}}(E^R_\tau) = (\ker \pi_{\mathbb{C}}) \cap \mathbb{C}^{E^R_\tau}$. In particular, there is an exact sequence

$$0 \to L_{\mathbb{C}}(E^R_\tau) \to \mathbb{C}^{E^R_\tau} \to \operatorname{span}_{\mathbb{C}} E^R_\tau \to 0\,.$$

Theorem: ([Al 2]) *Let $R \in M$. The above sequence for the corresponding complexes gives*

$$\begin{aligned} T^1_Y(-R) &= H^1((\text{span}_{\mathbb{C}} K^R_\bullet)^*) = H^1((\text{span}_{\mathbb{C}} E^R_\bullet)^*) \\ &\cong \Big[L_{\mathbb{C}}(\bigcup_{i=1}^m E^R_i)/\sum_{i=1}^m L_{\mathbb{C}}(E^R_i)\Big]^*. \end{aligned}$$

If, moreover, $\varphi : L_{\mathbb{C}}(E) \to \mathbb{C}$ induces some element of $T^1_Y(-R)$, then the A-linear map $I/I^2 \to A$ sending $\underline{z}^a - \underline{z}^b$ onto $\varphi(a-b) \cdot x^{\pi(a)-R}$ yields the same element via the T^1_Y-formula in (3.1).

(3.4) What do the above formulas mean for the simplest example, the two-dimensional cyclic quotient singularities? We use the notation of Example (2.4)(3), i.e. $\sigma = \langle (1,0); (-q,n) \rangle \subseteq \mathbb{R}^2$. We develop $\frac{n}{n-q}$ into a (negative) continued fraction

$$\frac{n}{n-q} = a_2 - \cfrac{1}{a_3 - \frac{1}{a_4 - \dots}} \qquad \text{(with } a_v \geq 2\text{)}.$$

Then $E \subseteq \sigma^\vee \cap \mathbb{Z}^2$ is given as the set $E = \{r^1, \dots, r^w\} \subseteq \langle [0,1], [n,q] \rangle \cap \mathbb{Z}^2$ with

- $r^1 = [0,1]$, $r^2 = [1,1]$, $r^w = [n,q]$;
- $r^{v-1} + r^{v+1} = a_v \cdot r^v$ ($v = 2, \dots, w-1$) (cf. [Ri] or [Od]).

Pinkham has already obtained that, if $w \geq 4$, $\dim T^1_Y = \sum_{v=2}^{w-1} a_v - 2$. The previous theorems should give the same result. Denoting $a^1 = (1,0)$ and $a^2 = (-q,n)$, they mean

$$T^1_Y(-R) = \Big((\text{span}_{\mathbb{C}} E^R_1) \cap (\text{span}_{\mathbb{C}} E^R_2) \Big/ \text{span}_{\mathbb{C}}(E^R_1 \cap E^R_2) \Big)^*.$$

There are four different cases for the multidegree $R \in M \cong \mathbb{Z}^2$; we assume $w \geq 3$, i.e. Y is not smooth:

(i) $R = r^2$ (or analogously $R = r^{w-1}$): We obtain $E^R_1 = \{r^1\}$ and $E^R_2 = \{r^3, \dots, r^w\}$, i.e. $\text{span}_{\mathbb{C}} E^R_1 = (a^1)^\perp$, $\text{span}_{\mathbb{C}} E^R_2 = \mathbb{C}^2$ (or $(a^2)^\perp$, if $w = 3$), and $\text{span}_{\mathbb{C}} E^R_{12} = 0$. Hence $T^1_Y(-R) \cong \mathbb{C}$ (or 0, if $w = 3$).

(ii) $R = r^v$ ($3 \leq v \leq w-2$): We obtain $E^R_1 = \{r^1, \dots, r^{v-1}\}$ and $E^R_2 = \{r^{v+1}, \dots, r^w\}$, hence $\text{span}_{\mathbb{C}} E^R_1 = \text{span}_{\mathbb{C}} E^R_2 = \mathbb{C}^2$, $\text{span}_{\mathbb{C}} E^R_{12} = 0$, and the theorem yields $T^1_Y(-R) \cong \mathbb{C}^2$.

(iii) $R = p \cdot r^v$ ($2 \leq v \leq w-1$, $2 \leq p < a_v$ for $w \geq 4$; or $v = 2 = w-1$, $2 \leq p \leq a_2$ for $w = 3$): We obtain $E^R_1 = \{r^1, \dots, r^v\}$ and $E^R_2 = \{r^v, \dots, r^w\}$, i.e. $\text{span}_{\mathbb{C}} E^R_1 = \text{span}_{\mathbb{C}} E^R_2 = \mathbb{C}^2$ and $\text{span}_{\mathbb{C}} E^R_{12} = \mathbb{C} \cdot R$. The theorem yields $T^1_Y(-R) \cong \mathbb{C}$.

(iv) For the remaining $R \in M$, either $E_1^R \subseteq E_2^R$ or $E_2^R \subseteq E_1^R$ or $\#(E_1^R \cap E_2^R) \geq 2$. In all these cases the theorem yields $T_Y^1(-R) = 0$.

Summing up, we get Pinkham's formula back. The vector space T_Y^2 may be obtained in a similar way.

(3.5) We would like to have an interpretation for the vector spaces $T_Y^p(-R)$ in terms of convex geometry. In fact, we do not know any for T_Y^2, but for T_Y^1 this was done in [Al 4]:
To any $R \in M$ we assign the crosscut $Q(R) := \sigma \cap [R = 1]$, cf. Example (2.2), and two $I\!R$-vector spaces $V(R)$, $W(R)$ defined as follows: $V(R) := V(Q(R))$ is the space of "generalized" (allowing negative dilatations) Minkowski summands of $Q(R)$ introduced in (2.3), and

$$W(R) := I\!R^{\#\{Q(R)\text{-vertices not belonging to } N\}}.$$

The latter provides coordinates s_i for each vertex $\bar{a}^i \in Q(R) \setminus N$, i.e. for each fundamental generator $a^i \in \sigma$ with $\langle a^i, R\rangle \geq 2$. To each compact edge $d^{ij} = \overline{\bar{a}^i \bar{a}^j}$ we associate a set of equations G_{ij} dealing with elements of $V(R) \oplus W(R)$. These sets are of one of the following three types:

(0) $G_{ij} = \emptyset$,

(1) $G_{ij} = \{s_i - s_j = 0\}$ provided both coordinates exist in $W(R)$, set $G_{ij} = \emptyset$ otherwise, or

(2) $G_{ij} = \{t_{ij} - s_i = 0,\ t_{ij} - s_j = 0\}$, dropping equations that do not make sense.

Restricting $V(R) \oplus W(R)$ to the (at most) three coordinates t_{ij}, s_i, s_j, the actual choice of G_{ij} is made such that these equations yield a subspace of dimension $1 + \dim T^1_{\langle a^i, a^j\rangle}(-R)$; since $\langle a^i, a^j\rangle$ is a two-dimensional cone, we do know $\dim T^1_{\langle a^i, a^j\rangle}(-R)$ from the previous section. Anyway, if Y is smooth in codimension two, then G_{ij} is always of type (2).

Theorem: ([Al 4]) *The infinitesimal deformations of Y_σ in degree $-R$ are given by $T_Y^1(-R) =$*

$$\{(\underline{t},\, \underline{s}) \in V_{\mathbb{C}}(R) \oplus W_{\mathbb{C}}(R) \mid (\underline{t},\, \underline{s}) \text{ fulfills all the equations } G_{ij}\} \;/\; \mathbb{C} \cdot (\underline{1},\, \underline{1})\,.$$

Corollary: *If Y is smooth in codimension two, then $T_Y^1(-R)$ is contained in $V_{\mathbb{C}}(R)/\,\mathbb{C} \cdot (\underline{1})$. It is built from those $\underline{t}$ such that $t_{ij} = t_{jk}$ whenever d^{ij}, d^{jk} are compact edges with a common non-lattice vertex $\bar{a}^j$ of $Q(R)$.*
Thus, $T_Y^1(-R)$ equals the set of equivalence classes of those Minkowski summands of $I\!R_{\geq 0} \cdot Q(R)$ that preserve up to homothety the stars of non-lattice vertices of $Q(R)$.

(3.6) In the last three sections we would like to mention some applications and examples. First, assume that Y_σ is smooth in codimension two. For the cone $\sigma \subseteq N_{\mathbb{R}}$ this assumption means that any two-dimensional face is generated by two elements of N forming a part of a $\mathbb{Z}$-basis of N.

Proposition: *There are only finitely many $R \in \sigma^\vee \cap M$ such that $T^1_Y(-R) \neq 0$. If, moreover, the three-dimensional faces of σ are simplicial (i.e. built from only three fundamental generators), then $T^1_Y(-R) = 0$ for each $R \in \operatorname{int}\sigma^\vee \cap M$.*

Degrees from $-(\operatorname{int}\sigma^\vee \cap M)$ somehow play the role of negative degrees when dealing with singularities with $(\mathbb{C})^*$-action. In contrast to the Gorenstein case below, there might be infinitesimal deformations in non-negative degrees, even for isolated singularities. In particular, it is not always true that cones which are smooth in codimension three provide rigid singularities:

Example: Let $\sigma = \langle a, b, x, y, z \rangle$ be the (four-dimensional) cone over some double tetrahedron.

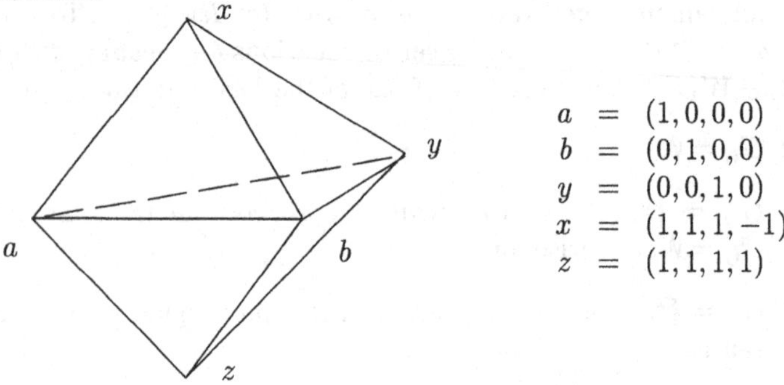

The two partial cones generated by a, b, y, x and a, b, y, z, respectively, are smooth. So are the proper faces of σ, i.e. Y_σ is an isolated singularity. Let $R = [0, 0, 1, 0]$; then the compact part of $Q(R)$ consists just of two edges containing the lattice-vertices $\bar{x}, \bar{y}, \bar{z}$.

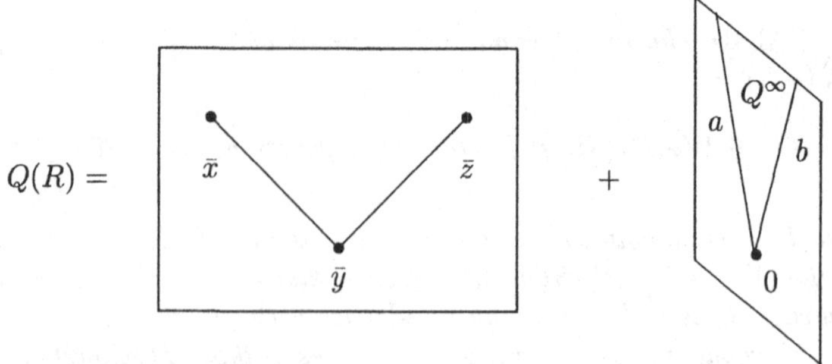

In particular, $\dim T^1_Y(-R) = 1$.

(3.7) If Y_σ is the affine cone over a projective toric variety Y_P (cf. (2.6)), then we may use the distinguished element $a^* \in N$ to transform M-multidegrees into ordinary integers. In particular, elements $R \in P \cap M \subseteq M$ yield $T^1_Y(-R) \subseteq T^1_Y(-1)$. For cones over smooth, projective surfaces, the reverse implication is also true:

Proposition: *Let P be a plane lattice polygon defining a smooth, projective surface. Except when P is isomorphic to the unit square (yielding $Y = [xy - zw = 0] \subseteq \mathbb{C}^3$), we have*

$$(T^1_Y)_- := \oplus_{k\geq 1} T^1_Y(-k) = T^1_Y(-1) = \oplus_{R\in P} T^1_Y(-R)\,.$$

Moreover, it is possible to list the series of polygons P provided $T^1_Y(-1) \neq 0$; they are at most hexagons such that the inner lattice points are contained in a line. These polygons are closely related to the so-called reflexive polygons occurring in [Bat].

(3.8) Finally, we consider $\mathbb{Q}$-Gorenstein singularities Y_σ provided by some lattice polytope $Q \subseteq M_{\mathbb{R}}$ sitting in height g (cf. (2.7)). Since $R^* \in M$ is a distinguished element, we may expect the homogeneous part $T^1_Y(-R^*)$ to be special.

Theorem: ([Al 1]) *Let σ = cone(Q) be a $\mathbb{Q}$-Gorenstein cone which is at least three-dimensional. Assume that T^1_Y is finite-dimensional (for instance, if Y_σ has an isolated singularity). Then,*

(1) Y_σ is rigid unless it is Gorenstein.

(2) Let Y_σ be Gorenstein, i.e. $g = 1$; then $T^1_Y = T^1_Y(-R^*) = V_{\mathbb{C}}(Q)/\mathbb{C}\cdot \underline{1}$. If, moreover, Y_σ is smooth in codimension three, then Y_σ is rigid, too.

If the restriction dim $T^1_Y < \infty$ is dropped, the situation will be more complicated. Examples are the three-dimensional, non-isolated Gorenstein singularities; they are given by plane lattice polygons Q. Each Q-edge $\overline{a^i a^{i+1}}$ (with $a^{m+1} := a^1$) of length n stands for a one-dimensional A_{n-1}-singularity inside Y_σ. Besides deformations with degree in $-\mathbb{N}\cdot R^*$, edges of length n provide $(n-1)$ infinite series of degrees $R \in M$ with dim $T^1_Y(-R) = 1$. Since

$$T^2_A(-R)^* = {\bigcap_{i=1}^m(\text{span}_{\mathbb{C}} K^R_{i,i+1})}\Big/{\text{span}_{\mathbb{C}}(\bigcap_{i=1}^m K^R_{i,i+1})}\,,$$

these deformations are, for each edge separately, unobstructed. They lift the versal deformation of A_{n-1}. Details may be found in [Al 4].

4 q-parameter families

(4.1) In the previous chapter we have seen that infinitesimal deformations of Y_σ in degree $-R$ are closely related to Minkowski summands of the crosscut $Q(R) = \sigma \cap [R = 1]$. Now, we would like to describe how certain Minkowski decompositions of $Q(R)$ give rise to genuine deformations living over smooth parameter spaces. We begin with negative degrees; let $R \in \sigma^\vee \cap M$. Assume we are given a Minkowski decomposition

$$Q(R) = Q_0 + Q_1 + \cdots + Q_q$$

meeting the following conditions:

(i) $Q_0 \subseteq [R = 1]$ and $Q_1, \ldots, Q_q \in [R = 0]$ are polyhedra with $Q(R)^\infty$ as their common cone of unbounded directions.

(ii) Each supporting hyperplane t of $Q(R)$ defines the faces $F(Q_0, t), \ldots, F(Q_q, t)$ of the indicated polyhedra; obviously, their Minkowski sum equals $F(Q(R), t)$. With at most one exception (depending on t), these faces should contain lattice vertices, i.e. vertices belonging to N.

Remark: If $\sigma = \operatorname{cone}(Q)$ defines a Gorenstein singularity as in (2.7), then for $R = R^*$ the above conditions become easier: Since $Q(R^*) = Q$ is a lattice polytope, all summands $Q_0, \ldots, Q_q$ have to be lattice polytopes, too.

These data provide a q-parameter deformation of Y_σ in degree $-R$ by the following *construction*:
Defining $\tilde{N} := N \oplus \mathbb{Z}^q$ (and $\tilde{M} := M \oplus \mathbb{Z}^q$), we embed the Minkowski summands as $(Q_0, 0)$, $(Q_1, e^1), \ldots, (Q_q, e^q)$ into the vector space $\tilde{N}_{\mathbb{R}}$; here $\{e^1, \ldots, e^q\}$ denotes the canonical basis of $\mathbb{Z}^q$. Together with $(Q(R)^\infty, 0)$, these polyhedra generate a cone $\tilde{\sigma} \subseteq \tilde{N}$ containing σ via $N \hookrightarrow \tilde{N}$, $a \mapsto (a; \langle a, R\rangle, \ldots, \langle a, R\rangle)$. Actually, σ equals $\tilde{\sigma} \cap N_{\mathbb{R}}$, and we obtain an inclusion $Y_\sigma \hookrightarrow X_{\tilde{\sigma}}$ between the associated toric varieties.
On the other hand, $[R, 0] : \tilde{N} \to \mathbb{Z}$ and $\mathrm{pr}_{\mathbb{Z}^q} : \tilde{N} \to \mathbb{Z}^q$ induce regular functions $g : X_{\tilde{\sigma}} \to \mathbb{C}$ and $(g^1, \ldots, g^q) : X_{\tilde{\sigma}} \to \mathbb{C}^q$, respectively. If restricted on Y_σ, they coincide.

Theorem: ([Al 1]) *The resulting map* $(g^1 - g, \ldots, g^q - g) : X_{\tilde{\sigma}} \to \mathbb{C}^q$ *is flat and has* $Y_\sigma \hookrightarrow X_{\tilde{\sigma}}$ *as special fiber.*

(4.2) Having both a description of T^1_Y and a recipe to construct certain deformations, one should ask for the Kodaira-Spencer map. Here it comes: Let $R \in \sigma^\vee \cap M$ and $Q(R) = Q_0 + \ldots + Q_q$ be a decomposition satisfying (i) and (ii) mentioned above. Denote by $(\bar{a}^i)_v$ the vertex of Q_v induced from $\bar{a}^i \in Q(R)$, i.e. $\bar{a}^i = (\bar{a}^i)_0 + \ldots + (\bar{a}^i)_q$.

Theorem: ([Al 4]) *The Kodaira-Spencer map of the corresponding toric deformation $X_{\tilde{\sigma}} \to \mathbb{C}^q$ is*

$$\varrho : \mathbb{C}^q = T_{\mathbb{C}^q,0} \longrightarrow T^1_Y(-R) \subseteq V_{\mathbb{C}}(R) \oplus W_{\mathbb{C}}(R)/\mathbb{C} \cdot (\underline{1}, \underline{1})$$

sending e^v onto the pair $[Q_v, \underline{s}^v] \in V(R) \oplus W(R)$ $(v = 1, \ldots, q)$ with

$$s_i^v := \begin{cases} 0 & \text{if the vertex } (\bar{a}^i)_v \text{ of } Q_v \text{ belongs to the lattice } N \\ 1 & \text{if } (\bar{a}^i)_v \text{ is not a lattice point.} \end{cases}$$

Remark: Setting $e^0 := -(e^1 + \cdots + e^q)$, we obtain $\varrho(e^0) = [Q_0, \underline{s}^0]$ with $\underline{s}^0$ defined similar to $\underline{s}^v$ in the previous theorem.

(4.3) If $R \notin \sigma^\vee \cap M$, then the previous construction fails. Moreover, in [Al 1] it was shown that deformations such that

- the total space together with the embedding of the special fiber still belong to the toric category and
- the Kodaira-Spencer map points injectively into some homogeneous piece $T^1_Y(-R) \subseteq T^1_Y$

are of the above type, including $R \in \sigma^\vee \cap M$. Nevertheless, in [Al 4] it has been shown that Minkowski decompositions of $Q(R)$ fulfilling (i) and (ii) yield q-parameter deformations in the general case, too. They are obtained from the previous construction for the adapted cone $\sigma' := \sigma \cap [R \geq 0]$ via blowing down. This causes the total space to lose its toric structure.

(4.4) Examples:

(4.4.1) *Cones over Del Pezzo surfaces:* Del Pezzo surfaces arising from $\mathbb{P}^2$ by blowing up not more than three points are toric. The affine cones over them are Gorenstein. Hence they fit in both patterns (2.6) and (2.7), and the polytopes $P \subseteq M_{\mathbb{R}}$ and $Q \subseteq N_{\mathbb{R}}$ are mutually dual.
Let Q be the hexagon from (2.2); it provides the cone over the Del Pezzo surface of degree 6. Identifying the hyperplanes $[R = 0]$ and $[R = 1]$, the two different Minkowski decompositions shown in (2.2) induce families with one and two parameters, respectively. The total spaces are the cones over $\mathbb{P}^1 \times \mathbb{P}^1 \times \mathbb{P}^1$ and $\mathbb{P}^2 \times \mathbb{P}^2$.

(4.4.2) *Pinkham's example:* The two-dimensional cone $\sigma := \langle (1,0), (-1,4) \rangle$ yields the cone over the rational normal curve of degree four; see Example (2.4)(3). T^1_Y consists of two one-dimensional parts in the degrees $-[1,1]$ and $-[1,3]$, respectively, and of a two-dimensional part in degree $-[1,2]$. Considering the latter, we obtain the real interval $[-1/2, 1/2] \subseteq \mathbb{R}^1$ as crosscut

$Q(R)$. According to our rules (i) and (ii), it allows two different Minkowski decompositions

$$[-1/2, 1/2] = [-1/2, 0] + [0, 1/2] = \{1/2\} + [-1, 0]$$

inside $I\!R^1 \cong [R = 0] \cong [R = 1]$. The associated one-parameter deformations yield two lines in the versal base space of Y_σ; they belong to the two different components of dimension one and three, respectively.

(4.4.3) The A_{n-1}-*singularity* from (1.4)(2) is given by the cone $\sigma = \langle(1,0), (1,n)\rangle$. The vector space T^1_Y is spread in the degrees $-[k,0]$ with $2 \le k \le n$, and for each k the previous method yields the perturbation $(z^n - xy) + t^{(k)} z^{n-k}$.
On the other hand, it is even possible to get all these families at one blow: despite T^1_Y vanishing in this degree, we consider $R^* = [1,0]$. The cross cut $Q(R^*)$ is an integer interval of length n. It splits into a Minkowski sum of n unit intervals causing an $(n-1)$-parameter family via the previous construction. However, since $T^1_Y(-R^*) = 0$, the Kodaira-Spencer map vanishes. The reason is that we have not obtained the versal deformation of A_{n-1}, but an $n!$-fold cover of it.

5 Versal deformations

(5.1) The *two-dimensional toric singularities*, i.e. the two-dimensional cyclic quotient singularities from Example (2.4)(3), have been investigated in [Ri], [KoSh], [Arn], [Ch], and [St 1]. Using Kollár/Shepherd-Barron's fundamental result about the one-to-one correspondence between so-called P-resolutions and components of the versal base space, Christophersen and Stevens have been able to give a detailed description of these components:

For a given two-dimensional cone σ let $a_2, \ldots, a_{w-1}$ be the positive integers introduced in (3.4). Since the A_{n-1}-singularities behave a little bit differently, assume $w \ge 4$ for simplicity. We define $K(\sigma)$ as the finite set

$$\begin{aligned} K(\sigma) \;:=\; \Big\{(k_2, \ldots, k_{w-1}) \in Z\!\!\!Z^{w-2} \,\Big|\, & 0 \le k_v \le a_v, \text{ and} \\ & \text{the continued fraction } k_2 - \tfrac{1}{k_3 - \ldots} \text{ is well defined and yields } 0\Big\}. \end{aligned}$$

Theorem: ([Ch], [St 1]) *The set $K(\sigma)$ parametrizes the irreducible components of the reduced, versal base space of Y_σ. Moreover, the component assigned to $\underline{k}$ is smooth and has dimension $\sum_{v=2}^{w-1}(a_v - 2k_v + 3) - 4$. Via the Kodaira-Spencer map it sits in the degrees $p \cdot r^v$ with $2 \le v \le w-1$ and $1 \le p \le a_v - k_v$ and, additionally, with the second dimension in some of the*

r^v with $3 \le v \le w-2$.
The element $(1,2,2,\ldots,2,1) \in K(\sigma)$ represents the Artin component.

In [Al 5] there is a straightforward description of how to find, to a given chain representing 0, the P-resolution as an explicit subdivision of the cone σ. The paper makes use of the fact that the singularities allowed on P-resolutions, the so-called T-singularities (cf. [KoSh]), may be easily described in combinatorial terms. They are the two-dimensional toric singularities arising from cones over intervals with integer length, i.e. over shifted lattice intervals.

(5.2) In higher dimensions, we have no correspondence relating deformation theory to certain partial resolutions which might be easier to describe. However, in [Al 3] the approach via Minkowski decompositions was used to obtain the versal deformation, even with its non-reduced structure, for isolated, three-dimensional, toric Gorenstein singularities. The concept works similarly in more general cases, i.e. for arbitrary dimensions and without the Gorenstein assumption, as long as the following two conditions are satisfied:

- Y_σ is smooth in codimension two (excluding the two-dimensional cyclic quotients) and
- T^1_Y is concentrated in one single multidegree $-R \in M$.

It is Theorem (3.8) that tells us that the isolated, three-dimensional Gorenstein singularities fulfill these assumptions (with $R = R^*$). For the sake of simplicity, we will use the rest of the paper to focus on this special case.

(5.3) Let $Q \subseteq \mathbb{R}^2$ be a lattice polygon such that the edges do not contain any interior lattice points. In particular, they are given by clockwise oriented primitive vectors $d^1, \ldots, d^m \in \mathbb{Z}^2$. The vector space $V(Q) \subseteq \mathbb{R}^m$ is given by the single equation $\sum_i t_i\, d^i = 0$; hence it is $(m-2)$-dimensional. Recalling from (2.3) that elements $\underline{t} \in C(Q) = V(Q) \cap \mathbb{R}^m_{\ge 0}$ represent Minkowski summands $Q_{\underline{t}}$ of $\mathbb{R}_{\ge 0} \cdot Q$, we may define the tautological cone as

$$\tilde{C}(Q) := \{(a, \underline{t}) \mid \underline{t} \in C(Q);\ a \in Q_{\underline{t}}\} \subseteq \mathbb{R}^2 \times V(Q) \subseteq \mathbb{R}^2 \times \mathbb{R}^m .$$

$\tilde{C}(Q)$ is, like $C(Q)$, a rational, polyhedral cone. By definition, there is a natural projection $\pi : \tilde{C}(Q) \twoheadrightarrow C(Q)$ yielding $\sigma := \mathrm{cone}(Q)$ as the pull back of $\mathbb{R}_{\ge 0} \cdot \underline{1} \subseteq C(Q)$. We obtain the following fiber product diagram of rational polyhedral cones:

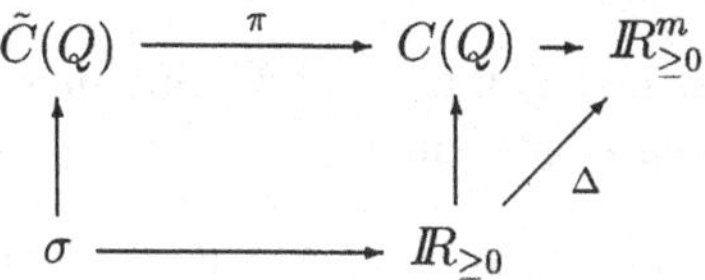

with Δ denoting the diagonal embedding. Applying the functor of building toric varieties from cones yields affine varieties $X_{\tilde{C}(Q)}$ and $S_{C(Q)}$ fitting into the fiber product diagram

$$\begin{array}{ccccccc}
X_{\tilde{C}(Q)} & \xrightarrow{\pi} & S_{C(Q)} & \longrightarrow & \mathbb{C}^m & \xrightarrow{\ell} & \mathbb{C}^m/\mathbb{C}\cdot\underline{1} \\
\uparrow & & \uparrow & \nearrow_{\Delta} & & & \uparrow \\
Y_\sigma & \longrightarrow & \mathbb{C} & & \longrightarrow & & \{0\}
\end{array}$$

with closed embeddings in the vertical directions. The map $S_{C(Q)} \to \mathbb{C}^m$ is finite, and, for the sake of simplicity, we will pretend in the following that it is a closed embedding, too. Denote by $\bar{\mathcal{M}} \subseteq \mathbb{C}^m/\mathbb{C}\cdot\underline{1}$ the largest closed subscheme such that $\mathcal{M} := \ell^{-1}(\bar{\mathcal{M}}) \subseteq \mathbb{C}^m$ is also contained in $S_{C(Q)}$. Theorem (3.8) says that both $\bar{\mathcal{M}}$ and T^1_Y sit in the same space $\mathbb{C}^m/\mathbb{C}\cdot\underline{1}$.

Theorem: ([Al 3]) *Restricting the map $X_{\tilde{C}(Q)} \to \mathbb{C}^m/\mathbb{C}\cdot\underline{1}$ onto $\bar{\mathcal{M}} \subseteq \mathbb{C}^m/\mathbb{C}\cdot\underline{1}$, the outer frame of the above commutative diagram yields the versal deformation of Y_σ.*

(5.4) What does $\bar{\mathcal{M}}$ look like? To answer this question, we define for each integer $k \geq 1$ the (vector-valued) polynomial

$$g_k(t_1,\ldots,t_m) := \sum_{i=1}^{m} t_i^k\, d^i$$

generating an ideal

$$\mathcal{J} := (g_k(\underline{t}) \mid k \geq 1) \subseteq \mathbb{C}[t_1,\ldots,t_m]\,.$$

Since $\sum_{i=1}^m d^i = 0$, this ideal may alternatively be generated by $g_k(t_1 - t_1,\ldots,t_m - t_1)$ with $k \geq 1$, i.e. $\mathcal{J}$ originally comes from the subring $\mathbb{C}[t_i - t_j \mid 1 \leq i,j \leq m]$.

Theorem: ([Al 3])

(1) $\mathcal{M}$ and $\bar{\mathcal{M}}$ are the affine schemes associated to the ideals $\mathcal{J} \subseteq \mathbb{C}[\underline{t}]$ and $\mathcal{J} \cap \mathbb{C}[t_i - t_j] \subseteq \mathbb{C}[t_i - t_j]$, respectively.

(2) If Q is contained in two different strips defined by pairs of parallel lines of lattice-distance $\leq k_0$ each, then the polynomials g_k with $k > k_0$ are not necessary for generating $\mathcal{J}$ or $\mathcal{J} \cap \mathbb{C}[t_i - t_j]$.

Since $g_1(\underline{t})$ are the linear forms defining $V(Q) \subseteq \mathbb{R}^m$, we see that $\mathcal{M}$ is not only contained in $\mathbb{C}^m$, but also in $V_{\mathbb{C}}(Q)$.

Example: Let Q be the hexagon from (2.2). Starting with $d^1 := \overline{(0,0)(1,0)}$, the anticlockwise oriented edges equal

$$d^1 = (1,0);\;\; d^2 = (1,1);\;\; d^3 = (0,1);$$

$$d^4 = (-1,0); \;\; d^5 = (-1,-1); \;\; d^6 = (0,-1).$$

Obviously, Q is contained in at least three different strips of thickness two. Hence, $\mathcal{J}$ is generated in degree ≤ 2. We obtain

$$\mathcal{J} = (t_1 + t_2 - t_4 - t_5,\;\; t_2 + t_3 - t_5 - t_6,\;\; t_1^2 + t_2^2 - t_4^2 - t_5^2,\;\; t_2^2 + t_3^2 - t_5^2 - t_6^2).$$

(5.5) Finally, we would like to mention the structure of the underlying reduced spaces of $\mathcal{M}$ and $\bar{\mathcal{M}}$. Let $Q = R_0 + \cdots + R_q$ be a decomposition of Q into a Minkowski sum of $q+1$ lattice polytopes. Then, denoting with $[R_v] \in C(Q)$ the point representing the summand R_v, the m-tuples $[R_0], \ldots, [R_q] \in I\!R^m$ have only 0 and 1 as entries and sum up to $(1, \ldots, 1)$. In particular, the $(q+1)$-plane $\mathbb{C} \cdot [R_0] + \ldots + \mathbb{C} \cdot [R_q] \subseteq \mathbb{C}^m$ (or its q-dimensional image via ℓ) is contained in $\mathcal{M}$ (in $\bar{\mathcal{M}}$, respectively). It is not difficult to see that there are no other points, i.e.

$\mathcal{M}_{red}$ and $\bar{\mathcal{M}}_{red}$ equal the union of those flats corresponding to maximal Minkowski decompositions of Q into lattice polytopes.

Example: Considering the hexagon again, the linear equations of $\mathcal{M}$ allow the substitution $t := t_1$, $s_1 := t_1 - t_3$, $s_2 := t_4 - t_2$, and $s_3 := t_1 - t_4$. The two quadratic equations transform into $s_1\, s_3 = s_2\, s_3 = 0$. Since $\mathcal{M} = \ell^{-1}(\bar{\mathcal{M}})$, the equations do not contain t; it is still a variable for $\mathcal{M}$, but not for $\bar{\mathcal{M}}$. We see that $\bar{\mathcal{M}}$ is the union of a line and a two-plane. The corresponding one- and two-parameter families are exactly those obtained in (4.4.1) corresponding to the Minkowski decompositions drawn in (2.2).

(5.6) In the same manner as we did with the hexagon, we may investigate the lattice rectangle $Q = \operatorname{conv}\{(0,0);\, (2,1);\, (2,2);\, (1,2)\}$ corresponding to the cone over the Del Pezzo surface of degree eight. Here the versal base space equals the fat point $\operatorname{Spec} \mathbb{C}[\varepsilon]/\varepsilon^2$.

References

[Al 1] Altmann, K.: Minkowski sums and homogeneous deformations of toric varieties. Tôhoku Math. J. **47** (1995), 151-184.

[Al 2] Altmann, K.: Obstructions in the deformation theory of toric singularities. E-print alg-geom/9405008; to appear in J. Pure Appl. Algebra.

[Al 3] Altmann, K.: The versal deformation of an isolated, toric Gorenstein singularity. E-print alg-geom/9403004; to appear in Invent. Math.

[Al 4] Altmann, K.: One-parameter families containing three-dimensional toric Gorenstein singularities. E-print alg-geom/9609006.

[Al 5] Altmann, K.: P-Resolutions of cyclic quotients from the toric viewpoint. E-print alg-geom/9602018.

[AlSl] Altmann, K.; Sletsjøe, A.B.: André-Quillen cohomology of monoid algebras. E-print alg-geom/9611014.

[Arn] Arndt, J.: Verselle Deformationen zyklischer Quotientensingularitäten. Dissertation, Universität Hamburg, 1988.

[Art] Artin, Michael: Lectures on deformations of singularities. Bombay: Tata Institute of Fundamental Research, 1976.

[Bat] Batyrev, V.V.: Dual polyhedra and mirror symmetry for Calabi-Yau hypersurfaces in toric varieties. J. Algebraic Geometry **3** (1994), 493-535.

[Ch] Christophersen, J.A.: On the components and discriminant of the versal base space of cyclic quotient singularities. In: Singularity Theory and its Applications, Warwick 1989, Part I: Geometric Aspects of Singularities, pp. 81-92, Springer-Verlag Berlin Heidelberg, 1991 (LNM 1462).

[Da] Danilov, V.I.: The geometry of toric varieties. Russian Math. Surveys **33**/2 (1978), 97-154.

[Fu] Fulton, W.: Introduction to toric varieties. Annals of Mathematics Studies **131**. Princeton, New Jersey, 1993.

[Ish] Ishida, M.-N.: Torus embeddings and dualizing complexes. Tôhoku Math. Journ. **32**, 111-146 (1980).

[Ke] Kempf, G., Knudsen, F., Mumford, D., Saint-Donat, B.: Toroidal embeddings I. Lecture Notes in Mathematics **339**, Springer-Verlag, Berlin-Heidelberg-New York, 1973.

[Kod] Kodaira, K.: Complex manifolds and deformation of complex structures. Grundlehren der mathematischen Wissenschaften **283**. Springer-Verlag 1986.

[KoSh] Kollár, J., Shepherd-Barron, N.I.: Threefolds and deformations of surface singularities. Invent. Math. **91**, 299-338 (1988).

[Lo] Loday, J.-L.: Cyclic homology. Grundlehren der mathematischen Wissenschaften **301**, Springer-Verlag 1992.

[Od] Oda, T.: Convex bodies and algebraic geometry. Ergebnisse der Mathematik und ihrer Grenzgebiete (3/15), Springer-Verlag 1988.

[Pa] Palamodov, V.P.: Deformations of complex spaces. Russian Math. Surveys **31** (1976), 129-197.

[Pi] Pinkham, H.: Deformations of algebraic varieties with G_m-action. Asterisque **20** (1974), 1-131.

[Ri] Riemenschneider, O.: Deformationen von Quotientensingularitäten (nach zyklischen Gruppen). Math. Ann. **209** (1974), 211-248.

[Sl] Sletsjøe, A.B.: Cohomology of monoid algebras. J. of Algebra **161** No. 1 (1993), 102-128.

[St 1] Stevens, J.: On the versal deformation of cyclic quotient singularities. In: Singularity Theory and its Applications, Warwick 1989, Part I: Geometric Aspects of Singularities, pp. 302-319, Springer-Verlag Berlin Heidelberg, 1991 (LNM 1462).

[St 2] Stevens, J.: Deformations of singularities. Habilitationsschrift, Universität Hamburg, 1995.

[Sm] Smilansky, Z.: Decomposability of polytopes and polyhedra. Geometriae Dedicata. **24** (1987), 29-49.

[Zi] Ziegler, G.M.: Lectures on polytopes. Graduate Texts in Mathematics **152** (1995), Springer-Verlag.

Institut für reine Mathematik
Humboldt-Universität zu Berlin
Ziegelstr. 13A
D-10099 Berlin
Germany

e-mail: altmann@mathematik.hu-berlin.de

Critical Points of Affine Multiforms on the Complements of Arrangements

James Damon*

To Terry Wall on his sixtieth birthday.

Introduction

We consider an affine hyperplane arrangement $A \subset \mathbb{C}^n$ which is a finite union of hyperplanes H_i defined by affine linear forms $\ell_i = 0$. In his work on the classical asymptotics of the KZ equation [Va], Varchenko conjectures

For generic complex values λ_i, the function (multiform) defined as a product of powers of affine linear forms $f = \prod \ell_i^{\lambda_i}$ has only nondegenerate critical points and the number of critical points equals the Euler characteristic of the complement of A.

He proves this result when A is the complexification of a real arrangement, extending an earlier result of Aomoto [Ao]. In [OT1], Orlik and Terao give a proof of this conjecture. They do so by relating both of these numbers to the Euler characteristic of a certain complex of logarithmic differential forms.

In this paper, we prove several results (Theorems 1 and 2) which also yield this conjecture as a corollary. These theorems give a topological proof of the conjecture and illuminate the direct relation between the Euler characteristic and the number of critical points. More generally, these results extend Varchenko's conjecture to nonlinear arrangements A on smooth complete intersections $X \subset \mathbb{C}^n$ so that f is now a product of higher degree polynomials restricted to X. We express the relative Euler characteristic as the sum of Milnor numbers of the critical points of $f|X\backslash A$ (allowing degenerate critical points). Moreover, we show this relative Euler characteristic equals a "relative singular Milnor number" of a "singular Milnor fibration" on the complete intersection X.

These results are valid when the λ_i take positive integer values. However, by applying an observation of Varchenko, and extending a lemma of Orlik-Terao, we still obtain the full conjecture. Moreover, in the case A is an "almost free arrangement", we can explicitly give a formula for the number of critical points in terms of the exponents of A and the degrees of the defining equations of X.

*Partially supported by a grant from the National Science Foundation

The basis for these results is the point of view used in [D1] that a central arrangement (in which all of the hyperplanes H_i pass through 0) defines a (highly) nonisolated singularity at 0, and any affine arrangement is the "singular Milnor fiber" of a generic hyperplane section of such a central arrangement. Then, local singular properties of the central arrangement translate into local properties of the "singular Milnor fiber" of the section, which in turn yield global properties of the affine arrangement. This applies as well to nonlinear arrangements on complete intersections. In particular, a weak version of Varchenko's conjecture for the sum of the Milnor numbers in the case where all $\lambda_i = 1$ already follows from the proof of proposition 5.2 of [D1] using the methods of [DM, 4]. To obtain the full strength of Varchenko's conjecture we allow a degeneracy in the section (which occurs when there are parallel flats in the affine arrangement), and then the singular Milnor fiber arises via translation of the section.

In §1, we recall the basic properties of algebraic transversality to singular varieties and introduce a version for multiforms. We also recall relevant properties of singular Milnor fibers and numbers. In §2, we state the local version of the main theorem and deduce the global results, including Varchenko's conjecture, giving several consequences and examples in §3. In §4, we establish the basic local result that relates the Euler characteristic of a nonisolated singularity Z defined by f on the Milnor fiber of a complete intersection V with the sum of the Milnor numbers of f restricted to the Milnor fiber of V. This extends the line of argument of Siersma [Si] and Looijenga [Lo], which in turn, extend the original argument of Lê [Lê2]. Then, in §5 we deduce the various results and their corollaries, including specific formulas for the number of critical points.

This author is especially grateful to both Alexander Varchenko and Hiroaki Terao for valuable conversations regarding these questions.

1 Transversality and Singular Milnor Fibers

We consider in this section the local singularity theory involving algebraic transversality conditions needed to define and determine the singular Milnor fibrations and numbers. We begin by recalling from [DM], [D1], and [D3] how transversality of singular varieties can be defined and how we can define singular Milnor fibers. Then we extend the transversality conditions to multiforms. *Because later we shall specifically relate local and global situations, we will find it convenient to work with local constructions in* $\mathbb{C}^{n+1}$ *(and global constructions in* $\mathbb{C}^n$*).*

Algebraic and Geometric Transversality :

Let $(V, 0) \subset \mathbb{C}^{n+1}$ be a germ of an analytic variety, and let $I(V)$ denote the ideal of holomorphic germs vanishing on V. If θ_{n+1} denotes the module of germs of holomorphic vector fields on $\mathbb{C}^{n+1}$ at 0, then, following Saito [Sa] we define

$$\mathrm{Derlog}(V) = \{\zeta \in \theta_{n+1} : \zeta(I(V)) \subset I(V)\}$$

which is an $\mathcal{O}_{\mathbb{C}^{n+1},0}$–module. Also if $(V,0)$ is a germ of a complete intersection, defined by $G = (G_1, \cdots, G_p) : \mathbb{C}^{n+1}, 0 \to \mathbb{C}^p, 0$, then

$$\mathrm{Derlog}(G) = \{\zeta \in \theta_{n+1} : \zeta(G) = 0\}.$$

Using these modules we define several analogues of the tangent space at singular points of V. For $z \in V$, we let $T_{log}V_{(z)}$ be the vector space spanned by $\{\zeta_{(z)}$: for ζ ranging over a set of generators of $\mathrm{Derlog}(V)\}$, and similarly for $T_{log}G_{(z)}$ using instead $\mathrm{Derlog}(G)$. Alternately we can consider the canonical Whitney stratification $\{V_i\}$ of V, with $V_0 = \mathbb{C}^{n+1}\backslash V$. Given $z \in V_i$, the geometric tangent space of V at z is $T_z V_i$. Then, as explained in [DM,§3] and [D1,§2],

$$T_{log}G_{(z)} \subseteq T_{log}V_{(z)} \subseteq T_z V_i.$$

If we have equality of the last two tangent spaces for all z in a neighborhood of 0, V is said to be *holonomic* [Sa, §3]. If all three are equal we say V is *h–holonomic*. At smooth points of V, $T_{log}G_{(z)} = T_{log}V_{(z)} = T_z V$; and off V, $T_{log}V_{(z)} = T_z\mathbb{C}^{n+1}$.

We use either of these last two tangent spaces to define the transversality of: smooth submanifolds to singular varieties, singular varieties to singular varieties, or even the transversality of map germs to singular varieties. If we use the algebraic tangent spaces, i.e. $T_{log}V$, then we obtain *algebraic transversality*, and with the geometric tangent spaces, *geometric transversality*.

By the relation of the tangent spaces, algebraic always implies geometric transversality. The converse holds for holonomic varieties, and then their intersection is again holonomic by [D1,§3 and 7]. If transversality holds in a punctured neighborhood of 0, we say if holds off 0. More generally a collection of varieties $V_1, \ldots, V_m$ are said to be in *(algebraic) general position off 0* if they are in general position using the (algebraic, resp.) geometric tangent spaces. The properties of these notions of transversality are given in [D1,§2,3,7]. For example, V is *algebraically transverse to W off 0* if for all z in a punctured neighborhood of 0

$$T_{log}V_{(z)} + T_{log}W_{(z)} = T_z\mathbb{C}^{n+1}.$$

Example 1.1 In the case of a central hyperplane arrangement $\mathcal{A}$ $(= \cup H_i)$ $\subset \mathbb{C}^{n+1}$, $\mathcal{A}$ can be thought of as a hypersurface singularity at 0 defined by $Q = \prod L_i = 0$, with L_i the defining linear equation for H_i, $i = 1, \ldots, s$. Then, for $z \in \mathcal{A}$, $T_{log}Q_{(z)} = T_{log}\mathcal{A}_{(z)}$ equals the flat of $\mathcal{A}$ containing z $(= \cap H_i$ for those H_i containing $z)$, which is the geometric tangent space. Thus, $\mathcal{A}$ is h–holonomic. Similarly, an isolated hypersurface singularity, or more generally an isolated complete intersection singularity (an ICIS) V is h–holonomic.

Thus, V is transverse (algebraically and geometrically) to the arrangement $\mathcal{A}$ off 0 if V is transverse (off 0) to each flat K, obtained by intersecting some subcollection of the H_i.

Definition 1.2 A nonlinear arrangement of hypersurfaces $\mathcal{A}, 0 \subset \mathbb{C}^{n+1}, 0$ based on a central hyperplane arrangement $A' \subset \mathbb{C}^m$ is defined by $\mathcal{A} = \Phi^{-1}(A')$ where $\Phi : \mathbb{C}^{n+1}, 0 \to \mathbb{C}^m, 0$ is a map which is (algebraically) transverse to A' off 0 (i.e. Φ is transverse to each flat of A' in a punctured neighborhood of 0 in $\mathbb{C}^{n+1}$).

Remark: By the transversality properties in [D1,§2,3], a nonlinear arrangement of hypersurfaces is h–holonomic.

Singular Milnor Fibers :

We next recall singular Milnor fibers and numbers, which play a key role in the various generalizations of Varchenko's conjecture. Suppose that $V' \subset \mathbb{C}^{p'}$ is a complete intersection of dimension k, not necessarily isolated. Let $f : \mathbb{C}^{n'}, 0 \to \mathbb{C}^{p'}, 0$ be (algebraically) transverse to V' off 0 and let $V = f^{-1}(V')$ as in (1.3).

$$
\begin{array}{ccc}
\mathbb{C}^{n'}, 0 & \xrightarrow{f} & \mathbb{C}^{p'}, 0 \\
\uparrow & & \uparrow \\
V = f^{-1}(V') & \longrightarrow & V'
\end{array}
\qquad (1.3)
$$

Then, we can consider a (topological) stabilization of f, $f_t : U \to \mathbb{C}^{p'}$ where U is a neighborhood of 0, f_0 is a representative of f, and f_t is geometrically transverse to a representative of V' on U. Then, using a theorem of Lê [Lê1], it can be proven, [DM, §4], [D1,§7] and see [D3], that on a sufficiently small ball B_ϵ the topological type of $V_t = f_t^{-1}(V') \cap B_\epsilon$ only depends on f and is homotopy equivalent to a bouquet of spheres of (real) dimension $k + n' - p'$. It is called the *singular Milnor fiber* of f. The number of spheres is called the *singular Milnor number* of f and denoted $\mu_{V'}(f)$, or just $\mu(V)$ if the map germ f is clear from the context.

In the case i denotes the inclusion of a sufficiently generic k-dimensional plane, then $\mu_{V'}(i)$ is independent of i and is denoted by $\mu_k(V')$, the k-th higher multiplicity. These multiplicities extend the higher multiplicities introduced by Teissier [Te] for hypersurfaces, and are closely related to those introduced by Lê–Teissier [Lê–T] for general singularities.

The singular Milnor numbers and higher multiplicities can be explicitly computed in the case that V' is a *free divisor* (which means, see [Sa], that Derlog(V') is a free $\mathcal{O}_{\mathbb{C}^{p'},0}$–module, necessarily of rank p'), then V is called an *almost free divisor* and $\mu_{V'}(f)$ can be computed as $\nu_{V'}(f)$, the length of a determinantal module, see [DM, thms 5,6] and [D1, §4]. This is further extended in [D1,§7,8] to *almost free complete intersections*, which are the transverse intersections off 0 of almost free divisors.

To see how singular Milnor fibers and numbers are relevant for Varchenko's conjecture, consider $\mathcal{V}$ an ICIS defined by $\mathcal{V} = G'^{-1}(0)$ for homogeneous $G' : \mathbb{C}^{n+1}, 0 \to \mathbb{C}^p, 0$ with $\mathcal{V}$ transverse to $\mathbb{C}^n = \mathbb{C}^n \times \{0\}$ off 0. Then $V = \mathcal{V} \cap \mathbb{C}^n$ is an ICIS defined by $G = (G', z_{n+1})$. Suppose $\mathcal{A}$ is a nonlinear arrangement of hypersurfaces and that V is algebraically transverse to $\mathcal{A}$ off 0. Then $V \cap \mathcal{A}$ is a complete intersection. For $|t|$ and $\epsilon > 0$ sufficiently small, $V_{t,\epsilon} = \mathcal{V} \cap (\mathbb{C}^n \times \{t\}) \cap B_\epsilon$ is the Milnor fiber of V; and moreover, $V_{t,\epsilon} \cap \mathcal{A}$ is the "singular Milnor fiber" of $V \cap \mathcal{A}$.

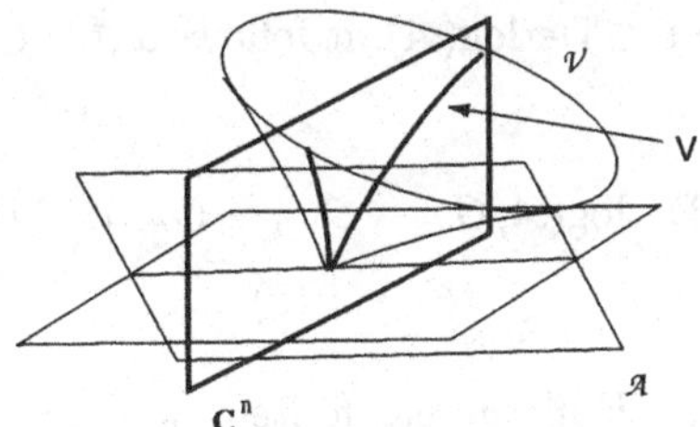

Figure 1.

As explained above, $V_{t,\epsilon} \cap \mathcal{A}$ is homotopy equivalent to a bouquet of spheres of real dimension $n - p - 1$. Furthermore, it follows from [D1] and [DM, §4] that the number of such spheres, which is the "singular Milnor number" of $V \cap \mathcal{A}$, equals the sum of the Milnor numbers of $Q|\, V_{t,\epsilon} \backslash \mathcal{A}$.

To extend these results to arbitrary affine arrangements (which may have parallel flats) we must allow a degeneracy in the condition on the section of $\mathcal{V}$. In addition, we show that the preceding results remain valid globally on $X = \mathcal{V} \cap (\mathbb{C}^n \times \{1\})$ even when Q is replaced by a multiform $F = \prod L_i^{\lambda_i}$, where now λ_i denote arbitrary positive integers.

Transversality to Multiforms :

This extension depends upon requiring "transversality to the multiform F". We follow Varchenko and Orlik-Terao and consider by logarithmic differentiation

$$(1.4) \qquad dF = F \cdot \left(\sum_{i=1}^{s} \lambda_i \frac{dL_i}{L_i} \right) = F \cdot \omega_\lambda \qquad \text{where} \qquad \omega_\lambda = \sum_{i=1}^{s} \lambda_i \frac{dL_i}{L_i}.$$

If $F(z) \neq 0$, then $\ker(dF(z)) = \ker(\omega_\lambda(z))$.

First, consider the case of a central hyperplane arrangement $\mathcal{A}$ with $F = Q$ (so all $\lambda_i = 1$ and we denote $\omega = \omega_{\mathbf{1}}$ where $\mathbf{1} = (1, \ldots, 1)$). Then, algebraic transversality of V to $\mathcal{A}$ off 0 actually implies a stronger result. For $z \in V \backslash \mathcal{A}$, $dQ(z)|T_zV \neq 0$ (see e.g. [DM,§3] or [D1,lemma 2.10]). This is equivalent to $\omega|T_zV \neq 0$ for any $z \in V \backslash \mathcal{A}$. Moreover, for $z \in V \cap \mathcal{A}$, if V is homogeneous then the Euler vector field e is tangent to V, and $\omega(e) = (1/Q)dQ(e) = s$. Thus, *$V$ is transverse to Q off 0* in the sense that $\omega|T_zV \neq 0$ in a (punctured) neighborhood of 0. This form of transversality has a natural extension to F (see also [OT1,§3]).

For an ICIS V defined by G and a (nonlinear) arrangement $\mathcal{A}$, we consider Derlog$(\mathcal{A}, G)$ defined to be Derlog$(\mathcal{A})\cap$Derlog(G). If we sheafify this module we obtain an object, also denoted by Derlog$(\mathcal{A}, G)$, which is still coherent.

Definition 1.5 We say that V is transverse to the multiform $F = \prod L_i^{\lambda_i}$ off 0 if for each $z \in V\backslash\{0\}$ in a punctured neighborhood of 0, there is $\zeta \in$ Derlog$(\mathcal{A}, G)$ such that $\omega_\lambda(\zeta) \neq 0$ at z.

By Saito [Sa,§1], since $\zeta \in$ Derlog$(\mathcal{A})$, it follows $\omega_\lambda(\zeta) \in \mathcal{O}_{\mathbb{C}^{n+1},0}$. We define the map

$$\begin{aligned} ev_F : \text{Derlog}(\mathcal{A}, G) &\to \mathcal{O}_{V,0} \simeq \mathcal{O}_{\mathbb{C}^{n+1},0}\backslash I(V) \\ \zeta &\mapsto \omega_\lambda(\zeta). \end{aligned} \tag{1.6}$$

Then, V is transverse to F iff ev_F has finite cokernel.

More generally, let $\mathcal{A} = \Phi^{-1}(A')$ be a nonlinear arrangement based on $A' \subset \mathbb{C}^m$, with A' defined by $Q = \prod L_i = 0$.

Definition 1.7 We say that V is transverse to $F = \prod(L_i \circ \Phi)^{\lambda_i}$ off 0 if for each $z \in V\backslash\{0\}$ in a punctured neighborhood of 0, there is $\zeta \in$ Derlog$(\mathcal{A}, G)$, such that $\Phi^*\omega_\lambda(\zeta) \neq 0$ at z.

As in (1.6) this can be expressed in terms of the finite dimension of the cokernel of ev_F except now we define $ev_F(\zeta) = \Phi^*\omega_\lambda(\zeta)$.

Example 1.8 As G is locally a submersion and V is smooth off 0, if V is transverse to $\mathcal{A}$ off 0, (1.6) reduces to the local case where (V, z) is a smooth subspace transverse to a subarrangement B of $\mathcal{A}$ at z. There is a local Euler vector field $e \in$ Derlog$(B \cap V)_{(z)}$. Then, $\omega_\lambda(e) = (1/F)dF(e) = \sum \lambda_i$, summed over those i for which H_i does not belong to B. For the special case of V a hyperplane through 0, not in $\mathcal{A}$, but which need not be transverse to $\mathcal{A}$ off 0, a calculation of Orlik–Terao [OT1,§4] shows that (1.6) is satisfied if for each subarrangement B with $r(B) \geq$ codim(V), $\sum \lambda_i \neq 0$ summed over those i for which H_i does not belong to B. The reduction to the local linear case shows that (1.6) also holds at z provided $\sum \lambda_i \neq 0$. Note this is always satisfied if all of the λ_i are positive.

2 The Number of Critical Points

We state in this section the main theorems for the critical points of a multiform f, relating the number of nondegenerate critical points (and more generally the sum of the Milnor numbers) of f on a smooth complete intersection X with the relative Euler characteristic of the induced arrangement on X. We deduce the global results, including Varchenko's conjecture, from the main local result.

Notation: *For local results we work in $\mathbb{C}^{n+1}$ and will denote maps and functions with capital letters; while for global results we work in $\mathbb{C}^n$ and use lower case letters.*

Let $G = (G_1, \ldots, G_{p+1}) : \mathbb{C}^{n+1}, 0 \to \mathbb{C}^{p+1}, 0$ define an ICIS V with Milnor fiber $V_t = G^{-1}(t) \cap B_\epsilon$. Also, let $Q = \prod L_i$ define the central hyperplane arrangement $A' \subset \mathbb{C}^m$. We suppose $\mathcal{A} = \Phi^{-1}(A')$ is a nonlinear arrangement of hypersurfaces based on A' (so that $\Phi : \mathbb{C}^{n+1}, 0 \to \mathbb{C}^m, 0$ is transverse to A' in a punctured neighborhood of 0). Moreover, we suppose that V is transverse to $F = \prod (L_i \circ \Phi)^{\lambda_i}$ (as in definition 1.7).

Theorem 1 (Local Version). *Let $F = \prod (L_i \circ \Phi)^{\lambda_i}$ where the λ_i are positive integers. Suppose that the ICIS V is transverse to both F and the (nonlinear) arrangement $\mathcal{A}$ off 0. Then, for ϵ and $|t|$ sufficiently small, F has only isolated singularities on $V_t \backslash \mathcal{A}$ and*

(2.1) (the sum of the Milnor numbers of $F|\, V_t \backslash \mathcal{A}) = |\chi(V_t, V_t \cap \mathcal{A})|$.

Moreover, the relative Euler characteristic is given by

$$|\chi(V_t, V_t \cap \mathcal{A})| = \mu(V) + \mu(V \cap \mathcal{A}). \tag{2.2}$$

Here $\mu(V)$ denotes the Milnor number of V (i.e. G) and $\mu(V \cap \mathcal{A})$ denotes the singular Milnor number of $V \cap \mathcal{A}$, i.e. $V_t \cap \mathcal{A}$ is homotopy equivalent to a bouquet of $n-p-1$-dimensional spheres and the number of such spheres equals $\mu(V \cap \mathcal{A})$.

The sum equals the relative singular Milnor number of $V \cap \mathcal{A}$ on V.

Remark 2.3 In the case that $\Phi = id$ (so $\mathcal{A} = A'$) then (2.1) and (2.2) of theorem 1 give a formula for the sum of the Milnor numbers of $F = \prod L_i^{\lambda_i}$ on $V_t \cap \mathcal{A}$, where $\prod L_i$ defines the central arrangement $\mathcal{A}$. The singular Milnor number $\mu(V \cap \mathcal{A})$ can be explicitly computed using the results of [D1,§6,8].

Next, we deduce from theorem 1 global versions of the results.

Relation between Local and Global Properties.

Again we let $Q = \prod L_i$ define a central arrangement $A' \subset \mathbb{C}^m$. Also, let $\Phi : \mathbb{C}^{n+1}, 0 \to \mathbb{C}^m, 0$ be a weighted homogeneous polynomial mapping with the property that $L_i \circ \Phi$ is homogeneous for each i, so that Q is homogeneous. Then, $\mathcal{A} = \Phi^{-1}(A')$ is a homogeneous nonlinear arrangement defined on all of $\mathbb{C}^{n+1}$. In addition we consider a homogeneous $G' : \mathbb{C}^{n+1}, 0 \to \mathbb{C}^p, 0$ which defines an ICIS $\mathcal{V}$. We introduce:

(2.4) Local transversality conditions :

1. Φ is transverse to A' off 0; and
2. $\mathcal{V}$, $\mathbb{C}^n = \mathbb{C}^n \times \{0\}$, $\mathcal{A} = \Phi^{-1}(A')$ in (algebraic) general position off 0.

We observe several consequences of (2.4). First, by (2) the intersection of any two of $\mathcal{V}$, $\mathbb{C}^n$ and $\mathcal{A}$ remains transverse to the third. Second, since $\mathbb{C}^n$ is transverse to $\mathcal{V}$ off 0, it follows that $\mathbb{C}^n \times \{t\}$ is transverse to $\mathcal{V}$ and $X = \mathcal{V} \cap (\mathbb{C}^n \times \{1\})$ is a smooth complete intersection in $\mathbb{C}^n \simeq \mathbb{C}^n \times \{1\}$. Likewise, $A = \mathcal{A} \cap (\mathbb{C}^n \times \{1\})$ is an affine arrangement of global hypersurfaces which is transverse to X. Third, let $\varphi = \Phi | \mathbb{C}^n \times \{1\}$. Then $q = \prod \ell_i$, where $\ell_i = L_i \circ \varphi$, defines the nonlinear affine arrangement A. We let $f = \prod (L_i \circ \varphi)^{\lambda_i}$.

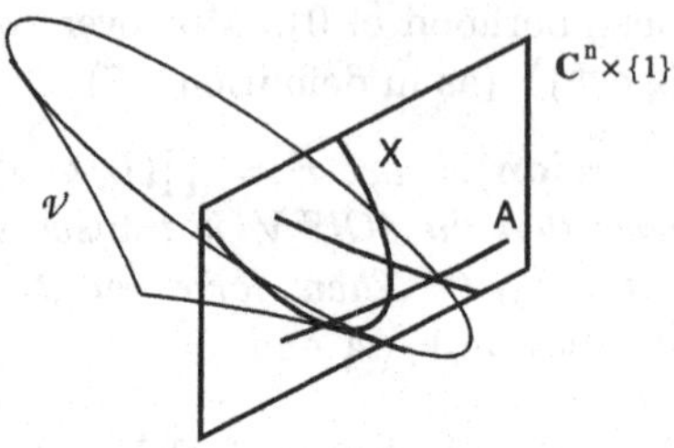

Figure 2.

These spaces are related to the (singular) Milnor fibers. Let $G_0(z) = z_{n+1}$. Then, $V = \mathcal{V} \cap \mathbb{C}^n$ is an ICIS which equals $G^{-1}(0)$ where $G = (G_0, G)$: $\mathbb{C}^{n+1}, 0 \to \mathbb{C}^{p+1}, 0$. Also, there are $\epsilon, \delta > 0$ so that if $|t| < \delta$ then $V_t = \mathcal{V} \cap (\mathbb{C}^n \times \{t\}) \cap B_\epsilon$ is the Milnor fiber of G, and $V_t \cap \mathcal{A}$ is the singular Milnor fiber of the complete intersection $V \cap \mathcal{A}$ (see [D1,§7]). The global spaces X and $X \cap A$ and the local Milnor fibers are related by the following.

Proposition 2.5 *Under the transversality assumptions (2.4), X is homeomorphic to the Milnor fiber V_t of G via a homeomorphism ψ which can be chosen to be the restriction of the $\mathbb{C}^*$-action on any compact subset of X (i.e. it is given by multiplication by t). Moreover, ψ can be chosen to send $X \cap A$ to $V_t \cap \mathcal{A}$ (see fig. 3).*

The proof of this result will be given in §6.

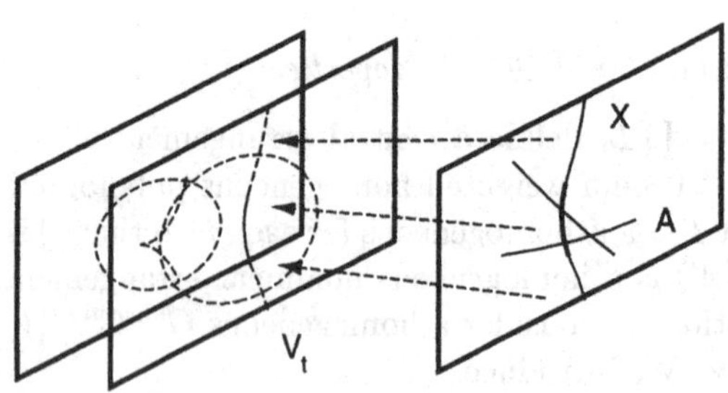

Figure 3.

Conversely, suppose we are given a smooth complete intersection $X \subset \mathbb{C}^n$ defined by g and a nonlinear arrangement $A = \varphi^{-1}(A') \subset \mathbb{C}^n$ for a central hyperplane arrangement $A' \subset \mathbb{C}^m$. If A' is defined by $q = \prod \ell_i$, then A is defined by $\ell_i = L_i \circ \varphi$. We can recover the local complete intersections $\mathcal{V}$ and $\mathcal{A}$ and the map germs G' and Φ by homogenization, e.g. from $g = (g_1, \ldots, g_p) : \mathbb{C}^n, 0 \to \mathbb{C}^p, 0$ defining X, we can recover $G' = (G_1, \ldots, G_p)$ where

$$G_i(z_1, \ldots, z_{n+1}) = (z_{n+1})^{d_i} \cdot g_i(z_1/z_{n+1}, \ldots, z_n/z_{n+1})$$

with $d_i = deg(g_i)$. Similarly we can recover Φ from φ.

In general $\mathcal{V}$ need not be an ICIS, nor need the local spaces we obtain satisfy the local transversality conditions (2.4). We distinguish the situations when they do.

Definition 2.6 We say that X is smooth including ∞ if G' defines an ICIS $\mathcal{V} = G'^{-1}(0)$ and $\mathcal{V}$ and $\mathbb{C}^n$ are transverse off 0 (so $V = \mathcal{V} \cap \mathbb{C}^n$ is also an ICIS).

Moreover, we then say that X is transverse to the nonlinear affine arrangement A including ∞ if Φ is transverse to A' off 0, and $\mathcal{V}$ and $\mathcal{A} = \Phi^{-1}(A')$ satisfy the local transversality conditions (2.4). Third, we say that X is transverse to the multiform f at ∞ if V is transverse to the multiform F.

Example 2.7 Affine Arrangements of Hyperplanes.
In the case $\Phi = id$, $A = \cup H_i$ is an affine arrangement of hyperplanes with H_i defined by $\ell_i = 0$, where now ℓ_i is an affine linear form on $\mathbb{C}^n$. Then, $q = \prod \ell_i$, and $f = \prod \ell_i^{\lambda_i}$ with λ_i positive integers. Then, L_i is the homogenization of ℓ_i (if $\ell_i = \sum a_{ij} z_j - c_i$ then $L_i = \sum a_{ij} z_j - c_i z_{n+1}$ and we let $a_{in+1} = -c_i$). Also, $Q = \prod L_i$ defines the central arrangement $A' \subset \mathbb{C}^{n+1}$ and $F = \prod L_i^{\lambda_i}$. Now if A has no parallel flats, then it is easily seen that $\mathbb{C}^n = \mathbb{C}^n \times \{0\}$ is transverse to A' off 0. The conditions (2.4) reduce to the smoothness of X at ∞ and the transversality of V to $A = A' \cap (\mathbb{C}^n \times \{0\})$ off 0.

The main general global result is the following.

Theorem 1 (Global Version). *Suppose that X is smooth and transverse to the affine (nonlinear) arrangement A including ∞. Let $q = \prod \ell_i$ define A and let $f = \prod \ell_i^{\lambda_i}$ where the λ_i are positive integers. If X is transverse to the multiform f at ∞, then f has only isolated singularities on $X \backslash A$ and*

$$\text{(the sum of the Milnor numbers of } f|\, X\backslash A) = |\chi(X, X \cap A)|.$$

Moreover, the relative Euler characteristic is given by

$$|\chi(X, X \cap A)| = \dim_{\mathbb{C}} H^{n-p}(X) + \dim_{\mathbb{C}} H^{n-p-1}(X \cap A)$$

(Moreover, these dimensions are given by the (singular) Milnor numbers $\mu(V)$ and $\mu(V \cap \mathcal{A})$.)

The Number of Nondegenerate Critical Points.

To give sufficient conditions that the critical points occurring in Theorem 1 are nondegenerate, we extend a proposition given by Orlik-Terao in [OT1,§4] for the case where $X = \mathbb{C}^n$. We suppose that the central arrangement A' is essential, which means that $\cap H_i = (0)$.

Definition 2.8 The nonlinear arrangement A is immersive on X including ∞ if the associated map $\tilde{\Phi} = (\Phi, z_{n+1}) : \mathbb{C}^{n+1}, 0 \to \mathbb{C}^{m+1}, 0$ restricts to $\mathcal{V}$ to be an immersion on a punctured neighborhood of 0.

Examples 2.9 If $\Phi = id$ and the central arrangement A' is essential then automatically A is immersive on X including ∞. Second, if A_1 is immersive on X including ∞ and A_2 is another nonlinear arrangement $(= \Phi_2^{-1}(A'_2))$ such that A_1 and A_2 are algebraically transverse off 0 (but A_2 is *not* required to be immersive), then $A_1 \cup A_2$ is a nonlinear arrangement which is also immersive on X including ∞. To see this, we recall from [D1,§2] that $A_1 \cup A_2 = (\Phi_1, \Phi_2)^{-1}(A'_1 \overset{\times}{\cup} A'_2)$ where $A'_2 \overset{\times}{\cup} A'_2$ denotes the "product-union" of A'_1 and A'_2, $= (A'_1 \times \mathbb{C}^{m_2}) \cup (\mathbb{C}^{m_1} \times A'_2)$ (in the parlance of linear arrangements, this is called the "product" of the arrangements A'_1 and A'_2). Then, by [D1, §2], A_1 and A_2 being algebraically transverse off 0 is equivalent to (Φ_1, Φ_2) being algebraically transverse off 0 to $A'_2 \overset{\times}{\cup} A'_2$. Then, the map $(\Phi_1, \Phi_2, z_{n+1})$ is immersive on $\mathcal{V}\backslash\{\prime\}$ as (Φ_1, z_{n+1}) already is.

Orlik and Terao proved [OT2, prop. 4.3] that for an (essential) affine hyperplane arrangement A, there is a nonempty Zariski open subset of λ such that f has only nondegenerate critical points, with the number of such points independent of λ. We extend this to multiforms on smooth complete intersections.

Proposition 2.10 *Suppose that X is smooth and transverse to the nonlinear affine arrangement A including ∞ and that A is immersive on X including ∞. Let $q = \prod \ell_i$ define A and let $f = \prod \ell_i^{\lambda_i}$, except now the λ_i are arbitrary complex numbers. Then, for a (nonempty) Zariski open subset $\Lambda(A, X)$ of $\lambda = (\lambda_1, \ldots, \lambda_k) \in \mathbb{C}^k$, X is transverse to f at ∞, $f | X\backslash A$ has only nondegenerate points, and the number of such points is independent of λ.*

Combining this proposition with Theorem 1 yields the global theorem.

Theorem 2 *Suppose that X is smooth and transverse to the affine (nonlinear) arrangement A including ∞ and A is immersive on X including ∞. Let $f = \prod \ell_i^{\lambda_i}$ where $\lambda = (\lambda_1, \ldots, \lambda_k) \in \Lambda(A, X)$. Then, f has only nondegenerate critical points on $X\backslash A$ and*

$$\text{(the number of critical points of } f | X\backslash A) = |\chi(X, X \cap A)|.$$

Again, this is a relative singular Milnor number.

In the special case $X = \mathbb{C}^n$, since $|\chi(\mathbb{C}^n, \mathbb{C}^n \cap A)| = |\chi(\mathbb{C}^n\backslash A)|$ (see e.g. §5), we obtain as a corollary of the proof of theorem 2,

Corollary 3 (Varchenko's conjecture).
Let $f = \prod \ell_i^{\lambda_i}$ where $\lambda = (\lambda_1, \ldots, \lambda_k) \in \Lambda(A, \mathbb{C}^n)$, Then, $f | \mathbb{C}^n\backslash A$ has only

nondegenerate critical points and

$$\text{the number of critical points} = |\chi(\mathbb{C}^n \backslash A)|.$$

The $\Lambda(A, \mathbb{C}^n)$ in the form of Varchenko's conjecture given here is that of Orlik–Terao [OT2]. We deduce the corollary from the proof of theorem 2 in §5.

3 Consequences using Singular Milnor numbers and Higher Multiplicities

We can explicitly compute the singular Milnor numbers appearing in Theorems 1 and 2 in a number of important special cases. First, for any essential central arrangement $A \subset \mathbb{C}^m$ the higher multiplicities (defined in §1) are related to the Poincare polynomial $P(A', t)$ of the complement of A' by

$$P(A', t) = (1 + t) \cdot \mu(A', t) \qquad \text{where} \qquad \mu(A', t) = \sum_{j=0}^{n-1} \mu_j(A')t^j$$

is the "multiplicity polynomial"of A' (see [D1,§5]). Thus, being given the higher multiplicities for an essential arrangement is equivalent to having the Poincare polynomial. Our corollaries will be stated in terms of the higher multiplicities of the associated central arrangement.

Second, in the case of free arrangements, these higher multiplicities are given in terms of the exponents. Let $A \subset \mathbb{C}^m$ be a free arrangement (recall this means that Derlog(A') is a free $\mathcal{O}_{\mathbb{C}^m,0}$–module, necessarily of rank m, see [To]). Such an arrangement has exponents $\exp(A') = (e_0, \ldots, e_{m-1})$ where $\{e_i - 1\}$ denotes the set of degrees of the generators of Derlog(A') with $e_0 = 1$ for the Euler vector field.

As earlier, let $A = \mathcal{A} \cap (\mathbb{C}^n \times \{1\})$ for $\mathcal{A} = \Phi^{-1}(A')$.

Definition 3.1 We say that A is an almost free affine hyperplane arrangement (based on A') if A' free and both Φ and $\Phi|(\mathbb{C}^n \times \{0\})$ are linear embeddings transverse to A' off 0. (We also refer to A as being A'–generic).

For example, if A' is a Boolean arrangement, then an A'–generic arrangement is usually referred to as a generic affine arrangement (or a general position arrangement).

The almost free affine arrangement A is the singular Milnor fiber of a generic hyperplane section of the almost free arrangement $\mathcal{A} = \Phi^{-1}(A')$, with singular Milnor number given by the higher multiplicity $\mu_n(\mathcal{A})$. By [D1,§4,E5], for an almost free arrangement $\mathcal{A}$ based on A', $\mu_k(\mathcal{A}) = \sigma_k(\exp'(A'))$ where $\exp'(A') = (e_1, \ldots, e_{m-1})$ $(\mu_0(A') = 1)$ and $\sigma_k(\mathbf{x})$ denote the k-th elementary symmetric polynomial in $\mathbf{x} = (x_1, \ldots, x_{m-1})$.

We shall also need a related function $s_k(\mathbf{x})$ which denotes the polynomial defined as the sum of all monomials of degree k in $\mathbf{x}$ (here $s_0(\mathbf{x}) = 1$ and $s_0(0) = 0$ for $k > 0$).

Corollary 4 *In the situation of theorem 2, suppose that A is an A'- generic affine hyperplane arrangement, with X smooth and transverse to A including ∞. If X is defined by $g = (g_1, \ldots, g_p)$ with $\deg(g_i) = d_i$, and $\lambda = (\lambda_1, \ldots, \lambda_k) \in \Lambda(A, X)$, then the critical points of f on $X \backslash A$ are nondegenerate and*

$$\text{(the number of critical points of } f|\, X\backslash A) = d \cdot \left(\sum_{j=0}^{n-p} \mu_{n-p-j}(A') s_j(\mathbf{d}-\mathbf{1}) \right)$$

where $d = \prod_{i=1}^{p} d_i$ and $\mathbf{d}-\mathbf{1} = (d_1 - 1, \ldots, d_p - 1)$.

In the special case of a smooth hypersurface we obtain

Corollary 5 *Suppose A is an A'-generic affine arrangement and that X is a smooth hypersurface of degree d which is smooth and transverse to A strongly at ∞. Let $f = \prod \ell_i^{\lambda_i}$ where $\lambda = (\lambda_1, \ldots, \lambda_k) \in \Lambda(A, X)$. Then, f has only nondegenerate critical points on $X \backslash A$ and*

$$\text{(3.2)} \qquad \text{(the number of critical points of } f|\, X\backslash A) = d^n \cdot P(A', d^{-1}).$$

Remark: It follows from results of Orlik-Terao [OT2] that when X is homogeneous, the relative Euler characteristic equals the RHS of (3.2) for arbitrary arrangements A. This suggests that Corollary 5 should hold without the condition on A. Also, there is a further relation between our notion of being an almost free (linear) arrangement and the notions in [OT2] of being pure. K.S. Lee has proven [Le] that an almost free arrangement is pure in the sense of Orlik-Terao [OT2]. Hence, several of the results of [D1,§5] on linear arrangements also now follow from results in [OT2].

Using the methods and results of [D1,§6 and 8], we obtain formulas for the number of critical points even if A is a nonlinear affine arrangement.

Corollary 6 (Varchenko's Conjecture for Nonlinear Arrangements). *Consider the situation of theorem 2 for $X = \mathbb{C}^n$, with A a nonlinear arrangement which is immersive on $\mathbb{C}^n$ including ∞. If $\lambda = (\lambda_1, \ldots, \lambda_k) \in \Lambda(A, \mathbb{C}^n)$, then $f|\, \mathbb{C}^n \backslash A$ has only nondegenerate critical points and*

$$\text{the number of critical points} = |\chi(\mathbb{C}^n \backslash A)|.$$

This number also equals the singular Milnor number of $\mathcal{A} \cap \mathbb{C}^n$, or alternately $\mu_n(\mathcal{A})$ (where $\mathcal{A} = \Phi^{-1}(A')$).

For special nonlinear arrangements we obtain two consequences.

Corollary 7 *As in corollary 6, with A a generic affine arrangement of m smooth hypersurfaces (defined by f_i of $\deg(f_i) = d_i$) which is immersive on $\mathbb{C}^n$ including ∞. Then, the number of critical points of $f|\,\mathbb{C}^n \backslash A$ equals*

(3.3)
$$s_n(\mathbf{d}-\mathbf{1}) + \binom{m-1}{1} \cdot s_{n-1}(\mathbf{d}-\mathbf{1}) + \cdots + \binom{m-1}{n-1} \cdot s_1(\mathbf{d}-\mathbf{1}) + \binom{m-1}{n}$$

where $\binom{\ell}{q} = 0$ if $q > \ell$ and $\mathbf{d} = (d_1, \ldots, d_m)$.

Corollary 8 *As in corollary 6 with A an A'-generic affine nonlinear arrangement of hypersurfaces each of degree $d+1$. Then*

$$\begin{aligned} &\text{the number of critical points} = |\chi(\mathbb{C}^n \backslash A)| \\ (3.4) \qquad &= \sigma_n(d, \ldots, d, e_1 \cdot (d+1), \ldots, e_{m-1} \cdot (d+1)) \end{aligned}$$

with n terms of d (where again $\exp'(A') = (e_1, \ldots, e_{m-1})$).

Examples 3.5 In the case with $m > n$, where $f_i = q_i$ are generic affine quadrics, then A is immersive on $\mathbb{C}^n$. Then, for example, for generic λ, $f = q_1^{\lambda_1} \cdot q_2^{\lambda_2} \cdot q_3^{\lambda_3}$ has only nondegenerate critical points on $\mathbb{C}^2 \backslash A$, and by corollary 8, the number of critical points $= \sigma_2(1,1,2,2) = 13$. We can see in fig 4 a) that there must be a critical point within each bounded region. This nonlinear affine arrangement is the singular Milnor fiber of $\mathcal{A} \cap \mathbb{C}^\epsilon$ and the bounded regions are bounded by the 13 singular vanishing cycles.

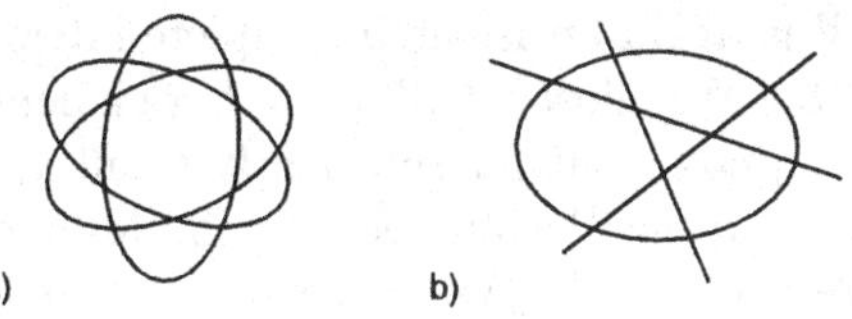

Figure 4.

Second, consider the case where $f_1 = g$ has degree $d+1$ and the f_i for $i > 1$ define a generic linear arrangement A_1 of $m-1$ hyperplanes so that the smooth hypersurface Y defined by g is smooth and transverse to A_1 including ∞. Provided $m - 1 > n$, then the nonlinear arrangement $A = Y \cup A_1$ is immersive on $\mathbb{C}^n$ including ∞ since A_1 is. Then, for generic α, λ, $f = g^\alpha \cdot \prod f_i^{\lambda_i}$ has only nondegenerate critical points on $\mathbb{C}^n \backslash A$ and by corollary 7, the number of critical points equals

$$(3.6) \qquad d^n + \binom{m-1}{1} \cdot d^{n-1} + \cdots + \binom{m-1}{n-1} \cdot d + \binom{m-1}{n}.$$

For example, on $\mathbb{C}^2$ when $m = 4$, and $\deg(g) = 2$, we obtain $1^2 + 3 + 3 = 7$ critical points. Again in the real picture fig. 4 b), there must be a critical point in each bounded region, with the 7 singular vanishing cycles in the singular Milnor fiber bounding these regions (see also §6 of [D1]).

Remark: Orlik and Terao have obtained a formula for the case of one nonlinear factor as above by relating it to the number of critical points of $\exp(G) \cdot \prod f_i^{\lambda_i}$. This correspondence should allow us to equally well obtain the number of critical points for this function by applying the relation in the reverse direction.

Remark 3.7 In all of the above corollaries, there are corresponding statements for the sum of the Milnor numbers for arbitrary positive λ provided only that transversality to f at ∞ holds.

4 Critical Points and the Topology of Singular Milnor Fibers

As a main step in proving the local version of theorem 1, we state and apply a generalization of the lemma of Siersma used in [DM]. In [Si] Siersma shows that the standard Morse theory type argument used by Looijenga in [Lo, chap. 5], and in its original form due to Lê [Lê2], can also be applied to nonisolated singularities defined by germs $g : \mathbb{C}^{n+1}, 0 \to \mathbb{C}, 0$. We show moreover that there is an analogue for nonisolated singularities defined on complete intersections. Let $\mathcal{X}, 0 \subset \mathbb{C}^{n+1}, 0$ be a k-dimensional isolated complete intersection and let $g, h : \mathbb{C}^{n+1}, 0 \to \mathbb{C}, 0$ define hypersurfaces singularities $Y = g^{-1}(0)$ and $Z = h^{-1}(0)$ so that g has an isolated singularity on $\mathcal{X}$. Thus, $Y \cap \mathcal{X}$ is again an isolated complete intersection with Milnor fiber $Y_{t,\epsilon} = Y_t \cap \mathcal{X} \cap B_\epsilon$. We allow $Y \cap \mathcal{X} \cap Z$ to have a nonisolated singularity at 0. However, we suppose that for sufficiently small γ, if $\gamma > |t|, |s| \geq 0$, then $Y_t \cap \mathcal{X} \cap Z_s$ is (stratified) transverse to the Milnor sphere $S_\epsilon = \partial B_\epsilon$. Moreover, we suppose that on $Y_{t_0,\epsilon} \backslash Z$, h has only a finite number of isolated singularities which approach 0 as t_0 approaches 0.

Proposition 4.1 *In the preceding situation, with $\gamma > |t_0| > 0$, $Y_{t_0,\epsilon} \cap \mathcal{X} \cap Z$ $(= \mathcal{X} \cap Y_{t_0} \cap Z \cap B_\epsilon)$ is homotopy equivalent to a bouquet of spheres of dimension $k-2$. If the number of such spheres is denoted by $\mu(\mathcal{X} \cap Y \cap Z)$ then*

$$\mu(\mathcal{X} \cap Y \cap Z) + \mu(\mathcal{X} \cap Y) = \sum \mu(h, z_j')$$

where the sum of the Milnor numbers on the right is over the isolated singularities of h on $Y_{t_0,\epsilon} \backslash Z$.

Remark: When $\mathcal{X}$, Y, and Z are in general position off 0, $\mu(\mathcal{X} \cap Y \cap Z)$ has an intrinsic meaning as a singular Milnor number.

To prove this result we must establish an analogue of a result of Looijenga [Lo, prop. 5.4] for this case.

Lemma 4.2 *In the above situation, there exist representatives $g : \mathcal{X} \cap \bar{B}_\epsilon \to B_{\delta'}$, $h : \mathcal{X} \cap \bar{B}_\epsilon \to B_\delta$, and $0 < \delta_1 < \delta$ so that:*

1. *the set of points* (t,s) *for which* $g^{-1}(t)$ *and* $h^{-1}(s)$ *are not transverse in* B_ϵ *is a curve in* $B_{\delta'} \times B_{\delta_1}$, *and*

2. *there is a homeomorphism* $\varphi : \mathcal{X}_1 \to \mathcal{X}_2$ *where* $\mathcal{X}_1 = \mathcal{X} \cap g^{-1}(B_{\delta'}) \cap \bar{B}_\epsilon$ *and* $\mathcal{X}_2 = \mathcal{X} \cap g^{-1}(B_{\delta'}) \cap h^{-1}(B_\delta) \cap \bar{B}_\epsilon$ *such that* $h \circ \varphi = h$.

The proof of the Lemma will be postponed until we complete the proof of the proposition.

Proof (of proposition 4.1). The proof of the proposition basically follows the line of proof of [Si] modifying the earlier argument of Looijenga [Lo]. To simplify notation we denote $h|Y_{t_0,\epsilon}$ by h_1. Also, we let $Z_{s,t_0,\epsilon} = \mathcal{X} \cap Y_{t_0} \cap Z_s \cap \bar{B}_\epsilon$. By our transversality assumption and the relative Ehresman fibration theorem, $h_1^{-1}(B_\delta)$ is a locally trivial fibration off 0 and the finite number of singular values w_j. We place small nonintersecting disks D_j around each w_j and D_0 around 0. We may connect a point $p_0 \in \partial D_0$ to points $p_j \in \partial D_j$ by a system of nonintersecting paths Γ (see fig. 5).

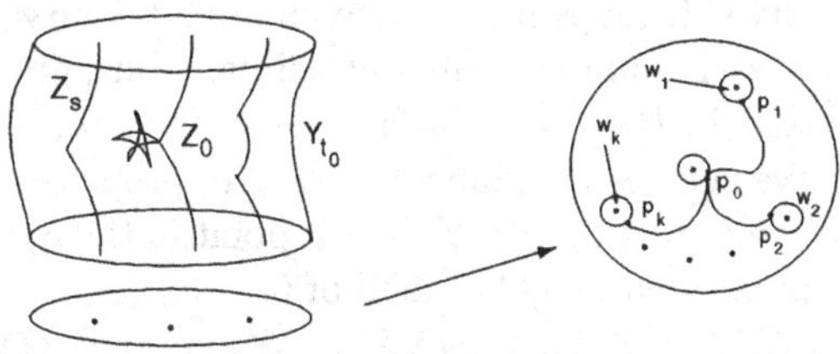

Figure 5.

Then, as in [Si], $h_1^{-1}(B_\delta)$ has $h_1^{-1}(D_0 \cup (\cup D_j) \cup \Gamma)$ as a deformation retract. Furthermore, each $h_1^{-1}(D_j)$ is obtained from $h_1^{-1}(p_j)$ by adjoining $\mu_{j'}$ cells of dimension $k-1$ for each singular point $z_{j'}$ lying over p_j. Thus, $h_1^{-1}(D_0 \cup (\cup D_j) \cup \Gamma)$ is obtained from $h_1^{-1}(D_0 \cup \Gamma)$ by adjoining $\sum \mu_{j'}$ cells of dimension $k-1$ and $h_1^{-1}(D_0 \cup \Gamma)$ has $h_1^{-1}(0)$ as a deformation retract. Thus, $h_1^{-1}(B_\delta)$ is homotopy equivalent to a complex obtained from $h_1^{-1}(0) = Z_{0,t_0,\epsilon}$ by adjoining $\sum \mu_{j'}$ cells of dimension $k-1$. Since $Y_{t_0,\epsilon}$ is a Milnor fiber, it is homotopy equivalent to a bouquet of spheres of real dimension $k-1$. Also, by lemma 4.2, there is a homeomorphism

$$(h_1^{-1}(B_\delta), Z_{0,t_0,\epsilon}) \simeq (Y_{t_0,\epsilon}, Z_{0,t_0,\epsilon}).$$

Thus, $h_1^{-1}(B_\delta)$ is $k-2$ connected. By a standard argument, as $h_1^{-1}(B_\delta)$ is obtained from $Z_{0,t_0,\epsilon}$ by attaching cells of dimension $k-1$, it follows that $Z_{0,t_0,\epsilon}$ is $(k-3)$–connected. Moreover, it is a Stein space of dimension $k-2$. Thus, $Z_{0,t_0,\epsilon}$ has the homotopy type of a CW–complex of dimension $k-2$. Thus, $Z_{0,t_0,\epsilon}$ is homotopy equivalent to a bouquet of spheres of dimension $k-2$. Then, $Z_{0,t_0,\epsilon}$ and $Y_{t_0,\epsilon}$ have zero homology except in dimensions $k-2$, resp. $k-1$. Since $Y_{t_0,\epsilon}$ is obtained (up to homotopy) from $Z_{0,t_0,\epsilon}$ by adjoining

$\sum \mu_{j'}$ cells of dimension $k-1$,

$$\dim_{\mathbb{C}} H^{k-1}(Y_{t_0,\epsilon}, Z_{0,t_0,\epsilon}) = \sum \mu_{j'}. \tag{4.3}$$

From the exact sequence of a pair we obtain

$$\begin{aligned} \dim_{\mathbb{C}} H^{k-1}(Y_{t_0,\epsilon}, Z_{0,t_0,\epsilon}) &= \dim_{\mathbb{C}} H^{k-1}(Y_{t_0,\epsilon}) + \dim_{\mathbb{C}} H^{k-1}(Z_{0,t_0,\epsilon}) \\ &= \mu(\mathcal{X} \cap Y) + \mu(\mathcal{X} \cap Y \cap Z). \end{aligned} \tag{4.4}$$

Since the intersections are topologically trivial over the Milnor sphere S_ϵ, for s sufficiently small, all singular points of $h_1 | (\mathcal{X} \cap Y_{t_0,\epsilon}) \backslash Z$ will be in B_ϵ. Combined with (4.3) and (4.4) this completes the proof. □

Proof (of Lemma 4.2). The proof of the lemma likewise follows the line of proof of Looijenga [Lo, prop. 5.4]. We may pick $\epsilon_0 > 0$ such that for $0 < \epsilon < \epsilon_0$, $\mathcal{X}$, $\mathcal{X} \cap Y$, and $\mathcal{X} \cap Y \cap Z$ are (stratified) transverse to S_ϵ. Next, the set of points in $\mathcal{X} \backslash Z$ at which dg and dh are linearly dependent form an analytic curve. It maps by (g, h) to an analytic curve α in $\mathbb{C}^2$ which intersects each$\{t\} \times \mathbb{C}$ in a finite number of points. Thus, given $\delta' > 0$, there are $0 < \delta_1 < \delta$ so that in $B_{\delta'} \times \mathbb{C}$, α in fact lies in $B_{\delta'} \times B_{\delta_1}$.

Third, by a curve selection argument as in [Lo, prop. 5.4], the gradients of $r = \|x\|^2 | \mathcal{X} \cap Y$ and $q = |h|^2 | \mathcal{X} \cap Y$ do not point in the opposite direction on $\mathcal{X} \cap Y \backslash Z$ in a punctured neighborhood of 0.

Then let $W = (\mathcal{X} \cap \bar{B}_\epsilon) \backslash \mathcal{X}_2$, where $\mathcal{X}_2 = \mathcal{X} \cap g^{-1}(B_{\delta'}) \cap h^{-1}(B_\delta) \cap \bar{B}_\epsilon$. By the compactness of $W \cap Y$, we may find δ and δ' so that the gradients of $r_t = \|x\|^2 | \mathcal{X} \cap Y_t$ and $q_t = |h|^2 | \mathcal{X} \cap Y_t$ do not point in opposite directions for $|t| < \delta'$. Then, following [Lo, prop. 5.4], we construct a C^∞ vector field ζ on W which is tangent to $W \cap Y_t$, the fibers of g, and so that $dr_t(\zeta)$ and $dq_t(\zeta) > 0$. Then, arguing as in [Lo], the integral curves of ζ yield the desired homeomorphism. □

5 Proofs of the Main Results

In this section we derive the main theorems and the corollaries by applying the results of §4.

Proof of Theorem 1 (Local Version). For the proof of the local version of theorem 1, we use the results of the preceding section. Recall that $G = (G_1, \ldots, G_{p+1}) : \mathbb{C}^{n+1}, 0 \to \mathbb{C}^{p+1}, 0$ defines the ICIS V. Let $\beta : \mathbb{C}, 0 \to \mathbb{C}^{p+1}, 0$ be a germ of a curve which only intersects the discriminant $D(G)$ at 0. Then, for $\epsilon > 0$ sufficiently small, $V_t = G^{-1}(t) \cap B_\epsilon$ is the Milnor fiber of V. We let $\mathcal{X} = \{(z, t) \in \mathbb{C}^{n+2} : G(z) = \beta(y)\}$ and let g denote the projection of $\mathcal{X}$ onto $\{0\} \times \mathbb{C}$. Then, $\mathcal{X}$ is an ICIS and g has an isolated singularity on $\mathcal{X}$. Then, we claim that we can apply proposition 4.1 to $\mathcal{X}$, g, and $h = F \circ pr_1$ (here $F = \prod (L_i \circ \Phi)^{\lambda_i}$ and $pr_1 : \mathbb{C}^{n+2}, 0 \to \mathbb{C}^{n+1}, 0$ denotes projection off the last factor). We apply proposition 4.1 using

(5.1)
$$\chi(V_t) = (-1)^{n-p}\mu(V) + 1 \quad \text{and} \quad \chi(V_t \cap \mathcal{A}) = (-1)^{n-p-1}\mu(V \cap \mathcal{A}) + 1.$$

By (5.1)

$$\chi(V_t, V_t \cap \mathcal{A}) = |\chi(V_t) - \chi(V_t \cap \mathcal{A})| = \mu(V) + \mu(V \cap \mathcal{A}).$$

By proposition 4.1, this equals the sum of the Milnor numbers of $F|\, V_t \backslash \mathcal{A}$. This completes the proof once we have verified that proposition 4.1 can be applied. To be able to apply proposition 4.1, we must show that there is a Milnor radius $\epsilon > 0$ so that for $|s|$ and $|t|$ sufficiently small, $V_t \cap F^{-1}(s)$ $(= \mathcal{X} \cap g^{-1}(t) \cap h^{-1}(s))$ is stratified transverse to S_ϵ, and that $F|\, V_t \backslash \mathcal{A}$ $(= h|\, \mathcal{X} \cap g^{-1}(t) \backslash h^{-1}(0)$ has only a finite number of singular points, all of which approach 0 as t approaches 0.

First, by assumption, $V = \mathcal{X} \cap \mathbb{C}^{n+1}$ is (algebraically) transverse to both $\mathcal{A} = \Phi^{-1}(A')$ and F in a punctured neighborhood of 0. We choose $\epsilon_0 > 0$ so that $\bar{B}_{\epsilon_0}$ lies in this neighborhood.

Step 1: Establishing the property of the critical points of $F|\, V_t \backslash \mathcal{A}$
As V is also algebraically transverse to F, for each $z \in V \backslash \{0\}$ in a punctured neighborhood of 0, there is $\zeta \in \mathrm{Derlog}(\mathcal{A}, G)$ such that $\Phi^*\omega_\lambda(\zeta) \neq 0$. As $\mathrm{Derlog}(\mathcal{A}, G)$ is a finitely generated $\mathcal{O}_{\mathbb{C}^{n+1},0}$-module, we can find a finite number of $\zeta_i \in \mathrm{Derlog}(\mathcal{A}, G)$ so that $\{\Phi^*\omega_\lambda(\zeta_i), i = 1, \dots, \ell, G_1, \dots, G_{p+1}\}$ do not simultaneously vanish in a punctured neighborhood (which we may assume contains $\bar{B}_{\epsilon_0}$).

Then, the equations $\Phi^*\omega_\lambda(\zeta_i) = 0, i = 1, \dots, \ell$, and $g = t$ define an analytic subset $E \subset \mathcal{X} \subset \mathbb{C}^{n+2}$ such that $E \cap (\mathbb{C}^{n+1} \times \{0\}) = \{0\}$. Thus, E is a curve in $\mathcal{X}$ which is finite over $\mathbb{C}$ with (by semi-continuity) at most $\dim_{\mathbb{C}}(\mathcal{O}_{\mathbb{C}^{n+1},0}/J)$ points in the fiber. Thus, there are γ and $\epsilon_1 > 0$ such that if $|t| < \gamma$, and $0 < \epsilon_1 \leq \epsilon \leq \epsilon_0$, then $E \cap g^{-1}(t) \cap S_\epsilon = \emptyset$. Thus, there are only a finite number of singular points of $h|\, V_t \backslash \mathcal{A}$ $(= E \cap g^{-1}(t))$. These lie in $\bar{B}_{\epsilon_1}$ and they approach 0 as t approaches 0.

Step 2: Stratified Transversality to S_ϵ
We must show that there is a Milnor radius $\epsilon > 0$ so that for all $|s|$ and $|t|$ sufficiently small, $V_t \cap F^{-1}(s)$ $(= \mathcal{X} \cap g^{-1}(t) \cap h^{-1}(s))$ is stratified transverse to S_ϵ.

By the parametrized transversality theorem and the algebraic criterion for transversality, it follows that (by possibly shrinking ϵ_0) we may assume that for $0 < \epsilon \leq \epsilon_0$, the following are (stratified) transverse to S_ϵ : $V = \mathcal{X} \cap g^{-1}(0)$, $V \cap \mathcal{A} = \mathcal{X} \cap g^{-1}(0) \cap h^{-1}(0)$, and $\mathcal{A}$.

Now $V \cap \mathcal{A} \cap S_\epsilon$ (or more generally $V \cap \mathcal{A} \cap (\bar{B}_{\epsilon_0} \backslash B_{\epsilon_1})$) is compact, and the condition for algebraic transversality to closed analytic subsets is an open condition. Hence, it is enough to show that $z \in V \cap \mathcal{A} \cap S_\epsilon$ has an open neighborhood U_z such that $\mathcal{X} \cap g^{-1}(t) \cap h^{-1}(s)$ is stratified transverse to each S_ϵ at all points of U_z.

We know that $\mathcal{X}$, $\mathbb{C}^{n+1} = g^{-1}(0)$, and $\mathcal{A} \times \mathbb{C} = h^{-1}(0)$ are in algebraic general position off 0 in $\bar{B}_{\epsilon_0}$. Moreover, $\mathcal{X}$ is an ICIS defined by $G_1(z,t) = G(z) - \beta(t)$, so it is h-holonomic, as is $\mathbb{C}^{n+1}$ for z_{n+2}.

We also claim that $\mathcal{A}$ is h-holonomic for F. Given this, it follows that the geometric tangent spaces to $\mathcal{X}$, $\mathbb{C}^{n+1}$, and $\mathcal{A} \times \mathbb{C}$ (off the axis $\{0\} \times \mathbb{C}$) are given by $T_{log}(G_1)_{(z)}$, $T_z\mathbb{C}^{n+1}$, and $T_{log}(F)_{(z)}$. These spaces are in general position and their intersection contains a vector transverse to T_zS_ϵ ($\epsilon = \|z\|$). By the coherence of the associated sheaves, it follows that for each $z' \in U_z$, a neighborhood of z, these tangent spaces at z' remain in general position and there is still a vector in the intersection of these tangent spaces at z' which is still transverse to $T_{z'}S_{\epsilon'}$ ($\epsilon' = \|z'\|$). Thus, there is $\gamma > 0$ so that for $|s|, |t| < \gamma$, $V_t = \mathcal{X} \cap g^{-1}(t)$ and $V_t \cap F^{-1}(s) = \mathcal{X} \cap g^{-1}(t) \cap h^{-1}(s))$ remain stratified transverse to S_ϵ for $0 < \epsilon_1 \le \epsilon \le \epsilon_0$.

Lastly, to see that $\mathcal{A}$ is h-holonomic for F, we note that if $F(z) = s \neq 0$ then by the coherence of Derlog(F), $T_{log}(F)_{(z)} = T_z(F^{-1}(s))$. If $z \in \mathcal{A}\backslash\{0\} = \Phi^{-1}(A')\backslash\{0\}$, since Φ is transverse to A' at z, there is a nonlinear change of coordinates so that

$$(\mathcal{A}, z) \simeq (A'' \times \mathbb{C}^k, (\Phi(z), 0))$$

where $A' \simeq A'' \times \mathbb{C}^r$ near $\Phi(z)$. However, A' (and hence A'') is h-holonomic for $F' = \prod L_i^{\lambda_i}$ at $\Phi(z)$ using the argument of [D1,lemma 2.1] since there is a local Euler vector field e vanishing at $\Phi(z)$ so that $e(F') = F'$. Thus, lifting e gives one at z so $\mathcal{A}$ is h-holonomic for F at z. □

Proof of Theorem 1 (Global Version). We use the notation of §2. In addition to the G_i, we also let $G_0 = z_{n+1}$, so $\deg(G_0) = d_0 = 1$. First we recall that X is defined by $G_i = 0$ for $i = 1, \ldots, p$ and $G_0 = 1$. We let $G = (G_0, G_1, \ldots, G_p) : \mathbb{C}^{n+1}, 0 \to \mathbb{C}^{p+1}, 0$. There is a $\gamma > 0$ so that if $|t| < \gamma$ then $V_t = \{z \in B_\epsilon : G_i(z) = 0, i = 1, \ldots, p, \text{and } G_0(z) = t\}$ is the Milnor fiber of G, and $V_t \cap \mathcal{A}$ is the singular Milnor fiber of the complete intersection $V \cap \mathcal{A}$ ($V = \mathcal{V} \cap \mathbb{C}^n$ where $\mathcal{V} = G'^{-1}(0)$).

We claim that the relation between the global result we want to establish regarding f and $X \cap A$ and the local Milnor fibers just described follows from proposition 2.5. By the preceding argument for the local version, the set of critical points of $F|\,(\mathcal{V} \cap (\mathbb{C}^n \times \{t\})\backslash\mathcal{A})$ as t varies forms an analytic curve without branches in $\mathbb{C}^n \times \{0\}$. Hence, we may choose ϵ' so that all of the critical points of $F|\,(\mathcal{V} \cap (\mathbb{C}^n \times \{t\})\backslash\mathcal{A})$ will be in $B_{\epsilon'}$ for sufficiently small t. Then, by the $\mathbb{C}^*$–action, the critical points of $F|\,X\backslash A$ will be in a compact set C. By proposition 2.5 for $|t|$ sufficiently small, there is a homeomorphism ψ from X to the Milnor fiber $V_t = \mathcal{V} \cap (\mathbb{C}^n \times \{t\}) \cap B_\epsilon$ which sends $A = X \cap A$ to $V_t \cap \mathcal{A}$ and is the restriction of the $\mathbb{C}^*$–action (and so holomorphic) on C.

Now we can see that proposition 2.5 together with the local version of Theorem 1 implies the global version. Since $\psi|C$ is a biholomorphic diffeomorphism, it is sufficient to prove

(5.1) $\quad$ (the sum of the Milnor numbers of $F|\, V_t \cap \mathcal{A}) = |\chi(V_t, V_t \cap \mathcal{A})|$

and that this relative Euler characteristic is given by

(5.2) $$|\chi(V_t, V_t \cap \mathcal{A})| = \dim_{\mathbb{C}} H^{n-p}(V_t) + \dim_{\mathbb{C}} H^{n-p-1}(V_t \cap \mathcal{A}).$$

However, V_t is the Milnor fiber of V and $V_t \cap \mathcal{A}$ is the singular Milnor fiber of $V \cap \mathcal{A}$. Hence,

$$\dim_{\mathbb{C}} H^{n-p}(V_t) = \mu(V) \qquad \text{and} \qquad \dim_{\mathbb{C}} H^{n-p-1}(V_t \cap \mathcal{A}) = \mu(V \cap \mathcal{A}).$$

Thus, both (5.1) and (5.2) will follow provided we may apply the local version of Theorem 1 to F, $\mathcal{V}$, and $\mathcal{A}$. However, the global conditions exactly ensure that this is so. □

Proof of Theorem 2. By proposition 2.10, there is a Zariski open subset $\Lambda(A, X) \subset \mathbb{C}^k$ such that for $\lambda = (\lambda_1, \dots, \lambda_k) \in \Lambda(A, X)$ and $f = \prod \ell_i^{\lambda_i}$, then $f|\, X \backslash A$ has only nondegenerate critical points and the number is constant. Thus, to prove the theorem it is enough to determine the number for some $\lambda \in \Lambda(A, X)$.

We use an observation of Varchenko that if $\mathbb{N}$ denotes the positive integers then $\mathbb{N}^k$ is not contained in a proper Zariski closed subset of $\mathbb{C}^k$. (This can be seen by a naive argument involving induction on k. If $k = 1$ then the only proper Zariski closed subsets of $\mathbb{C}$ are finite. By induction, if $\mathbb{N}^k \subset B \subsetneq \mathbb{C}^k$ for a Zariski closed B, then for $\lambda_1 \in \mathbb{N}$, $\{\lambda_1\} \times \mathbb{N}^{k-1} \subset B \cap (\{\lambda_1\} \times \mathbb{C}^{k-1})$ which is Zariski closed. By induction $\{\lambda_1\} \times \mathbb{C}^{k-1} \subset B$ for all $\lambda_1 \in \mathbb{N}$. Thus, given $\lambda \in \mathbb{C}^k$, $(m, \lambda_2, \dots, \lambda_k) \in B$ for all $m \in \mathbb{N}$. Hence, $\mathbb{N} \times \{(\lambda_2, \dots, \lambda_k)\} \subset B \cap (\mathbb{C} \times \{(\lambda_2, \dots, \lambda_k)\})$, which is Zariski closed so again $\mathbb{C} \times \{(\lambda_2, \dots, \lambda_k)\} \subset B \cap (\mathbb{C} \times \{(\lambda_2, \dots, \lambda_k)\})$, i.e. $\lambda \in B$.)

Thus, there are $\lambda \in \Lambda(A, X) \cap \mathbb{N}^k$. Then, we can apply theorem 1. Now, each singular point is nondegenerate so its Milnor number is 1, so the sum of the Milnor numbers is just the number of critical points. This yields the result. □

Proofs of the corollaries.
All of the corollaries except Corollary 3, Varchenko's conjecture, follow immediately from Theorems 1 and 2.

Corollary 4: We note that by proposition 2.5,

$$|\chi(X, X \cap A)| = |\chi(V_t, V_t \cap \mathcal{A})|.$$

However $(V_t, V_t \cap \mathcal{A})$ is the relative singular Milnor fiber of $(V \cap \mathcal{A})$ on the Milnor fiber of V. This equals by Theorem 8.19 of [D1], the expression given in corollary 4. □

Corollary 5: Corollary 5 is a special case of corollary 4 where the first part follows from theorem 1 and (3.2) is a special case of the formula in Corollary 4 (see also [D1,§8]). □

Corollary 6: We first may apply theorem 2 to obtain that the number of critical points $= |\chi(\mathbb{C}^n, A)|$ with $A = (\mathbb{C}^n \times \{1\}) \cap \mathcal{A})$. Also, for an affine variety $A \subset \mathbb{C}^n$,

$$|\chi(\mathbb{C}^n \backslash A)| = |1 - \chi(A)| = |\chi(\mathbb{C}^n, A)|$$

giving the first part of corollary 6. In addition,

$$|\chi(\mathbb{C}^n \backslash A)| = \dim_{\mathbb{C}} H^n(\mathbb{C}^n) + \dim_{\mathbb{C}} H^{n-1}(A).$$

The first dimension is 0. By lemma 4.1 applied in the case $X = \mathbb{C}^n$, the second equals the singular Milnor number $\mu(\mathbb{C}^n \cap \mathcal{A})$. As $\mathbb{C}^n$ is algebraically transverse to $\mathcal{A}$ off 0, $\mu(\mathbb{C}^n \cap \mathcal{A}) = \mu_n(\mathcal{A})$. □

Corollaries 7 and 8: Then, (3.3) follows from proposition 6.10 of [D1,§6], giving corollary 7, and for (3.4), we apply proposition 6.12 of [D1,§6]. □

Proof of Varchenko's conjecture.
Because $\mathbb{C}^n = \mathbb{C}^n \times \{0\}$ need not be transverse to the associated central arrangement $\mathcal{A}$, the general form of Varchenko's conjecture does not follow directly from theorem 2, However, we claim it does follow from the proof. We choose $\lambda \in \Lambda(A, \mathbb{C}^n)$ with all λ_i positive integers.

First, let $\mathbb{C}^n_t = \mathbb{C}^n \times \{t\}$, $\mathcal{A}_t = \mathcal{A} \cap \mathbb{C}^n \times \{t\}$, $B_{R,t} = B_R \times \{t\}$, for B_R the ball of radius R, and similarly for the sphere $S_{R,t}$. We also let $F_t = F|\mathbb{C}^n_t$ and $r_t = ||z||^2|\mathbb{C}^n_t$. Note $dr_t = dr_0$ under the natural identification $\mathbb{C}^n_t \simeq \mathbb{C}^n$.

Then, there is an $R_0 > 0$ such that if $R \geq R_0$, then $S_{R,t}$ is transverse $\mathcal{A}_t$ in $\mathbb{C}^n_t$ for all $|t| \leq 1$. This is separately true for a single $\mathcal{A}_t$ because the distance–squared function only has a single critical point, a minimum, on any affine subspace. Then, we can use a common R_0 for $t = 0, 1$ and extend to $|t| \leq 1$ using the $\mathbb{C}^*$–action. From this it follows by stratification theory that for $R \geq R_0$ and $|t| \leq 1$,

$$(B_{R,t}, (B_{R,t} \cap \mathcal{A}_t)) \simeq (\mathbb{C}^n_t, \mathcal{A}_t). \tag{5.3}$$

Second, the critical set of F_t for $|t| \leq 1$ is compact (because there are only a finite number of branches to the curve of critical points as in theorem 1). Thus, we can increase R_0 and find an $s_0 > 0$ such that the critical points of F_t for $|t| \leq 1$ lie in B_{R_0}, and the critical values lie in B_{s_0}.

Third, we claim

Lemma 5.4 *There is an $s_1 \geq s_0$, $0 < t_0 < 1$ arbitrarily small, and an $R_1 \geq R_0$ so that for any t with $t_0 \leq |t| \leq 1$*

1. *$F^{-1}(s) \cap \mathbb{C}^n_t$ is transverse to $S_{R,t}$ (in $\mathbb{C}^n_t$) for all $|s| \leq s_1$ and $R \geq R_1$;*
2. *the differentials $d|F_t|^2$ and dr_t do not point in opposite directions for any $z \in \bar{B}_R \backslash F^{-1}(B_{s_1})$.*

Let s_1, t_0 and R_1 denote the values given by Lemma 5.4. For a specific t_1 with $t_0 < |t_1| \leq 1$, we can argue as in §4. First, by (2) of Lemma 5.4 and the argument of lemma 4.2, $F_{t_1}^{-1}(\bar{B}_{s_1}) \cap B_{R_1,t_1}$ is a strong deformation retract of B_{R_1,t_1}. Furthermore, by the proof of Lemma 4.1, $F_{t_1}^{-1}(0) \cap B_{R_1,t_1} = \mathcal{A} \cap B_{R_1,t_1}$ is $n-2$–connected and $F_{t_1}^{-1}(\bar{B}_{s_1}) \cap B_{R_1,t_1}$ is obtained from $F_{t_1}^{-1}(0) \cap B_{R_1,t_1}$ by attaching n–cells, with the number of cells equal to the number of critical points of F_{t_1} on $\mathbb{C}_t^n \backslash \mathcal{A}_{t_1}$ (all of which have Milnor number 1). Hence, as in the proof of theorem 2 and using (5.3),

$$\begin{aligned}|\chi(\mathbb{C}^n, A)| &= |\chi(\mathbb{C}^n_{t_1}, \mathcal{A}_{t_1})| \\ &= |\chi(F_{t_1}^{-1}(\bar{B}_{s_1}) \cap B_{R_1,t_1}, \mathcal{A} \cap B_{R_1,t_1})| \\ &= \text{number of critical points of } f \text{ on } \mathbb{C}^n \backslash A.\end{aligned}$$

Note: If A contains no pairs of parallel flats, then $\mathbb{C}^n$ is transverse to $\mathcal{A}$ off 0, so Theorem 2 does yield that the number of critical points equals $\mu_n(\mathcal{A})$ (which equals $\mu(A)$, the Mobius function of A by proposition 5.6 of [D1]).

Proof of Lemma 5.4. We will argue by contradiction using the following two lemmas.

Lemma 5.5 *Let $z' \neq 0$ be contained in a minimal flat K of $\mathcal{A}$ such that $K \subset \mathbb{C}^n (= \mathbb{C}^n \times \{0\})$.*

1. *There is a tubular neighborhood Z (in $\mathbb{C}^{n+1}$) of the line ℓ through 0 containing z' so that given $s_1 > 0$ there is an $R_1 > 0$ depending on s_1, so that $F_t^{-1}(s) \cap Z$ is transverse to $S_{R,t}$ in $\mathbb{C}_t^n$ for all $|s| \leq s_1 \cdot |t|^N$ where $N = \deg(F)$ and all $R \geq R_1$; and*

2. *there is a neighborhood U of z' in $\mathbb{C}^{n+1}$ so that $d|F_t|^2$ and dr_t do not point in opposite directions for all $(z,t) \in U$.*

Lemma 5.6 *Let $\mathcal{N}$ be a subset of $\mathbb{C}^n \times \bar{B}_1 \backslash \{0\} \subset \mathbb{C}^n \times \mathbb{C}$ which is preserved by the (standard) $\mathbb{R}_+$–action on $\mathbb{C}^{n+1}$. Suppose for some $0 < |t| \leq 1$, $\mathcal{N} \cap \mathbb{C}_t^n$ is unbounded. Then, there is a nonzero point $(z, 0)$ in the closure of $\mathcal{N}$.*

We first deduce Lemma 5.4 from these lemmas. First we find s_1. Let

$$\mathcal{W}_1 = \{(z,t) \in \mathbb{C}^n \times (\bar{B}_1 \backslash \{0\}) : \text{there is a } \lambda < 0 \text{ such that } d|F_t|^2 = \lambda dr_t\}$$

Then, as F is a homogeneous polynomial, $\mathcal{W}_1$ is a subset of $\mathbb{C}^n \times (\bar{B}_1 \backslash \{0\})$ which is preserved by the (usual) $\mathbb{C}^*$–action. First, suppose $\mathcal{W}_1 \cap \mathbb{C}_t^n$ is unbounded for some $0 < |t| \leq 1$, then by Lemma 5.6, there is a nonzero point $(z', 0)$ in the closure of $\mathcal{W}_1$. However, $d|F_0|^2$ and dr_0 do not point in opposite directions for z in a punctured neighborhood of 0 in $\mathbb{C}^n$ (see e.g. [Lo, chap. 2]), and hence on all of $\mathbb{C}^n \backslash \{0\}$ by the $\mathbb{C}^*$–action. Then, it follows that if either $(z', 0) \notin \mathcal{A}$ or $\mathcal{A}$ is transverse to $\mathbb{C}^n$ at $(z', 0)$, then there is a neighborhood of $(z', 0)$ in $\mathbb{C}^{n+1}$ on which $d|F_0|^2$ and dr_0 do not point in opposite directions. If instead $\mathcal{A}$ is not transverse to $\mathbb{C}^n$ at $(z', 0)$, then such a neighborhood exists by (2) of Lemma 5.5. This contradicts $(z', 0)$ being in the closure of $\mathcal{W}_1$.

As $\mathcal{W}_1 \cap \mathbb{C}^n_t$ is bounded for each $0 < |t| \leq 1$, then by the $\mathbb{C}^*$–action, $|F|$ is bounded on $\mathcal{W}_1$, by say $s_1 \geq s_0$, and hence on $\mathcal{W}_1 \cap \mathbb{C}^n_t$ by $s_1|t|^N$. Hence, $d|F_t|^2$ and dr_t will not point in opposite directions for $(z,t) \in (\mathbb{C}^n \backslash F^{-1}(\bar{B}_{s_1})) \times (\bar{B}_1 \backslash \{0\})$ yielding (2) of Lemma 5.4.

Choose $t_0 > 0$ with $(0, t_0) \in Z$ (from (1) of Lemma 5.5), let $s_2 = s_1/t_0^N$, and define

$$\begin{aligned}\mathcal{W}_2 =&\{(z,t) \in \mathbb{C}^n \times (\bar{B}_1 \backslash \{0\}) : \|z\| = R, F(z,t) = s \text{ with } |s| \leq s_2|t|^N, \\ &\text{and } F_t^{-1}(s) \text{ is not transverse to } S_{R,t} \text{ at } (z,t)\}.\end{aligned}$$

Again, $\mathcal{W}_2$ is a subset of $\mathbb{C}^n \times (\bar{B}_1 \backslash \{0\})$ preserved by the $\mathbb{C}^*$–action. If $\mathcal{W}_2 \cap \mathbb{C}^n_t$ is unbounded for some $0 < |t| \leq 1$, then by lemma 5.6, there is a nonzero $(z', 0)$ in the closure of $\mathcal{W}_2$. By continuity, $F(z', 0) = 0$. Also, using the $\mathbb{R}_+$–action, if $R > 0$,we can find such a $(z', 0)$ in the closure with $\|z'\| \geq R$.

There are $\epsilon, \delta > 0$ so that $F_0^{-1}(s)$ is (stratified) transverse to $S_{\epsilon,0}$ for all $|s| \leq \delta$. Using the $\mathbb{C}^*$–action, we obtain for any $s_1 > 0$ an $R_1 \geq R_0$ such that $F_0^{-1}(s)$ is transverse to $S_{R,0}$ for all $|s| \leq s_1$ and all $R \geq R_1$. We also suppose R_1 is large enough using s_2 in (1) of Lemma 5.5 and that $\|z'\| > R_1$. Thus, if $\mathcal{A}$ is transverse to $\mathbb{C}^n$ at $(z', 0)$, then there is a neighborhood U of $(z', 0)$ in $\mathbb{C}^{n+1}$ so that if $(z,t) \in U$ with $F(z,t) = s$ and $\|z\| = R > R_1$ then $F^{-1}(s)$ is transverse to $S_{R,0}$ at (z,t). If $\mathcal{A}$ is not transverse to $\mathbb{C}^n$ at $(z', 0)$, then such a neighborhood exists by (1) of Lemma 5.5. Again, this contradicts $(z', 0)$ being in the closure of $\mathcal{W}_2$.

Thus, $\mathcal{W}_2 \cap \mathbb{C}^n_t$ is bounded for each $0 < |t| \leq 1$, and hence for all such t by the $\mathbb{C}^*$–action. Thus, given $0 < |t| \leq 1$, there is an $R_1 \geq R_0$ such that $\mathcal{W}_2 \cap \mathbb{C}^n_t \subset B_{R_1,t}$. Hence, (1) of Lemma 5.4 follows since $t_0 \leq |t|$ implies $s_2|t|^N \geq s_1$. □

Proof of Lemma 5.6. Suppose for some $0 < |t'| \leq 1$, $\mathcal{N} \cap \mathbb{C}^n_{t'}$ is unbounded. Hence, for each k there is a $(w_k, t') \in \mathcal{N}$ with $s_k = \|w_k\| \geq k$. Then, by the $\mathbb{R}_+$–action, $(z_k, t_k) = (\frac{R}{s_k} \cdot w_k, \frac{R}{s_k} \cdot t') \in \mathcal{N}$ and $\|z_k\| = R$ for all k. Also, $t_k \to 0$. By the compactness of S_R, there is a convergent sequence of $(z_k, t_k) \in \mathcal{N} \cap S_{R,t_k}$ with limit $(z', 0) \in$ (closure of $\mathcal{N}$) $\cap S_{R,0}$. □

Proof of Lemma 5.5. Let K be the smallest flat of $\mathcal{A}$ containing z' and contained in $\mathbb{C}^n$. We let $I = \{i : L_i(K) = 0\}$ and $I' = \{1, \ldots, n\} \backslash I$. We may choose coordinates so that the line through z' is the z_k–axis and the L_i for $i \in I$ only depend on $(z_{k+1}, \ldots, z_{n+1})$. By the minimality of K, all of the L_i for $i \in I'$ depend on z_k, which we shall single out and denote by y. Furthermore, we may assume that y is chosen to equal 1 for z'.

We let $\omega_1 = \sum_{i \in I} \lambda_i \frac{dL_i}{L_i}$ and $\omega_2 = \sum_{i \in I'} \lambda_i \frac{dL_i}{L_i}$ so that $\omega_\lambda = \omega_1 + \omega_2$. Likewise, we let $F_1 = \prod_{i \in I} \lambda_i \frac{dL_i}{L_i}$ and $F_2 = \prod_{i \in I'} \lambda_i \frac{dL_i}{L_i}$, so we may factor $F = F_1 \cdot F_2$ and so $F_i \omega_i = dF_i$. Since F_1 and ω_1 don't involve y, both are constant along lines parallel to the y–axis.

If $y = u + iv$, then $\frac{\partial}{\partial u} = \frac{\partial}{\partial y} + \frac{\partial}{\partial \bar{y}}$. We see

$$\begin{aligned} \frac{\partial}{\partial \bar{y}}(|F|^2) &= |F_1|^2 \cdot \frac{\partial}{\partial y}(|F_2|^2) = |F_1|^2 \cdot |F_2|^2 \omega_2(\frac{\partial}{\partial y}) \\ &= |F|^2 \cdot g(y, z) \end{aligned} \tag{5.7}$$

where $g(y, z) = \sum_{i \in I'} \frac{\lambda_i a_i}{L_i}$ for $L_i = a_i y + \ell_i(z)$ with $a_i \neq 0$ for $i \in I'$. Then,

$$\frac{\partial}{\partial u}(|F|^2) = |F|^2 \left(2\mathrm{Re}(g(y, z))\right) \tag{5.8}$$

and

$$g(1, 0) = \sum_{i \in I'} \frac{\lambda_i a_i}{a_i} = \sum_{i \in I'} \lambda_i > 0. \tag{5.9}$$

Thus, in a neighborhood of $z' = (1, 0)$, $\mathrm{Re}(g(y, z) > 0$ so by (5.8) and (5.9), $d|F_t|^2$ and dr_t are both positive on $\frac{\partial}{\partial u}$ and so do not point in opposite directions.

For (1), we may view F_1 as a function on $\mathbb{C}^{n+1-k}$. By (1.8), its restriction to slices $z_{n+1} = t$ is singular off $F_1^{-1}(0)$ on a finite number of lines $\mathcal{L}_i$ in the complement of $F_1^{-1}(0)$ where the coefficients of $dz_{k+1}, \ldots, dz_n$ in ω_1 vanish. On such a line $\mathcal{L}_i$, $F_1 = c_i t^{N_1}$ ($c_i \neq 0$), where $N_1 = \deg(F_1)$. As each factor of F_2 contains y, if we are in a sufficiently small closed tubular neighborhood Z of the y–axis, $|F_2| \geq C_2 |y|^{N_2}$ for $|y| \geq R_1 (\geq R_0)$. Here $N_2 = \deg(F_2)$. Furthermore, we may assume on Z with $|y| \geq R_1$, that $|F|$ is increasing with increasing $|y|$ (by (5.8) and (5.9)) and that lines parallel to the y–axis are transverse to $S_{R,t}$. Let $Z_{\geq R}$, resp. Z_R, denote the subset of Z with $|y| \geq R$, resp. $|y| = R$. Then, in $Z_{\geq R_1}$ and on $M_i = \mathbb{C}^k \times \mathcal{L}_i$,

$$|F| \geq C_2 |c_i| |t|^{N_1} |y|^{N_2}. \tag{5.10}$$

From (5.10), if $C_1 = \min\{|c_i|\}$, then $|F| > s_1 |t|^N$ on $M_i \cap Z_R$ provided

$$\frac{C_1 C_2}{s_1} > \left(\frac{|t|}{R}\right)^{N_2/N_1}. \tag{5.11}$$

As $|z_{n+1}| = |t|$ is bounded on Z, given $s_1 > 0$, we may increase R_1 so that (5.11) holds on $Z_{\geq R_1}$. Let $\mathcal{C} = \{z \in \mathbb{C}^{n+1} : \text{for } t = z_{n+1}, |F_t| \leq s_1 |t|^N\}$. Choosing Z of the form $\bar{B}' \times \mathbb{C} \times \bar{B}''$, then $\mathcal{C}' = \mathcal{C} \cap (\bar{B}' \times \{y : |y| = R_1\} \times \partial \bar{B}'')$ is compact and misses all the M_i. Then $\|\omega_1\|$ is nonvanishing on $\mathcal{C}'$ and $\to \infty$ as $z \to 0$ along $\mathbb{R}_+$ orbits of $\mathbb{C}^{n+1-k}$ on which ω_1 is not zero. Since $|F|$ increases with $|y|$ on $Z_{\geq R_1}$, we conclude $\|\omega_1\| \geq C > 0$ on $\mathcal{C} \cap Z_{\geq R_1}$. As ω_1 contains $\frac{\partial}{\partial y}$ in its kernel, $\ker(\omega_1)$ will be transverse to all $S_{R,t}$ on $\mathcal{C} \cap Z_{\geq R_1}$. Since $\|\omega_2\| \to 0$ uniformly as $|y| \to \infty$ on $Z_{\geq R_1}$, by further increasing R_1, $\ker(\omega_1 + \omega_2)$ will remain transverse to $S_{R,t}$ on $\mathcal{C} \cap Z_{\geq R_1}$, completing the proof. □

6 Proofs of Propositions 2.5 and 2.10

Proof of Proposition 2.5.

Step 1: Constructing the Homeomorphism
There are two steps for constructing the homeomorphism. By the $\mathbb{C}^*$–action $t \cdot (z_1, \ldots, z_{n+1}) \mapsto (t \cdot z_1, \ldots, t \cdot z_{n+1})$, we obtain a biholomorphic diffeomorphism ψ_1 between $X = \mathcal{V} \cap (\mathbb{C}^n \times \{1\})$ with $\mathcal{V} \cap (\mathbb{C}^n \times \{t\})$ such that $\psi_1(X \cap A)(= \psi_1(X \cap \mathcal{A})) = \mathcal{V} \cap \mathcal{A} \cap (\mathbb{C}^n \times \{t\})$. Thus, ψ_1 has all of the required properties. Second, we shall construct for $|t|$ sufficiently small, a homeomorphism ψ_2 from $\mathcal{V} \cap (\mathbb{C}^n \times \{t\})$ to the Milnor fiber $V_t = \mathcal{V} \cap (\mathbb{C}^n \times \{t\} \cap B_\epsilon)$ which sends $\mathcal{A} \cap V \cap (\mathbb{C}^n \times \{t\})$ to $\mathcal{A} \cap V_t$ and is the identity on $B_{\epsilon'}$ for $0 < \epsilon' < \epsilon$, with ϵ' arbitrarily close to ϵ. Furthermore, given a compact subset C of X, we may choose $|t|$ small enough so that $t \cdot C \subset B_{\epsilon'}$. Then, $\psi_2 \circ \psi_1$ will be the desired homeomorphism (see fig. 6).

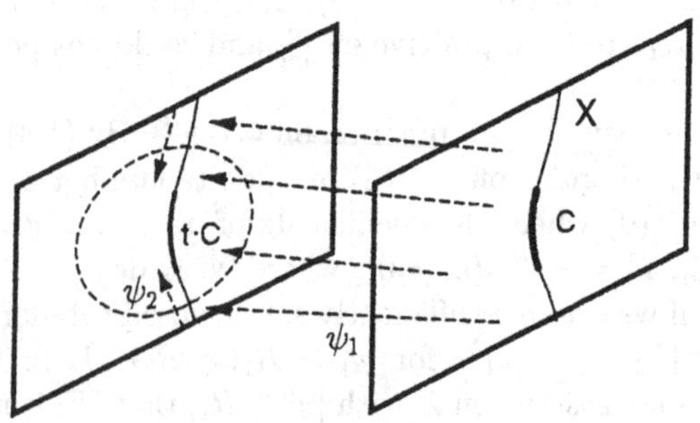

Figure 6.

To construct ψ_2 we shall show that there is a $\delta > 0$ so that if $|t| < \delta$, then $V_t = V \cap (\mathbb{C}^n \times \{t\})$ is transverse to $\mathcal{A} \cap (\mathbb{C}^n \times \{t\})$, and both V_t and $V_t \cap \mathcal{A}$ are (stratified) transverse to S_r for $r \geq \epsilon$. Hence, by the openness of transversality (to closed Whitney stratified sets) they are transverse for $r \geq \epsilon'$, where $\epsilon' < \epsilon$ but sufficiently close. Then, we can stratify V_t by $V_t \backslash \mathcal{A}$ and $V_t \cap W_i$ where W_i ranges over the strata of a Whitney stratification of $\mathcal{A}$. By the transversality condition, $||z||^2 | V_t$ is proper and has no critical points on any strata for $||z|| \geq \epsilon'$. Thus, by stratification theory, there is a homeomorphism of pairs

$$\begin{aligned}(V_t \backslash B_{\epsilon'}, (V_t \cap \mathcal{A}) \backslash B_{\epsilon'}) &\simeq (V_t \backslash S_{\epsilon'}, V_t \cap \mathcal{A} \cap S_{\epsilon'}) \times [\epsilon', \infty) \\ &\simeq (V_t \cap (\bar{B}_\epsilon \backslash B_{\epsilon'}), V_t \cap \mathcal{A} \cap (\bar{B}_\epsilon \backslash B_{\epsilon'}))\end{aligned}$$

where the second is the identity on $V_t \cap S_{\epsilon'}$. Thus, the homeomorphism may be combined with the identity on $V_t \cap \bar{B}_{\epsilon'}$ to give the desired homeomorphism.

Lastly it remains to establish the transversality. For this we shall first show that because the transversality is equivalent to algebraic transversality, the set of points where transversality fails is given by an ideal with (real) homogeneous generators $\{H_1, \ldots, H_s\}$ of real degrees $\deg(H_i) = b_i$. Then, this ideal only vanishes at 0 when $z_{n+1} = 0$, The generators H_i, when viewed as deformations of polynomials in $z' = (z_1, \ldots, z_n)$ with respect to the parameter z_{n+1} are deformations of negative weight, i.e. by terms in z' of degree $< b_i$. Second, we may then apply lemma 6.2, to conclude that for $z_{n+1} = t$ sufficiently small, there is an $R > 0$ so that the H_i do not simultaneously vanish for $||z'|| \geq R$. Third, using the $\mathbb{C}^*$–action we, in fact, show that given $\epsilon > 0$, for t small enough this is true even for $|z'| \geq \epsilon$.

Step 2: Algebraic Condition for Transversality

By assumption, $\mathcal{V}$, $\mathcal{A}$, and $\mathbb{C}^n$ are in algebraic general position off 0 (in a neighborhood of 0, and since they are homogeneous, on all of $\mathbb{C}^{n+1}\backslash\{0\}$). Also, they are holonomic in the sense of Saito [Sa], and by [D1,§2], their intersections are also holonomic. It follows by [D1,§2] that for them, geometric transversality (to strata of the Whitney stratification) is equivalent to algebraic transversality, and hence, can be expressed by an algebraic condition which we give next.

We may choose homogeneous generators $\{\eta_i : 1 \leq i \leq m\}$ for Derlog(V) and $\{\zeta_j : 1 \leq i \leq \ell\}$ for Derlog$(V \cap \mathcal{A})$. We may write $\eta_j = \eta'_j + i \cdot \eta''_j$ for real vector fields η'_j and η''_j, and similarly for $\zeta_j = \zeta'_j + i \cdot \zeta''_j$. Also, we let e denote the real Euler vector field on $\mathbb{C}^n$. Then, for example, the transversality of $V_t \cap \mathcal{A} = \mathcal{V} \cap \mathcal{A} \cap (\mathbb{C}^n \times \{t\})$ to $S_r \subset \mathbb{C}^n \times \{t\}$ at a point $z \in S_r$ is equivalent to the nonvanishing of at least one of the real homogeneous polynomials

$$\{\langle \zeta'_{j(z)}, e_{(z)} \rangle, \langle \zeta''_{j(z)}, e_{(z)} \rangle, \quad 1 \leq j \leq \ell\} \tag{6.1}$$

with a similar condition for V_t using η'_j and η''_j. Here, $\langle\, ,\, \rangle$ denotes the inner product.

We let $\{H_1(z, z_{n+1}), \ldots, H_s(z, z_{n+1})\}$ denote either set of real homogeneous polynomials. Thus, the $H_i(z, z_{n+1})$ are real homogeneous of say degree d_i. Then, transversality will fail for either V or $V \cap \mathcal{A}$ at the points where all of the H_i vanish. Since $V = \mathcal{V} \cap \mathbb{C}^n$ and $\mathcal{A} \cap \mathbb{C}^n$ are both homogeneous and (algebraically) transverse, V and $V \cap \mathcal{A}$ are transverse to S_r for $0 < r < \epsilon$ (and hence for all $r > 0$ by homogeneity). Thus, at least one of $H_1(z, 0), \ldots, H_s(z, 0)$ will be nonzero at any point in a punctured neighborhood of 0.

Step 3: A Nonvanishing Lemma for Transversality

Let $H_i(z, t)$ be real polynomial germs of degrees $d_i > 0$, for $i = 1, \ldots, s$, with z denoting coordinates for $\mathbb{C}^n$ and t for $\mathbb{C}$. Furthermore, we suppose that $h_i(z) = H_i(z, 0)$ is homogeneous of degree d_i. Then, we think of $(H_1(z, t), \ldots, H_s(z, t))$ as a "deformation of negative weight of the weighted homogeneous germ "$(h_1, \ldots, h_s)$. Then, the following lemma is the dual statement to: a deformation of nonnegative weight of a weighted homogeneous germ with isolated zero has an isolated zero [D2, lemma 12.9].

Lemma 6.2 *Consider the polynomial germ $(H_1, \ldots, H_s)$ which is a deformation of negative weight (as described above) of $h(z) = (h_1(z), \ldots, h_s(z))$. If $h(z)$ does not vanish in a punctured neighborhood of 0 then there are $R, \epsilon > 0$ such that if $|t| < \epsilon$ and $\|z\| > R$, then not all of the $H_i(z,t) = 0$.*

Proof. The argument basically follows that given in [D2, lemma 12.9] but for the dual case of behavior at infinity.

First $h(z) = (h_1(z), \ldots, h_s(z))$ does not vanish in a punctured neighborhood of 0. Since all of the h_i are homogeneous, by the $\mathbb{C}^*$–action, h does not vanish off 0. Choose b_i so that for all i, $b_i \cdot d_i = N$. Then, $\rho = \sum |h_i|^{2b_i}$ is real weighted homogeneous of degree $2N$. If $z^{\boldsymbol{\alpha}}$ is a monomial of degree k then there is a constant $C > 0$, so that

$$|z^{\boldsymbol{\alpha}}| \leq C \cdot \rho^{k/2N} \tag{6.3}$$

Now each H_i has the form

$$H_i(z,t) = h_i(z) + \sum c_{\boldsymbol{\alpha}} t^{m_{\boldsymbol{\alpha}}} \cdot z^{\boldsymbol{\alpha}} \qquad \text{summed over } |\boldsymbol{\alpha}| < d_i$$

where all $m_{\boldsymbol{\alpha}} > 0$. Thus, by (6.3), there is a $C_i > 0$ so that if $|t| < \epsilon$ and $\|z\| > 1$ then

$$|\sum c_{\boldsymbol{\alpha}} t^{m_{\boldsymbol{\alpha}}} \cdot z^{\boldsymbol{\alpha}}| \leq \epsilon C_i \rho^{(d_i - 1)/2N}.$$

Let $H = \sum |H_i|^{2b_i}$. If $|t| < \epsilon$ and $\|z\| > 1$ then

$$H \geq \rho - \epsilon C \rho^{(1-1/2N)} = \rho^{(1-1/2N)}(\rho^{1/2N} - \epsilon \cdot C).$$

Then, there is a constant $C' > 0$ so that $\rho \geq C'$ for $\|z\| \geq 1$. By choosing ϵ smaller still so that $\epsilon \cdot C < (1/2)C'^{1/2N} = C''$, we obtain

$$H \geq C'' \cdot \rho^{(1-1/2N)} \qquad \text{for } \|z\| \geq 1 \text{ and } |t| < \epsilon \text{ (for the smaller } \epsilon).$$

Thus, $H \neq 0$ for $\|z\| \geq 1$, implying for any given z, some $H_i(z,t) \neq 0$. □

Step 4: Strengthening the Conclusion of the Lemma
The lemma asserts that there exist R and $\epsilon > 0$ so that if $|t| < \epsilon$ and $\|z\| \geq R$, then some $H_i(z,t) \neq 0$. However, we can apply the $\mathbb{R}_+$–action, and conclude that if $|t| < \delta\epsilon$ and $\|z\| \geq \delta R$, some $H_i(z,t) \neq 0$. Now choose δ sufficiently small so that $\delta R < \epsilon'$, so we have the desired conclusion when $|t| < \delta\epsilon$. This completes the proof of proposition 2.5. □

Proof of Proposition 2.10. This proposition is basically an application of a version for multivalued functions [OT1,§4] of a well–known result about the "catastrophe map" for a parametrized family of holomorphic germs $F(z, \lambda) : \mathbb{C}^{n+k}, 0 \to \mathbb{C}, 0$, together with special properties of linear affine arrangements. We recall in the holomorphic case, if the partial differential map $D_z : (z, \lambda) \mapsto d_z F(0)$ is a submersion at 0 then $Z = D_z^{-1}(0)$ is a smooth germ at 0 of dimension k and the restriction of projection from $\mathbb{C}^{n+k}$ to $\mathbb{C}^k$ gives the "catastrophe map" $p : Z, 0 \to \mathbb{C}^k, 0$. The regular points of this map correspond to the points where D_z, viewed as a function of z, is transverse to 0. These are the Morse singularities of F viewed as a function of z (see e.g. [Lo,chap 4]). Thus, the critical points of p are the points where F does not have a Morse singularity. Then, by Sard's theorem, for almost all λ, i.e. off the discriminant of p, $F(\cdot, \lambda) : \mathbb{C}^{n+k}, 0 \to \mathbb{C}, 0$ has only Morse singularities.

Then, the argument in [OT1,§4] can be seen as an extension of this to multivalued functions F, and combining the local results to obtain a global result. Rather than choosing holomorphic representatives of branches of F where $F \neq 0$,and applying the holomorphic version, one can argue directly (as is done for linear arrangements in [OT1,§4]) by noting that $d_z F(z, \lambda) = F \cdot \Phi^* \omega_\lambda$. Where $F(z, \lambda) \neq 0$, the submersion condition at (z, λ) becomes $d_{(z,\lambda)} \Phi^* \omega_\lambda(z)$ has rank n. First, just as we argued in the proof of the global version of theorem 1, the global result for $f = \prod \ell_i^{\lambda_i}$ reduces by proposition 2.5 to the local case for $F = \prod (L_i \circ \Phi)^{\lambda_i}$. Then, the image of $d_\lambda(\Phi^* \omega_\lambda)$ is spanned by

$$(6.4) \qquad \{\Phi^* \frac{dL_1}{L_1}, \ldots, \Phi^* \frac{dL_k}{L_k}\}.$$

We claim that in a punctured neighborhood of 0, the restriction of (6.4) spans the cotangent space $T^* V_t$; or alternately that $T^* V$ is spanned by the restriction to $T^* V$ of dz_{n+1} together with (6.4). As $F \neq 0$, it is enough to use $\{\Phi^* dL_i, 1 \leq i \leq k, dz_{n+1}\}$.

Let $\tilde{\Phi} = (\Phi, z_{n+1}) : \mathbb{C}^{n+1}, 0 \to \mathbb{C}^m, 0$. By the immersive assumption, $d(\tilde{\Phi}|V)(z)$ is injective if $z \in V \backslash \{0\}$. Hence, a sufficient condition for $\{\Phi^* dL_i, 1 \leq i \leq k, dz_{n+1}\}$ to span $T_z^* V$ is that $\{L_i, 1 \leq i \leq k, dz_{n+1}\}$ span the dual space $(\mathbb{C}^{m+1})^*$. Since A' is essential, $\cap H_i = (0)$ in $\mathbb{C}^m$. Thus, $\{L_i, 1 \leq i \leq k\}$ span $(\mathbb{C}^m)^*$, hence $\{L_i, 1 \leq i \leq k, dz_{n+1}\}$ span $(\mathbb{C}^{m+1})^*$ as required.

Thus, we can apply the preceding argument to the multivalued map

$$\begin{aligned} \tilde{F}(z, \lambda) : V \backslash \mathcal{A} \times \mathbb{C}^k &\to \mathbb{C} \\ (z, \lambda) &\mapsto F(z) = \prod (L_i \circ \Phi)^{\lambda_i}. \end{aligned}$$

Then, by the preceding, $d_\lambda \Phi^* \omega_\lambda$ has maximal rank on $V \backslash \{0\}$, and thus as a map,

$$\Psi = \Phi^* \omega_\lambda : V \backslash \mathcal{A} \times \mathbb{C}^k \to T^* V$$

lies over projection on the first factor and is transverse to the zero-section Z' of $T^* V$. Thus, $Z = \Psi^{-1}(Z')$ is smooth of dimension k.

Denote the restriction of the projection map by $p : Z \to \mathbb{C}^k$. By the $\mathbb{C}^*$–action, p is determined by its behavior in a neighborhood of 0. Then, the fiber of p over $\lambda \in \mathbb{C}^k$ consists of the critical points of F (with exponents given by λ). This is finite so p is a finite–to–one mapping. It follows that the critical set $C(p)$ is a proper algebraic subset of Z and hence constructible of positive codimension. Thus, the discriminant $D(p) = p(C(p))$ is also a constructible subset of $\mathbb{C}^k$ of positive codimension. Hence, it is contained in a proper Zariski closed subset of $\mathbb{C}^k$, with complement denoted by $\Lambda_1(X)$. As p is a covering off $\Lambda_1(X)$, the fibers are constant. Hence, the number of critical points is independent of $\lambda \in \Lambda_1(X)$.

Lastly, we must show that X is transverse to the multiform f at ∞. Consider in (1.6) the ideal in $\mathcal{O}_{\mathbb{C}^{n+1},0}$ generated by $I(V)$ and $ev_F(\mathrm{Derlog}(\mathcal{A}, G))$, except as in (1.7) $ev_F(\zeta) = \Phi^*\omega_\lambda(\zeta)$. We temporarily denote this ideal by J_λ. Then, if for some value of λ, $\dim_{\mathbb{C}} \mathcal{O}_{\mathbb{C}^n,0}/J_\lambda = \ell < \infty$, then by a standard argument using Nakayama's lemma the set of λ for which $\dim_{\mathbb{C}} \mathcal{O}_{\mathbb{C}^{n+1},0}/J_\lambda < \infty$ is a non-empty Zariski open subset $\Lambda_2(X) \subset \mathbb{C}^k$. However, as remarked in §1, by [DM,§4] or [D1,lemma 2.10] the algebraic transversality of $\mathcal{A}$ to V implies that V is algebraically transverse to $Q_1 = Q \circ \Phi$ which defines the nonlinear arrangement $\mathcal{A}$. Thus, $\Lambda_2(X)$ is nonempty as it contains $\mathbf{1} = (1, \ldots, 1)$. We let $\Lambda(A, X) = \Lambda_1(X) \cap \Lambda_2(X)$ to complete the proof. □

References

[Ao] V.I.Aomoto, On the vanishing cohomology attached to certain many valued meromorphic forms, *Math. Soc. Japan*, 27 (1975), 248-255.

[D1] J.Damon, *Higher Multiplicities and Almost Free Divisors and Complete Intersections*, 589, Memoirs Amer. Math. Soc., 1996.

[D2] J.Damon, *Topological Triviality and Versality for subgroups of $\mathcal{A}$ and $\mathcal{K}$*, 389, Memoirs Amer. Math. Soc., 1988.

[D3] J.Damon, Singular Milnor fibers and higher multiplicities for non-isolated complete intersections, Int. Sem. Sing. and Complex Geom., ed. Q. K. Lu et al., *AMS–IP Studies Adv. Math.* 5 (1997) ,28-53.

[DM] J.Damon and D.Mond, $\mathcal{A}$–codimension and the vanishing topology of discriminants, *Invent. Math.*, 106 (1991), 217-242.

[Lê] Lê D.T., Le concept de singularité isolée de fonction analytique, *Adv. Studies in Pure Math.*, 8 (1986), 215-227.

[Lê2] Lê D.T., Calculation of Milnor number of an isolated singularity of complete intersection, *Funct. Anal. and Appl.*, 8 (1974),127-131.

[LêT] Lê D.T. and B.Teissier, Cycles évanescents, sections planes, et conditions de Whitney II, *Proc. Sym. Pure Math.* 44, Part II (1983), 65-103.

[Le] K.S.Lee, *On logarithmic forms and arrangements of hyperplanes*, Thesis, University of Wisconsin 1995.

[Lo] E.J.N.Looijenga, *Isolated singular points on complete intersections*, Lecture Notes in Math., London Math. Soc., Cambridge University Press, 1984.

[OT] P.Orlik and H.Terao, *Arrangements of Hyperplanes*, Grundlehren der Math. Wiss. 300, Springer Verlag, 1992.

[OT2] P.Orlik and H.Terao, The number of critical points of a product of powers of linear functions, *Invent. Math.*, 120, (1995), 1-14.

[OT3] P.Orlik and H.Terao, Arrangements and Milnor Fibers, *Math. Ann.*, 301 (1995), 211-235.

[Sa] K.Saito, Theory of logarithmic differential forms and logarithmic vector fields, *J. Fac. Sci. Univ. Tokyo Sect. Math.*, 27 (1980), 265-291.

[Si] D.Siersma, Vanishing cycles and special fibres, *Singularity Theory and its Applications: Warwick , Part I,* Springer Lecture Notes, 1462 (1991), 292-301.

[Te] B.Teissier, Cycles évanescents, sections planes, et conditions de Whitney, *Singularités à Cargèse, Astérisque,* 7, 8 (1973), 285-362.

[To1] H.Terao, Arrangements of hyperplanes and their freeness I,II *J. Fac. Sci. Univ. Tokyo Sect. Math.*, 27 (1980), 293-320.

[V] A.N.Varchenko, Critical points of the product of powers of linear functions and families of bases of singular vectors, *Compositio Math.* 97 (1995), 385-401.

Department of Mathematics
University of North Carolina
Chapel Hill, NC 27599
USA.

Strange Duality, Mirror Symmetry, and the Leech Lattice

Wolfgang Ebeling

Dedicated to Terry Wall.

Abstract

We give a survey of old and new results concerning Arnold's strange duality. We show that most of the features of this duality continue to hold for the extension discovered by C.T.C. Wall and the author. The results include relations to mirror symmetry and the Leech lattice.

Introduction

More than 20 years ago, V.I. Arnold [Ar] discovered a strange duality among the 14 exceptional unimodal hypersurface singularities. A beautiful interpretation of this duality was given by H. Pinkham [P1] and independently by I.V. Dolgachev and V.V. Nikulin [DN, D3]. I. Nakamura related this duality to the Hirzebruch-Zagier duality of cusp singularities [Na1, Na2].

In independent work in early 1982, C.T.C. Wall and the author discovered an extension of this duality embracing on the one hand series of bimodal singularities and on the other, complete intersection surface singularities in $\mathbb{C}^4$ [EW]. We showed that this duality also corresponds to Hirzebruch-Zagier duality of cusp singularities.

Recent work has aroused new interest in Arnold's strange duality. It was observed by several authors (see [D4] and the references there) that Pinkham's interpretation of Arnold's original strange duality can be considered as part of a two-dimensional analogue of the mirror symmetry of families of Calabi-Yau threefolds. Two years ago, K. Saito [S] discovered a new feature of Arnold's strange duality involving the characteristic polynomials of the monodromy operators of the singularities and he found a connection with the characteristic polynomials of automorphisms of the famous Leech lattice. Only shortly after, M. Kobayashi [Kob] found a duality of the weight systems associated to the 14 exceptional unimodal singularities which corresponds to Arnold's strange duality. He also related it to mirror symmetry.

In this paper we first review these results. Then we consider our extension of this duality and examine which of the newly discovered features continue to hold. It turns out that with a suitable construction, Pinkham's interpretation can be extended to a larger class of singularities. In this way, one obtains many new examples of mirror symmetric families of K3 surfaces. We also associate characteristic polynomials to the singularities involved in our extension of the duality and show that Saito's duality continues to hold. Moreover, in this way we can realize further characteristic polynomials of automorphisms of the Leech lattice. The connection with the Leech lattice seems to be rather mysterious. We discuss some facts which might help to understand this connection. We conclude with some open questions.

We thank the referee for his useful comments.

1 Arnold's Strange Duality

We first discuss Arnold's original strange duality among the 14 exceptional unimodal hypersurface singularities.

We recall Dolgachev's construction [D1, D2] (see also [L1]) of these singularities. Let $b_1 \leq b_2 \leq b_3$ be positive integers such that $\frac{1}{b_1} + \frac{1}{b_2} + \frac{1}{b_3} < 1$. Consider the upper half plane $\mathbb{H} = \{x + iy \in \mathbb{C} \mid y > 0\}$ with the hyperbolic metric $\frac{1}{y^2}(dx^2 + dy^2)$ and a solid triangle $\Delta \subset \mathbb{H}$ with angles $\frac{\pi}{b_1}, \frac{\pi}{b_2}, \frac{\pi}{b_3}$. Let Σ be the subgroup of the group of isometries of $\mathbb{H}$ generated by the reflections in the edges of Δ, and let Σ_+ be the subgroup of index 2 of orientation preserving isometries. Then $\Sigma_+ \subset \mathrm{PSL}_2(\mathbb{R})$ and acts linearly on the total space $T\mathbb{H}$ of the tangent bundle of $\mathbb{H}$. The inclusion $\mathbb{H} \subset T\mathbb{H}$ as zero section determines an inclusion $\mathbb{H}/\Sigma_+ \subset T\mathbb{H}/\Sigma_+$ of orbit spaces. Collapsing $\mathbb{H}/\Sigma_+$ to a point yields a normal surface singularity (X, x_0). This singularity is called a *triangle singularity*. The numbers b_1, b_2, b_3 are called the *Dolgachev numbers* $\mathrm{Dol}(X)$ of the singularity. Scalar multiplication in the fibres of the tangent bundle $T\mathbb{H}$ induces a good $\mathbb{C}^*$-action on X. A resolution of the singularity (X, x_0) can be obtained by the methods of [OW]. A minimal good resolution consists of a rational curve of self-intersection number -1 and three rational curves of self-intersection numbers $-b_1$, $-b_2$, and $-b_3$ respectively intersecting the exceptional curve transversely.

By [D1], for exactly 14 triples (b_1, b_2, b_3) the singularity (X, x_0) is a hypersurface singularity. Thus it can be given by a function germ $f : (\mathbb{C}^3, 0) \to (\mathbb{C}, 0)$ where f is weighted homogeneous with weights w_1, w_2, w_3 and degree N. The corresponding weighted homogeneous functions, weights and degrees are indicated in Table 1. It turns out that these singularities are unimodal and one gets in this way exactly the 14 exceptional unimodal hypersurface singularities in Arnold's classification [Ar]. (The equations in Table 1 are obtained by setting the modulus equal to zero.)

Let (X, x_0) be one of the 14 hypersurface triangle singularities, and denote by X_t and μ its Milnor fibre and Milnor number respectively. We denote by $\langle\ ,\ \rangle$ the intersection form on $H_2(X_t, \mathbb{Z})$ and by $H = (H_2(X_t, \mathbb{Z}), \langle\ ,\ \rangle)$ the Milnor lattice. A.M. Gabrielov [G] has shown that there exists a weakly distinguished basis of vanishing cycles of H with a Coxeter-Dynkin diagram

Name	Equation	N	Weights	Dol	Gab	μ	d	Dual
E_{12}	$x^7+y^3+z^2$	42	6 14 21	2 3 7	2 3 7	12	1	E_{12}
E_{13}	$x^5y+y^3+z^2$	30	4 10 15	2 4 5	2 3 8	13	−2	Z_{11}
E_{14}	$x^8+y^3+z^2$	24	3 8 12	3 3 4	2 3 9	14	3	Q_{10}
Z_{11}	$x^5+xy^3+z^2$	30	6 8 15	2 3 8	2 4 5	11	−2	E_{13}
Z_{12}	$x^4y+xy^3+z^2$	22	4 6 11	2 4 6	2 4 6	12	4	Z_{12}
Z_{13}	$x^6+xy^3+z^2$	18	3 5 9	3 3 5	2 4 7	13	−6	Q_{11}
Q_{10}	$x^4+y^3+xz^2$	24	6 8 9	2 3 9	3 3 4	10	3	E_{14}
Q_{11}	$x^3y+y^3+xz^2$	18	4 6 7	2 4 7	3 3 5	11	−6	Z_{13}
Q_{12}	$x^5+y^3+xz^2$	15	3 5 6	3 3 6	3 3 6	12	9	Q_{12}
W_{12}	$x^5+y^4+z^2$	20	4 5 10	2 5 5	2 5 5	12	5	W_{12}
W_{13}	$x^4y+y^4+z^2$	16	3 4 8	3 4 4	2 5 6	13	−8	S_{11}
S_{11}	$x^4+y^2z+xz^2$	16	4 5 6	2 5 6	3 4 4	11	−8	W_{13}
S_{12}	$x^3y+y^2z+xz^2$	13	3 4 5	3 4 5	3 4 5	12	13	S_{12}
U_{12}	$x^4+y^3+z^3$	12	3 4 4	4 4 4	4 4 4	12	16	U_{12}

Table 1: The 14 exceptional unimodal singularities.

of the form of Fig. 1. The author [E1] has shown that this diagram even corresponds to a distinguished basis of vanishing cycles (cf. also [E5]). (For the notions of a distinguished and weakly distinguished basis of vanishing cycles see e.g. [AGV]). The numbers p_1, p_2, p_3 are called the *Gabrielov numbers* $\mathrm{Gab}(X)$ of the singularity. Here each vertex represents a sphere of self-intersection number -2, two vertices connected by a single solid edge have intersection number 1, and two vertices connected by a double broken line have intersection number -2. Using the results of K. Saito (see [E3, Theorem 3.4.3]), one can see that the Gabrielov numbers are uniquely determined by the singularity. We denote by d the discriminant of H, i.e. the determinant of an intersection matrix with respect to a basis of H.

Arnold has now observed: there exists an involution $X \mapsto X^*$ on the set of the 14 exceptional unimodal singularities, such that

$$\mathrm{Dol}(X) = \mathrm{Gab}(X^*), \quad \mathrm{Gab}(X) = \mathrm{Dol}(X^*), \quad N = N^*, \quad \mu + \mu^* = 24.$$

This is called *Arnold's strange duality.* Note that also $d = d^*$.

H. Pinkham [P1] has given the following interpretation of this duality. (This was independently also obtained by I.V. Dolgachev and V.V. Nikulin [DN, D3].) The Milnor fibre X_t can be compactified in a weighted projective space to a surface with three cyclic quotient singularities on the curve at infinity; a minimal resolution of these singularities yields a K3 surface S. Denote by $G(p_1, p_2, p_3)$ the subgraph of the graph of Fig. 1 which is obtained by omitting the vertices with indices $\mu - 1$ and μ. Let $M(p_1, p_2, p_3)$ be the lattice (the free abelian group with an integral quadratic form) determined by the graph $G(p_1, p_2, p_3)$. Then $H = M(p_1, p_2, p_3) \oplus U$, where U is a unimodular hyperbolic plane (the lattice of rank 2 with a basis $\{e, e'\}$ such that $\langle e, e' \rangle =$

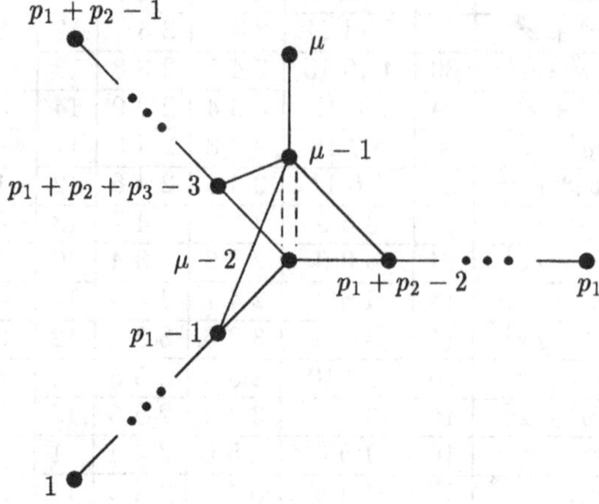

Figure 1: Coxeter-Dynkin diagram of an exceptional unimodal singularity.

1, $\langle e, e \rangle = \langle e', e' \rangle = 0$) and $\oplus$ denotes orthogonal direct sum. The dual graph of the curve configuration of S at infinity is given by $G(b_1, b_2, b_3)$. The inclusion $X_t \subset S$ induces a primitive embedding $H_2(X_t, \mathbb{Z}) \hookrightarrow H_2(S, \mathbb{Z})$ and the orthogonal complement is just the lattice $M(b_1, b_2, b_3)$. By [Nik] the primitive embedding of $M(p_1, p_2, p_3) \oplus U$ into the unimodular K3 lattice $L := H_2(S, \mathbb{Z})$ is unique up to isomorphism.

In this way, Arnold's strange duality corresponds to a duality of K3 surfaces. This is a two-dimensional analogue of the mirror symmetry between Calabi-Yau threefolds. This has recently been worked out by Dolgachev [D4]. We give an outline of his construction. Let M be an even non-degenerate lattice of signature $(1, t)$. An M*-polarized* K3 surface is a pair (S, j) where S is a K3 surface and $j : M \hookrightarrow \text{Pic}(S)$ is a primitive lattice embedding. Here $\text{Pic}(S)$ denotes the Picard group of S. An M-polarized K3 surface (S, j) is called *pseudo-ample* if $j(M)$ contains a pseudo-ample divisor class. We assume that M has a unique embedding into the K3 lattice L and the orthogonal complement $M^\perp$ admits an orthogonal splitting $M^\perp = U \oplus \check{M}$. (Dolgachev's construction is slightly more general.) Then we consider the complete family $\mathcal{F}$ of pseudo-ample M-polarized K3 surfaces and define its *mirror family* $\mathcal{F}^*$ to be any complete family of pseudo-ample $\check{M}$-polarized K3 surfaces. It is shown in [D4] that this is well defined and that there is the following relation between $\mathcal{F}$ and $\mathcal{F}^*$: The dimension of the family $\mathcal{F}$ is equal to the rank of the Picard group of a general member of the mirror family $\mathcal{F}^*$. In particular, this can be applied to $M = M(b_1, b_2, b_3)$ and $\check{M} = M(p_1, p_2, p_3)$ for one of the 14 Dolgachev triples (b_1, b_2, b_3). See [D4] for further results and references.

It was observed by I. Nakamura [Na1, Na2] that Arnold's strange duality corresponds to Hirzebruch-Zagier duality of hyperbolic (alias cusp) singularities. For details see [Na1, Na2, EW].

2 Kobayashi's Duality of Weight Systems

In his paper [Kob], M. Kobayashi has observed a new feature of Arnold's strange duality which we now want to explain.

A quadruple $W = (w_1, w_2, w_3; N)$ of positive integers with $N \in \mathbb{N}w_1 + \mathbb{N}w_2 + \mathbb{N}w_3$ is called a *weight system.* The integers w_i are called the weights and N is called the degree of W. A weight system $W = (w_1, w_2, w_3; N)$ is called *reduced* if $\gcd(w_1, w_2, w_3) = 1$.

Let $W = (w_1, w_2, w_3; N)$ and $W' = (w'_1, w'_2, w'_3; N')$ be two reduced weight systems. An 3×3- matrix Q whose elements are non-negative integers is called a *weighted magic square* for (W, W'), if

$$(w_1, w_2, w_3)Q = (N, N, N) \quad \text{and} \quad Q\begin{pmatrix} w'_1 \\ w'_2 \\ w'_3 \end{pmatrix} = \begin{pmatrix} N' \\ N' \\ N' \end{pmatrix}.$$

(In the case $w_1 = w_2 = w_3 = w'_1 = w'_2 = w'_3 = 1$, Q is an ordinary magic square.) Q is called *primitive,* if $|\det Q| = N = N'$. We say that the weight systems W and W' are *dual* if there exists a primitive weighted magic square for (W, W').

Kobayashi now proves:

Theorem 1 (M. Kobayashi) *Let $W = (w_1, w_2, w_3; N)$ be the weight system of one of the 14 exceptional unimodal singularities. Then, up to permutation, there exists a unique dual weight system W^*. The weight system W^* belongs to the dual singularity in the sense of Arnold.*

Moreover, Kobayashi shows that there is a relation between this duality of weight systems and the polar duality between certain polytopes associated to the weight systems. Such a polar duality was considered by V. Batyrev [Ba] in connection with the mirror symmetry of Calabi-Yau hypersurfaces in toric varieties. We refer to [E6] for a more precise discussion of this relation.

3 Saito's Duality of Characteristic Polynomials

Let $f : (\mathbb{C}^3, 0) \to (\mathbb{C}, 0)$ be a germ of an analytic function defining an isolated hypersurface singularity (X, x_0). A characteristic homeomorphism of the Milnor fibration of f induces an automorphism $c : H_2(X_t, \mathbb{Z}) \to H_2(X_t, \mathbb{Z})$ called the *(classical) monodromy operator* of (X, x_0). It is a well known theorem (see e.g. [Br]) that the eigenvalues of c are roots of unity. This means that the characteristic polynomial $\phi(\lambda) = \det(\lambda I - c)$ of c is a monic polynomial whose roots are roots of unity. Such a polynomial can be written in the form

$$\phi(\lambda) = \prod_{m \geq 0} (\lambda^m - 1)^{\chi_m} \quad \text{for } \chi_m \in \mathbb{Z},$$

where all but finitely many of the integers χ_m are equal to zero. We note some useful formulae.

Proposition 1 (i) $\mu = \deg \phi = \sum_{m>0} m\chi_m$.

(ii) *If* $\sum_{m>0} \chi_m = 0$ *then*

$$\phi(1) = \prod_{m>0} m^{\chi_m}.$$

(iii) $\phi(1) = (-1)^\mu d$.

(iv) $\operatorname{tr} c^k = \sum_{m|k} m\chi_m$*; in particular* $\operatorname{tr} c = \chi_1$.

Proof. (i) is obvious. For the proof of (ii) we use the identity

$$(\lambda^m - 1) = (\lambda - 1)(\lambda^{m-1} + \cdots + \lambda + 1).$$

To prove (iii), let A be the intersection matrix with respect to a distinguished basis $\{\delta_1, \ldots, \delta_\mu\}$ of vanishing cycles. Write A in the form $A = V + V^t$ where V is an upper triangular matrix with -1 on the diagonal. Let C be the matrix of c with respect to $\{\delta_1, \ldots, \delta_\mu\}$. Then $C = -V^{-1}V^t$ (see e.g. [E3, Proposition 1.6.3]). Therefore

$$\phi(1) = \det(I - C) = \det V^{-1}(V + V^t) = (-1)^\mu d.$$

Finally, (iv) is obtained as in [AC2] using the identity

$$\det(I - tc) = \exp(\operatorname{tr}(\log(I - tc))) = \exp(-\sum_{k \geq 1} \frac{t^k}{k} \operatorname{tr} c^k).$$

This proves Proposition 1.

By A'Campo's theorem [AC1]

$$\operatorname{tr} c = -1.$$

We assume that c has finite order h. This is true, for example, if f is a weighted homogeneous polynomial of degree N. In this case $h = N$. Then $\chi_m = 0$ for all m which do not divide h. K. Saito [S] defines a *dual polynomial* $\phi^*(\lambda)$ to $\phi(\lambda)$:

$$\phi^*(\lambda) = \prod_{k|h} (\lambda^k - 1)^{-\chi_{h/k}}.$$

He obtains the following result.

Theorem 2 (K. Saito) *If $\phi(\lambda)$ is the characteristic polynomial of the monodromy of an exceptional unimodal singularity X, then $\phi^*(\lambda)$ is the corresponding polynomial of the dual singularity X^*.*

For $\phi(\lambda) = \prod_{m|h}(\lambda^m - 1)^{\chi_m}$ we use the symbolic notation

$$\pi := \prod_{m|h} m^{\chi_m}.$$

In the theory of finite groups, this symbol is known as a *Frame shape* [F, CN]. For, if one has a rational finite-dimensional representation of a finite group, then the zeros of the characteristic polynomials of each element of the group are also roots of unity. For a given rational representation, one can thus assign to each conjugacy class of the group its Frame shape. The number

$$\deg(\pi) = \sum m\chi_m$$

is called the *degree* of the Frame shape π.

Let us denote the Frame shape of the dual polynomial $\phi^*(\lambda)$ by π^*. The Frame shapes of the monodromy operators of the 14 exceptional unimodal singularities are listed in Table 2.

Name	π	π^*	Dual
E_{12}	$2\cdot3\cdot7\cdot42/1\cdot6\cdot14\cdot21$	$2\cdot3\cdot7\cdot42/1\cdot6\cdot14\cdot21$	E_{12}
E_{13}	$2\cdot3\cdot30/1\cdot6\cdot15$	$2\cdot5\cdot30/1\cdot10\cdot15$	Z_{11}
E_{14}	$2\cdot3\cdot24/1\cdot6\cdot8$	$3\cdot4\cdot24/1\cdot8\cdot12$	Q_{10}
Z_{12}	$2\cdot22/1\cdot11$	$2\cdot22/1\cdot11$	Z_{12}
Z_{13}	$2\cdot18/1\cdot6$	$3\cdot18/1\cdot9$	Q_{11}
Q_{12}	$3\cdot15/1\cdot5$	$3\cdot15/1\cdot5$	Q_{12}
W_{12}	$2\cdot5\cdot20/1\cdot4\cdot10$	$2\cdot5\cdot20/1\cdot4\cdot10$	W_{12}
W_{13}	$2\cdot16/1\cdot4$	$4\cdot16/1\cdot8$	S_{11}
S_{12}	$13/1$	$13/1$	S_{12}
U_{12}	$4\cdot12/1\cdot3$	$4\cdot12/1\cdot3$	U_{12}

Table 2: Frame shapes of the 14 exceptional unimodal singularities.

To two Frame shapes $\pi = \prod m^{\chi_m}$ and $\pi' = \prod m^{\chi'_m}$ of degree n and n' respectively one can associate a Frame shape $\pi\pi'$ of degree $n + n'$ by concatenation

$$\pi\pi' := \prod m^{\chi_m} \prod m^{\chi'_m} = \prod m^{\chi_m+\chi'_m}.$$

If π and π' are the Frame shapes of the operators $c : H \to H$ and $c' : H' \to H'$ respectively, $\pi\pi'$ is the Frame shape of the operator $c \oplus c' : H \oplus H' \to H \oplus H'$.

In the appendix of [S], Saito notes the following observation: If π is the Frame shape of the monodromy operator of an exceptional unimodal singularity, then the symbol $\pi\pi^*$ of degree 24 is a Frame shape of a conjugacy class of the automorphism group of the Leech lattice. The Leech lattice is a 24-dimensional even unimodular positive definite lattice which contains no roots (see e.g. [E4]). It was discovered by J. Leech in connection with the search for densest sphere packings. Its automorphism group G is a group of order $2^{22}3^9 5^4 7^2 11 \cdot 13 \cdot 23$. The quotient group $\mathrm{Co}_1 := G/\{\pm 1\}$ is a famous sporadic simple group discovered by J. Conway. The Frame shapes of the 164 conjugacy classes of G have been listed by T. Kondo [Kon].

4 An Extension of Arnold's Strange Duality

C.T.C. Wall and the author [EW] have found an extension of Arnold's strange duality. In order to consider this extension, we have to enlarge the class of singularities which we want to discuss.

On the one hand, instead of restricting to the hypersurface case, we can also look at isolated complete intersection singularities (abbreviated ICIS in the sequel). Pinkham has shown [P2] that for exactly 8 triples (b_1, b_2, b_3) the triangle singularities with these Dolgachev numbers are ICIS, but not hypersurface singularities. They are given by germs of analytic mappings $(g, f) : (\mathbb{C}^4, 0) \to (\mathbb{C}^2, 0)$. They are $\mathcal{K}$-unimodal singularities and appear in Wall's classification [Wa]. For certain values of the modulus, the equations are again weighted homogeneous. The corresponding 8 triples (b_1, b_2, b_3), Wall's names and weighted homogeneous equations with weights (w_1, w_2, w_3, w_4) and degrees (N_1, N_2) are shown in Table 3.

Name	Equations	N	Weights	Dol	Gab	μ	d	Dual
J'_9	$xw + y^2$ $x^3 + yw + z^2$	16 18	6 8 9 10	2 3 10	2 2 2 3	9	−4	$J_{3,0}$
J'_{10}	$xw + y^2$ $x^2y + yw + z^2$	12 14	4 6 7 8	2 4 8	2 2 2 4	10	8	$Z_{1,0}$
J'_{11}	$xw + y^2$ $x^4 + yw + z^2$	10 12	3 5 6 7	3 3 7	2 2 2 5	11	−12	$Q_{2,0}$
K'_{10}	$xw + y^2$ $x^3 + z^2 + w^2$	10 12	4 5 6 6	2 6 6	2 3 2 3	10	12	$W_{1,0}$
K'_{11}	$xw + y^2$ $x^2y + z^2 + w^2$	8 10	3 4 5 5	3 5 5	2 3 2 4	11	−20	$S_{1,0}$
L_{10}	$xw + yz$ $x^3 + yw + z^2$	11 12	4 5 6 7	2 5 7	2 2 3 3	10	11	$W_{1,0}$
L_{11}	$xw + yz$ $x^2y + yw + z^2$	9 10	3 4 5 6	3 4 6	2 2 3 4	11	−18	$S_{1,0}$
M_{11}	$2xw + y^2 + z^2$ $x^3 + 2yw$	8 9	3 4 4 5	4 4 5	2 3 3 3	11	−24	$U_{1,0}$

Table 3: The 8 triangle ICIS.

By [H1], the notion of Milnor fibre can also be extended to ICIS. We assume that (g, f) are generically chosen such that $(X', 0) = (g^{-1}(0), 0)$ is an isolated hypersurface singularity of minimal Milnor number μ_1 among such choices of g. Then the *monodromy operator* of $(X, 0)$ is defined to be the monodromy operator of the function germ $f : (X', 0) \to (\mathbb{C}, 0)$. By [E3] there exists a distinguished set of generators consisting of $\nu := \mu + \mu_1$ vanishing cycles, where μ is the rank of the second homology group of the Milnor fibre. Again the monodromy operator is the Coxeter element of this set, i.e. the product of the ν reflections corresponding to the vanishing cycles of the distinguished set of generators. For the 8 triangle ICIS, a Coxeter-Dynkin diagram corresponding to such a distinguished set is depicted in Fig. 2 (cf. [E3]). Let us call the characteristic numbers p_1, p_2, p_3, p_4 of these graphs the *Gabrielov numbers* Gab(X) of the singularity. They are also indicated in Table 3. Again, using [E3, Theorem 3.4.3] one can see that these numbers are uniquely defined.

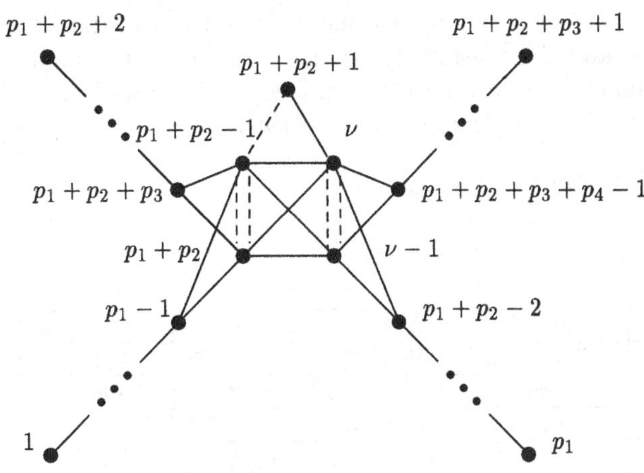

Figure 2: Coxeter-Dynkin diagram of a triangle ICIS.

On the other hand, instead of starting with a hyperbolic triangle, we can start with a hyperbolic quadrilateral. Let b_1, b_2, b_3, b_4 be positive integers such that

$$\frac{1}{b_1} + \frac{1}{b_2} + \frac{1}{b_3} + \frac{1}{b_4} < 2.$$

One can perform the same construction as above with a solid quadrilateral with angles $\frac{\pi}{b_1}$, $\frac{\pi}{b_2}$, $\frac{\pi}{b_3}$, $\frac{\pi}{b_4}$ instead of a triangle. The resulting normal surface singularities are called *quadrilateral singularities* (with *Dolgachev numbers* b_1, b_2, b_3, b_4) (cf. [L2]). Again these singularities admit a natural good $\mathbb{C}^*$-action. A minimal good resolution consists of a rational curve of self-intersection number -2 together with four rational curves of self-intersection numbers $-b_1$, $-b_2$, $-b_3$, and $-b_4$ respectively intersecting the first curve transversely.

For 6 quadruples (b_1, b_2, b_3, b_4) these singularities are isolated hypersurface singularities. They are bimodal in the sense of Arnold. They can again be defined by weighted homogeneous equations. By [EW] they have a distinguished basis with a Coxeter-Dynkin diagram of the following form: it is obtained by adding a new vertex with number 0 to the graph of Fig. 1 in one of the following two ways:

(1) It is connected to the vertex μ and to the vertex $p_1 + p_2$ by ordinary lines. We refer to this diagram by the symbol $(p_1, p_2, \underline{p_3})$.

(2) It is connected to the vertex μ and to the vertices p_1 and $p_1 + p_2 - 1$ by ordinary lines. We refer to this diagram by the symbol $(p_1, \underline{p_2}, \underline{p_3})$.

The corresponding data are indicated in Table 4. The numbers p_1, p_2, p_3 defined by this procedure are unique except for the singularities $W_{1,0}$ and $S_{1,0}$, where we have two triples each. This follows from [E2, Supplement to Theorem 4.1]. One can also obtain an easier proof by considering the discriminants (and in one case the discriminant quadratic form) of the singularities and the possible determinants of the graphs (and in one case the discriminant quadratic form determined by the graph). The absolute values of the determinants in the respective cases are

(1) $|d| = 4(p_1 p_2 - p_1 - p_2)$,

(2) $|d| = (p_1 - 1)(p_2 + p_3)$.

Name	Equation	N	Weights	Dol	Gab	μ	d	Dual
$J_{3,0}$	$x^9 + y^3 + z^2$	18	2 6 9	2 2 2 3	$2\ 3\ \underline{10}$	16	4	J'_9
$Z_{1,0}$	$x^7 + xy^3 + z^2$	14	2 4 7	2 2 2 4	$2\ 4\ \underline{8}$	15	-8	J'_{10}
$Q_{2,0}$	$x^6 + y^3 + xz^2$	12	2 4 5	2 2 2 5	$3\ 3\ \underline{7}$	14	12	J'_{11}
$W_{1,0}$	$x^6 + y^4 + z^2$	12	2 3 6	2 2 3 3	$2\ 5\ \underline{7}$ $2\ \underline{6}\ \underline{6}$	15	-12	K'_{10} L_{10}
$S_{1,0}$	$x^5 + y^2z + xz^2$	10	2 3 4	2 2 3 4	$3\ 4\ \underline{6}$ $3\ \underline{5}\ \underline{5}$	14	20	K'_{11} L_{11}
$U_{1,0}$	$x^3y + y^3 + z^3$	9	2 3 3	2 3 3 3	$4\ \underline{4}\ \underline{5}$	14	27	M_{11}

Table 4: The 6 quadrilateral hypersurface singularities.

For another 5 quadruples (b_1, b_2, b_3, b_4) the quadrilateral singularities with these Dolgachev numbers are ICIS. They are also $\mathcal{K}$-unimodal and appear in Wall's lists [Wa]. They can also be given by weighted homogeneous equations. These equations and Wall's names are listed in Table 5.

The singularities $J'_{2,0}$, $L_{1,0}$, $K'_{1,0}$, and $M_{1,0}$ can be given by equations (g, f) where g has Milnor number $\mu_1 = 1$. Coxeter-Dynkin diagrams of these singularities are computed in [E3]. Using transformations as in the proof of [E3, Proposition 3.6.1], these graphs can be transformed to the following graphs. A Coxeter-Dynkin diagram corresponding to a distinguished set of generators is obtained by adding a new vertex to the graph of Fig. 2. It gets the number $p_1 + p_2 + 2$ and the indices of the old vertices with numbers

Name	Equations	Restrictions	N	Weights
$J'_{2,0}$	$xw + y^2$ $ax^5 + xy^2 + yw + z^2$	$a \neq 0, -\frac{4}{27}$	8 10	2 4 5 6
$L_{1,0}$	$xw + yz$ $ax^4 + xy^2 + yw + z^2$	$a \neq 0, -1$	7 8	2 3 4 5
$K'_{1,0}$	$xw + y^2$ $ax^4 + xy^2 + z^2 + w^2$	$a \neq 0, \frac{1}{4}$	6 8	2 3 4 4
$M_{1,0}$	$2xw + y^2 - z^2$ $x^2y + ax^2z + 2yw$	$a \neq \pm 1$	6 7	2 3 3 4
$I_{1,0}$	$x^3 + w(y - z)$ $ax^3 + y(z - w)$	$a \neq 0, 1$	6 6	2 3 3 3

Name	Dol	Gab	μ	d	Dual
$J'_{2,0}$	2 2 2 6	$2\ 2\ 2\ \underline{6}$	13	-16	$J'_{2,0}$
$L_{1,0}$	2 2 3 5	$2\ 2\ 3\ \underline{5}$ $2\ 2\ \underline{4}\ \underline{4}$	13	-28	$L_{1,0}$ $K'_{1,0}$
$K'_{1,0}$	2 2 4 4	$2\ 3\ 2\ \underline{5}$ $2\ \underline{4}\ 2\ \underline{4}$	13	-32	$L_{1,0}$ $K'_{1,0}$
$M_{1,0}$	2 3 3 4	$2\ 3\ \underline{3}\ \underline{4}$ $2\ \underline{3}\ 3\ \underline{4}$	13	-42	$M_{1,0}$
$I_{1,0}$	3 3 3 3	$\underline{3}\ \underline{3}\ \underline{3}\ \underline{3}$	13	-54	$I_{1,0}$

Table 5: The 5 quadrilateral ICIS.

$p_1 + p_2 + 2, p_1 + p_2 + 3, \ldots, \nu$ are shifted by 1. New edges are introduced in one of the following ways:

(1) The new vertex is connected to the vertex $p_1 + p_2 + 1$ and to the vertex with new index $p_1 + p_2 + p_3 + 3$ (old index $p_1 + p_2 + p_3 + 2$) by ordinary lines. We refer to this diagram by the symbol $(p_1, p_2, p_3, \underline{p_4})$.

(2) The new vertex is connected to the vertex $p_1 + p_2 + 1$ and to the vertices with new indices $p_1 + p_2 + 3$ and $p_1 + p_2 + p_3 + 2$ (old indices $p_1 + p_2 + 2$ and $p_1 + p_2 + p_3 + 1$ respectively) by ordinary lines. We refer to this diagram by the symbol $(p_1, p_2, \underline{p_3}, \underline{p_4})$.

(3) The new vertex is connected to the vertex $p_1 + p_2 + 1$, to the vertex p_1, and to the vertex with new index $p_1 + p_2 + p_3 + 2$ (old index $p_1 + p_2 + p_3 + 1$) by ordinary lines. We refer to this diagram by the symbol $(p_1, \underline{p_2}, p_3, \underline{p_4})$.

The absolute values of the determinants of the respective graphs are

(1) $|d| = 4(p_1 p_2 p_3 - p_1 - p_3)$,

(2) $|d| = p_1 p_2 (p_3 + p_4 + 2) - p_1 - p_2 - p_3 - p_4$,

(3) $|d| = p_1 p_3 (p_2 + p_4)$.

Comparing the values of these determinants with the discriminants of the above 4 quadrilateral ICIS, we find that the graphs listed in Table 5 are the only possible graphs of the types (1), (2), or (3) for these singularities. Again, in two cases the numbers p_1, p_2, p_3, p_4 are not uniquely defined.

For the remaining singularity $I_{1,0}$, $\mu_1 = 2$. This singularity can be given by the following equations:

$$\begin{aligned} g(z) &= z_1^2 + z_2^2 + z_3^2 + z_4^3, \\ f(z) &= a_1 z_1^2 + a_2 z_2^2 + a_3 z_3^2 + a_4 z_4^3, \end{aligned}$$

where $a_i \in \mathbb{R}$, $a_1 < a_2 < a_3 < a_4$. For such a singularity H. Hamm [H2] has given a basis of the complexified Milnor lattice $H_{\mathbb{C}} = H \otimes \mathbb{C}$. As in [E3, Sect. 2.3], one can show that the cycles he constructs are in fact vanishing cycles and that there exists a distinguished set $\{\delta_1, \ldots, \delta_\nu\}$ of generators for this singularity with the following intersection numbers:

$$\langle \delta_{i+1}, \delta_{i+2} \rangle = -1, \quad i = 0, 2, 4, 6, 8, 10,$$

$$\langle \delta_{i+1}, \delta_{i+3} \rangle = \langle \delta_{i+1}, \delta_{i+4} \rangle = 0, \langle \delta_{i+2}, \delta_{i+3} \rangle = \langle \delta_{i+2}, \delta_{i+4} \rangle = 0, \quad i = 0, 4, 8,$$

$$\langle \delta_{i+1}, \delta_j \rangle = \langle \delta_{i+3}, \delta_j \rangle = -1, \quad i = 0, 4, 8, \ 1 \le j \le 12, j \ne i+1, i+2, i+3, i+4,$$

$$\langle \delta_{i+2}, \delta_{2k} \rangle = \langle \delta_{i+4}, \delta_{2k} \rangle = -1, \quad i = 0, 4, 8, \ 1 \le k \le 6, 2k \ne i+2, i+4,$$

$$\langle \delta_{i+2}, \delta_{2k-1} \rangle = \langle \delta_{i+4}, \delta_{2k-1} \rangle = 0, \quad i = 0, 4, 8, \ 1 \le k \le 6, 2k-1 \ne i+1, i+3,$$

$$\langle \delta_{2k-1}, \delta_{13} \rangle = -1, \langle \delta_{2k-1}, \delta_{15} \rangle = 0, \langle \delta_{2k}, \delta_{13} \rangle = 0, \langle \delta_{2k}, \delta_{15} \rangle = -1, \quad 1 \le k \le 6,$$

$$\langle \delta_i, \delta_{14} \rangle = -1, \quad 1 \le i \le 12,$$

$$\langle \delta_{13}, \delta_{14} \rangle = 0, \langle \delta_{13}, \delta_{15} \rangle = 0, \langle \delta_{14}, \delta_{15} \rangle = 0.$$

By the following sequence of braid group transformations (for the notation see e.g. [E3]) the distinguished set $\{\delta_1, \ldots, \delta_\nu\}$ can be transformed to a distinguished set $\{\delta'_1, \ldots, \delta'_\nu\}$ where the Coxeter-Dynkin diagram corresponding to the subset $\{\delta'_1, \ldots, \delta'_{12}\}$ is the graph of Fig. 3:

$$\beta_5, \beta_4, \beta_3, \beta_2; \alpha_6, \alpha_7, \alpha_8, \alpha_9, \alpha_{10}, \alpha_{11}; \alpha_5, \alpha_6; \alpha_3, \alpha_4, \alpha_5;$$

$$\beta_8, \beta_7, \beta_6; \beta_{10}, \beta_9, \beta_8, \beta_7; \kappa_4, \kappa_5, \kappa_6, \kappa_7, \kappa_8, \kappa_9.$$

We refer to the Coxeter-Dynkin diagram corresponding to $\{\delta'_1, \ldots, \delta'_\nu\}$ by the symbol $(\underline{3}, \underline{3}, \underline{3}, \underline{3})$. This notation is motivated by Fig. 3. We admit, however, that it is somewhat arbitrary.

The corresponding symbols are indicated in Table 5.

If one compares the Dolgachev and Gabrielov numbers of Tables 3 and 4 and of Table 5, then one observes a correspondence between the 8 triangle ICIS and the 6 quadrilateral hypersurface singularities and between the 5 quadrilateral ICIS. The corresponding "dual" singularities are indicated in the last column of each table. Note that this correspondence is not always a duality in the strict sense. For the quadrilateral hypersurface singularity

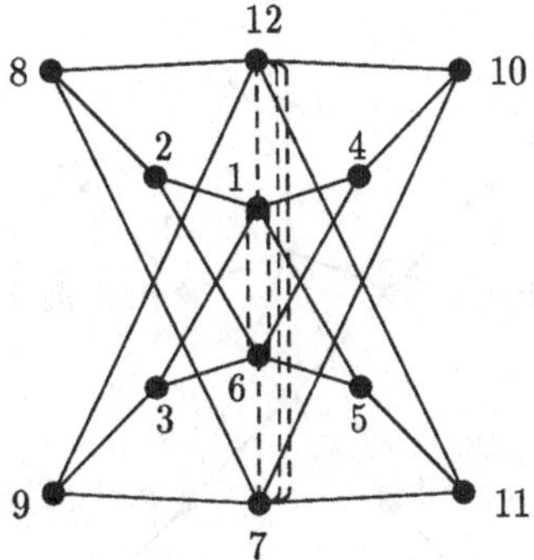

Figure 3: Subgraph of a Coxeter-Dynkin diagram of the singularity $I_{1,0}$.

$W_{1,0}$ we have two corresponding ICIS K'_{10} and L_{10} and for $S_{1,0}$ we have the corresponding ICIS K'_{11} and L_{11}. The quadrilateral ICIS $L_{1,0}$ and $K'_{1,0}$ are both self-dual and dual to each other. In the other cases the correspondence is one-to-one. Pinkham also defined "Gabrielov numbers" (in a weaker sense) for the triangle ICIS [P2] and he already made part of this observation (unpublished). This duality also corresponds to the Hirzebruch-Zagier duality of cusp singularities (see [Na2, EW]).

If one now compares the Milnor numbers of dual singularities, one finds

- for the triangle ICIS versus quadrilateral hypersurface singularities: $\mu + \mu^* = 25$.
- for the quadrilateral ICIS: $\mu + \mu^* = 26$.

(Note that also d and d^* do not coincide in each case.) So one still has to alter something. There are two alternatives:

(1) subtract 1 for quadrilateral.

(2) subtract 1 for ICIS.

In [EW] we considered the first alternative. The quadrilateral singularities are first elements ($l = 0$) of series of singularities indexed by a non-negative integer l. We showed that to each such series one can associate a virtual element $l = -1$. We defined for these Milnor lattices, Coxeter-Dynkin diagrams, and monodromy operators, and showed that all features of Arnold's strange duality including Pinkham's interpretation continue to hold. For more details see below.

A new discovery is that the second alternative works as well, and this also leads to an extension of Saito's duality. Recall that the triangle or quadrilateral ICIS with $\mu_1 = 1$ have a Coxeter-Dynkin diagram D which is either the graph of Fig. 2 or an extension of it. By similar transformations as in the proof of [E3, Proposition 3.6.2], this graph can be transformed to

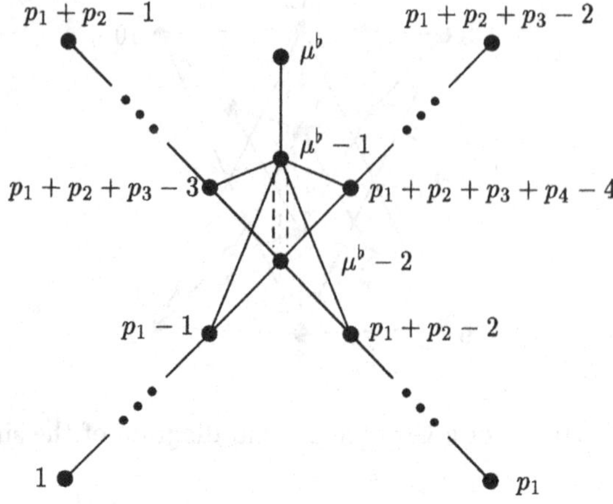

Figure 4: Reduced Coxeter-Dynkin diagram $D^\flat$.

a graph containing the subgraph $D^\flat$ depicted in Fig. 4. (Unfortunately, this proof has to be modified slightly. The correct sequence of transformations is

$$\beta_{\rho-1}, \beta_{\rho-2}, \beta_{\rho-3}, \beta_{\rho-3}, \beta_{\rho-2}, \kappa_{\rho-1}$$

and one has to consider the vertices $\lambda'_{\rho-2}$ and λ'_ρ instead of $\lambda'_{\rho-1}$ and λ'_ρ.) By [E3, Remark 3.6.5], the passage from D to $D^\flat$ can be considered as a kind of "de-suspension". For the singularity $I_{1,0}$, let $D^\flat$ be the graph of Fig. 3. In each case, the new Coxeter-Dynkin diagram $D^\flat$ has $\mu - 1$ instead of $\mu + \mu_1$ vertices. Denote the corresponding Coxeter element (product of reflections corresponding to the vanishing cycles) by $c^\flat$. For the ICIS with $\mu_1 = 1$ one can compute (see [E3, Proposition 3.6.2]) that

$$\pi(c^\flat) = \pi(c)/1.$$

This means that $c^\flat$ has the same eigenvalues as c but the multiplicities of the eigenvalue 1 differ by 1. For the singularity $I_{1,0}$ one has $\pi(c) = 6^3/1 \cdot 2^2$ (cf. [H2]), whereas $\pi(c^\flat) = 3^2 6^2/1^2 2^2$. Note that in any case

$$\operatorname{tr} c^\flat = -2.$$

The passage from c to $c^\flat$ corresponds to the passage from the Milnor lattice H of rank μ to a sublattice $H^\flat$ of rank $\mu^\flat = \mu - 1$. The corresponding discriminants and discriminant quadratic forms of the lattices $H^\flat$ are listed in Table 6. Here we use the notation of [EW].

The Frame shapes of the corresponding operators $c^\flat$ are listed in Table 7. It turns out that the substitution

$$\mu \mapsto \mu^\flat, \quad c \mapsto c^\flat, \quad H \mapsto H^\flat$$

Name	$\mu^\flat$	$d^\flat$	$(H^\flat)^*/H^\flat$	dual form	dual
J'_9	8	4	q_{D_4}	q_{D_4}	$J_{3,0}$
J'_{10}	9	−8	$q_{D_4} + q_{A_1}$	$q_{D_4} + w^1_{2,1}$	$Z_{1,0}$
J'_{11}	10	12	$q_{D_4} + q_{A_2}$	$q_{D_4} + w^{-1}_{3,1}$	$Q_{2,0}$
K'_{10}	9	−12	$w^1_{2,2} + w^{-1}_{3,1}$	$w^{-1}_{2,2} + w^1_{3,1}$	$W_{1,0}$
L_{10}	9	−12	$w^1_{2,2} + w^{-1}_{3,1}$	$w^{-1}_{2,2} + w^1_{3,1}$	$W_{1,0}$
K'_{11}	10	20	$w^1_{2,1} + w^1_{2,1} + w^{-1}_{5,1}$	$w^{-1}_{2,1} + w^{-1}_{2,1} + w^{-1}_{5,1}$	$S_{1,0}$
L_{11}	10	20	$w^1_{2,1} + w^1_{2,1} + w^{-1}_{5,1}$	$w^{-1}_{2,1} + w^{-1}_{2,1} + w^{-1}_{5,1}$	$S_{1,0}$
M_{11}	10	27	$w^{-1}_{3,2} + w^{-1}_{3,1}$	$w^1_{3,2} + w^1_{3,1}$	$U_{1,0}$
$J'_{2,0}$	12	16	$v_1 + v_1$	$v_1 + v_1$	$J'_{2,0}$
$L_{1,0}$ $K'_{1,0}$	12	32	$w^1_{2,2} + w^{-1}_{2,3}$	$w^1_{2,2} + w^{-1}_{2,3}$	$L_{1,0}$ $K'_{1,0}$
$M_{1,0}$	12	49	$w^1_{7,1} + w^{-1}_{7,1}$	$w^1_{7,1} + w^{-1}_{7,1}$	$M_{1,0}$
$I_{1,0}$	12	81	$w^1_{3,1} + w^1_{3,1}$ $+w^{-1}_{3,1} + w^{-1}_{3,1}$	$w^{-1}_{3,1} + w^{-1}_{3,1}$ $+w^1_{3,1} + w^1_{3,1}$	$I_{1,0}$

Table 6: Discriminants and discriminant quadratic forms of the lattices $H^\flat$.

for the ICIS yields

$$\mu + \mu^* = 24, \quad d = d^*, \quad \pi_{X^*} = \pi^*_X.$$

Moreover, the lattice H admits an embedding into the even unimodular lattice

$$K_{24} = (-E_8) \oplus (-E_8) \oplus U \oplus U \oplus U \oplus U$$

of rank 24. This lattice can be considered as the full homology lattice of a K3 surface,

$$K_{24} = H_0(S, \mathbb{Z}) \oplus H_2(S, \mathbb{Z}) \oplus H_4(S, \mathbb{Z}),$$

where the inner product on $H_0(S,\mathbb{Z}) \oplus H_4(S,\mathbb{Z})$ is defined in such a way that this lattice corresponds to a unimodular hyperbolic plane U. The orthogonal complement of H is the lattice $\check{H}$ of the singularity X^* (cf. Table 6).

Let us consider Pinkham's interpretation in the new cases. The Milnor fibre of a triangle or quadrilateral isolated hypersurface or complete intersection singularity can be compactified in such way that after resolving the singularities one gets a K3 surface S [P3]. We consider the dual graph of the curve configuration at infinity in each case. Let $G(p_1, p_2, p_3, p_4)$ and $\tilde{G}(p_1, p_2, p_3, p_4)$ be the subgraphs of the graphs of Fig. 4 and Fig. 2 respectively obtained by omitting the vertices $\mu^\flat - 1$ and $\mu^\flat$, and $p_1 + p_2 - 1$, $p_1 + p_2 + 1$, and ν respectively. Denote by $M(p_1, p_2, p_3, p_4)$ and $\tilde{M}(p_1, p_2, p_3, p_4)$ the corresponding lattices. Recall that the homology lattice $H_2(S, \mathbb{Z})$ of the K3 surface is denoted by L.

First start with a triangle ICIS (X, x_0) with Dolgachev numbers (b_1, b_2, b_3). Then the dual graph is the graph $G(b_1, b_2, b_3)$. This yields an embedding $M(b_1, b_2, b_3) \subset L$ and the orthogonal complement is the Milnor lattice $H =$

Name	π	π^*	Dual
$J_{3,0}$	$2\cdot 3\cdot 18^2/1\cdot 6\cdot 9^2$	$2^2 3\cdot 18/1^2 6\cdot 9$	J'_9
$Z_{1,0}$	$2\cdot 14^2/1\cdot 7^2$	$2^2 14/1^2 7$	J'_{10}
$Q_{2,0}$	$3\cdot 12^2/1\cdot 6^2$	$2^2 12/1^2 4$	J'_{11}
$W_{1,0}$	$2\cdot 12^2/1\cdot 4\cdot 6$	$2\cdot 3\cdot 12/1^2 6$	K'_{10} L_{10}
$S_{1,0}$	$10^2/1\cdot 5$	$2\cdot 10/1^2$	K'_{11} L_{11}
$U_{1,0}$	$9^2/1\cdot 3$	$3\cdot 9/1^2$	M_{11}
$J'_{2,0}$	$2^2 10^2/1^2 5^2$	$2^2 10^2/1^2 5^2$	$J'_{2,0}$
$L_{1,0}$ $K'_{1,0}$	$2\cdot 8^2/1^2 4$	$2\cdot 8^2/1^2 4$	$L_{1,0}$ $K'_{1,0}$
$M_{1,0}$	$7^2/1^2$	$7^2/1^2$	$M_{1,0}$
$I_{1,0}$	$3^2 6^2/1^2 2^2$	$3^2 6^2/1^2 2^2$	$I_{1,0}$

Table 7: Frame shapes of the triangle ICIS and quadrilateral singularities.

$\tilde{M}(p_1, p_2, p_3, p_4) \oplus U$. By alternative (1) (cf. [EW]) the dual of (X, x_0) is a bimodal series; the Milnor lattice of the "virtual" $l = -1$ element of the corresponding series is $M(b_1, b_2, b_3) \oplus U$. One can even associate a Coxeter element to the dual "virtual" singularity; it has order lcm (N_1, N_2) where N_1, N_2 are the degrees of the equations of (X, x_0) [EW], whereas the monodromy operator of (X, x_0) has order N_2. There is no Saito duality of characteristic polynomials. The dual singularities and orders of the monodromy operators are listed in Table 8.

Name	μ	Dol	Gab	d	h	h^*	μ^*	Dual
J'_9	9	2 3 10	2 2 2 3	−4	18	144	15	$J_{3,-1}$
J'_{10}	10	2 4 8	2 2 2 4	8	14	84	14	$Z_{1,-1}$
J'_{11}	11	3 3 7	2 2 2 5	−12	12	60	13	$Q_{2,-1}$
K'_{10}	10	2 6 6	2 3 2 3	12	12	60	14	$W_{1,-1}$
K'_{11}	11	3 5 5	2 3 2 4	−20	10	40	13	$S_{1,-1}$
L_{10}	10	2 5 7	2 2 3 3	11	12	132	14	$W^\sharp_{1,-1}$
L_{11}	11	3 4 6	2 2 3 4	−18	10	90	13	$S^\sharp_{1,-1}$
M_{11}	11	4 4 5	2 3 3 3	−24	9	72	13	$U_{1,-1}$

Table 8: The duality: 8 triangle ICIS versus 8 bimodal series.

On the other hand, we can start with a quadrilateral hypersurface singularity (X, x_0) with Dolgachev numbers (b_1, b_2, b_3, b_4). Then the dual graph is the graph $G(b_1, b_2, b_3, b_4)$. We obtain an embedding $M(b_1, b_2, b_3, b_4) \subset L$ and the orthogonal complement is the Milnor lattice of (X, x_0) described in Table 4. Here we use alternative (2) for the duality. The reduced Milnor lattice $H^\flat$ of the dual triangle ICIS according to Table 4 is the lattice $M(b_1, b_2, b_3, b_4) \oplus U$.

Finally, let (X, x_0) be one of the 5 quadrilateral ICIS. Then the dual graph is again the graph $G(b_1, b_2, b_3, b_4)$. One has an embedding $M(b_1, b_2, b_3, b_4) \subset L$ and the orthogonal complement is the Milnor lattice of (X, x_0) described in Table 5. Combining both alternatives (1) and (2), the lattice $M(b_1, b_2, b_3, b_4) \oplus U$ can be interpreted as follows: The 5 quadrilateral ICIS are the initial $l = 0$ elements of 8 series of ICIS. To each such series one can again associate a virtual $l = -1$ element with a well-defined Milnor lattice. Then $M(b_1, b_2, b_3, b_4) \oplus U$ is the reduced Milnor lattice $\check{H}^\flat$ of the dual virtual singularity. This correspondence is indicated in Table 9. There is also a duality between the 8 virtual singularities as indicated in [EW] (see also [E3, Table 3.6.2]). There is no Saito duality of characteristic polynomials in both cases. But as we have seen above, using alternative (2) we get a third correspondence, for which Saito's duality of characteristic polynomials holds.

Name	μ	Dol	Gab	d	μ^*	Dual
$J'_{2,0}$	13	2 2 2 6	$2\ 2\ 2\ \underline{6}$	-16	11	$J'_{2,-1}$
$L_{1,0}$	13	2 2 3 5	$2\ 2\ 3\ \underline{5}$ $2\ 2\ \underline{4}\ \underline{4}$	-28	11	$L^\sharp_{1,-1}$ $K^\flat_{1,-1}$
$K'_{1,0}$	13	2 2 4 4	$2\ 3\ 2\ \underline{5}$ $2\ \underline{4}\ 2\ \underline{4}$	-32	11	$L_{1,-1}$ $K'_{1,-1}$
$M_{1,0}$	13	2 3 3 4	$2\ 3\ \underline{3}\ \underline{4}$ $2\ \underline{3}\ 3\ \underline{4}$	-42	11	$M^\sharp_{1,-1}$ $M_{1,-1}$
$I_{1,0}$	13	3 3 3 3	$\underline{3}\ \underline{3}\ \underline{3}\ \underline{3}$	-54	11	$I_{1,-1}$

Table 9: The duality between the quadrilateral ICIS.

By Dolgachev's construction [D4], to each case of Pinkham's construction there corresponds a pair of mirror symmetric families of K3 surfaces. Moreover, also to each case where we only have a pair of lattices embedded as orthogonal complements to each other in the lattice K_{24} (cf. Table 6) there corresponds such a mirror pair.

One can also investigate Kobayashi's duality of weight systems for our extension of Arnold's strange duality. As already observed by Kobayashi [Kob], only some of the weight systems of the quadrilateral hypersurface singularities have dual weight systems, the dual weight systems are in general not unique and they correspond again to isolated hypersurface singularities. Since an ICIS has two degrees N_1 and N_2, it is not quite clear how to generalize the notion of a weighted magic square. One possibility would be to work with the sum of the degrees $N := N_1 + N_2$ and to use also 3×4 and 4×4 matrices instead of 3×3 matrices. Then one finds again that in some cases there does not exist a dual weight system, the dual weight systems are in general not unique, most cases are self-dual, and only in the cases $J'_{10} \leftrightarrow Z_{1,0}$, $K'_{11} \leftrightarrow S_{1,0}$, $J'_{2,0} \leftrightarrow J'_{2,0}$, and $M_{1,0} \leftrightarrow M_{1,0}$ of our duality do there exist weighted magic squares giving a duality of the corresponding weight systems.

However, there is a relation between our extended duality and a polar duality between the Newton polytopes generalizing Kobayashi's observation for Arnold's strange duality. This can be used to explain Saito's duality of characteristic polynomials. For details see the forthcoming paper [E6].

5 Singular Moonshine

Let us consider the symbols $\pi\pi^*$ of Table 7. It turns out that they all occur in the list of Kondo, too. These pairs and the pairs from the original Arnold duality correspond to self-dual Frame shapes of the group G with trace -2, -3, or -4. By examining Kondo's list one finds that there are 22 such Frame shapes and all but 3 occur. They are listed in Table 10. Here we use the ATLAS notation [ATL] for the conjugacy classes. For each value of the trace one symbol is missing.

Special elements of G correspond to the deep holes of the Leech lattice. A *deep hole* of the Leech lattice Λ_{24} is a point in $\mathbb{R}^{24}$ which has maximal distance from the lattice points. It is a beautiful theorem of J.H. Conway, R.A. Parker, and N.J.A. Sloane ([CPS], see also [E4]) that there are 23 types of deep holes in Λ_{24} which are in one-to-one correspondence with the 23 isomorphism classes of even unimodular lattices in $\mathbb{R}^{24}$ containing roots, which were classified by H.-V. Niemeier [Nie]. These lattices are characterized by the root systems which they contain. The Frame shape of the Coxeter element of such a root system is also the Frame shape of an automorphism of the Leech lattice. We have indicated in Table 10 the type of the root system of the Niemeier lattice, if a Frame shape corresponds to the Coxeter element of such a root system.

The automorphism group G of the Leech lattice contains the Mathieu group M_{24} (see e.g. [E4]). S. Mukai [M] has classified the finite automorphism groups of K3 surfaces (automorphisms which leave the symplectic form invariant) and shown that they admit a certain embedding into the Mathieu group M_{24}. He gives a list of 11 maximal groups such that every finite automorphism group imbeds into one of these groups. A table of the centralizers of the conjugacy classes of G can be found in [Wi]. For an element $g \in G$, denote its centralizer by $Z(g)$ and the finite cyclic group generated by g by $\langle g \rangle$. We have marked by $(*)$ in Table 10 the cases where there is an obvious inclusion of $Z(g)/\langle g \rangle$ in one of Mukai's groups. It follows that in these cases there is a K3 surface with an operation of $Z(g)/\langle g \rangle$ on it by symplectic automorphisms.

To a Frame shape

$$\pi = \prod_{m|N} m^{\chi_m}$$

ATLAS	Frame		Niemeier	Duality
21A	$2^2 3^2 7^2 42^2/1^2 6^2 14^2 21^2$	$*$		$E_{12} \leftrightarrow E_{12}$
15E	$2^2 3 \cdot 5 \cdot 30^2/1^2 6 \cdot 10 \cdot 15^2$	$*$	$D_{16} \oplus E_8$	$E_{13} \leftrightarrow Z_{11}$
24B	$2 \cdot 3^2 4 \cdot 24^2/1^2 6 \cdot 8^2 12$	$*$		$E_{14} \leftrightarrow Q_{10}$
11A	$2^2 22^2/1^2 11^2$	$*$	D_{12}^2	$Z_{12} \leftrightarrow Z_{12}$
18B	$2 \cdot 3 \cdot 18^2/1^2 6 \cdot 9$	$*$	$A_{17} \oplus E_7$	$Z_{13} \leftrightarrow Q_{11}$
15B	$3^2 15^2/1^2 5^2$	$*$		$Q_{12} \leftrightarrow Q_{12}$
20A	$2^2 5^2 20^2/1^2 4^2 10^2$	$*$		$W_{12} \leftrightarrow W_{12}$
16B	$2 \cdot 16^2/1^2 8$	$*$	$A_{15} \oplus D_9$	$W_{13} \leftrightarrow S_{11}$
13A	$13^2/1^2$	$*$	A_{12}^2	$S_{12} \leftrightarrow S_{12}$
12E	$4^2 12^2/1^2 3^2$	$*$		$U_{12} \leftrightarrow U_{12}$
9C	$2^3 3^2 18^3/1^3 6^2 9^3$		$D_{10} \oplus E_7^2$	$J'_9 \leftrightarrow J_{3,0}$
7B	$2^3 14^3/1^3 7^3$	$*$	D_8^3	$J'_{10} \leftrightarrow Z_{1,0}$
12K	$2^2 3 \cdot 12^3/1^3 4 \cdot 6^2$	$*$	$A_{11} \oplus D_7 \oplus E_6$	$J'_{11} \leftrightarrow Q_{2,0}$ $K'_{10} \leftrightarrow W_{1,0}$ $L_{10} \leftrightarrow W_{1,0}$
10E	$2 \cdot 10^3/1^3 5$	$*$	$A_9^2 \oplus D_6$	$K'_{11} \leftrightarrow S_{1,0}$ $L_{11} \leftrightarrow S_{1,0}$
9A	$9^3/1^3$		A_8^3	$M_{11} \leftrightarrow U_{1,0}$
5B	$2^4 10^4/1^4 5^4$		D_6^4	$J'_{2,0} \leftrightarrow J'_{2,0}$
8C	$2^2 8^4/1^4 4^2$		$A_7^2 \oplus D_5^2$	$L_{1,0} \leftrightarrow L_{1,0}$ $L_{1,0} \leftrightarrow K'_{1,0}$ $K'_{1,0} \leftrightarrow K'_{1,0}$
7A	$7^4/1^4$		A_6^4	$M_{1,0} \leftrightarrow M_{1,0}$
6A	$3^4 6^4/1^4 2^4$			$I_{1,0} \leftrightarrow I_{1,0}$
10A	$5^2 10^2/1^2 2^2$	$*$		
15A	$2^3 3^3 5^3 30^3/1^3 6^3 10^3 15^3$	$*$	E_8^3	
12A	$2^4 3^4 12^4/1^4 4^4 6^4$		E_6^4	

Table 10: Self-dual Frame shapes of G with trace -2, -3 or -4.

one can associate a modular function [Kon]. Let

$$\eta(\tau) = q^{1/24} \prod_{n=1}^{\infty} (1 - q^n), \quad q = e^{2\pi i \tau}, \tau \in \mathbb{H},$$

be the Dedekind η-function. Then define

$$\eta_\pi(\tau) = \prod_{m/N} \eta(m\tau)^{\chi_m}.$$

Saito [S] has proved the identity

$$\eta_\pi \left(-\frac{1}{N\tau} \right) \eta_{\pi^*}(\tau) \sqrt{d} = 1,$$

where $d = \prod m^{\chi_m}$ and π^* is the dual Frame shape. From this it follows that $\eta_{\pi\pi^*}$ is a modular function for the group

$$\Gamma_0(N) = \left\{ \begin{pmatrix} a & b \\ c & d \end{pmatrix} \in \mathrm{SL}_2(\mathbb{Z}) \mid c \equiv 0(N) \right\}.$$

Question 1 Let $\pi\pi^*$ be one of the self-dual Frame shapes of Table 10. Is there any relation of $\eta_{\pi\pi^*}$ to the analogue of Dedekind's eta function for K3 surfaces considered in [JT] ?

The Frame shape $\pi\pi^*$ is the Frame shape of the operator $c \oplus c^*$ which can be considered as an automorphism of a sublattice of finite index of the even unimodular 24-dimensional lattice K_{24}, which is the full homology lattice of a K3 surface. The lattice K_{24} has the same rank as the Leech lattice, but contrary to the Leech lattice it is indefinite and has signature $(4, 20)$.

Question 2 Is there an explanation for this strange correspondence between operators of different lattices?

Is it only a purely combinatorial coincidence? One can try to classify finite sequences $(\chi_1, \chi_2, \ldots, \chi_N)$ with the following properties:

(1) $\chi_m \in \mathbb{Z}$ for all $m = 1, \ldots, N$,

(2) $\sum m\chi_m = 24$,

(3) $\chi_m = 0$ for $m \nmid N$,

(4) $\chi_m = -\chi_{N/m}$ for $m|N$,

(5) $\prod m^{\chi_m} \in \mathbb{N}$,

(6) $\chi_1 \in \{-2, -3, -4\}$,

(7) $|\chi_m| \leq |\chi_N|$ for $m|N$.

By a computer search, one finds for $N \leq 119$ in addition to the 22 Frame shapes of Table 10 only the following Frame shapes:

$$3^2 6 \cdot 12^2/1^2 2 \cdot 4^2, \quad 2^4 3^4 4^4 24^4/1^4 6^4 8^4 12^4, \quad 2^2 4^2 5^2 40^2/1^2 8^2 10^2 20^2.$$

These Frame shapes also appear in Kondo's tables, namely in the tables of certain transforms of the Frame shapes of G [Kon, Table III, 4C; Table II, 12A; Table II, 20A].

References

[AC1] N. A'Campo: Le nombre de Lefschetz d'une monodromie. *Indag. Math.* **35**, 113–118 (1973).

[AC2] N. A'Campo: La fonction zêta d'une monodromie. *Comment. Math. Helvetici* **50**, 233–248 (1975).

[Ar] V.I. Arnold: Critical points of smooth functions and their normal forms. *Usp. Math. Nauk.* **30**:5, 3–65 (1975) (Engl. translation in *Russ. Math. Surv.* **30**:5, 1–75 (1975)).

[AGV] V.I. Arnold, S.M. Gusein-Zade, A.N. Varchenko: *Singularities of Differentiable Maps, Volume II.* Birkhäuser, Boston Basel Berlin 1988.

[Ba] V.V. Batyrev: Dual polyhedra and the mirror symmetry for Calabi-Yau hypersurfaces in toric varieties. *J. Alg. Geom.* **3**, 493–535 (1994).

[Br] E. Brieskorn: Die Monodromie der isolierten Singularitäten von Hyperflächen. *Manuscripta math.* **2**, 103–161 (1970).

[ATL] J.H. Conway, R.T. Curtis, S.P. Norton, R.A. Parker, R.A. Wilson: *ATLAS of Finite Groups.* Oxford Univ. Press, London New York 1985.

[CN] J.H. Conway, S.P. Norton: Monstrous moonshine. *Bull. Lond. Math. Soc.* **11**, 308–339 (1979).

[CPS] J.H. Conway, R.A. Parker, N.J.A. Sloane: The covering radius of the Leech lattice. *Proc. R. Soc. Lond. A* **380**, 261–290 (1982).

[D1] I.V. Dolgachev: Quotient-conical singularities on complex surfaces. *Funkt. Anal. Jego Prilozh.* **8**:2, 75–76 (1974) (Engl. translation in *Funct. Anal. Appl.* **8**, 160–161 (1974)).

[D2] I.V. Dolgachev: Automorphic forms and weighted homogeneous singularities. *Funkt. Anal. Jego Prilozh.* **9**:2, 67–68 (1975) (Engl. translation in *Funct. Anal. Appl.* **9**, 149–151 (1975)).

[D3] I.V. Dolgachev: Integral quadratic forms: applications to algebraic geometry. In: *Sem. Bourbaki, 1982/83, Exposé* n° *611*, Astérisque, Vol. **105/106**, Soc. Math. France 1983, pp. 251–275.

[D4] I.V. Dolgachev: Mirror symmetry for lattice polarized K3 surfaces. Preprint, University of Michigan, Ann Arbor, 1995 (alg-geom/9502005).

[DN] I.V. Dolgachev, V.V. Nikulin: Exceptional singularities of V.I. Arnold and K3 surfaces. In: *Proc. USSR Topological Conference in Minsk*, 1977.

[E1] W. Ebeling: Quadratische Formen und Monodromiegruppen von Singularitäten. *Math. Ann.* **255**, 463–498 (1981).

[E2] W. Ebeling: Milnor lattices and geometric bases of some special singularities. *L'Enseignement Math.* **29**, 263–280 (1983).

[E3] W. Ebeling: The Monodromy Groups of Isolated Singularities of Complete Intersections. *Lect. Notes in Math.*, Vol. **1293**, Springer-Verlag, Berlin etc., 1987.

[E4] W. Ebeling: *Lattices and Codes.* Vieweg, Wiesbaden 1994.

[E5] W. Ebeling: On Coxeter-Dynkin diagrams of hypersurface singularities. *J. Math. Sciences* **82** (5), 3657–3664 (1996).

[E6] W. Ebeling: Strange duality and polar duality. In preparation.

[EW] W. Ebeling, C.T.C. Wall: Kodaira singularities and an extension of Arnold's strange duality. *Compositio Math.* **56**, 3–77 (1985).

[F] J.S. Frame: Computation of the characters of the Higman-Sims group and its automorphism group. *J. Algebra* **20** (2), 320–349 (1972).

[G] A.M. Gabrielov: Dynkin diagrams of unimodal singularities. *Funkt. Anal. Jego Prilozh.* **8**:3, 1–6 (1974) (Engl. translation in *Funct. Anal. Appl.* **8**, 192–196 (1974)).

[H1] H. Hamm: Lokale topologische Eigenschaften komplexer Räume. *Math. Ann.* **191**, 235–252 (1971).

[H2] H. Hamm: Exotische Sphären als Umgebungsränder in speziellen komplexen Räumen. *Math. Ann.* **197**, 44–56 (1972).

[JT] J. Jorgenson, A. Todorov: A conjectured analogue of Dedekind's eta function for K3 surfaces. *Math. Res. Letters* **2**, 359–376 (1995).

[Kob] M. Kobayashi: Duality of weights, mirror symmetry and Arnold's strange duality. Preprint Tokyo Institute of Technology 1995 (alg-geom/9502004).

[Kon] T. Kondo: The automorphism group of Leech lattice and elliptic modular functions. *J. Math. Soc. Japan* **37**, 337–361 (1985).

[L1] E. Looijenga: The smoothing components of a triangle singularity. I. *Proc. Symp. Pure Math.* Vol. **40**, Part 2, 173–183 (1983).

[L2] E. Looijenga: The smoothing components of a triangle singularity. II. *Math. Ann.* **269**, 357–387 (1984).

[M] S. Mukai: Finite groups of automorphisms of K3 surfaces and the Mathieu group. *Invent. math.* **94**, 183–221 (1988).

[Na1] I. Nakamura: Inoue-Hirzebruch surfaces and a duality of hyperbolic unimodular singularities I. *Math. Ann.* **252**, 221–235 (1980).

[Na2] I. Nakamura: Duality of cusp singularities. In: *Complex Analysis of Singularities* (R.I.M.S. Symposium, K. Aomoto, organizer), Kyoto 1981, pp. 1–18.

[Nie] H.-V. Niemeier: Definite quadratische Formen der Dimension 24 und Diskriminante 1. *J. Number Theory* **5**, 142–178 (1973).

[Nik] V.V. Nikulin: Integral symmetric bilinear forms and some of their applications. *Izv. Akad. Nauk. SSSR Ser. Mat.* **43**, 111–177 (1979) (Engl. translation in *Math. USSR Izv.* **14**, No. 1, 103–167 (1980)).

[OW] P. Orlik, P. Wagreich: Isolated singularities of algebraic surfaces with $\mathbb{C}^*$-action. *Ann. of Math.* **93**, 205–228 (1971).

[P1] H. Pinkham: Singularités exceptionelles, la dualité étrange d'Arnold et les surfaces K-3. *C. R. Acad. Sc. Paris, Serie A*, **284**, 615–618 (1977).

[P2] H. Pinkham: Groupe de monodromie des singularités unimodulaires exceptionelles. *C. R. Acad. Sc. Paris, Serie A*, **284**, 1515–1518 (1977).

[P3] H. Pinkham: Deformations of normal surface singularities with $\mathbb{C}^*$ action. *Math. Ann.* **232**, 65–84 (1978).

[S] K. Saito: On a duality of characteristic polynomials for regular systems of weights. Preprint RIMS, Kyoto University 1994.

[Wa] C.T.C. Wall: Classification of unimodal isolated singularities of complete intersections. *Proc. Symp. Pure Math. Vol.* **40**, *Part 2*, 625–640 (1983).

[Wi] R.A. Wilson: The maximal subgroups of Conway's group Co_1. *J. Algebra* **85**, 144–165 (1983).

Institut für Mathematik,
Universität Hannover,
Postfach 6009,
D-30060 Hannover,
Germany.

e-mail: ebeling@math.uni-hannover.de

Geometry of Equisingular Families of Curves

Gert-Martin Greuel Eugenii Shustin *

Dedicated to C.T.C. Wall on the occasion of his 60th birthday.

Contents

---

*The authors were supported by Grant No. G 039-304.01/95 from the German Israeli Foundation for Scientific Research and Development, by the DFG–Schwerpunkt "Algorithmische Zahlentheorie und Algebra", and by the Minerva center for Geometry at the Tel Aviv University.

Introduction

Singular algebraic curves, their existence, deformation, families (from the local and global point of view) have attracted the continuous attention of algebraic geometers since the last century. The aim of our paper is to give an account of results, new trends and bibliography related to the geometry of equisingular families of algebraic curves on smooth algebraic surfaces over an algebraically closed field of characteristic zero. This theory is founded in basic works of Plücker, Severi, Segre, Zariski, and has close links with, and finds important applications in singularity theory, topology of complex algebraic curves and surfaces, and in real algebraic geometry.

We shall concentrate on the following traditional questions. Given some classification of isolated curve singularities we call an equivalence class a *type*. Then the discriminant in the linear system $|D|$ on a smooth algebraic surface Σ is decomposed into the set of non-reduced or reducible curves and the sets $V(D; S_1, \ldots, S_r)$ of (irreducible) curves in $|D|$ having r singular points of types $S_1, \ldots, S_r$ as their only singular points (in the sequel these sets are referred to as *equisingular families* – ESF). We ask:

- Is $V(D; S_1, \ldots, S_r)$ non-empty ?
- Is $V(D; S_1, \ldots, S_r)$ smooth and does it have the "regular", or "expected", codimension in $|D|$ (expressed in terms of $S_1, \ldots, S_r$) ?
- What is the adjacency relation of ESF in the given $|D|$ (i.e. what are the deformations of a curve $C \in |D|$) ? Is this adjacency regular (i.e. do deformations of C depend only on the local structure of its singular points) ?
- Is $V(D; S_1, \ldots, S_r)$ irreducible ?

Among the results related to the problems stated, we discuss basically *sufficient conditions* of a numerical kind (i.e. expressed via numerical invariants of surfaces, curves, singularities), which ensure affirmative answers to the questions listed above for the given ESF. The case of *plane curves* is central in the theory and to this paper. In fact, results on curves on other surfaces appear as extensions and generalizations of the corresponding ones for plane curves.

As one of the most important set of applications we mention various constructions based on deformation theory. This especially concerns Hilbert's 16th problem, the classification of real algebraic curves. First of all, the real version of Severi's theorem on the independence of deformations of nodes of plane curves, given by Brusotti [12], was the basis of classical constructions of real non-singular algebraic curves by Harnack, Hilbert, Brusotti, Wiman [37]. Using generalizations of this method on certain surfaces, Gudkov [37] classified real algebraic curves of small degree on quadrics, and recently Pecker [54] proved that any g between zero and the Eisenbud-Harris upper bound

$\pi_d(m, \alpha)$ is the genus of a non-singular curve in $\mathbb{P}^m$ of degree d, not lying on a hypersurface of small degree. The Viro method [84], [85], [86], being invented for the construction of real non-singular algebraic curves, brought new ideas to deformation theory. Combining it with the independence of singular point deformations and criteria for the smoothness of ESF, one can

- construct curves (real or complex) with a prescribed collection of singularities [67], [29], [72] (see Section 1 below for more details) and prescribed topology [57] [62];
- classify deformations of certain singular points [63], [65], [64];
- construct interesting topological objects (see, for instance, in [21] counterexamples to one of Kirby's conjecture).

The material is organized in four sections: 1. Existence of curves with given singularities (or, equivalently, non-emptiness of ESF), 2. Smoothness of ESF and independence of singular point deformations, 3. Irreducibility of ESF, 4. Open questions.

The authors would like to thank C. Lossen for his help during the preparation of this article.

1 Existence of plane curves with given singularities

1.1 Statement of the problem

The question, what are the possible singularities of a complex plane algebraic curve of a given degree, is classical and can be traced back to Euler, Plücker, Segre, Severi. We consider two aspects of this problem: analytic and topological.

Let (C, z), (C', z') be germs of complex plane algebroid curves with isolated singular points z, z'. They are called analytically (topologically) equivalent if there is a local analytic diffeomorphism (homeomorphism) of neighborhoods of z, z' in the plane taking (C, z) into (C', z'). Analytic equivalence can be expressed as an isomorphism of the complete local rings

$$\hat{O}_{C,z} \cong \hat{O}_{C',z'} \ .$$

Topological equivalence is completely characterized by the following discrete invariants (see [92], [88], [77], [9]): the embedded resolution tree of (C, z), the multiplicities of the strict transforms of (C, z) at infinitely near points (including z) and the intersection numbers of the strict transforms of C with the corresponding exceptional divisors.

Analytic or topological equivalence classes of isolated singular points are called analytic or topological singularities. So, the general problem is: given an integer $d > 0$ and singularities (analytic or topological) $S_1, \ldots, S_r$, does there exist an irreducible curve of degree d having r singular points of types $S_1, \ldots, S_r$, respectively, as its only singularities ?

An important particular case is the same problem for one singularity. Namely, let S be an analytic or topological singularity. What is the minimal degree $d(S)$ of a curve having a singular point of type S ? In other words, we ask about a normal form of minimal degree of the given singularity.

1.2 Restrictions

Severi's theorem [61] that the inequality

$$n \leq \frac{(d-1)(d-2)}{2} \tag{1.1}$$

is necessary and sufficient for the existence of an irreducible curve of degree d with exactly n nodes, is the only complete answer of this type of problem. For other singularities, starting with ordinary cusps, the problem is still open except the case of degrees ≤ 6, see [36], [89], [90], [16], [17], [79], [80], [81].

There are various known restrictions of the type

$$\sum_{i=1}^{r} \alpha(S_i) \leq ad^2 + o(d^2)$$

as necessary conditions for the existence of a curve of degree d with r singular points of types (topological or analytic) $S_1, \ldots, S_r$, where α is some positive invariant of the singularities, $a = \text{const} > 0$ (see a number of such inequalities both for arbitrary singularities, or for certain classes of them in [10], [42], [44], [56], [82]). We mention here the classical ones:

$$\sum_{i=1}^{r} \delta(S_i) \leq \frac{(d-1)(d-2)}{2} ,$$

$$\sum_{i=1}^{r} \mu(S_i) \leq (d-1)^2, \tag{1.2}$$

where δ is the δ-invariant and μ is the Milnor number.

In the case of one singularity there is a lower bound on the degree $d(S)$ of a normal form of the singularity S, coming from (1.2):

$$d(S) \geq \sqrt{\mu(S)} + 1 , \tag{1.3}$$

and an upper bound coming from Tougeron's theorem on the sufficiency of the $(\mu(S)+1)$-jet [78]:

$$d(S) \leq \mu(S) + 1 .$$

Note the asymptotically large gap between the bounds.

1.3 Some previously known constructions

The other side of the problem is the construction of curves with prescribed singularities. The following two approaches are classical.

The first is to somehow construct a curve of the given degree, which is degenerate with respect to the requested curve, and then deform it in order to get the prescribed singularities. For example, Severi's condition (1.1) is obtained as follows: take the union of d generic straight lines, show that all $d(d-1)/2$ nodes of such a curve can be smoothed or preserved independently, then smooth $d(d-1)/2 - n$ nodes and get the requested irreducible curve. Nodal curves on the blown up projective plane can be constructed using the same idea [2]. Another example, a construction of curves with nodes, cusps and ordinary triple points, was considered in [23]. Note that the number of cusps and triple points in the latter case was bounded from above by $2d/3$, whereas all known restrictions, in particular (1.2), bound the number of arbitrary singularities by a quadratic function of d. Natural difficulties, appearing this way, do not allow, actually, to extend the range of application. Namely, we have to understand the deformations of singularities of the initial degenerate curve, and that the deformations required can be realized simultaneously in the space of curves of a given degree.

The second way consists of a construction especially adapted to the given degree and given collection of singularities. It may be based on a sequence of rational transformations of the plane applied to a more or less simple initial curve in order to get a curve of a given degree with given singularities. Or it may consist of the invention of a polynomial defining the required curve. This can be illustrated by the construction of singular curves of degrees ≤ 6, cited above. The construction of cuspidal curves in [40], [44], [93] is of the same type. Namely, Ivinskis [44] constructs curves of degree $d = 6q$ with $9q^2 = d^2/4$ cusps from a curve $F_6(x_0, x_1, x_2) = 0$ of degree 6 with 9 cusps, by a q^2-sheeted branched covering of the plane:

$$F_6(x_0^q, x_1^q, x_2^q) = 0.$$

Hirano [40] improves Ivinskis' procedure and obtains, in particular, curves of degree $d = 2 \cdot 3^k$ with $9(9^k - 1)/8$ cusps. Finally Zariski-type constructions (see [93], chapter 8, and [76])

$$F_{2q}^3 + G_{3q}^2 = 0$$

where $F_{2q} = 0$, $G_{3q} = 0$ are generic curves of degrees $2q$, $3q$, respectively, gives a curve of degree $d = 6q$ with $6q^2 = d^2/6$ cusps. This approach does not, however, lead to any general answer for the following reasons. First, for any collection of singularities we have to invent a new construction, and second, obtaining a curve with a certain number of singularities, we cannot guarantee that there are curves with smaller numbers of these singularities; for instance, in the above constructions of cuspidal curves, the cusps are dependent, which does not allow us to smooth them independently.

Another idea, based on a modification of Viro's method of gluing polynomials (see the original method in [84], [85], [86]) and on the independence of singular point deformations, was suggested in [67], [43]. It was shown, for instance, that if

$$n + 2k \leq \frac{d^2}{2} + O(d)\ ,$$

then there exists an irreducible curve of degree d with exactly n nodes and k cusps. Below we describe some details of the procedure but here we only say that it requires us first to have initial curves of small degrees having given singularities, and second, to check that the ESF of curves of these small degrees are smooth and have "regular" dimensions. Such requirements are quite restrictive, but once we have constructed a singular curve, we can remove any singularity while retaining the others, i.e. any smaller collection of singularities is realizable as well. One further advantage is that the procedure works over the reals.

Finally, we emphasize a common defect of all these constructions: they are not applicable to the problem of finding a normal form of a single singularity.

1.4 Asymptotically optimal sufficient existence conditions

Here we describe a new approach to the problem of the existence of curves of a given degree with given topological singularities, that gives an asymptotically optimal sufficient existence condition for arbitrary singularities [28]. The case of analytic singularities remains open; we discuss it in the last section.

Theorem 1 *If an integer $d > 0$ and topological singularities $S_1, \ldots, S_r$ satisfy*

$$\sum_{i=1}^{r} \mu(S_i) \leq \frac{d^2}{396}\ , \tag{1.4}$$

then there exists an irreducible curve of degree d with r singular points of types $S_1, \ldots, S_r$ as its only singularities.

Theorem 2 *For any topological singularity S there exists an irreducible curve of degree*

$$d \leq 14\sqrt{\mu(S)} \tag{1.5}$$

having a singular point of type S as its only singularity.

These conditions are optimal with respect to the exponents of d and $\mu(S)$, as compared with the necessary conditions (1.2), (1.3).

The proof is done in three main steps:

Step 1. Existence of a curve of degree d with a singular point of the given topological type S is reduced to a relation

$$h^1(\mathbb{P}^2, \mathcal{J}_X(d)) = 0$$

for the ideal sheaf $\mathcal{J}_X \subset \mathcal{O}_{\mathbb{P}^2}$ of a certain zero-dimensional subscheme $X \subset \mathbb{P}^2$, associated with S, such that

$$\deg X \leq a \cdot \mu(S), \quad a = \text{const} > 0. \tag{1.6}$$

Step 2. Given a scheme $X \in \mathcal{GS}$ (cf. Section 1.5), it may happen that

$$h^1(\mathbb{P}^2, \mathcal{J}_X(d)) > 0$$

for $d \sim \deg X$. Due to the semi-continuity of h^1 we reduce this parameter by choosing a generic element X_0 in the corresponding Hilbert scheme. In addition, we show that

(1.7)
$$h^1(\mathbb{P}^2, \mathcal{J}_{X_0}(d)) = 0 \quad \text{as} \quad d \geq b\sqrt{\deg X_0}, \quad b = \text{const} > 0\ .$$

Step 3. The final stage is a construction of curves with many singular points. This is done by means of a version of the Viro method.

Below we describe each step in more detail.

1.5 Zero-dimensional schemes associated with topological singularities

Let z be a point of a reduced curve $C \subset \mathbb{P}^2$. It is called essential if C is singular at z. If $q \neq z$ is infinitely near to z, we denote by $C_{(q)}$ (respectively, $\widehat{C}_{(q)}$) the strict (respectively, total) transform under the composition of blowing-ups, $\pi_{(q)} : S_{(q)} \to S$ defining q. We call q *essential* if it is not a node (ordinary double point) of the union of $C_{(q)}$ with reduced exceptional divisor.

For $z \in C$ let $T(C, z)$ denote the (infinite) complete embedded resolution tree of (C, z) with vertices the points infinitely near to z. It is naturally oriented, inducing a partial ordering on its vertices such that $z < q$ for all $q \in T(C, z)\backslash\{z\}$. If $z \notin C$ we define $T(C, z)$ to be the empty tree. Moreover, let

$$T^*(C, z) := \{q \in T(C, z) \mid q \text{ is essential}\}$$

denote the tree of essential points of (C, z) which is a finite subtree of $T(C, z)$.

Let $T^* \subset T(C, z)$ be a finite, connected tree, containing $T^*(C, z)$. For any point $q \in T^*$ and any $f \in \mathcal{O}_{S,z}$ let $f_{(q)}$, respectively $\hat{f}_{(q)}$, be the strict, respectively total, transform under the modification $\pi_{(q)}$ defining q. Set $m_q := \mathrm{mt}(C_{(q)}, q)$, $\widehat{m}_q := \mathrm{mt}(\widehat{C}_{(q)}, q)$, where mt denotes multiplicity, and define the ideal

$$J := J(C, T^*) := \{f \in \mathcal{O}_{\mathbb{P}^2,z} \mid \mathrm{mt}(\hat{f}_{(q)}, q) \geq \widehat{m}_q, \ \ q \in T^*\} \subset \mathcal{O}_{\mathbb{P}^2,z}$$

and the subscheme of S defined by J,

$$X := X(C, T^*) = Z(J), \quad \mathcal{O}_{X,z} := \mathcal{O}_{\mathbb{P}^2,z}/J,$$

which is concentrated on $\{z\}$. The scheme X is called a *generalized singularity scheme* and the class of zero-dimensional subschemes of $\mathbb{P}^2$, constructed in this way, is denoted by $\mathcal{GS}$. The subclass of schemes $X \in \mathcal{GS}$ with $T^* = T^*(C, z)$ is denoted by $\mathcal{S}$, $X \in \mathcal{S}$ is called a *singularity scheme.*

A scheme $X \in \mathcal{GS}$ defines uniquely a topological singularity, and almost all germs $f \in J$, where $J \subset \mathcal{O}_{\mathbb{P}^2,z}$ is the ideal of X, belong to this topological type.

The following statement reduces the existence problem for one singularity to an h^1-vanishing for ideal sheaves of zero-dimensional schemes in the plane. Let (C, z) be a reduced germ of an algebroid curve, L be a generic straight line (a smooth germ) through z, and let the scheme $X \in \mathcal{GS}$ be defined by the germ (CL, z) and the tree, which contains $T^*(C, z)$ and the first non-essential point of any local branch of (C, z) (by abuse of notation, C denotes the curve as well as an equation for it).

Lemma 1 *If*

$$h^1(\mathbb{P}^2, \mathcal{J}_X(d-1)) = 0, \tag{1.8}$$

where $\mathcal{J}_X \subset \mathcal{O}_{\mathbb{P}^2}$ is the ideal sheaf of X, then there exists an irreducible curve of degree d having only one singular point, and this point is topologically equivalent to (C, z).

We illustrate the proof in the case of an ordinary singularity of multiplicity m (represented by a transversal intersection point of m smooth branches). We take $X' \in \mathcal{GS}$, $X' \subset X$, concentrated at the point z and defined by the ideal $\mathfrak{m}_z^{m+1} \subset \mathcal{O}_{\mathbb{P}^2,z}$. The relation

$$h^1(\mathbb{P}^2, \mathcal{J}_{X'}(d-1)) = 0$$

means that the linear conditions defining X', i.e. the vanishing of the m-jet of the Taylor series at z, are independent in the space of curves of degree $d-1$. Hence there exists a curve C of degree $d-1$ whose m-jet at z is a non-degenerate m-form, that determines the given singularity at z. Since C may not be irreducible with many singularities, we consider CL and use similar reasoning to show for each singular point $z_i \neq z$ of CL the existence of a curve C_i of degree d having the given singularity at z but not z_i. Finally, using Bertini's theorem, a generic element of the linear system $\lambda CL + \sum \lambda_i C_i$ gives the required irreducible curve with z as its only singular point.

1.6 h^1-vanishing criteria for zero-dimensional schemes in the plane

One can easily show that X and d in Lemma 1 satisfy (1.6). Hence, for asymptotically optimal results (1.4), (1.5), one has to prove (1.8) under condition (1.7).

Given a scheme $X \in \mathcal{GS}$, it may happen that

$$h^1(\mathbb{P}^2, \mathcal{J}_X(d)) > 0$$

for $d \sim \deg X$. Due to the semi-continuity of h^1 we reduce this parameter by choosing a generic element in the corresponding Hilbert scheme $\mathcal{H}(X)$. Namely we show that $\mathcal{H}(X)$ is irreducible, and any two elements $X_0, X_1 \in \mathcal{H}(X)$ can be connected by a family $X_t \in \mathcal{H}(X)$, $t \in [0,1]$, with underlying families of topologically equivalent germs (C_t, z_t) and trees $T_t \subset T(C_t, z_t)$, $t \in [0,1]$, defining these schemes. Then we prove

Lemma 2 *If $X \in \mathcal{GS}$ is defined by a germ (C,z) and satisfies*

$$\deg X < \frac{(d - \mathrm{mt}X - \mathrm{mt}_s X)^2}{10}, \tag{1.9}$$

where $\mathrm{mt}X$ is the multiplicity of C at z and $\mathrm{mt}_s X$ is the total multiplicity of the singular branches of C at z, then there exists $X_0 \in \mathcal{H}(X)$ such that

$$h^1(\mathbb{P}^2, \mathcal{J}_{X_0}(d)) = 0.$$

Since $\mathrm{mt}_s X \le \mathrm{mt}X \le \sqrt{\mu(X)} + 1$, the estimate (1.9) can be reduced to (1.7).

To find such a generic scheme X_0, we follow Hirschowitz's ideas [41], which he applied to the same problem for schemes of generic fat points in the plane. First, a good "approximation" of a generic scheme can be obtained by requiring general position with respect to straight lines, that means, the intersection with any straight line should be not too big. The second idea consists of using an inductive procedure based on

Lemma 3 *Let L be a straight line, $X \subset \mathbb{P}^2$ be a zero-dimensional scheme. Then*

$$h^1(\mathbb{P}^2, \mathcal{J}_{X:L}(d-1)) = 0, \quad \deg(X \cap L) \le d+1,$$

(where $X : L$ is the residue of X with respect to L, and $X \cap L$ is the scheme-theoretic intersection) implies

$$h^1(\mathbb{P}^2, \mathcal{J}_X(d)) = 0 \, .$$

In our case we perform the same procedure, using the main observation that the class $\mathcal{GS}$ is closed with respect to the residue operation (which does not hold for the more natural subclass $\mathcal{S}$):

Lemma 4 *In the previous notation, if $X \in \mathcal{GS}$ then $X : L \in \mathcal{GS}$.*

1.7 Gluing singular points

Given topological singularities $S_1, \ldots, S_r$ and an integer $d > 0$, we want to construct an irreducible curve of degree d with exactly r singular points of types $S_1, \ldots, S_r$ respectively.

First, with any S_i, $1 \le i \le r$, we associate the zero-dimensional scheme X_i as in Lemma 1, an integer d_i such that

$$h^1(\mathbb{P}^2, \mathcal{J}_{X_i}(d_i - 1)) = 0, \quad i = 1, \ldots, r,$$

and an irreducible curve C_i of degree d_i, having one singular point of type S_i. Then we take a zero-dimensional scheme $Y \subset \mathbb{P}^2$ of r generic fat points of multiplicities $d_1 + 1, \ldots, d_r + 1$, satisfying

$$h^1(\mathbb{P}^2, \mathcal{J}_Y(d)) = 0 \; . \tag{1.10}$$

By Hirschowitz's theorem [41], the latter relation holds if

$$\sum_{i=1}^{r} \frac{(d_i + 1)(d_i + 2)}{2} < \left[\frac{(d + 3)^2}{4} \right] . \tag{1.11}$$

Finally, we take a curve $C \in |\mathcal{J}_Y(d)|$ and simultaneously deform its singular points in order to produce a curve with singularities $S_1, \ldots, S_r$. Here we use a Viro-type one-parameter deformation, which essentially replaces the zero d_i-jet of the curve C at a point $z_i \in Y$ by an equation of a curve $\widetilde{C}_i$ of degree d_i, close to C_i belonging to the same ESF.

2 Smoothness of equisingular and equianalytic families

In this section we discuss results on the smoothness of the variety of projective curves on a smooth surface Σ with fixed singularity type, considered as a subscheme of the Hilbert scheme of Σ. We actually consider a stronger condition, the so-called T-property, which states that the conditions imposed by the singularities are independent (or transversal), which is more natural than just smoothness.

Starting with methods and results for arbitrary surfaces, we concentrate on the most important case $\Sigma = \mathbb{P}^2$, and then finish with a few remarks about rational and $K3$-surfaces and surfaces of general type.

2.1 *T*-property

Let Σ be a smooth projective surface (embedded in some projective space) and let Hilb_h^Σ be the Hilbert scheme of Σ, parametrizing closed subschemes of Σ with given Hilbert polynomial h. Let $V = V_h(S_1, \ldots, S_r)$ denote the locally closed subscheme of reduced curves on Σ, having exactly r singularities of (analytic or topological) types $S_1, \dots, S_r$ (cf. [87] for nodes and cusps and [26], [27] in general for the existence of V). Intuitively, each singularity imposes open and closed conditions and all conditions for all singular points define a certain locally closed subvariety of Hilb_h^Σ. We say that V has the T-property if all closed conditions for all singular points are independent, that is, the corresponding subvarieties of Hilb_h^Σ are smooth hypersurfaces and intersect transversally. Hence, V is smooth of the "expected codimension", which is equal to the number of closed conditions.

Already the classical Italian geometers [61], [59], [60] noticed that it is possible to express the T-property infinitesimally for families of plane curves with nodes and cusps. The classical term for the T-property of such families is "completeness of the linear characteristic series of families with nodes and cusps" (cf. [93], appendix to Chapter V). Best known is certainly Severi's result [61] saying that nodal curves in $\mathbb{P}^2$ satisfy the T-property.

To give a precise definition, recall that $H^0(C, \mathcal{N}_{C/\Sigma})$ is the Zariski tangent space to Hilb_h^Σ at C, as well as the space of first order embedded deformations of C in Σ, where $\mathcal{N}_{C/\Sigma} = \mathcal{O}_\Sigma(C) \otimes \mathcal{O}_C$ is the normal sheaf of C in Σ. It appears in the exact sequence

$$0 \to \Theta_C \to \Theta_\Sigma \otimes_{\mathcal{O}_\Sigma} \mathcal{O}_C \to \mathcal{N}_{C/\Sigma} \to \mathcal{T}_C^1 \to 0, \tag{2.1}$$

where $\mathcal{T}_C^1$ is a skyscraper sheaf concentrated on the singularities of C with $\mathcal{T}_{C,z}^1 \cong \mathcal{O}_{\Sigma,z}/I_z^{ea}$ and

$$I_z^{ea} = (f, f_x, f_y)\mathcal{O}_{\Sigma,z}$$

for $f(x,y) = 0$ a local equation of C at z. Moreover, we have Wahl's equisingularity ideal [88],

$$I_z^{es} = \{g \in \mathcal{O}_{\Sigma,z} \mid f + \varepsilon g \text{ is equisingular over } \mathrm{Spec}(\mathbb{C}[\varepsilon]/\varepsilon^2)\}$$

and inclusions

$$I_z^{ea} \subset I_z^{es}.$$

Let J^{ea}, J^{es} be the corresponding ideal sheaves defining 0-dimensional schemes $X^{ea} \supset X^{es}$ on Σ . Note that X^{ea} and X^{es} are actually subschemes of C. For simplicity we use the superscript $'$ and mean ea (respectively es) if we consider families of curves with fixed analytic (respectively topological) singularity types $S_1, \dots, S_r$. To be consistent with notation we sometimes write $V' = V_h'(S_1, \dots, S_r)$ for $V = V_h(S_1, \dots, S_r)$.

We set

$$J_{X'/C}(C) := J'\mathcal{O}_C(C)|_C = \mathrm{Ker}\,(\mathcal{O}_\Sigma(C) \otimes \mathcal{O}_C \to \mathcal{O}_{X'})\,|_C,$$

and then, from (2.1) we deduce the exact sequence (noting that $\mathcal{O}_{X^{ea}}$ surjects onto $\mathcal{O}_{X^{es}}$)

$$0 \to \mathcal{J}_{X'/C}(C) \to \mathcal{O}_\Sigma(C) \otimes \mathcal{O}_C \to \mathcal{O}_{X'} \to 0. \tag{2.2}$$

It follows from general deformation theory (cf. [26], [27]) that

- $H^0\big(C, J_{X'/C}(C)\big)$ = Zariski tangent space to V' at C;
- $H^1\big(C, J_{X'/C}(C)\big) = 0 \Leftrightarrow V'$ is smooth at C of the expected dimension $C^2 + 1 - p_a(C) - \deg(X')$.

Thus, we may *define* the T-property of V' by the vanishing of $H^1(C, J_{X'/C}(C))$ for all C in V'. This holds, of course, for an empty V', hence the T-property has nothing to do with existence.

We remark that the expected codimensions are

$$\deg(X^{ea}) = \tau(C), \quad \text{respectively} \quad \deg(X^{es}) = \tau(C) - \operatorname{mod}(C),$$

where

$$\tau(C) = \sum_{x\in C} \dim_{\mathbb{C}} T^1_{C,x}$$

is the total Tjurina number and

$$\operatorname{mod}(C) = \sum_{x\in C} \operatorname{mod}(C, x)$$

the total modality in the sense of Arnold, that is, with respect to right equivalence ([3]).

Note that fixing the Hilbert polynomial of C is equivalent to fixing the arithmetic genus $p_a(C)$ and, either the degree of the embedding or the self-intersection C^2 or the intersection with the canonical divisor, $C \cdot K_\Sigma$. Classically, instead of Hilb_h^Σ, the complete linear system $|C|$ on Σ is considered. By the remark above, the incidence variety is flat over $|C|$ and hence, there exists a unique (injective) morphism $|C| \to \mathrm{Hilb}^\Sigma_{h(C)}$, which, on the tangent level, corresponds to

$$H^0\big(\Sigma, \mathcal{O}_\Sigma(C)\big)/H^0(\Sigma, \mathcal{O}_\Sigma) \hookrightarrow H^0\big(\Sigma, \mathcal{O}_\Sigma(C) \otimes \mathcal{O}_C\big).$$

Hence, we may consider $|C|$ as a subscheme of Hilb. If Σ is a regular surface $\big(h^1(\mathcal{O}_\Sigma) = 0\big)$, then $|C|$ coincides with Hilb.

2.2 Curves on general surfaces

Basically, only one sufficient condition for smoothness for general surfaces Σ is known. The following result is proven in [76] for curves with nodes and cusps and in [26], respectively [27], for arbitrary analytic, respectively topological, singularity types and gives the following condition for the T-property.

If C is irreducible then $H^1\big(C, J_{X'/C}(C)\big) = 0$ if

$$-K_\Sigma C > \deg X' - \varepsilon', \tag{2.3}$$

where $\varepsilon' \geq 0$ is the "isomorphism defect" (cf. [26], [27]).

If C is reducible and $C = C_1 \cup \cdots \cup C_s$, the decomposition into irreducible components, then $H^1\big(C, J_{X'/C}(C)\big) = 0$ if for $i = 1, \ldots, s$

$$-K_\Sigma C_i > \deg(X' \otimes \mathcal{O}_{C_i}) - \varepsilon_i' \tag{2.4}$$

where $\varepsilon_i' \geq 0$ is a local invariant with $\varepsilon_i^{ea} \geq 1$ and $\deg(X^{ea} \otimes \mathcal{O}_{C_i}) = \Sigma_{j \neq i} C_j C_i + \tau(C_i)$.

Since nodes do not contribute to the right-hand side of (2.3) and (2.4) we deduce that the variety of nodal curves on Σ with a fixed number of nodes has the T-property at C if $-K_\Sigma C_i > 0$ for all irreducible components C_i of C.

(2.3) and (2.4) may be considered as the "standard" inequalities, except for the isomorphism defects ε' and ε_i'.

These local isomorphism defects come from a refinement of the Riemann-Roch formula for rank 1 sheaves which are not locally free (cf. [26]) and can be quite big for $J_{X^{es}/C}$. For example,

$$\varepsilon^{es} = \frac{m(m-1)}{2} - 2$$

for an ordinary m-tuple point, $m \geq 3$ (cf. [27]).

As a result we deduce that, if $S_1, \ldots, S_r$ are topological types of ordinary singularities, $V_h^{es}(S_1, \ldots, S_r)$ has the T-property at $C = C_1 \cup \cdots \cup C_s$ if

$$-K_\Sigma C_i > \sum_{\substack{x \in C_i \cap \operatorname{Sing}(C) \\ \operatorname{mt}_x(C) > 2}} \operatorname{mt}_x(C_i), \qquad i = 1, \ldots, s, \tag{2.5}$$

where mt_x denotes the multiplicity at x. For $\Sigma = \mathbb{P}^2$, the left-hand side is equal to $3 \cdot \deg(C_i)$ and we obtain a result of Giacinti-Diebolt [22].

The vanishing of $H^1\big(C, J_{X'/C}(C)\big)$ implies $H^1\big(C, \mathcal{O}_\Sigma(C) \otimes \mathcal{O}_C\big) = 0$ and hence, the smoothness of $\operatorname{Hilb}_h^\Sigma$ at C. But it has another important consequence. If $X' = X^{ea}$ then $H^0\big(C, \mathcal{O}_\Sigma(C) \otimes \mathcal{O}_C\big) \to H^0(C, T_C^1)$ is surjective and any infinitesimal deformation of the singularities $S_1, \ldots S_r$ can be lifted to an infinitesimal deformation of the global curve C as a subscheme of Σ. Moreover, since $H^1\big(C, \mathcal{O}_\Sigma(C) \otimes \mathcal{O}_C\big) = 0$ these deformations are unobstructed and hence, any deformation of $S_1, \ldots, S_r$ is induced by a subfamily of curves in Σ close to C. We call this the *independence of deformations of* $S_1, \ldots, S_r$ (or *regular adjacency*) in $\operatorname{Hilb}_h^\Sigma$.

More generally, let $\{S_1, \ldots, S_t\}$ be analytic and $\{S_{t+1}, \ldots, S_r\}$ topological singularity types and let X' denote the corresponding mixed analytic-topological scheme. Then the vanishing of $H^1\big(C; J_{X'/C}(C)\big)$ implies the independence of deformations of $S_1, \ldots, S_t$ without changing the topological type of $S_{t+1}, \ldots, S_r$.

2.3 Curves in $\mathbb{P}^2$

Severi's result [61], that the family of plane curves of degree d having a fixed number of nodes is a T-variety, does not generalize to more complicated singularities. Wahl [87] showed the existence of an ESF of curves of degree 104 with 900 ordinary cusps and 3636 nodes without the T-property; the reduction of his family is, however, smooth (of dimension bigger than the expected one).

Luengo constructed in [52] an irreducible plane curve of degree 9 having one singularity of type A_{35} such that the corresponding ESF has the expected dimension but is not smooth, nor even its reduction. Further examples were found in [71].

The classical result [93] is that the family $V_d(n,k)$ of curves of degree d with n nodes and k cusps is a T-variety at an irreducible curve C if

$$k < 3d. \tag{2.6}$$

For $\mathbb{P}^2$ and arbitrary singularity types $S_1, \ldots, S_r$ (analytic or topological), the general approach from Section 2.2 gives that $V_d(S_1, \ldots, S_r)$ has the T-property at an irreducible curve C, if

$$\deg(X') - \varepsilon' < 3d, \tag{2.7}$$

(cf. [26], [27]) which is of the same nature. Indeed we cannot improve this inequality substantially if we consider X' as a subscheme of the given curve C and use vanishing theorems on C.

Progress has been made by considering X' not as a subscheme of C but as a subscheme of $\mathbb{P}^2$ and looking for an irreducible curve C' of degree $d' < d$ containing X' as a subscheme and then use vanishing theorems on C'. This idea appears for the first time in [63] with C' a generic polar of C, which is irreducible if C is not the union of lines through one point. This led to the sufficient condition (cf. [63], [27])

$$\deg X' - \tilde{\varepsilon}' < 4d - 4 \tag{2.8}$$

where $\tilde{\varepsilon}' \geq 0$, which can be computed or estimated by local methods (cf. [27]). Note that C need not to be irreducible.

The use of C' instead of C is possible, since for $\mathbb{P}^2$ we have

$$h^1\big(\mathbb{P}^2, \mathcal{O}_{\mathbb{P}^2}(k)\big) = h^2\big(\mathbb{P}^2, \mathcal{O}_{\mathbb{P}^2}(k)\big) = 0 \quad (k \geq 0),$$

hence

$$H^1\big(C, J_{X'/C}(C)\big) = 0 \Leftrightarrow H^1\big(\mathbb{P}^2, J_{X'}(C)\big) = 0 \Leftrightarrow H^1\big(C', J_{X'/C'}(C)\big) = 0$$

for C' a curve of degree $d' \leq d = \deg C$.

The above inequalities (2.6), (2.7), (2.8) have a right–hand side which is linear in d while the left–hand side may be quadratic in d (for example, for d lines through one point). Other generalizations of the same kind were obtained in [66], [83].

In [68] and [70] the T-property (and irreducibility) was shown when the total Milnor number was bounded by a quadratic function in d. These bounds have been improved in [71], using ideas from [28], stating that $V_d(S_1,\ldots,S_r)$ has the T-property at C and is of codimension $\sum_{i=1}^{r}\tau'(S_i)$, if

$$\sum_{i=1}^{r}(\mu(S_i)+2)^4 \le (d+6)^2, \tag{2.9}$$

where μ is the Milnor number, $\tau'=\tau$ (the Tjurina number) for an analytic type and $\tau'=\tau$-mod for a topological type. For curves with n nodes and k cusps $V_d(n,k)$ is a T-variety at an irreducible C if

$$\frac{4}{5}n+2k \le \frac{(d+6)^2}{18}. \tag{2.10}$$

Moreover, in [71] it was shown that the variety of curves of degree $2pq$ having q^2 singularities of type A_{6p-1} has a non-smooth irreducible component if $p \ge 5$, $q \ge 3$ (cf. also Section 4).

2.4 Curves on special surfaces

Besides $\mathbb{P}^2$, few special surfaces have been considered where the results are better than for the general case discussed in 2.2.

So far, only nodal curves on rational surfaces, $K3$-surfaces and surfaces of general type have been considered (cf. [74], [75], [53], [29], [14]).

Let Σ be a smooth surface, D a divisor of Σ and $V(D,n)$ the variety of nodal curves in the complete linear system $|D|$ with exactly n nodes. Tannenbaum [74] derives for a rational surface Σ the condition

$$-K_\Sigma\cdot C_i > 0, \qquad i=1,\ldots,s, \tag{2.11}$$

for the T-property of $V(D,n)$ at $C=C_1\cup\cdots\cup C_s$, C_i irreducible, if the generic member of $|D|$ is irreducible. Note that for a rational surface the result follows directly from (2.4) of Section 2.2.

However, for $\Sigma=\mathbb{P}^2_m$, the projective plane blown-up in m generic points $p_1,\ldots,p_m$, we have better results which are, for large m, asymptotically (in d) nearly optimal. Let $E_0\subset\Sigma$ be the preimage of a generic straight line in $\mathbb{P}^2$ and $E_1,\ldots,E_m$ the exceptional divisors. Then, for an irreducible nodal curve $C\in\left|dE_0-d_1E_1-\ldots-d_mE_m\right|$, (2.11) reads

$$3d > \sum_{i=1}^{m} d_i. \tag{2.12}$$

This already shows that $V=V(d;d_1,\ldots,d_m;n)$, the variety of nodal curves in $\left|dE_0-d_1E_1-\ldots-d_mE_m\right|$ with n nodes, is smooth at C for $m\le 9$.

For $m \geq 10$ it is shown in [29] that V has the T-property at C if the following two conditions are fulfilled:

$$\left[\sqrt{2n}\right] < \frac{d}{2} + 3 - \frac{\sqrt{2}}{2}\sqrt{\sum_{i=1}^{m}(d_i+2)^2}\ ,$$

$$\left[\sqrt{2n}\right] < d + 3 - \sqrt{2}\sqrt{2 + \sum_{i=1}^{m}(d_i+2)(d_i+1)}.$$

If C is reducible, the T-property holds at C if analogous conditions are fulfilled for each component of C.

For a $K3$-surface, K_Σ is trivial, hence the above condition is never fulfilled. Tannenbaum shows in [75] that the T-property holds for $V(D,n)$ at $C \in V(D,n)$ if and only if C is irreducible. Also, the independence of deformations of the nodes fails if C is reducible. Nevertheless, $V(D,n)$ is always smooth at $C = C_1 \cup \cdots \cup C_s$ of dimension $\dim|D| - n + (s-1)$.

In [14] Chiantini and Sernesi consider nodal curves C on surfaces of general type Σ. They study a rank 2 bundle on Σ associated to the nodes of C, and use an inequality for the Bogomolov instability of this bundle, in order to prove that $V(C,n)$ has the T-property at an irreducible C. Namely, if C is numerically equivalent to pK_Σ, $p \in \mathbb{Q}$, $p > 2$, then $V(D,n)$ has the T-property, provided

$$n < \frac{1}{4}p(p-2)(K_\Sigma)^2,$$

or if

$$n < \frac{1}{4}(p-1)^2(K_\Sigma)^2$$

p odd, and the Neron-Severi group is $NS(\Sigma) = \mathbb{Z} \cdot K_\Sigma$. These results seem to be asymptotically nearly optimal (see also Section 4).

3 Irreducibility of equisingular families

The question about the irreducibility of ESF is more delicate than the existence and smoothness problems, especially concerning sufficient conditions. Rather difficult from the algebraic-geometric point of view, it is of special topological interest, being connected with the problem of the fundamental group of the complement of a plane complex algebraic curve [93] (and through this, with the classification of complex algebraic surfaces). Even the case of plane nodal curves appeared to be very hard: Severi started to handle it [61], but failed (see details in [2], [39]), then Arbarello and Cornalba [2] made some progress. Finally Harris [39] proved that the family $V_d(n)$ of irreducible plane curves of degree d with n nodes as their only singularities is irreducible for any

$$n \leq \frac{(d-1)(d-2)}{2} .$$

The classical Zariski examples of reducible ESF and other such examples [22], [68], show that for more complicated singularities, starting with ordinary cusps, possible numerical sufficient conditions for the irreducibility of ESF should be different from the necessary existence conditions for curves with a given collection of singularities.

The results in this direction basically concern plane curves and we will pay most attention to this case.

3.1 Plane curves

There are three main approaches to the problem about irreducibility of ESF.

(1) One possible approach consists of building a connecting path between two equisingular curves, using explicit equations of the curves, or projective transformations. This method works for small degrees only. Besides the classical case of conics and cubics, it is proven that all ESF of quartic curves [11] and of quintic curves [16], [90] are irreducible. For degrees > 5 this is not true, and the method is no longer efficient, except for some very special cases [79], [80], [81].

(2) Another approach, suggested by Arbarello and Cornalba [2], consists of relating an ESF to the moduli space of curves of a given genus, known to be irreducible. This gave some particular results on families of plane nodal curves [2], and plane curves with nodes and cusps. Namely Kang [45] proved that the family $V_d(n,k)$ of plane irreducible curves of degree d with n nodes and k cusps as their only singularities is irreducible if

$$k \leq \frac{d+1}{2}, \quad \frac{(d-1)(d-2)}{2} \geq n \geq \frac{d^2-4d+1}{2} .$$

(3) A new idea of Harris [39] completed the case of plane nodal curves. It consists of proceeding inductively from rational nodal curves (whose family is classically known to be irreducible) via study of a degeneration of nodal curves to any family of curves of a given degree with a given number of nodes. Further development of this idea lead to new results by Ran [55]: an ESF of plane irreducible curves of a given degree with a given number of nodes and one ordinary singularity is irreducible, and by Kang [46]: an ESF of irreducible curves of a given degree with a given number of nodes and a given number $k \leq 3$ of cusps is irreducible. Note that the requirement to study all possible deformations of the curves considered does not allow us to extend such an approach to more complicated singularities or to a large number of any singularities different from nodes.

(4) The following approach is applicable to ESF of curves (and even projective hypersurfaces of any dimension) with any quantities of arbitrary singularities and provides the most general results. So we explain it in more

detail. Namely, let V be an ESF of plane curves of degree d with a given collection of (topological or analytic) singularities. With an isolated singular point z of a plane curve C we can associate zero-dimensional schemes $X^s(z)$ and $X^a(z)$ such that

- for almost all elements f in the local ideal $J_{X^s(z)} \subset \mathcal{O}_{\mathbb{P}^2,z}$ of the scheme $X^s(z)$ (respectively in the local ideal $J_{X^a(z)} \subset \mathcal{O}_{\mathbb{P}^2,z}$ of the scheme $X^a(z)$) C and $C+f$ have the same topological (respectively analytic) singularity at z,
- $X^s(z)$ (respectively $X^a(z)$) varies in an irreducible flat family $\mathcal{X}(S)$ as z runs through all singular points of topological (respectively analytic) type S.

For instance, $X^s(z)$ can be defined as $X(C, T^*(C,z))$ in the notation of Section 1.5, and $X^a(z)$ can be defined by the ideal

$$J_{X^a(z)} = (\mathfrak{m}_z)^{\mu(z)+2} \subset \mathcal{O}_{\mathbb{P}^2,z}$$

(see [83], Theorem 1.2.1). If a curve $C \in V$ has singularities $S_1, \ldots, S_r$, then we define a natural morphism

$$\Phi : \widetilde{V} \to \mathcal{X}(S_1) \times \cdots \times \mathcal{X}(S_r) ,$$

where $\widetilde{V}$ is a finite covering of V, consisting of curves $C \in V$ with ordered singular points. By the construction of $\mathcal{X}(S)$, for the irreducibility of V it is enough to show that Φ is surjective and its fibers are irreducible (open sets of projective spaces). Both conditions follow from

$$h^1(\mathbb{P}^2, \mathcal{J}_{X(C)}(d)) = 0, \quad \text{for each } C \in V, \tag{3.1}$$

where $X(C)$ is the union of schemes $X^s(z)$ (or $X^a(z)$, according as we consider topological or analytic singularities) taken over all singular points $z \in C$. The inequality, derived easily from [26], Proposition 5.2,

$$\deg X(C) < 3d \tag{3.2}$$

is sufficient for (3.1). The right-hand side here is linear in d, whereas the left-hand side, being the sum of certain positive invariants of $S_1, \ldots, S_r$, may be of order d^2. We observe also that (3.1) may fail for some curves $C \in V$ as (3.2) fails, even in the case of nodal curves [66]. To improve the result, one can weaken the requirement to

$$V_{reg} = \{C \in V \mid h^1(\mathbb{P}^2, \mathcal{J}_{X(C)}(d)) = 0\}$$

is open dense in V. In other words

$$\dim(V \backslash V_{reg}) < \dim V, \tag{3.3}$$

where $\dim V$ is the virtual "regular" dimension of V (in fact, the minimal possible actual dimension).

In [68], [69], [70] the failure of (3.1) was interpreted as: the singular points of $C \in V$ lie on a curve of a relatively small degree. Then, using an irreducible curve of degree $n < \alpha d$ passing through the scheme $X^s(C)$ or $X^a(C)$, one can estimate the dimension of the set of such curves C and get the following sufficient condition for the irreducibility of V:

$$n + 2k < \frac{d^2}{225}, \tag{3.4}$$

if $V = V_d(n, k)$, and

$$\sum_{i=1}^{r} \mu(S_i) \leq \min_{1 \leq i \leq r} f(S_i) \cdot d^2, \tag{3.5}$$

$$f(S) = \frac{2}{(\mu(S) + \mathrm{mt}(S) - 1)^2 (3\mu(S) - \mathrm{mt}(S)^2 + 3 \cdot \mathrm{mt}(S) + 2)^2}.$$

The latter conditions have a right-hand side quadratic in d.

D. Barkats [7] approached the problem (3.3) from the viewpoint of the theory of zero-dimensional subschemes of the plane [15] and obtained inequalities with bigger coefficients of d^2,

$$3n + 5k < \frac{7}{40} d^2 - \frac{1}{8} d - 5$$

to be sufficient for the irreducibility of $V_d(n, k)$, and

$$\sum_{i=1}^{r} b(S_i) < \frac{2b - 3}{2b(b-1)} d^2 - \frac{2b - 9}{2(b-1)} d \tag{3.6}$$

to be sufficient for the irreducibility of V, when $S_1, \ldots, S_r$ are topological singularities, $b(S) = \deg X^s(z)$ as z is of type S, and

$$b = \max_{i=1,\ldots,r} b(S_i).$$

3.2 Curves on rational surfaces

The methods described above were extended to some rational surfaces.

So, Ran [55] showed that, for any Hirzebruch surface (fan) Σ, a family of irreducible curves of a given bidegree with a given number of nodes is irreducible. He used both Harris' method and the theory of deformations of fans. The reasons mentioned above show it is dificult to generalize this method to any surface and curves with arbitrary singularities.

On the other hand, the approach of [68], [69], [70] was applied to $\Sigma = \mathbb{P}^2_m$, the plane blown up at m generic points, and the families $V(d; d_1, \ldots, d_m; n)$

of irreducible curves with n nodes, belonging to the linear system $|dE_0 - d_1E_1 - \ldots - d_mE_m|$. As a result, in [29] the inequalities

$$\left[\sqrt{2k}\,\right] < \frac{d}{4}+1-\frac{1}{4}\sqrt{\sum_{i=1}^{m} d_i^2}\,, \tag{3.7}$$

$$\left[\sqrt{2k}\,\right] < \frac{d}{2}+1-\frac{\sqrt{2}}{2}\sqrt{\sum_{i=1}^{m}(d_i+2)^2}\,, \tag{3.8}$$

are shown to be sufficient for the irreducibility of $V(d; d_1, \ldots, d_m; n)$. In contrast to the irreducibility problem for plane curves, discussed in the previous subsection, here we need to have a certain irreducible curve going through the singularities of an arbitrary curve $C \in V = V(d; d_1, \ldots, d_m; n)$. But this strongly depends on the fact that only nodal curves are considered.

4 Perspectives and problems

In this section we indicate some problems naturally appearing in the field, possible new results, and other relevant ideas.

(1) The problem of the existence of plane curves of a given degree with given analytic (non-simple) singularities is still open. In general, we only have the upper bounds mentioned above. The zero-dimensional scheme approach used in the topological situation does not work here, because the residue operation does not respect the analytic type of singular points. Another approach is based on the relation between topological and analytic singularities. Let us call the quotient $CS(S) = V_h(S)/\sim$ (with respect to analytic equivalence) a *classifying space* of a topological singularity S. In the case of a simple singularity S, $CS(S)$ is a point, while in general it is merely a (connected) topological space (not even Hausdorff). In order to obtain a (coarse) moduli space as an algebraic variety, we have to stratify $V_h(S)$ into invariant strata and call the disjoint union of these strata, $M(S)$, the moduli space of a topological singularity S (cf. [33]). What is a "good" compactification of $M(S)$? This may be important in view of

Theorem 3 *There is an open dense subset $U \subset V_h(S)$ such that for any analytic singularity $\mathcal{A} \in U$ there exists an irreducible curve of degree*

$$d \leq 14\sqrt{\mu(S)}$$

having one singular point of type $\mathcal{A}$ as its only singularity.

We conjecture that for all topological singularities S, and $\mathcal{A} \in M(S)$, the minimal degree $d = d(\mathcal{A})$ as in Theorem 3 satisfies

$$d(\mathcal{A}) \leq a \cdot d(S)$$

for some absolute constant a. This would imply an asymptotically optimal sufficient condition for the existence of curves with given analytic singularities, similar to that for topological singularities.

(2) In connection with the previous problem it seems to be relevant to develop the theory of zero-dimensional schemes in the plane of general type and of types associated with singularities, as, for instance, of class $\mathcal{GS}$. The principal problem is the computation of $h^1(\mathbb{P}^2, \mathcal{J}_X(d))$ (or the so-called Castelnuovo function [15]). From our point of view it is important to estimate the minimal d such that

$$h^1(\mathbb{P}^2, \mathcal{J}_X(d)) = 0 \tag{4.1}$$

for X a generic element in the corresponding Hilbert scheme. We conjecture that there exists an absolute constant $a > 0$ so that if

$$d \geq a\sqrt{\deg X}$$

then (4.1) holds, which would generalize Hirschowitz's theorem [41], mentioned in Section 1.7.

It is interesting to strengthen Hirschowitz's inequality (1.11) itself, to improve the constants in the sufficient existence conditions involving such inequalities. In particular, in [73], (1.11) is strengthened to

$$\max_{i<j}(d_i + d_j) < d, \quad \sum_{i=1}^{r} \frac{(d_i+1)(d_i+2)}{2} < \frac{2d^2 + 8d - 15}{5} .$$

The Harbourne-Hirschowitz conjecture [41], [38] states that

$$\max_{i<j<k}(d_i + d_j + d_k) \leq d - 3, \quad \sum_{i=1}^{r} \frac{(d_i+1)(d_i+2)}{2} \leq \frac{(d+1)(d+2)}{2}$$

is sufficient for (1.10).

(3) In a similar way we can state the problem of the existence of hypersurfaces in $\mathbb{P}^m$, $m \geq 3$, of a given degree with isolated singular points of given types. Our conjecture is that the inequality

$$\sum_{i=1}^{r} \mu(S_i) \leq a_m d^m$$

with a certain positive a_m, depending only on m, suffices for the existence of an irreducible hypersurface of degree d in $\mathbb{P}^m$ having r isolated singular points of types $S_1, \ldots, S_r$, respectively as its only singularities. Such an estimate would be asymptotically optimal in view of Bruce's upper bounds to the total Milnor number of projective hypersurfaces [10].

We would like to point out two aspects of this problem. The problem, in fact, consists of two problems – that for topological singularities and for analytic ones. A possible approach may be based on the study of zero-dimensional schemes in $\mathbb{P}^m$. Namely, it might consist of a description of zero-dimensional schemes associated with topological or analytic isolated singularities, and the development of an h^1-vanishing criteria for generic (in some sense) zero-dimensional schemes of certain classes, which conjecturally should look like

$$h^1(\mathbb{P}^m, \mathcal{J}_X(d)) = 0 \quad \text{as} \quad \deg X \le ad^m, \quad a = \text{const} > 0.$$

(4) In the smoothness and irreducibility problem for ESF of plane algebraic curves, the question of optimality of the numerical sufficient conditions for these properties remains open. We recall that the sufficient condition for the existence (1.4) discussed in Section 1.4, is asymptotically optimal, i.e. it differs from the necessary condition (1.2) only in the coefficient of d^2 in the right-hand side. Another situation is observed for the smoothness and the irreducibility: there are sufficient conditions which are "linear in d", like those from [26], [27]:

$$\sum_{i=1}^r (\tau(S_i) - 1) < 3d, \tag{4.2}$$

$$\sum_{i=1}^r \tau(S_i) < 4d - 4, \tag{4.3}$$

and ones "quadratic in d", like (3.5) in [70] and (3.6) in [8]. The latter conditions are weaker than any asymptotically optimal one, because the coefficient of d^2 in the right-hand side depends on the "worst singularity" S_i and tends to zero as $\mu(S_i) \to \infty$. The following statement shows that, unlike the situation for existence, there cannot be an asymptotically optimal sufficient condition for the smoothness of ESF [71]:

Theorem 4 *The family $V_{2pq}(q^2 \cdot A_{6p-1})$, where A_{6p-1} is the singularity given by $y^2 + x^{6p} = 0$, has a non-smooth irreducible component if $p \ge 5$, $q \ge 3$.*

Indeed, here we have a sequence of non-smooth ESF as the number of some singularities S becomes $\sim 9d^2/\mu(S)^2$, where d and $\mu(S)$ tend to ∞. In Figure 1, we give a graphical illustration of the conditions described. Here $x = d^2$ is the squared degree of a curve, $y = \sum_{i=1}^r \mu(S_i)$ is the total complexity of the singularities. The straight lines $y = x$ and $y = x/396$ indicate the necessary and sufficient conditions for the existence of a curve with given singularities. The parabola $y = 9\sqrt{x}$ goes through points corresponding to non-smooth ESF, and the parabola $y = 4\sqrt{x} - 4$ indicates a sufficient condition for the smoothness of ESF (4.3).

We conjecture that there cannot be an asymptotically optimal sufficient condition for the irreducibility of ESF.

One more conjecture concerns a possible form of sufficient conditions for the smoothness and irreducibility of ESF:

Conjecture 1 *There exists a positive invariant β of topological (analytic) singularities and an absolute constant $a > 0$ such that*

$$\sum_{i=1}^r \beta(S_i) \le ad^2 \tag{4.4}$$

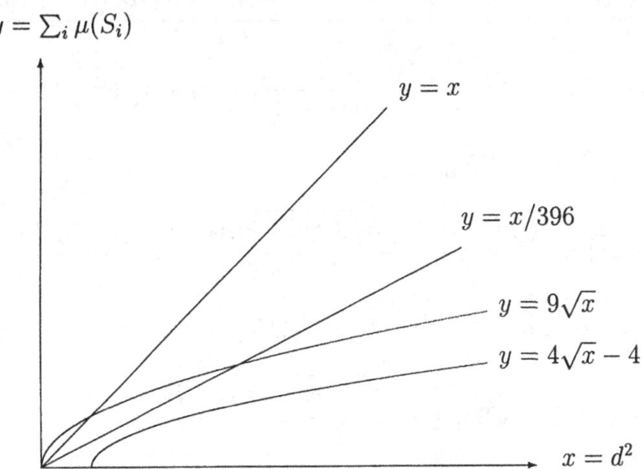

Figure 1: Necessary and sufficient conditions for existence and smoothness.

is sufficient for the smoothness and the irreducibility of the corresponding topological (respectively analytic) ESF, and, in addition, there exists a sequence of curves of degrees $d \to \infty$ with the total β-invariant, growing faster than d^2.

Condition (4.4) is weaker than any asymptotically optimal one, but is stronger that any condition of type (3.5) in the following sense: any condition of type (4.4) can be reduced to a condition of type (3.5), for instance,

$$\sum_{i=1}^{r} \mu(S_i) \leq \min_{1 \leq i \leq r} \frac{\mu(S_i)}{\beta(S_i)} d^2,$$

but no condition of type (3.5) can be reduced to any condition of type (4.4).

(5) Given a smooth algebraic surface Σ and a divisor D on Σ, the problem about the geometry of ESF of curves in the linear system $|D|$ can be stated as for the plane. What are numerical necessary and sufficient conditions for the non-emptiness, smoothness and irreducibility of such an ESF? The general sufficient condition for the smoothness of ESF

$$\sum_{i=1}^{r} (\tau(S_i) - 1) < -DK_\Sigma, \tag{4.5}$$

where K_Σ is the canonical divisor, seems not to be optimal, since on the plane it turns into a "linear" condition (4.2). New results on families of nodal curves on $\mathbb{P}^2_m$ the projective plane blown up at $m > 9$ generic points [29], and on nodal curves on surfaces of general type [14], cover a much wider range than (4.5). These new conditions can be formulated in invariant form as follows. Assume that $\Sigma = \mathbb{P}^2_m$. Then the inequalities

$$n \le \frac{A_1 + B_1}{2} + 1 - \frac{(m+9)B_1^2 + 9(m-9)A_1 + 6\sqrt{m}B_1\sqrt{B_1^2 + (m-9)A_1}}{2(m-9)^2} - \frac{3}{2(m-9)}(\sqrt{m}B_1 + 3\sqrt{B_1^2 + (m-9)A_1}),$$

where $A_1 = D^2$, $B_1 = DK_\Sigma$,

$$[\sqrt{2n}] < \frac{(\sqrt{m} - 3\sqrt{2})\sqrt{B_2^2 + (m-9)A_2} - (\sqrt{2m} - 3)B_2}{2(m-9)},$$

where $A_2 = (D - 2K_\Sigma)^2$, $B_2 = (D - 2K_\Sigma)K_\Sigma$,

$$[\sqrt{2n}] \le \frac{\sqrt{B_3^2 + (m-9)A_3} - B_3}{4(\sqrt{m} + 3)},$$

where $A_3 = D^2$, $B_3 = DK_\Sigma$, suffice for the family of curves with n nodes in the linear system $|D|$ to be non-empty, smooth and irreducible, respectively. Similarly, if Σ is a surface of general type, then the Chiantini-Sernesi sufficient condition for the smoothness and the T-property of the family of curves with n nodes in the linear system $|D|$ (see [14]) is

$$n < \frac{D^2 - 2DK_\Sigma + 1}{4}.$$

Comparing these results with the general necessary condition for the existence of an irreducible curve in $|D|$ on an arbitrary surface Σ (for stronger restrictions for special surfaces see [13], [91]) namely

$$\sum_{i=1}^{r} \delta(S_i) \le \frac{D^2 + DK_\Sigma}{2} + 1,$$

we conjecture that an asymptotically optimal form of a sufficient condition for the properties considered may be

$$\sum_{i=1}^{r} \alpha(S_i) \le \Phi(D^2, DK_\Sigma),$$

with some positive invariant α of (topological or analytic) singularities and a function Φ, linear in D^2 and quadratic in DK_Σ.

(6) Problems on the geometry of ESF of algebraic curves can be formulated in a similar way for equisingular families of hypersurfaces of a given degree in $\mathbb{P}^m$, $m \geq 3$, with given isolated singularities. In contrast to the existence problem, which conjecturally admits an approach coming from the study of curves discussed above, for the smoothness and irreducibility we do not have yet an adequate approach. The first step in this direction was made by Du Plessis and Wall [19]. They used the theory of discriminant matrices and obtained, in particular, the following result:

Lemma 5 *If*

$$\frac{1}{m} + \frac{1}{d} > \frac{1}{2} \tag{4.6}$$

and

$$\sum_{i=1}^{r} \tau(S_i) < (d-1)^{m-1}(2m + 2d - md),$$

then the family of hypersurfaces of degree d *in* $\mathbb{P}^m$ *with given analytic singularities* $S_1, \ldots, S_r$ *is smooth and has dimension*

$$\binom{d+m}{m} - 1 - \sum_{i=1}^{r} \tau(S_i) \ .$$

The inequality (4.6) restricts the range of application to finitely many pairs d, $m \geq 3$. So the question of how to handle the general case, with arbitrary d, m, still remains open and the most important in the problem discussed.

References

[1] Alexander J. and Hirschowitz A. Le methode d'Horace éclatée: application à l'interpolation en degrée quatre. *Invent. Math.* **107** (1992), 585-602.

[2] Arbarello E. and Cornalba M. A few remarks about the variety of irreducible plane curves of given degree and genus. *Ann. sci. Ec. norm. super.* **16** (1983), no. 3, 467-488.

[3] Arnol'd V.I., Gusein-Zade S.M., and Varchenko A.N. *Singularities of differentiable maps, vol.1.* Boston, Basel, Stuttgart: Birkhäuser Verlag, 1985.

[4] Artin M. Versal deformations and algebraic stalks. *Invent. Math.* **27** (1974), 165-189.

[5] Artin M. *Deformations of singularities.* Tata Institute of Fundamental Research, Bombay, 1976.

[6] Barkats D. *Non vacuité des variétés des courbes planes a noeuds et a cusps.* Preprint no. 363, Université de Nice-Sophia-Antipolis, December 1993.

[7] Barkats D. *Irreducibilité des variétés des courbes planes a noeuds et a cusps.* Preprint no. 364, Université de Nice-Sophia-Antipolis, December 1993.

[8] Barkats D. *Private communication.*

[9] Brieskorn E. and Knörrer H. *Plane algebraic curves.* Birkhäuser-Verlag, 1986.

[10] Bruce J. W. Singularities on a projective hypersurface. *Bull. Lond. Math. Soc.* **13** (1981), 47-50.

[11] Bruce J. W. and Giblin P. J. A stratification of the space of plane quartic curves. *Proc. London Math. Soc.* **42** (1981), 270-298.

[12] Brusotti L. Sulla "piccola variazione" di una curva piana algebrica reali. *Rend. Rom. Ac. Lincei (5)* **30** (1921), 375-379.

[13] Chang M. C. and Ran Z. Divisors on some generic hypersurfaces. *J. Diff. Geom.* **38** (1993), 671-678.

[14] Chiantini L. and Sernesi E. Nodal curves on surfaces of general type. *Math. Ann.* **307** (1997), 41-56.

[15] Davis E. D. 0-dimensional subschemes of $\mathbb{P}^2$: new applications of Castelnuovo's function. *Ann. Univ. Ferrara* **32** (1986), 93-107.

[16] Degtyarev A. *Isotopy classification of complex plane projective curves of degree five.* Preprint P-3-87, Steklov Math. Inst., Leningrad branch, 1987 (Russian).

[17] Degtyarev A. *Quintics in $\mathbb{C}^2$ with non-abelian fundamental group.* Preprint, Max-Planck-Institut für Mathematik, 1995.

[18] Diaz S., and Harris J. Ideals associated to deformations of singular plane curves. *Trans. Amer. Math. Soc.* **309** (1988), no. 2, 433-468.

[19] Du Plessis A. A. and Wall C. T. C. *Versal deformations in spaces of polynomials of fixed weight.* Preprint no. 16, University of Aarhus, July 1996.

[20] Du Plessis A. A. and Wall C. T. C. *The Geometry of Topological Stability.* Clarendon Press, Oxford, 1995.

[21] Finashin S. and Shustin E. On imaginary plane curves and Spin quotients of complex surfaces by complex conjugation. In: *Topics in Singularity Theory. V. I. Arnold's 60th Anniversary Collection (AMS Translations, Ser. 2, vol. 180)/* A. Khovanskii etc eds., AMS, Providence R. I., 1997, pp. 93-101.

[22] Giacinti-Diebolt C. Variétés des courbes projectives planes de degrée et lieu singulier donnés. *Math. Ann.* **266** (1984), no. 3, 321-350.

[23] Gradolato M. and Mezzetti E. Curves with nodes, cusps and ordinary triple points. *Ann. Univ. Ferrara, sez. 7,* **31** (1985), 23-47.

[24] Gradolato M.A. and Mezzetti E. Families of curves with ordinary singular points on regular surfaces. *Ann. mat. pura ed appl.* **150** (1988), 281-298.

[25] Greuel, G.-M. *A remark on the paper of A. Tannenbaum.* Composito Math. **51**, 185-187 (1984).

[26] Greuel G.-M. and Karras U. Families of varieties with prescribed singularities. *Compositio Math.* **69** (1989), no. 1, 83-110.

[27] Greuel G.-M. and Lossen C. Equianalytic and equisingular families of curves on surfaces. *Manuscr. Math.* **91** (1996), 323-342.

[28] Greuel G.-M., Lossen C. and Shustin E. Plane curves of minimal degree with prescribed singularities. *Invent. Math.* **133** (1998), no. 3, 539-580.

[29] Greuel G.-M., Lossen C. and Shustin E. Families of nodal curves on the blown-up projective plane. *Trans. Amer. Math. Soc.* **350** (1998), no. 1, 251-274.

[30] Greuel, G.-M., Pfister, G., Hertling, C. Moduli spaces of semiquasihomogeneous singularities with fixed principal part. *J. Alg. Geom.* **6** (1997), no. 1, 169-199.

[31] Greuel G.-M., and Pfister G. Moduli for singularities. In: *Singularities/London Math. Soc. Lect. Notes 201*, J.-P. Brasselet, ed., Cambridge Univ. Press, Cambridge, 1994, pp. 119-146.

[32] Greuel, G.-M., and Pfister, G. Geometric quotients of unipotent group actions. *Proc. London Math. Soc.* (3) **67** (1993), 75-105.

[33] Greuel, G.-M. and Pfister, G. On moduli spaces of semiquasihomogeneous singularities. In: *Progress in mathematics* **134**, Birkhäuser, NY, 1996, 174-185.

[34] Greuel, G.-M. and Pfister, G. *Geometric quotients of unipotent group actions II.* Preprint, Univ. of Kaiserslautern 1995. To be published in "Singularities", Festband on the occasion of Brieskorn's 60th birthday.

[35] Greuel, G.-M., Pfister, G. and Schönemann, H. *SINGULAR, A Computeralgebra system for Algebraic Geometry and Singularity theory.* Available via ftp from helios.mathematik.uni-kl.de.

[36] Gudkov D. A., Tai M. L., and Utkin G. A. Complete classification of irreducible curves of the 4th order. *Mat. Sbornik* **69** (1966), no. 2, 222-256 (Russian).

[37] Gudkov D.A. Topology of real projective algebraic varieties. *Rus. Math. Surveys* **29** (1974), no. 4, 3-79.

[38] Harbourne B. Complete linear systems on rational surfaces. *Trans. Amer. Math. Soc.* **289** (1985), 213-226.

[39] Harris J. On the Severi problem. *Invent. Math.* **84** (1985), 445-461.

[40] Hirano A. Constructions of plane curves with cusps. *Saitama Math. J.* **10** (1992), 21-24.

[41] Hirschowitz A. Une conjecture pour la cohomologie des diviseurs sur les surfaces rationelles génériques. *J. reine angew. Math.* **397** (1989), 208-213.

[42] Hirzebruch F. Singularities of algebraic surfaces and characteristic numbers. *Contemp. Math.* **58** (1986), 141-155.

[43] Itenberg I. and Shustin E. Real algebraic curves with real cusps. *Amer. Math. Soc. Transl. (2)* **173** (1996), 97-109.

[44] Ivinskis K. *Normale Flächen und die Miyaoka-Kobayashi Ungleichung.* Diplomarbeit, Bonn, 1985.

[45] Kang P.-L. On the variety of plane curves of degree d with δ nodes and k cusps. *Trans. Amer. Math. Soc.* **316** (1989), no. 1, 165-192.

[46] Kang P.-L. A note on the variety of plane curves with nodes and cusps. *Proc. Amer. Math. Soc.* **106** (1989), no. 2, 309-312.

[47] Korchagin A.B. and Shustin E. Affine curves of degree 6 and smoothings of the ordinary 6th order singular point. *Math. USSR Izvestia* **33**, N.3, 501-520 (1989).

[48] Laudal A. *Formal moduli of algebraic structures. (Lecture Notes Math., vol. 754)*, Springer, Berlin etc., 1979.

[49] Laudal O.A., Martin B., and Pfister G. Moduli of plane curve singularities with $\mathbb{C}^*$-action. In: *Banach Cent. Publ. Vol. 20. Singularities.* Warszawa, 1988, pp. 255-278.

[50] Lindner M. Über Mannigfaltigkeiten ebener Kurven mit Singularitäten. *Arch. Math.* **28** (1977), 603-610.

[51] Loeser I. Deformations des courbes planes. *Seminaire Bourbaki*, 1986-87, no. 679.

[52] Luengo I. The μ-constant stratum is not smooth. *Invent. Math.* **90** (1987), 139-152.

[53] Nobile A. Families of curves on surfaces. *Math. Zeitschrift* **187** (1984), no. 4, 453-470.

[54] Pecker D. *On the geometric genus of projective curves.* Preprint, Université Pierre & Marie Curie, 1996.

[55] Ran Z. Families of plane curves and their limits: Enriques' conjecture and beyond. *Annals of Math.* **130** (1989), no. 1, 121-157.

[56] Sakai F. Singularities of Plane Curves. In: *Geometry of Complex Projective Varieties, Seminars and Conferences 9*, Mediterranian Press, Rende, 1993, pp. 257-273.

[57] Santos F. Construction of real algebraic plane curves. *Proc. Real Algebraic and Analytic Geometry, Trento, Sept. 1992*, Berlin, Walther de Gruyter, 1993, pp. 213-228.

[58] Schlessinger M. Functors of Artin rings. *Trans. AMS* **130** (1968), 208-222.

[59] Segre B. Dei sistemi lineari tangenti ad un qualunque sistema di forme. *Atti Acad. naz. Lincei Rendiconti serie 5*, **33** (1924), 182-185.

[60] Segre B. Esistenza e dimensione di sistemi continui di curve piane algebriche con dati caraterri. *Atti Acad. naz. Lincei Rendiconti serie 6*, **10** (1929), 31-38.

[61] Severi, F. *Vorlesungen über Algebraische Geometrie (Anhang F).* Leipzig, Teubner, 1921.

[62] Shustin E. New M-curve of the 8th degree. *Math. Notes of Acad. Sci. USSR* **42** (1987), 606-610.

[63] Shustin E. Versal deformations in the space of plane curves of fixed degree. *Function. Anal. Appl.* **21** (1987), 82-84.

[64] Shustin E. Geometry of discriminant and topology of algebraic curves. *Proc. Internat. Congr. of Math., Kyoto, Japan, Aug. 21-29, 1990*, vol. 1, Springer, 1991, pp. 559-567.

[65] Shustin E. Hyperbolic and minimal smoothings of singular points. *Selecta Math. Sov.* **10** (1991), no.1, 19-25.

[66] Shustin E. On manifolds of singular algebraic curves. *Selecta Math. Sov.* **10** (1991), no.1, 27-37.

[67] Shustin E. Real plane algebraic curves with prescribed singularities. *Topology* **32** (1993), no. 4, 845-856.

[68] Shustin E. Smoothness and irreducibility of varieties of algebraic curves with nodes and cusps. *Bull. SMF* **122** (1994), 235-253.

[69] Shustin E. *Smoothness and irreducibility of families of plane algebraic curves with ordinary singularities.* In: *Israel Mathematical Conference Proceedings, vol. 9,* AMS, Providence Rh. I., 1996, pp. 393-416.

[70] Shustin E. *Geometry of equisingular families of plane algebraic curves. J. Algebraic Geometry* **5** (1996), no. 2, 209-234.

[71] Shustin E. Smoothness of equisingular families of plane algebraic curves. *International Math. Research Notices* **2** (1997), 67-82.

[72] Shustin E. Gluing of singular and critical points. *Topology* **37** (1998), no. 1, 195-217.

[73] Shustin E. and Tyomkin I. *Linear systems of plane curves with generic multiple points.* Preprint 4-96, Tel Aviv University, May 1996.

[74] Tannenbaum A. Families of algebraic curves with nodes. *Compositio Math.* **41** (1980), 107-126.

[75] Tannenbaum A. Families of curves with nodes on K3-surfaces. *Math. Ann.* **260** (1982), 239-253.

[76] Tannenbaum A. On the classical characteristic linear series of plane curves with nodes and cuspidal points: two examples of Beniamino Segre. *Compositio Math.* **51** (1984), 169-183.

[77] Teissier B. The hunting of invariants in the geometry of discriminants. In: *Real and Complex Singularities./* P.Holm, editor. Oslo: Sijthoff-Noordhoff Publ., Alphen aan den Rijn, 1977, p. 565-677.

[78] Tougeron J. C. Ideaux des fonctions différentiables. *Ann. Inst. Fourier (Grenoble)* **18** (1968), no. 1, 177-240.

[79] Urabe T. On quartic surfaces and sextic curves with certain singularities. *Proc. Japan. Acad. Ser. A.* **59** (1983), 434-437.

[80] Urabe T. On quartic surfaces and sextic curves with singularities of type $\widetilde{E}_8$, $T_{2,3,7}$, $\widetilde{E}_{12}$. *Publ. Inst. Math. Sci. Kyoto Univ.* **80** (1984), 1185-1245.

[81] Urabe T. Singularities in a certain class of quartic surfaces and sextic curves and Dynkin graphs. *Proc. Vancouver Conf. Algebraic Geom., July 2-12, 1984.* Providence, 1986, p. 477-497.

[82] Varchenko A.N. Asymptotic of integrals and Hodge structures. In: *Modern Problems of Math., Vol. 22 (Itogi nauki i tekhniki VINITI),* Moscow, 1983, pp. 130-166 (Russian).

[83] Vassiliev V.A. Stable cohomology of complements to the discriminant of deformations of singularities of smooth functions. *J. Soviet Math.* **52** (1990), no. 4, 3217-3230.

[84] Viro O. Ya. Gluing of algebraic hypersurfaces, smoothing of singularities and construction of curves. *Proc. Leningrad Int. Topological Conf., August 1983,* Nauka, Leningrad, 1983, pp. 149-197 (Russian).

[85] Viro O. Ya. Gluing of plane real algebraic curves and construction of curves of degrees 6 and 7. *Lect. Notes Math.* **1060**, Springer, Berlin etc., 1984, pp. 187-200.

[86] Viro O.Ya. Real algebraic plane curves: constructions with controlled topology. *Leningrad Math. J.* **1** (1990), 1059-1134.

[87] Wahl J. Deformations of plane curves with nodes and cusps. *Amer. J. Math.* **96** (1974), 529-577.

[88] Wahl J. Equisingular deformations of plane algebroid curves. *Trans. Amer. Math. Soc.* **193** (1974), 143-170.

[89] Wall C. T. C. Geometry of quartic curves. *Math. Proc. Camb. Phil. Soc.* **117** (1995), 415-423.

[90] Wall C. T. C. Highly singular quintic curves. *Math. Proc. Camb. Phil. Soc.* **119** (1996), 257-277.

[91] Xu G. Subvarieties of general hypersurfaces in projective spaces. *J. Diff. Geom.* **39** (1994), 139-172.

[92] Zariski O. Studies in equisingularity I-III. *Amer. J. Math.* **87** (1965), 507-536, *Amer. J. Math.* **87** (1965), 972-1006, and *Amer. J. Math.* **90** (1968), 961-1023.

[93] Zariski O. *Algebraic surfaces.* 2nd ed. Heidelberg, Springer, 1971.

Added in proof. Recently the following result was proven by E. Shustin and I. Tyomkin ("Versal deformations of projective hypersurfaces with isolated singularities", Preprint, Tel Aviv University, May 1997) and independently by A. du Plessis and C. T. C. Wall (personal communication): the inequality

$$\sum_{i=1}^{r} \tau(S_i) < 4d - 4$$

provides the smoothness and the T-property of the family of hypersurfaces of degree d with isolated singular points of analytic types $S_1, \ldots, S_r$ in $\mathbb{P}^n$, $n \geq 3$.

Gert-Martin Greuel
Universität Kaiserslautern
Fachbereich Mathematik
Erwin-Schrödinger-Strasse
D-6750 Kaiserslautern
Germany.

Eugenii Shustin
School of Mathematical Sciences
Tel Aviv University
Tel Aviv 69978
Israel.

Arrangements, KZ Systems and Lie Algebra Homology

Eduard Looijenga

To Terry Wall, for his 60th birthday.

Introduction

The Knizhnik–Zamolodchikov equations were originally defined in terms of a local system associated to tuples of finite dimensional irreducible representations of SU_2, but were soon afterwards generalized to a Kac–Moody setting. The natural question that arises is whether these local systems admit a topological interpretation. A paper by Varchenko–Schechtman [8] comes close to answering this affirmatively and it is this article and related work that we intend to survey here.

Our presentation deviates at certain points from the original sources. First, we felt it worthwhile to introduce the notion of a Knizhnik–Zamolodchikov system (in (1.4)), whose value is enhanced by the simple criterion (1.5). Such systems also occur in the theory of root systems (both in a linear and in an exponential setting) and we took the occasion to discuss them briefly from our point of view. A KZ system leads to what is perhaps the most natural class of local systems on hyperplane complements endowed with a given extension over the whole space as vector bundle. That may already be sufficient reason for this notion to merit a more thorough investigation than we can give here.

Second, we treated the cohomology of hyperplane complements (due Arnol'd, Brieskorn and Orlik–Solomon) using the methods of sheaf theory, an approach we advocated on an earlier occasion for its effectiveness. Furthermore, we kept the statements free of genericity assumptions regarding the exponents; this applies in particular to the discussion in section 4. Finally, we avoided the use of Lie bialgebras structures and comodules in [8].

This is a slightly edited version of a manuscript written in the Spring of 1994. It is based on two talks given one year earlier in a seminar on hypergeometric functions in Utrecht.

1 Knizhnik–Zamolodchikov Systems

(1.1) Let U be a connected analytic manifold and F a complex vector space. The trivial vector bundle $F \times U \to U$ has $\mathcal{F} := \mathcal{O}_U \otimes F$ as its sheaf of holomorphic sections. A holomorphic connection on this vector bundle is given by an $\mathrm{End}(F)$-valued holomorphic 1-form $E = \sum_i \omega_i \otimes E_i \in \Omega^1(U) \otimes \mathrm{End}(F)$:

$$D_E : \mathcal{O}_U \otimes F \to \Omega^1_U \otimes F, \quad \phi \otimes v \mapsto d\phi \otimes v + \sum_i \phi\omega_i \otimes E_i(v).$$

It is customary to extend D_E to a derivation of degree 1 in $\Omega^\bullet_U \otimes F$ by

$$D_E : \Omega^k_U \otimes F \to \Omega^{k+1}_U \otimes F, \quad \omega \otimes v \mapsto d\omega \otimes v + \sum_i \omega_i \wedge \omega \otimes E_i(v).$$

Then

$$D_E D_E(\omega \otimes v) = \sum_i d\omega_i \wedge \omega \otimes E_i(v) + \sum_{i,j} \omega_i \wedge \omega_j \wedge \omega \otimes E_i E_j(v).$$

So $D_E D_E$ is $\mathcal{O}_U$-linear and is equal to wedging with $dE + E \wedge E \in \Omega^2(U) \otimes \mathrm{End}(F)$. The last expression is called the *curvature* of the connection D_E. If it is constant zero, we say that D_E is *flat*; then the local sections of the vector bundle annihilated by D_E define a local system $\mathbb{F} \subset \mathcal{O}_U \otimes F$ and $(\Omega^\bullet_U, D_E)$ becomes a complex of coherent sheaves which resolves $\mathbb{F}$; this is the De Rham complex of $\mathbb{F}$.

In case U is an algebraic variety there is defined a special class of holomorphic differentials on U, the logarithmic forms: If $\overline{U} \supset U$ is a smooth completion of U which adds to U a normal crossing divisor D, then a regular differential ω on U is said to be *logarithmic* if it defines a section of $\Omega_{\overline{U}}(\log D)$; it is well-known that this property is independent of the choice of the completion. According to Deligne [3] such a differential is always closed. If instead of being a normal crossing divisor, D is *arrangement-like*, that is, in local-analytic coordinates given by a product of linear forms $f_1, \dots, f_k$, then a regular differential on U is logarithmic if and only if in any such coordinate patch it is locally a linear combination combination of the logarithmic forms $f_1^{-1}df_1, \dots, f_k^{-1}df_k$ with analytic coefficients.

We say that the connection E (on the trivial vector bundle over U with fiber F) is *logarithmic* if its coefficients are. For such connections there is the following flatness criterion:

(1.2) Proposition. *Suppose that $\overline{U} \supset U$ is a smooth completion of U which adds to U an arrangement-like divisor D whose irreducible components D_α are smooth. Suppose that $\overline{U}$ has no nonzero regular 2-forms and that and any irreducible component D_α has no nonzero regular 1-forms. Then a logarithmic connection E on U is flat if and only if for every for every intersection I of two distinct irreducible components of D, the sum $\sum_{D_\alpha \supset I} \mathrm{Res}_{D_\alpha} E$ commutes with each of its terms $\mathrm{Res}_{D_\alpha} E$ ($D_\alpha \supset I$).*

Proof. A straightforward local calculation shows that the last property is equivalent to the vanishing of any double residue of $E \wedge E$ along any I. So it is certainly a necessary condition for flatness, but it is also sufficient: if this is the case for every I, then the residue of $E \wedge E$ along any D_α has as coefficients regular differentials on D_α. Since there are no nonzero everywhere regular differentials on any D_α it follows that $E \wedge E$ is regular everywhere. As there are no nonzero regular 2-forms on $\overline{U}$, we have in fact that $E \wedge E = 0$.

We remark that if we add to a flat logarithmic connection a logarithmic differential (which acts on $\mathcal{O}_U \otimes F$ via the first factor), then the resulting connection is still logarithmic and flat.

The examples we shall deal with involve a projective space P and a finite collection $\mathcal{C}$ of hyperplanes in P. Our base manifold U is then $P - \cup_{H \in \mathcal{C}} H$. From (1.2) we immediately conclude:

(1.3) Corollary. *Let E define a logarithmic connection D_E on $\mathcal{O}_U \otimes F$. Then this connection is flat if and only if for every codimension two intersection I of members of $\mathcal{C}$, the sum $\sum_{H \in \mathcal{C}, H \supset I} \operatorname{Res}_H E$ commutes with each of its terms $\operatorname{Res}_H E$ $(H \in \mathcal{C}, H \supset I)$.*

(1.4) If the equivalent conditions of the corollary are fulfilled we say that the quadruple $(P, \mathcal{C}, F, E)$ (or simply E) defines a *Knizhnik–Zamolodchikov (KZ) system*. A flat local section of a KZ system (F, E) is given by an F-valued holomorphic map on an open subset of U; the composite of such a map with a linear form on F is called a *hypergeometric function.*

Usually a KZ system will be given in terms of an affine space V and a finite collection $\mathcal{C}$ of hyperplanes in V with $U = V - \cup_{H \in \mathcal{C}} H$. Then by adding to V the hyperplane at infinity we find ourselves in the previous case. Every $H \in \mathcal{C}$ determines a logarithmic differential

$$\omega_H := \frac{df_H}{f_H},$$

where f_H is an affine-linear form whose zero set is H. This is a closed holomorphic form on U which does not depend on the choice of f_H. The ω_H's make up a basis of the logarithmic differentials on U, so that any logarithmic connection on $\mathcal{O}_U \otimes F$ is of the form $E = \sum_{H \in \mathcal{C}} \omega_H \otimes E_H$ with $E_H \in \operatorname{End}(F)$. So:

(1.5) Corollary. *The connection D_E is flat if and only if for every codimension two intersection I of members of $\mathcal{C}$, the sum $\sum_{H \in \mathcal{C}, H \supset I} E_H$ commutes with each of its terms E_H $(H \in \mathcal{C}, H \supset I)$.*

(1.6) Here are two important classes of examples.

1. (Heckman–Opdam–Dunkl) Let V and $\mathcal{C}$ be as above and let G be a finite subgroup of $GL(V)$ which preserves $\mathcal{C}$. (A case in point is when

G is a finite complex reflection group and $\mathcal{C}$ is the collection of fixed point hyperplanes of complex reflections in G.) We take for F the group algebra $\mathbb{C}[G]$ of G (acting on itself by left multiplication) and for every $H \in \mathcal{C}$ we choose $E_H \in \mathbb{C}[Z_G(H)]$ (where $Z_G(H)$ is the group of $g \in G$ which fix H pointwise) such that $E_{g(H)} = gE_Hg^{-1}$ for all $H \in \mathcal{C}$ and $g \in G$. It is clear that then for a codimension two intersection I of reflection hyperplanes the element $\sum_{H \in \mathcal{C}, H \supset I} E_H$ is central in $\mathbb{C}[Z_G(I)]$, so that $E := \sum_H \omega_H \otimes E_H$ defines a KZ system.

The monodromy group of this local system will take values in the group of units of $\mathbb{C}[G]$. So if F is a finite dimensional complex representation of G, then F together with the image of E in $\Omega_U \otimes \mathrm{End}(F)$ defines another such system. Notice that these local systems come with a natural G-action and therefore descend to local systems on a Zariski open subset of the G-orbit space of V.

2. Let $\mathfrak{g}$ be a finite dimensional complex Lie algebra and B a nondegenerate symmetric bilinear form on $\mathfrak{g}$. (Such a form exists if and only if $\mathfrak{g}$ is reductive.) Let $B^\vee \in \mathfrak{g} \otimes \mathfrak{g}$ represent the inverse form on the dual of $\mathfrak{g}$ (so if $(e_k)_k$ is an orthonormal basis of $\mathfrak{g}$, then $B^\vee = \sum_k e_k \otimes e_k$). The Lie bracket is defined by a map $\mathfrak{g} \otimes \mathfrak{g} \to \mathfrak{g}$. If we denote its adjoint by $u : \mathfrak{g} \to \mathfrak{g} \otimes \mathfrak{g}$, then we have the following identities in $U\mathfrak{g} \otimes U\mathfrak{g}$:

$$-[X \otimes 1, B^\vee] = [1 \otimes X, B^\vee] = u(X).$$

For $1 \leq k, l \leq 3$, $k \neq l$, let $E_{k,l} \in \mathfrak{g}^{\otimes 3}$ be obtained by inserting $B^\vee$ in the slots k and l. The $\mathfrak{g}$-invariance of B amounts to the property that the expression $B(X, [Y, Z])$ is anti-symmetric in its three arguments. If we dualize, we find that $[E_{\sigma(1),\sigma(2)}, E_{\sigma(2),\sigma(3)}]$ is anti-symmetric in $\sigma \in \mathfrak{S}_3$. This implies that the sum $E_{1,2} + E_{2,3} + E_{1,3}$ commutes with each of its terms.

Now for $n \geq 2$, let V_n be the subspace of $\mathbb{C}^n$ defined by $z_1 + \cdots + z_n = 0$, let $\mathcal{C}_n$ be the collection of hyperplanes $z_k - z_l = 0$ $(1 \leq k < l \leq n)$ and let U_n be the complement in V_n of the union of these. Denote the logarithmic form associated to $z_k - z_l = 0$ by $\omega_{k,l}$ and let $E_{k,l} \in U\mathfrak{g}^{\otimes n}$ have the obvious meaning. Then it follows from the preceding remarks that $E := \sum_{1 \leq k < l \leq n} \omega_{k,l} \otimes E_{k,l} \in B^\vee_{U_n} \otimes U\mathfrak{g}^{\otimes n}$ defines a KZ system. If $M_1, \ldots, M_n$ are representations of $\mathfrak{g}$, then the same is true for the image of E in $B^\vee_U \otimes \mathrm{End}(M_1 \otimes \cdots \otimes M_n)$. The equations defining the flat sections of such systems are known as the *Knizhnik–Zamolodchikov equations.*

Since $B^\vee$ is a $\mathfrak{g}$-invariant element (of $\mathfrak{g}^{\otimes 2}$), so are the $E_{k,l}$'s. This implies that the $\mathfrak{g}$-action on the fibres sends flat sections to flat sections. So if

$$M_1 \otimes \cdots \otimes M_n = \oplus_S F_S \otimes S$$

is a decomposition into isotypical components (with S running over a system of mutually inequivalent irreducible representations of $\mathfrak{g}$), then by Schur's lemma the connection actually takes values in $\oplus_S \mathrm{End}(F_S)$. Hence for each

S, we have a KZ system on the trivial vector bundle with fiber $F_S = \mathrm{Hom}_{\mathfrak{g}}(S, M_1 \otimes \cdots \otimes M_n)$.

An interesting case is $\mathfrak{g} = \mathfrak{gl}(W)$ (where W is a finite dimensional vector space) and $M_i = W$ for all i. A classical theorem of Weyl asserts that $W^{\otimes n}$ decomposes as a direct sum of irreducible $GL(W) \times \mathfrak{S}_n$-modules with multiplicity one. This implies that each F_S is an irreducible $\mathfrak{S}_n$-module. It can be shown that its KZ system is of the type discussed in example 1.

2 Intermezzo: The Toric Logarithmic System Associated to a Root System

(2.1) There is also a toric version of the KZ connection, which is due to Cherednik and Matsuo. Although this will not play a role here we cannot resist the temptation to give it a similar treatment. Let R be a finite, possibly nonreduced root system in a real vector space V with Weyl group W and $L \subset V$ a W-invariant lattice containing the roots. Then $T := \mathrm{Hom}(L, \mathbb{C}^*)$ is an algebraic torus with W-action. Every $\alpha \in R$ determines a character $\chi_\alpha : T \to \mathbb{C}^*$. We denote the kernel of this character by D_α. Our logarithmic connection will be defined on $U := T - \cup_{\alpha\in R} D_\alpha$ and will have values in $\mathbb{C}[W]$. In order to be able to apply (1.2) we need a completion of T. We take (as the most natural choice) the torus embedding $T \subset \hat{T}$ defined by the decomposition of $\mathrm{Hom}(\mathbb{C}^*, T) \otimes_{\mathbb{Z}} \mathbb{R} \cong V^\vee$ into chambers. This completion need not be smooth, but it admits a smooth Galois cover and that is good enough for our purpose. It adds to T a normal crossing divisor (or rather an orbit space of such a divisor under a finite group action) whose irreducible components are bijectively indexed by the one-dimensional facets in $V^\vee$. Every one-dimensional facet contains a unique indivisible element of L, so if $\Pi \subset L$ denotes the collection of such elements, then for every $p \in \Pi$ we have an irreducible component D^p of $\hat{L} - L$. Then D_α, $\alpha \in R$ meets D^p, $p \in \Pi$ if and only if $\alpha(p) = 0$ and in that case no other irreducible component of $\hat{T} - U$ passes through $D_\alpha \cap D^p$. For every root $\alpha \in R$ we define two logarithmic differentials on U:

$$\eta_\alpha := \chi_\alpha^* \left(\frac{dt}{t}\right), \quad \omega_\alpha := \chi_\alpha^* \left(\frac{(t+1)dt}{(t-1)t}\right).$$

A simple calculation shows that these elements have the following residues:

$$\mathrm{Res}_{D^p} \eta_\alpha = \alpha(p), \quad \mathrm{Res}_{D^p} \omega_\alpha = -|\alpha(p)|, \quad \mathrm{Res}_{D_\alpha} \omega_\alpha = 2;$$

all other residues vanish. We fix a chamber C in $V^\vee$ so that we have system of positive roots R^+ and we define for $\alpha \in R$, $\epsilon_\alpha \in \mathrm{End}\,\mathbb{C}[W]$ by

$$\epsilon_\alpha(w) = \begin{cases} w & \text{if } w^{-1}\alpha \in R^+, \\ -w & \text{otherwise.} \end{cases}$$

These endomorphisms possess the following properties:

(i) $\epsilon_\alpha + \epsilon_{-\alpha} = 0$,

(ii) for $w \in W$, $w\epsilon_\alpha w^{-1} = \epsilon_{w(\alpha)}$.

(2.2) Proposition. *If $k \in R \to k_\alpha \in \mathbb{C}$ is a W-invariant function, then*

$$E := \sum_{\alpha\in R} k_\alpha(\omega_\alpha \otimes s_\alpha + \eta_\alpha \otimes s_\alpha\epsilon_\alpha)$$

defines a flat logarithmic W-equivariant connection on the trivial bundle over U with fiber $\mathbb{C}[W]$.

Proof. The W-invariance is clear from property *(ii)* above. Since $\hat{T}$ is rational, it has no regular 2-form. For the same reason, no irreducible component of $\hat{T} - U$ has a regular 1-form. We verify the criterion (1.2) along every intersection I of two distinct irreducible components of $\hat{T} - U$. First note that

$$\mathrm{Res}_{D_\alpha} E = 2k_\alpha s_\alpha,$$
$$\mathrm{Res}_{D^p} E = \sum_{\beta\in R} k_\beta(-|\beta(p)|s_\beta + \beta(p)s_\beta\epsilon_\beta).$$

If I is of the form $D_\alpha \cap D_\beta$, then the argument of example 1 applies. If I is of the form $D_\alpha \cap D^p$, then we need to show that $\mathrm{Res}_{D_\alpha} E = s_\alpha$ commutes with $\mathrm{Res}_{D^p} E$. But this follows from the W-invariance of E and the fact that $s_\alpha(p) = p$. There remains the case when I is of the form $D^p \cap D^{p'}$. This intersection is nonempty only if p and p' are in the closure of a single chamber; since E is W-invariant, we may as well assume that this chamber is C. Then

$$\mathrm{Res}_{D^p} E = \sum_{\beta\in R^+} -2k_\beta\beta(p)s_\beta(1-\epsilon_\beta).$$

and there is a similar expression for $\mathrm{Res}_{D^{p'}} E$. Now, if β is a positive root, then $(1-\epsilon_\beta)(w) = 0$ unless $w^{-1}\beta \in R^-$. We can express this in a better way if we write $w \in W$ as the reduced expression in the simple reflections $w = s_{\alpha_1}s_{\alpha_2}\cdots s_{\alpha_l}$, where $\alpha_1, \dots, \alpha_l$ are simple roots relative C. Then $(1 - \epsilon_\beta)(w) = 0$ unless there is a (unique) $i \in \{1, \dots, l\}$ with β a positive multiple of $s_{\alpha_1}\dots s_{\alpha_{i-1}}(\alpha_i)$ and in that case $(1-\epsilon_\beta)(w) = 2s_{\alpha_1}\cdots\widehat{s_{\alpha_i}}\cdots s_{\alpha_l}$. Hence

$$\mathrm{Res}_{D^p} E(w) = \sum_{i=1}^{l} -4\tilde{k}_{\alpha_i}s_{\alpha_1}\dots s_{\alpha_{i-1}}(\alpha_i)(p)s_{\alpha_1}\cdots\widehat{s_{\alpha_i}}\cdots s_{\alpha_l}.$$

where $\tilde{k}_{\alpha_i}$ denotes the sum of the k_β's for which β is a positive multiple of α_i. This formula gives us for the commutator $[\mathrm{Res}_{D^{p'}} E, \mathrm{Res}_{D^p} E]$ applied to w a linear combination of the elements $s_{\alpha_1}\cdots\widehat{s_{\alpha_i}}\cdots\widehat{s_{\alpha_j}}\cdots s_{\alpha_k}$; the coefficient of that element is proportional to the value in (p, p') of the 2-form

$$s_{\alpha_1}\cdots s_{\alpha_{i-1}}\left(\alpha_i \wedge (1-s_{\alpha_i})s_{\alpha_{i+1}}\cdots s_{\alpha_{j-1}}\alpha_j\right).$$

Since the image of $1 - s_{\alpha_i}$ is spanned by α_i, this 2-form is clearly zero. Hence Res_{D^p} and $\mathrm{Res}_{D^{p'}}$ commute. This completes the proof.

(2.3) Remarks.

1. Matsuo actually proves the flatness of a more general connection: he allows an additional term in $\Omega(T) \otimes \operatorname{End} \mathbb{C}[W] = \operatorname{Hom}(\mathbb{C}[W], \Omega(T) \otimes \mathbb{C}[W])$ which depends on a choice of $\lambda \in V \otimes \mathbb{C}$: if $d\lambda$ is the translation invariant differential on T corresponding to λ, then it is given by $w \in W \mapsto wd\lambda \otimes w \in \Omega(T) \otimes \mathbb{C}[W]$. In a sense this generalization is only apparent, for on the universal cover $\operatorname{Hom}_{\mathbb{R}}(V, \mathbb{C})$ of T this new connection is equivalent to the old one by means of the endomorphism of $\mathcal{O}_T \otimes \mathbb{C}[W]$ which multiplies w by $w\lambda$.

2. Since the algebra $\mathbb{C}[W]$ does not contain the endomorphims ϵ_α, the monodromomy group will no longer take values in the group of units of $\mathbb{C}[W]$. So there is no obvious way to descend the connection to any representation of W. Heckman and Opdam have however identified the monodromy group in terms of an affine Hecke algebra and so one expects the connection to descend to certain representations of this algebra.

3 Cohomology of Arrangements

(3.1) Let V be a finite dimensional affine space over the complex numbers and let $\mathcal{C}$ be a collection of affine hyperplanes in V. We denote

$$D := \cup_{H \in \mathcal{C}} H, \quad U := V - D$$

and name their inclusions in V $i : D \subset V$ and $j : U \subset V$. A *facet* of $\mathcal{C}$ is a nonempty intersection of members of $\mathcal{C}$ (this includes V); the collection of codimension p-facets is denoted by $\mathcal{C}^p$. We give a resolution $\mathcal{F}^\bullet$ of the sheaf $i_*\mathbb{Z}_D$ which generalizes a standard construction in the normal crossing case: denote by $D^{[p]}$ the disjoint union of the p-facets of $\mathcal{C}$, by $i_p : D^{[p]} \to V$ be the obvious map and by $U^{[p]}$ the difference of $i_p(D^{[p]})$ and $i_p(D^{[p+1]})$ (but we regard $U^{[p]}$ also as a subset of $D^{[p]}$). Let $\mathcal{F}^1 := \mathbb{Z}_{D^{[1]}}$. Then we have an inclusion $i_*\mathbb{Z}_D \subset i_{1*}\mathcal{F}^1$ whose cokernel $\mathcal{F}^{2'}$ is supported by the image of $D^{[2]}$. This cokernel is constant on each connected component of $U^{[2]}$ and so the direct image of $\mathcal{F}^{2'}|U^{[2]}$ on $D^{[2]}$ is a sheaf $\mathcal{F}^2$ which is constant on every component L of $D^{[2]}$. There is an obvious morphism $i_{1*}\mathcal{F}^1 \to i_{2*}\mathcal{F}^2$ whose kernel is $i_*\mathbb{Z}_D$ and whose cokernel is supported by the image of i_3. It is clear that we can continue in this way.

Here is a more precise description. Define a *p-flag* of $\mathcal{C}$ as a chain

$$L^\bullet = (L^p \subset L^{p-1} \subset \cdots \subset L^0 = V), \quad p = 0, 1, \ldots$$

with $L^i \in \mathcal{C}^i$. Let F^p be the free abelian group generated by the p-flags of $\mathcal{C}$ modulo the relations

$$\sum_{L \in \mathcal{C}^i, L^{i+1} \subset L \subset L^{i-1}} [L^p \subset \cdots \subset L^{i+1} \subset L \subset L^{i-1} \subset \cdots \subset L^0] = 0, \quad 0 < i < p.$$

We turn this into a complex by putting

$$d([L^p \subset \cdots \subset L^0]) := \sum_{L \in \mathcal{C}^{p+1}, L \subset L^p} [L \subset L^p \subset \cdots \subset L^0].$$

If we restrict ourselves to those hyperplanes that contain a given $L \in \mathcal{C}^p$, then we shall write $F_L^\bullet$ instead. It is easy to see that the complex $\mathcal{F}^p|L$ is the constant sheaf $\mathbb{Z}_L \otimes F_L^p$.

(3.2) Proposition. *There is a natural isomorphism*

$$H^\bullet(V, D; \mathbb{Z}) \cong H^\bullet(F^\bullet, d).$$

Proof. If $\mathcal{I}$ is any sheaf on V, then there is a standard exact sequence

$$0 \to j_! j^* \mathcal{I} \to I \to i_* i^* \mathcal{I} \to 0.$$

If $i^*\mathcal{I}$ has the property that every local section is a sum of local sections with support contained in a single member of $\mathcal{C}$, then we can use the resolution $\mathcal{F}^\bullet$ to resolve $i_* i^* \mathcal{I}$, and we get an exact complex

$$0 \to j_! j^* \mathcal{I} \to \mathcal{I} \to i_{1*}(i_1^* \mathcal{I} \otimes \mathcal{F}^1) \to i_{2*}(i_2^* \mathcal{I} \otimes \mathcal{F}^2) \cdots.$$

If we regard this as a resolution of $j_! j^* \mathcal{I}$, then there is an associated spectral sequence

$$E_1^{p,q} = H^q(D^{[p]}, i^* \mathcal{I} \otimes \mathcal{F}^p) = \bigoplus_{L \in \mathcal{C}^p} H^q(L; \mathcal{I}|L) \otimes F_L^p \Rightarrow H^{p+q}(j_! j^* \mathcal{I}),$$

where by convention $i_0 : D^{[0]} = V$ and $\mathcal{F}_0 = \mathbb{Z}_V$. This applies in particular to the Godement resolution $\mathcal{I}^\bullet$ of the constant sheaf $\mathbb{Z}_V$. Then this spectral sequence becomes

$$E_1^{p,q} = \bigoplus_{L \in \mathcal{C}^p} H^q(L; \mathbb{Z}) \otimes F_L^p \Rightarrow H^{p+q}(V, D; \mathbb{Z}),$$

Since $H^q(L; \mathbb{Z}) = 0$ for $q \neq 0$, the spectral sequence degenerates at E_2 and the proposition follows.

We have also a dual use for the F^p's, which at the same time leads to a homological interpretation of F_L^p. Let U_L denote the complement in V of the union of hyperplanes in $\mathcal{C}$ that contain L.

(3.3) Proposition. *There is a natural isomorphism $H^p(U; \mathbb{Z}(p)) \cong (F^p)^\vee$. If $L \in \mathcal{C}^p$, then $(F^p)^\vee$ is naturally isomorphic to $H^p(U_L; \mathbb{Z}(p))$, so that the canonical map $H^p(U; \mathbb{Z}) \to \oplus_{L \in \mathcal{C}^p} H^p(U_L; \mathbb{Z})$ is an isomorphism.*

Proof. Let $\mathcal{I}^\bullet$ be the Godement resolution of $\mathbb{Z}_V$. Then a similar argument as in the proof of (3.2) shows that we have an exact complex of graded differential sheaves

$$\cdots \to i_{2*}(\mathrm{Hom}(\mathcal{F}^2, i_2^!\mathcal{I}^\bullet) \to i_{1*}(\mathrm{Hom}(\mathcal{F}^1, i_1^!\mathcal{I}^\bullet) \to \mathcal{I}^\bullet \to j_*j^*\mathcal{I}^\bullet \to 0.$$

The Thom isomorphism shows that $i_p^!\mathcal{I}^\bullet$ is quasi-isomorphic to the shifted (and twisted) constant sheaf $\mathbb{Z}_{D^{[p]}}[-2p](-p)$. This gives the spectral sequence

$$\begin{aligned} E_1^{-p,q} &= H^{-2p+q}(D^{[p]}, \mathrm{Hom}(\mathcal{F}^p, \mathbb{Z}(-p)) \\ &= \bigoplus_{L\in\mathcal{C}^p} \mathrm{Hom}(F_L^p, H^{-2p+q}(L;\mathbb{Z})(-p)) \Rightarrow H^{-p+q}(U;\mathbb{Z}). \end{aligned}$$

Since $E_1^{-p,q} = 0$ for $-2p+q \neq 0$, the spectral sequence degenerates at the E_1-term, and it follows that $H^p(U;\mathbb{Z}) \cong \oplus_{L\in\mathcal{C}^p}(F_L^p)^\vee(-p) = (F^p)^\vee(-p)$. If we apply this to the collection of hyperplanes passing through $L \in \mathcal{C}^p$, we find that $H^p(U_L;\mathbb{Z}) \cong (F_L^p)^\vee(-p)$.

This proposition shows that the cohomology of U is torsion free. By the universal coefficient theorem, the homology of U is then also torsion free. It can be shown that each F_L^p is torsion free, so that $F_L^p \cong H_p(U_L;\mathbb{Z}(-p))$.

(3.4) Just as in the normal crossing case, there is another description of $H^p(U)$ in terms of logarithmic differentials. It goes as follows. The logarithmic differential ω_H defined in (1.4) is a closed form on $U - H$ whose period over a circle generating $H_1(V-H;\mathbb{Z})$ is equal to $\pm 2\pi\sqrt{-1}$. Hence ω_H defines a generator of $H^1(V-H, \mathbb{Z}(-1))$. Let $A^\bullet$ be the $\mathbb{Z}$-algebra of meromorphic forms generated by the ω_H's. It consists of closed forms which are holomorphic on U so that we have a natural algebra homomorphism

$$A^\bullet \to \oplus_p H^p(U;\mathbb{Z}(p)).$$

According to Brieskorn this is an isomorphism (we shall give the proof below). The two descriptions are related through an iterated residue map: any p-flag $L^\bullet$ defines a linear form $\mathrm{Res}_{L^\bullet}$ on A^p by taking iterated residues along the flag: for $p \geq 1$ we let

$$\mathrm{Res}_{L^\bullet} : \omega \in A^p \mapsto \mathrm{Res}_{L^p}\mathrm{Res}_{L^{p-1}}\cdots\mathrm{Res}_{L^1}(\omega) \in \mathbb{Z},$$

and by convention Res_V is the identity map of $\mathbb{Z}$. A straightforward calculation shows that this factors through a linear map

$$\mathrm{Res}^p : A^p \to (F^p)^\vee.$$

Its composite with the isomorphism $(F^p)^\vee \cong H^p(U;\mathbb{Z}(p))$ yields the map $A^p \to H^p(U;\mathbb{Z}(p))$ above.

(3.5) Proposition. *For every p the maps $A^p \to H^p(U;\mathbb{Z}(p))$ and Res^p : $A^p \to (F^p)^\vee$ are isomorphisms.*

Proof. The two assertions are equivalent so that it is enough to prove one of them. We proceed with induction on $\dim V$; so we assume that $\dim V \geq 1$ and the assertions proved for arrangements in affine spaces of smaller dimension.

For $L \in \mathcal{C}^p$, let $A^\bullet_L$ be the algebra associated to the collection of $H \in \mathcal{C}$ that contain L. So if $p = \operatorname{codim} L$, then A^p_L is the subgroup of A^p spanned by the $\omega_{H1} \wedge \cdots \wedge \omega_{H_p}$ with $H_1 \cap \cdots \cap H_p = L$. Clearly, A^p is spanned by the A^p_L with $L \in \mathcal{C}^p$. Since A^p_L maps to $(F^p_L)^\vee$, it is enough to show that $A^p_L \to (F^p_L)^\vee$ is an isomorphism. This follows from our induction hypothesis unless $p = \dim V$. We therefore concentrate on the case that $p = \dim V$ and $\cap_{H \in \mathcal{C}} H$ is a singleton $\{0\}$. Choose $H \in \mathcal{C}$ and let $f_H : V \to \mathbb{C}$ be an affine-linear form whose zero set is H. Put $V_1 := f^{-1}(1)$. With an obvious interpretation of notation we have an isomorphism $U_1 \times (\mathbb{C} - \{0\}) \to U$, $(z,t) \mapsto tz$. By inductive assumption, there are is a natural isomorphism $A^{p-1}_1 = (F^p_1)^\vee$. We have an exact sequence

$$0 \to F^p_1 \to F^p \to F^{p-1}_1 \to 0,$$

where the last map assigns to a p-flag $\{0\} = L^p \subset \cdots \subset L^0$ the $(p-1)$-flag $L^{p-1} \cap V_1 \subset \cdots L^0 \cap V_1$. This parallels the sequence

$$0 \to A^{p-1}_1 \xrightarrow{\omega_H \wedge} A^p \to A^p_1 \to 0,$$

where the last map is given by restriction. It is a simple exercise to show that this sequence is exact also. There is an obvious morphism from this exact sequence to the dual of the previous one. By induction hypothesis this is an isomorphism on the extremal terms. The five lemma implies that we have an isomorphism on the middle term.

4 Twisted Coefficients

If $(V, \mathcal{C}, F, E)$ defines a KZ system, then $A^\bullet(\mathcal{C}) \otimes F$ is a distinguished subcomplex of the De Rham complex of $\mathbb{F}$ with the differential D_E operating as wedging with E. It is possible to give a topological interpretation of the cohomology of this subcomplex. As we will only need this when F of dimension one, we shall restrict ourselves to that case. In that case E is given by a logarithmic differential for which we shall write η instead.

It will be best to work on the projective completion of V and so we may as well start out this way: until further notice P is a complex projective space, $\mathcal{C}$ a finite collection of hyperplanes of P, D the union of the hyperplanes $H \in \mathcal{C}$, $U := P - D$, F a one-dimensional complex vector space and η a logarithmic differential which is regular on U. The the associated rank one local system on $U = P - D$ is denoted by $\mathbb{F}_\eta$.

(4.1) There is a natural way to blow up P such that D transforms into a normal crossing divisor. Let us say that a facet L of $\mathcal{C}$ is *abnormal* if the projective space of hyperplanes containing L has a projective basis of $(1+\operatorname{codim} L)$ members of $\mathcal{C}$. It is clear that the set of abnormal subspaces is closed under intersection. Such subspaces are the centers of a blow-up $\pi: \tilde{P} \to P$ which has the property that the pre-image of D is a normal crossing divisor: blow up first the abnormal points; then the proper transforms of the abnormal lines become separated, blow these up and notice that the abnormal planes get separated and continue in this way. We end up with a blow-up $\pi: \tilde{P} \to P$ that resolves D to a normal crossing divisor $\tilde{D}$. The following proposition is implicit in a paper by Esnault–Schechtman–Viehweg [4].

(4.2) Proposition. *Let $\Phi(\eta)$ be the family of the sets $Z \subset U$ that are closed in U and have the property that their closure in P is disjoint from any abnormal facet L with the property that $\sum_{H\supset L; H\in\mathcal{C}} \operatorname{Res}_H \eta$ is a positive integer. This is a family of supports and there is natural isomorphism*

$$H^\bullet(A^\bullet, E\wedge) \to H^\bullet_{\Phi(\eta)}(V; \mathbb{F}_\eta).$$

Proof. Let $\Omega^\bullet_{\tilde{P}}(\log Y)$ be the logarithmic holomorphic De Rham complex of $(\tilde{P}, Y)$. This is a complex of free $\mathcal{O}_{\tilde{P}}$-modules which represents the total direct image of the sheaf of complex constants on $\tilde{P} - Y \cong U$. Notice that the elements of A^p define global sections of this complex. Following Deligne [3], the spectral sequence defined by the logarithmic De Rham resolution

$$E_1^{p,q} = H^q(\tilde{P}, \Omega^p_{\tilde{P}}(\log \tilde{D})) \Rightarrow H^{p+q}(U; \mathbb{C})$$

degenerates at the E_1-term. This means that we have a decreasing filtration $F^\bullet$ (the Hodge filtration) on $H^\bullet(U;\mathbb{C})$ such that

$$Gr^p_F H^n(U;\mathbb{C}) = H^{n-p}(\tilde{P}, \Omega^p_{\tilde{P}}(\log \tilde{D})).$$

By (3.5), $A^p \otimes \mathbb{C}$ maps isomorphically onto $H^p(U;\mathbb{C})$. As $H^0(\tilde{P}, \Omega^p_{\tilde{P}}(\log \tilde{D})) = H^0(P, \Omega^p_P(\log D)) = A^p \otimes \mathbb{C}$ it follows that $H^q(\tilde{P}, \Omega^p_{\tilde{P}}(\log \tilde{D})) = 0$ for $q \neq 0$ and $H^0(\tilde{P}, \Omega^p_{\tilde{P}}(\log \tilde{D})) = A^p \otimes \mathbb{C}$. We now invoke another result of Deligne [1] which we state as follows:

Proposition. *Let X be a complex manifold, $Y \subset X$ a normal crossing divisor, and η a closed global section of $\Omega^1_X(\log Y)$. Let $Y' \subset Y$ be the union of irreducible components Y_i of Y for which $\operatorname{Res}_{Y_i} \eta$ is a positive integer and denote the inclusions $k: X - Y \to X - Y'$ and $k': X - Y' \subset X$. If $\mathbb{F}_\eta$ denotes the local system over $X - Y$ defined by the structure sheaf with flat connection η, then $(\Omega^\bullet_X(\log Y), d + \eta\wedge)$ represents $R^\bullet k'_! R^\bullet k_* \mathbb{F}_\eta$.*

This is not quite the way it is stated, but this is what is proved.

So if in our case, $\tilde{D}'$, $\tilde{k}: U \subset \tilde{P} - \tilde{D}'$ and $\tilde{k}': \tilde{P} - \tilde{D}' \subset \tilde{P}$ have corresponding meanings, then we have a spectral sequence whose E_1-term is the same as in the constant case, but whose differentials are different:

$$E_1^{p,q} = H^q(\tilde{P}, \Omega^p_{\tilde{P}}(\log \tilde{D})) \Rightarrow H^{p+q}(R^\bullet \tilde{k}'_! R^\bullet \tilde{k}_* \mathbb{F}_\eta).$$

From the preceding calculation we deduce that this sequence degenerates at the E_2-term and that $E_2^{0,p} = H^p(A^\bullet, \eta\wedge) = H^p(R^\bullet \tilde{k}'_! R^\bullet \tilde{k}_* \mathbb{F}_\eta)$. The image D' of $\tilde{D}'$ in P is the union of abnormal facets L for which E_L is a positive integer; if $k : U \subset P - D'$ and $k' : P - D' \subset P$ are the inclusions, then $H^p(R\tilde{k}'_! R\tilde{k}_* \mathbb{F}_\eta) \cong H^p(Rk'_! R\tilde{k}_* \mathbb{F}_\eta) \cong H^p_{\Phi(E)}(U; \mathbb{F}_\eta)$.

(4.3) We now return to the affine case and assume that P arises from a projective completion of an affine space V.

Given $L \in \mathcal{C}^p$, set

$$\tilde{s}_\eta(L) := \sum_{H \supset L, H \in \mathcal{C}} \mathrm{Res}_H(\eta)\omega_H \in A^1 \otimes \mathbb{C}$$

and define for a flag $L^\bullet$ of positive length

$$\tilde{s}_\eta(L^\bullet) = \tilde{s}_\eta(L^1) \wedge \tilde{s}_\eta(L^2) \wedge \cdots \wedge \tilde{s}_\eta(L^p) \in A^p \otimes \mathbb{C}.$$

We complete this definition by putting $\tilde{s}_\eta(L^0) = 1$. One verifies easily that $\tilde{s}_\eta$ factors through a homomorphism $s_\eta : F^p \to A^p \otimes \mathbb{C}$ and that this factor is a chain map

$$s_\eta : (F^\bullet, d) \to (A^\bullet \otimes \mathbb{C}, \eta\wedge).$$

So s_η induces a homomorphism $H^\bullet(V, D; \mathbb{Z}) \to H^\bullet_{\Phi(\eta)}(U; \mathbb{F}_\eta)$. It is desirable to have a reasonable topological interpretation of this map, especially for its image.

If we identify $A^\bullet$ with the dual of $F^\bullet$, we end up with a bilinear form

$$S_\eta : F^p \otimes F^p \to \mathbb{C}.$$

Concretely, if $L^\bullet$ and $L'^\bullet$ are p-flags, then

$$S_\eta(L^\bullet, L'^\bullet) = \sum_{(H_1,\ldots,H_p),\sigma} \mathrm{sign}(\sigma)\eta_{H_1} \cdots \eta_{H_p},$$

where the sum is over all ordered p-tuples $(H_1, \ldots, H_p)$ in $\mathcal{C}$ and $\sigma \in \mathfrak{S}_p$ with the property that $L^k = H_1 \cap \cdots \cap H_k$ and $L'^k = H_{\sigma(1)} \cap \cdots \cap H_{\sigma(k)}$ for $k = 1, \cdots, p$. (Notice that given $(H_1, \ldots, H_p)$, there can be at most one $\sigma \in \mathfrak{S}_p$ with these properties, so that the sum is really over a set of p-tuples in $\mathcal{C}$.) From this formula we see that S_η is symmetric.

5 Higher Direct Images of a KZ System

(5.1) We continue with the situation of the previous section. In addition we assume given a linear subspace T of the translation space of V. We denote the quotient affine space $T\backslash V$ by V' and the projection map $V \to V'$ by π. The members of $\mathcal{C}$ that are T-invariant define a a collection $\mathcal{C}'$ of hyperplanes of W. We shall make the assumption that every facet of $\mathcal{C}$ projects onto a facet of C'; this assumption can always be made to hold by adding to C' the hyperplanes that are images of facets of C.

We have a corresponding algebra $A'^{\bullet}$ of logarithmic forms which we often regard as a subalgebra of $A^{\bullet}$. We form the algebra of relative forms:

$$A^{\bullet}_{\pi} := A^{\bullet}/(A'^{1} \wedge A^{\bullet-1}).$$

Lemma. *The sequence*

$$0 \to A'^{1} \otimes (A^{\bullet-1}_{\pi}, \eta\wedge) \xrightarrow{\wedge} (A^{\bullet}, \eta\wedge) \to (A^{\bullet}_{\pi}, \eta\wedge) \to 0$$

is a short exact sequence of complexes.

Proof. (Sketch) We have to show that the map $\wedge : A'^{1} \otimes A^{\bullet-1}_{\pi} \to A'^{1} \wedge A^{\bullet-1}_{\pi}$ is injective. So let $\sum_{H \in C'} \omega_H \wedge \zeta_H = 0$, with $\zeta_H \in A^p$. Fix $H_0 \in C'$. Taking the residue along H_0 yields an an identity

$$\zeta_{H_0}|H_0 - \sum_{H \in C' - H_0} \omega_H \wedge \operatorname{Res}_{H_0} \zeta_H = 0.$$

For $H \neq H_0$, $\operatorname{Res}_{H_0} \zeta_H$ is the restriction of some $\xi_H \in A^{p-1}$. The difference of ζ_{H_0} and $\sum_{H \in C' - \{H_0\}} \omega_H \wedge \xi_H$ is then an element of A^p with trivial restriction to H_0. It is then not hard to show that this expression is in the ideal of $A^{\bullet}$ spanned by the ω_H with H parallel to H_0. In particular, $\zeta_{H_0} \in A'^{1} \wedge A^{p-1}$.

The long exact cohomology sequence for this short exact sequence has a differential

$$H^p(A^{\bullet}_{\pi}, \eta\wedge) \to A'^{1} \otimes H^p(A^{\bullet}_{\pi}, \eta\wedge),$$

which, when regarded as an element of $A'^{1} \otimes \operatorname{End}(H^p(A^{\bullet}_{\pi}))$, is denoted by E_η. Since the square of the differential is zero, we have $E_\eta \wedge E_\eta = 0$, so that E_η defines a KZ system; we will refer to this as the *direct image KZ system defined by* η.

The underlying local system has a topological interpretation. For this we projectively complete V in the T-direction. This gives a (trivial) bundle of projective spaces $\tilde{\pi} : \tilde{V} \to V'$. The closures of the members of C plus the $\mathbb{P}(T)$-bundle at infinity define a "projective arrangement $\tilde{C}$ over V'".

Much of what we did in the absolute case carries over this relative case: Let U' denote the complement of the union of the members of C' in V' and let $\mathbb{H}^p_\eta$ denote the sheaf on U' associated to the presheaf which assigns to every open $B \subset U''$ the cohomology group $H^p_{\Phi(\eta)|\pi^{-1}B \cap U}(\pi^{-1}B \cap U; \mathbb{F}_\eta)$. This sheaf is locally constant and the underlying holomorphic vector bundle $\mathcal{O}_{U'} \otimes \mathbb{H}^p_\eta$ has the following algebraic description: Let $\mathcal{A}^p_\pi$ denote the sheaf on U' of relative logarithmic p-forms; (so that a section of $\mathcal{A}^p_\pi$ over B is a relative meromorphic p-form on $\tilde{\pi}^{-1}B$, with logarithmic poles along the members of $\tilde{C}$). Both the exterior derivative and wedging with η makes $\mathcal{A}^{\bullet}_\pi$ a complex of sheaves. The two operations anti-commute and so we have a complex $(\mathcal{A}^{\bullet}_\pi, d + \eta\wedge)$.

(5.2) Proposition. *There is a natural isomorphism*

$$\mathcal{H}^p(\mathcal{A}^\bullet_\pi, d+\eta\wedge) \cong \mathcal{O}_{U'} \otimes \mathbb{H}^p_\eta.$$

The proof is similar to that of (4.2).

Notice that every element of $A^\bullet_\pi$ defines a global section of $\mathcal{A}^\bullet_\pi$. This restriction homomorphism

$$r^\bullet : (A^\bullet_\pi, \eta\wedge) \to (H^0(U'; \mathcal{A}^\bullet_\pi), d+\eta\wedge)$$

is a morphism of complexes and so induces homomorphisms

$$H^p(r^\bullet) : H^p(A^\bullet_\pi, \eta\wedge) \to H^0(U'; \mathcal{H}^p(\mathcal{A}^\bullet_\pi, d+\eta\wedge)) \cong H^0(U'; \mathcal{O}_{U'} \otimes \mathbb{H}^p_\eta).$$

(5.3) Proposition. *The maps $r^\bullet$ and $H^p(r^\bullet)$ are injections.*

Proof. (Sketch) Let $\zeta \in A^p$ map to an element of the kernel of r^p. This implies that for every $y \in U''$, the restriction of η to the fiber $\tilde{V}_y$ is exact. This restriction is a logarithmic form on a projective variety. According to Deligne [3] such a form can be exact only when it is zero. So locally over U', ζ is in the ideal generated by A'^1. A computation shows that this this even true locally over all of V'. Then η can be written as $\sum_{H\in C'} \omega_H \wedge \zeta_H$ with ζ_H a linear combination of elements of A^{p-1} with coefficients polynomial functions on V'. If we replace these polynomial functions by their constant term then the corresponding linear combination is still equal to ζ (we are in effect taking the degree zero part of an equation), and so $\zeta \in A'^1 \wedge A^{p-1}$. This shows that $r^\bullet$ is injective. For the injectivity of $H^p(r^\bullet)$ one proceeds in the same way: if $\zeta \in A^p$ can be written as $\eta\wedge\xi + \sum_{H\in C'} \omega_H \wedge \zeta_H$ with ξ and ζ_H linear combinations of elements of A^{p-1} with polynomial functions on V' as coefficients, then by taking the degree zero part we find a solution inside A^{p-1} itself.

Note that in this discussion the residues of η along the members of C' do not matter.

It is likely that a weak Lefschetz theorem holds which asserts that $H^p(r^\bullet)$ is an isomorphism onto $H^0(U'; \mathcal{H}^p_\eta)$ for $p < \dim T$.

(5.4) Proposition. *The natural flat connection on $\mathcal{O}_{U'} \otimes \mathbb{H}^p_\eta$ induces the KZ connection E_η on $H^p(A^\bullet_\pi, \eta)$.*

Proof. The flat connection $\mathcal{O}_{U'} \otimes \mathbb{H}^p_\eta$ transported to $\mathcal{H}^p(\mathcal{A}^\bullet_\pi, d+\eta\wedge)$ is given by a map

$$D : \mathcal{H}^p(\mathcal{A}^\bullet_\pi, d+\eta\wedge) \to \Omega^1_{U'} \otimes \mathcal{H}^p(\mathcal{A}^\bullet_\pi, d+\eta\wedge)$$

which can be described as follows (see for instance [2]): a section of $\mathcal{H}^p(\mathcal{A}^\bullet_\pi, d+\eta\wedge)$ over a Stein coordinate patch $(B; z_1, z_2, \ldots, z_k)$ of U' is representable by a logarithmic form ζ on $\tilde{\pi}^{-1}B$ with the property that $d\zeta + \eta\wedge\zeta = \sum_{i=1}^k dz_i \wedge$

ζ_i for certain logarithmic forms ζ_i on $\tilde{\pi}^{-1}B$. Then ζ_i maps to a cocycle of $(\mathcal{A}^p_\pi, d + \eta\wedge)$ and

$$D([\zeta]) = \sum_{i=1}^{k} dz_i \otimes [\zeta_i].$$

The proposition now follows easily.

6 Embedding a Lie Algebra Complex in a Flag Complex

In this section U_r is the free associative algebra over $\mathbb{C}$ on r variables $f_1, \ldots, f_r$ and $L_r \subset U_r$ the (free) complex Lie algebra generated by these elements. The algebra U_r is multigraded by $\mathbb{Z}^r_{\geq 0}$ and L_r inherits this grading.

(6.1) Let $\tilde{V}_r$ be a complex vector space of dimension r with coordinates $t_1, \ldots, t_r$. Put $z = -\sum_i t_i$, and let V_r denote the quotient of $\tilde{V}_r$ by the line $t_1 = \cdots = t_r$. Let $\tilde{\mathcal{C}}_r$ be the collection of hyperplanes in $\tilde{V}$ having an equation $z = t_i$ or $t_i = t_j$, and $\mathcal{C}_r$ those of V_r having an equation by $t_i = t_j$. We identify p-flags of $\mathcal{C}_r$ with p-flags of $\tilde{\mathcal{C}}_r$ so that $F^\bullet(\mathcal{C}_r)$ can be regarded as a graded submodule of $F^\bullet(\tilde{\mathcal{C}}_r)$. For a facet L of $\tilde{\mathcal{C}}_r$ and $i \in \{1, \ldots r\}$, define $\epsilon(i, L)$ to be 1 resp. -1 if the number of hyperplanes of the form $z = t_j, j > i$ which contain L is even resp. odd. We define an action of U_r on $F^\bullet(\tilde{\mathcal{C}}_r)$ by letting f_i act as "left multiplication" with $\{z = t_i\}$ up to sign:

$$f_i[L^p \subset \cdots \subset L^0] := \begin{cases} 0 & \text{if } L^p \subset (z = t_i), \\ \epsilon(i, L^p)[(z = t_i) \cap L^p \subset L^p \subset \cdots \subset L^0] & \text{otherwise.} \end{cases}$$

Then we have a linear mapping

$$\tilde{\psi} : U_r \to F^\bullet(\tilde{\mathcal{C}}_r), \quad u \mapsto u[L_0]$$

which preserves the (total) degree. Notice that for $i > j$, $[f_i, f_j]$ is left multiplication with $-[(z = t_i = t_j) \subset (t_i = t_j)]$. With induction we see that a k-fold commutator $[f_{i_{k+1}}, [f_{i_k}, [\cdots, f_{i_1}] \cdots]$ with $i_{k+1} > i_k > \cdots > i_1$ acts as left multiplication with $(-1)^k$ times the $(k+1)$-flag

$$[(z = t_{i_{k+1}} = \cdots = t_{i_1}) \subset (t_{i_{k+1}} = \cdots = t_{i_1}) \subset (t_{i_k} = \cdots = t_{i_1}) \subset \cdots \subset (t_{i_2} = t_{i_1})],$$

and a permutation σ of these elements has the effect of multiplying the flag by the sign of σ. So if we omit the smallest term we obtain an element of $F^k(\mathcal{C}_r)$. This defines a linear map

$$\psi : L_r \to F^\bullet(\mathcal{C}_r) \otimes \mathbb{C}$$

of degree -1.

(6.2) Let $V_{r,n}$ be a complex vector space of dimension $r+n-1$ and $\{z_1,\dots,z_n;t_1,\dots,t_r\}$ a set of generators of its dual whose sum is 0. Let $\mathcal{C}_{r,n}$ be the collection of hyperplanes of the form $z_\nu = t_i$ and $t_i = t_j$. There is an obvious map $\pi_{r,n}: V_{r,n} \to V_r$ and a corresponding embedding $\mathcal{C}_r \subset \mathcal{C}_{r,n}$ and consequently a linear map $\psi : L_r \to F^{r-p}(\mathcal{C}_{r,n})\otimes\mathbb{C}$. Substituting z_ν for z defines an embedding $\tilde{\mathcal{C}}_r \subset \mathcal{C}_{r,n}$ and hence a linear map $\tilde{\psi}_\nu : U_r \to F^\bullet(\mathcal{C}_{r,n})\otimes\mathbb{C}$. Now consider the linear map

$$(L_r^{\otimes p}\otimes U_r^{\otimes n})_{(1,1,\dots,1)} \to F^{r-p}(\mathcal{C}_{r,n})\otimes\mathbb{C}$$

defined by the rule

$$g_p\otimes\cdots\otimes g_1\otimes u_1\otimes\cdots\otimes u_n \mapsto \psi(g_p)\cdots\psi(g_1)\tilde{\psi}_1(u_1)\cdots\tilde{\psi}_n(u_n).$$

If the $g_i's$ are homogeneous, then interchanging g_i and g_{i+1} makes the image acquire the sign $(-1)^{(\deg(g_i)-1)(\deg(g_{i+1})-1)}$. We wish to modify the definition of this mapping so that this change in sign becomes -1 instead, thus making the modified map factorize through $(\wedge^p L_r\otimes U_r^{\otimes n})_{(1,1,\dots,1)}$. To this end, assume that $g_p,\dots,g_1$ resp. $u_1,\dots,u_n$ are Lie-monomials resp. noncommutative monomials in $f_1,\dots,f_r$. If $(g_p\otimes\cdots\otimes g_1\otimes u_1\otimes\cdots\otimes u_n)$ has multidegree $(1,\dots,1)$ and $f_1,\dots,f_n$ appear in the order $f_{\sigma(r)},\dots,f_{\sigma(1)}$, then we multiply the above expression with $\mathrm{sign}(\sigma)(-1)^{\sum_i i(\deg(g_i)-1)}$. This has indeed the desired effect with respect to interchanging adjacent g_i's so that we now have a well-defined map

$$\Psi_{r,n} : (\wedge^\bullet L_r\otimes U_r^{\otimes n})_{(1,1,\dots,1)} \to F^{r-\bullet}(\mathcal{C}_{r,n})\otimes\mathbb{C}.$$

(6.3) Claim. *This is a $\mathfrak{S}_r$-equivariant isomorphism of graded vector spaces.*

Proof. This is left to the reader. For equivariance, one checks that $\Psi_{r,n}$ commutes with interchanging the indices i and $i+1$.

Recall that for any L_r-module M one has the standard chain complex

$$C_\bullet(L_r,M) = (\wedge^\bullet L_r\otimes M,\partial),$$

whose differential is defined by

$$\partial(g_p\wedge\cdots\wedge g_1\otimes m) = \sum_{i=1}^{p}(-1)^{i-1}g_p\wedge\cdots\wedge\widehat{g_i}\wedge\cdots\wedge g_1\otimes g_i m$$
$$+\sum_{1\le i<j\le p}^{p}(-1)^{i+j}g_p\wedge\cdots\wedge\widehat{g_j}\wedge\cdots\wedge\widehat{g_i}\wedge\cdots\wedge g_1\wedge[g_j,g_i]\otimes m.$$

If M is multigraded by $\mathbb{Z}^r_{\ge 0}$ in a way that is compatible with the L_r-action (think of $U_r^{\otimes n}$), then the differential respects the multigrading of the complex. So both source and target of $\Psi_{r,n}$ are complexes. One can check that $\Psi_{r,n}$ commutes with the differentials. The proof is a matter of careful bookkeeping of the signs and so we omit it. Thus:

(6.4) Proposition. *We have an $\mathfrak{S}_r$-equivariant isomorphism of complexes*

$$\Psi_{r,n} : (C_\bullet(L_r, U_r^{\otimes n})_{(1,1,\ldots,1)} \to F^{r-\bullet}(\mathcal{C}_{r,n}) \otimes \mathbb{C}.$$

With this isomorphism constructed, we can identify the other multigraded pieces of $C_\bullet(L_r, U_r^{\otimes n})$ with flag complexes as well. Given $\lambda = (\lambda_1, \ldots, \lambda_r) \in \mathbb{Z}^r_{\geq 0}$, let U_λ be the free associative algebra on generators $f_{i,j}$, $i = 1, \ldots, r$, $j = 1, \ldots, \lambda_i$. We define an embedding $U_r \to U_\lambda$ by mapping the generator f_i to $\sum_{j=1}^{\lambda_i} f_{i,j}$. This identifies U_r with the part of U_λ that is fixed under the obvious action of $\mathfrak{S}_\lambda := \mathfrak{S}_{\lambda_1} \times \cdots \times \mathfrak{S}_{\lambda_r}$. Similarly, we have an isomorphism

$$(C_\bullet(L_r, U_r^{\otimes n})_\lambda \cong (C_\bullet(L_\lambda, U_\lambda^{\otimes n})^{\mathfrak{S}_\lambda}_{(1,1,\ldots,1)}.$$

In view of (6.4) and the results of section 3 we get:

(6.5) Theorem. *For every multidegree $\lambda \in \mathbb{Z}^r_{\geq 0}$ we have a natural isomorphism of complexes*

$$\Psi_{r,\lambda} : (C_\bullet(L_r, U_r^{\otimes n})_\lambda \cong F^{|\lambda|-\bullet}(\mathcal{C}_{\lambda,n})^{\mathfrak{S}_\lambda} \otimes \mathbb{C},$$

(where $\mathcal{C}_{\lambda,n}$ has the obvious interpretation) and (dually) an isomorphism of complexes

$$\Psi^\vee_{r,\lambda} : A^\bullet(\mathcal{C}_{\lambda,n})^{\mathfrak{S}_\lambda} \otimes \mathbb{C} \cong (C^\bullet(L_r, ((U_r)^\vee)^{\otimes n})_{-\lambda}.$$

The preceding construction is somewhat unsatisfactory since it appears to come out of the blue. One would like to see a more conceptual interpretation of this isomorphism.

7 A KZ System Attached to a Class of Lie Algebras

We begin by briefly reviewing the beginnings of the theory of Kac–Moody Lie algebras (in fact, of an even more general class of Lie algebras). The reader will find more information in the book by Kac [5].

First a remark concerning graded duals: if $V = \oplus_{a \in A} V_a$ is a vector space graded by an abelian group A, then its *graded dual* $V^\vee$ is defined as the direct sum of the duals of its homogeneous components with the understanding that the dual of V_a has degree $-a$. So if the homogeneous components V_a are finite dimensional, then the double graded dual of V is canonically isomorphic to V.

(7.1) We start out with a finite dimensional complex vector space $\mathfrak{h}$, a nondegenerate symmetric bilinear form B on $\mathfrak{h}$ and linearly independent forms $\alpha_1, \ldots, \alpha_r \in \mathfrak{h}^\vee$ (called the *fundamental roots*). Let $h_i \in \mathfrak{h}$ be characterized by the property that $B(h_i, h) = \alpha_i(h)$ for all $h \in \mathfrak{h}$ and let $\tilde{\mathfrak{g}}$ be the Lie algebra generated by $\mathfrak{h}$ and indeterminates $e_1^+, \dots e_r^+, e_1^-, \dots e_r^-$ subject to the relations

$$[e_i^+, e_j^-] = \delta_{ij} h_i,$$
$$[h, e_i^\epsilon] = \epsilon\alpha_i(h) e_i^\epsilon,$$
$$[h, h'] = 0$$

for all $h, h' \in \mathfrak{h}$ and $\epsilon \in \{\pm\}$. The elements $e_1^\epsilon, \dots, e_n^\epsilon$ generate a free Lie subalgebra $\tilde{\mathfrak{n}}_\epsilon$ and we have a *triangular decomposition*

$$\tilde{\mathfrak{g}} = \tilde{\mathfrak{n}}_- \oplus \mathfrak{h} \oplus \tilde{\mathfrak{n}}_+.$$

The adjoint action of $\mathfrak{h}$ on $\tilde{\mathfrak{g}}$ is semi-simple. Clearly, the weights that occur in $\mathfrak{n}_-$ resp. $\mathfrak{n}_+$ are negative resp. positive integral linear combinations of the α_i's so that the triangular decomposition is refined by the *weight decomposition*

$$\tilde{\mathfrak{g}} = \oplus_\lambda \tilde{\mathfrak{g}}_\lambda.$$

Notice that if we replace B by a nonzero scalar multiple we get an isomorphic Lie algebra.

(7.2) The form B extends (uniquely) to $\tilde{\mathfrak{g}}$ as a $\tilde{\mathfrak{g}}$-invariant symmetric bilinear form $\tilde{B}$. In particular, $\tilde{\mathfrak{g}}_\lambda$ and $\tilde{\mathfrak{g}}_\mu$ are perpendicular unless $\lambda + \mu = 0$. One checks easily that $\tilde{B}(e_i^+, e_j^-) = \delta_{ij}$. The *Chevalley involution* $\tilde{\tau} : \tilde{\mathfrak{g}} \to \tilde{\mathfrak{g}}$ is defined as minus the identity on $\mathfrak{h}$ and sends $e_i^\pm$ to $-e_i^\mp$. So $\tilde{\tau}$ interchanges $\mathfrak{g}_\lambda$ and $\mathfrak{g}_{-\lambda}$. It is symmetric with respect to $\tilde{B}$, so that

$$\tilde{S}(g, g') := -\tilde{B}(\tilde{\tau} g, g')$$

is a symmetric bilinear form on $\tilde{\mathfrak{g}}$. Now the weight decomposition $\tilde{\mathfrak{g}} = \oplus_\lambda \tilde{\mathfrak{g}}_\lambda$ is orthogonal with respect to $\tilde{S}$. Observe that $\tilde{S}$ is not $\mathfrak{g}$-invariant but satisfies

$$\tilde{S}([g, g_1], g_2) + \tilde{S}(g_1, [\tilde{\tau}(g), g_2]) = 0.$$

We put

$$\tilde{\mathfrak{b}}_+ := \mathfrak{h} \oplus \tilde{\mathfrak{n}}_+.$$

The nilspace of $\tilde{S}$ is also the nilspace of $\hat{B}$ and is therefore a graded ideal. The quotient algebra, denoted $\mathfrak{g}$, inherits the grading of $\tilde{\mathfrak{g}}$:

$$\mathfrak{g} := \oplus_\lambda \mathfrak{g}_\lambda,$$

the forms $\tilde{B}$ resp. $\tilde{S}$ descend to a nondegenerate forms B resp. S on $\mathfrak{g}$ and $\tilde{\tau}$ induces an involution τ of $\mathfrak{g}$. The grading is orthogonal with respect S and the triangular decomposition of $\tilde{\mathfrak{g}}$ determines one of $\mathfrak{g}$:

$$\mathfrak{g} := \mathfrak{n}_- \oplus \mathfrak{h} \oplus \mathfrak{n}_-.$$

Notice that B induces a pairing $\mathfrak{n}_- \times \mathfrak{n}_+ \to \mathbb{C}$. This pairing is perfect in the graded sense, so that $\mathfrak{n}_+$ can be identified with the graded dual of $\mathfrak{n}_-$.

Remark. This class of algebras contains the Kac–Moody algebras associated to a symmetrizable Cartan matrix, in particular it contains the finite dimensional complex reductive Lie algebras. To see this, assume that $B(h_i, h_i) \neq 0$ for all i and consider the matrix $(n_{ij} := 2B(h_i, h_j)/B(h_i, h_i))_{i,j}$. If this is a *generalized Cartan matrix* in the sense that its off-diagonal entries are nonnegative integers with $n_{ij} = 0 \Leftrightarrow n_{ji} = 0$, then $\mathfrak{g}$ is one of its associated Kac-Moody algebras. This follows from a theorem of Gabber and Kac which states that in this situation the nilspace of $\tilde{S}$ is generated by the "Serre elements" $(\mathrm{ad} e_i^\epsilon)^{-n_{ij}+1}(e_j^\epsilon)$ $(i \neq j)$. Every symmetrizable generalized Cartan matrix arises this way. If (n_{ij}) is a Cartan matrix, then $\mathfrak{g}$ is a finite dimensional reductive Lie algebra and the image of $\mathfrak{h}$ in it is a Cartan subalgebra.

The projection $\tilde{\mathfrak{b}}_+ \to \mathfrak{h}$ is a Lie algebra homomorphism and so every $\mu \in \mathfrak{h}^*$ defines a one-dimensional representation $\mathbb{C}_\mu$ of $\tilde{\mathfrak{b}}_+$. The induced module

$$\tilde{M}(\mu) := U\tilde{\mathfrak{g}} \otimes_{U\tilde{\mathfrak{b}}_+} \mathbb{C}_\mu$$

is called the Verma module of $\tilde{\mathfrak{g}}$ with highest weight μ. Since we have a factorization $U\tilde{\mathfrak{g}} = U\tilde{\mathfrak{n}}_- \otimes U\tilde{\mathfrak{b}}_+$, $M(\mu)$ is canonically isomorphic to $U(\tilde{\mathfrak{n}}_-)$ as a $\tilde{\mathfrak{n}}_-$-module. Notice that $U(\tilde{\mathfrak{n}}_-)$ is a free associative algebra on $e_1^-, \ldots, e_r^-$. Since $\mathfrak{h}$ acts semi-simply on $U\mathfrak{g}$, it does so on $\tilde{M}(\mu)$. Whence an eigen space decomposition:

$$\tilde{M}(\mu) = \oplus_{\pi \in \mathfrak{h}^*} \tilde{M}(\mu)_\pi$$

with $\tilde{M}(\mu)_\pi = 0$ unless $\pi \in \mu + \mathbb{Z}_{\leq 0}\alpha_1 + \cdots \mathbb{Z}_{\leq 0}\alpha_r$. We equip $M(\mu)$ with a bilinear form, also denoted by $\tilde{S}$, which is characterized by the property that $\tilde{S}(1 \otimes 1, 1 \otimes 1) = 1$ and $\tilde{S}(gx, y) + \tilde{S}(x, \tilde{\tau}(g)y) = 0$ for all $g \in \mathfrak{g}$ and $x, y \in \tilde{M}(\mu)$. One verifies that $\tilde{S}$ is symmetric and that the grading of $\tilde{M}(\mu)$ is orthogonal with respect to $\tilde{S}$. There is an associated adjoint map

$$\tilde{s} : \tilde{M}(\mu) \to \tilde{M}(\mu)^\vee,$$

which maps each homogeneous summand to its dual. If we twist the contragradient representation on $\tilde{M}(\mu)^\vee$ with the Chevalley involution, then this map becomes a $\tilde{\mathfrak{g}}$-homomorphism. The image of $\tilde{s}$ may be identified with the Verma module for $\mathfrak{g}$ with highest weight μ:

$$M(\mu) := U\mathfrak{g} \otimes_{U\mathfrak{b}_+} \mathbb{C}_\mu,$$

in other words, $\tilde{s}$ induces a nondegenerate form s on $M(\mu)$.

(7.3) We now fix weights $\mu_1, \ldots, \mu_n \in \mathfrak{h}^*$, and consider the $\tilde{\mathfrak{g}}$-module

$$\tilde{M} := \tilde{M}(\mu_1) \otimes_{\mathbb{C}} \cdots \otimes_{\mathbb{C}} \tilde{M}(\mu_n).$$

We restrict the module structure to $\tilde{\mathfrak{n}}_-$ and form the standard chain complex $(\wedge^\bullet \tilde{\mathfrak{n}}_- \otimes \tilde{M}, \partial)$. The symmetric bilinear forms $\tilde{S}$ defined on $\tilde{n}_-$ and the $\tilde{M}(\mu_i)$'s combine to give such a form (also denoted by $\tilde{S}$) on $\wedge^\bullet \tilde{\mathfrak{n}}_- \otimes \tilde{M}$. The associated adjoint map is

$$\tilde{s} : \wedge^\bullet \tilde{\mathfrak{n}}_- \otimes \tilde{M} \to (\wedge^\bullet \tilde{\mathfrak{n}}_- \otimes \tilde{M})^\vee.$$

(7.4) Lemma. *The image of* $\tilde{s} : \wedge^\bullet\tilde{\mathfrak{n}}_- \otimes \tilde{M} \to (\wedge^\bullet\tilde{\mathfrak{n}}_- \otimes \tilde{M})^\vee$ *is a quotient complex of of* $(\wedge^\bullet\tilde{\mathfrak{n}}_- \otimes \tilde{M}, \partial)$ *whose graded dual is canonically isomorphic to the standard cochain complex of* $\mathfrak{n}_+$ *with values in* M, $\mathrm{Hom}(\wedge^\bullet\mathfrak{n}_+, M)$. *This isomorphism multiplies the weights with* -1. *In particular, the weight* λ *piece of the pth homology group of the image of* $\wedge^\bullet\tilde{s} \otimes s$ *is dual to the weight* λ *piece of* $H^p(\mathfrak{n}_+, M)$.

Proof. From the preceding discussion it is clear that $\tilde{s}$ factorizes as

$$\wedge^\bullet\tilde{\mathfrak{n}}_- \otimes \tilde{M} \to \wedge^\bullet\mathfrak{n}_- \otimes M \cong \wedge^\bullet\tilde{\mathfrak{n}}_-^\vee \otimes M^\vee \to (\wedge^\bullet\tilde{\mathfrak{n}}_- \otimes \tilde{M})^\vee,$$

where the first map is surjective, the middle map is given by s and the last one is injective. So the image can be identified with $\wedge^\bullet\mathfrak{n}_- \otimes M$, which is indeed a quotient complex of $\wedge^\bullet\tilde{\mathfrak{n}}_- \otimes \tilde{M}$. Since B induces a perfect pairing between $\mathfrak{n}_-$ and $\mathfrak{n}_+$, the third term may be identified with $\wedge^\bullet\mathfrak{n}_+ \otimes M^\vee$. The resulting isomorphism $\wedge^\bullet\mathfrak{n}_- \otimes M \cong \wedge^\bullet\mathfrak{n}_+ \otimes M^\vee$ is an isomorphism of complexes if we give $M^\vee$ the (untwisted) contragradient representation. Notice that this isomorphism multiplies the weights with -1. The lemma follows from this.

(7.5) Let $\alpha \in \mathfrak{h}^*$ be a *root* of $\mathfrak{h}$ in $\mathfrak{g}$, that is, a nonzero weight for which $\mathfrak{g}_\lambda$ is nontrivial. The form B establishes a perfect pairing between the finite dimensional vector spaces $\mathfrak{g}_\alpha$ and $\mathfrak{g}_{-\alpha}$ and thus determines an element $B_\alpha^\vee \in \mathfrak{g}_\alpha \otimes \mathfrak{g}_{-\alpha}$. Since B is symmetric, interchanging the factors transforms $B_\alpha^\vee$ in $B_{-\alpha}^\vee$. We denote by $B_0^\vee \in \mathfrak{h} \otimes \mathfrak{h}$ the symmetric tensor that represents the inverse of $B|\mathfrak{h} \times \mathfrak{h}$ and we put

$$B^\vee := B_0^\vee + \sum_{\alpha \text{ a root}} B_\alpha^\vee,$$

considered as an element of a completed tensor product of $\mathfrak{g}$ with itself. In that sense it is $\mathfrak{g}$-invariant, just as in example 2 of (1.6) and for the same reason.

For $1 \leq k < l \leq n$, and α a root or zero, we let $(B_\alpha^\vee)_{k,l}$ denote the action of $B_\alpha^\vee$ on M via the kth and lth factor. Notice that if this action is nontrivial, then $\alpha \in \mu_k + \mathbb{Z}_{\leq 0}\{\alpha_1, \dots, \alpha_r\}$ and $-\alpha \in \mu_l + \mathbb{Z}_{\leq 0}\{\alpha_1, \dots, \alpha_r\}$. So this is the case for only finitely many α's. Therefore, there is a well-defined action of $B_{k,l}^\vee := \sum_\alpha (B_\alpha^\vee)_{k,l}$ on M. By letting $\wedge^\bullet\mathfrak{n}_+$ be inert, we extend this to an action on the standard cochain complex $\mathrm{Hom}(\wedge^\bullet\mathfrak{n}_+, M)$. Notice that $B_{k,l}^\vee$ is $\mathfrak{g}$-equivariant, respects the weights and commutes with the differential. So for every weight $\lambda \in \mathfrak{h}^*$, $B_{k,l}^\vee$ acts on $\mathrm{Hom}(\wedge^\bullet\mathfrak{n}_+, M)_\lambda$ and on $H^p(\mathfrak{n}_+, M)_\lambda$. Since $H^0(\mathfrak{n}_+, M)_\lambda$ can be identified with $\mathrm{Hom}_\mathfrak{g}(M(\lambda), M)$, this provides us with a generalization of example 2 of (1.6):

(7.6) Proposition. *The quadruple* $(V_n, \mathcal{C}_n, H^p(\mathfrak{n}_+, M)_\lambda, \sum_{1\leq k<l\leq n} \omega_{k,l} \otimes B_{k,l}^\vee)$ *defines a KZ system.*

The proof is similar. Our aim is to identify the KZ system defined by $H^\bullet(\mathfrak{n}_+, M)_\lambda$ with a summand of a direct image KZ system as discussed in section 5. This we do in the next section.

8 A KZ System as a Gauss-Manin System

We continue with the situation of the previous section. We put $\mu := \mu_1 + \cdots + \mu_n$ and we fix a nonnegative integral weight $\lambda = \lambda_1\alpha_1 + \cdots + \lambda_r\alpha_r$, $\lambda_i \in \mathbb{Z}_{\geq 0}$.

Define a logarithmic differential η on $V_{\lambda,n}$ relative to the collection $\mathcal{C}_{\lambda,n}$ by letting its residue along $(t_{i_p} = t_{j_q})$ resp. $(z_\nu = t_{i_p})$ be $B_0^\vee(\alpha_i, \alpha_j)$ resp. $-B_0^\vee(\mu_\nu, \alpha_{i_p})$. According to (4.3) this defines a symmetric bilinear form S_η on $F^\bullet(\mathcal{C}_{\lambda,n})^{\mathfrak{S}_\lambda} \otimes \mathbb{C}$. Recall that $\tilde{\mathfrak{n}}_-$ is a free Lie algebra on r generators, and note that we have a $\tilde{\mathfrak{n}}_-$-module isomorphism $U\tilde{\mathfrak{n}}_- \cong \tilde{M}(\mu)$ of weight μ. So (6.5) gives an isomorphism of complexes

$$\Psi_{r,\lambda} : C_\bullet(\tilde{n}_-, \tilde{M})_{\mu-\lambda} \cong F^{|\lambda|-\bullet}(\mathcal{C}_{\lambda,n})^{\mathfrak{S}_\lambda} \otimes \mathbb{C}.$$

(8.1) Proposition. *The pull-back of S_η under this isomorphism is $(-1)^{|\lambda|}S$.*

The proof is straightforward and so we omit it.

(8.2) Corollary. *The map $\Psi_{r,\lambda}$ induces an isomorphism of complexes between the image of s_η (which is a subcomplex of $(A^{|\lambda|-\bullet}(\mathcal{C}_{\lambda,n})^{\mathfrak{S}_\lambda} \otimes \mathbb{C}, \eta\wedge)$) and the dual of the Lie algebra cochain complex* $\mathrm{Hom}(\wedge^\bullet \mathfrak{n}_+, M)$. *In particular, we have isomorphisms*

$$H^{|\lambda|-p}(\mathrm{image}(s_\eta)) \cong (H^p(\mathfrak{n}_+, M)_{\mu-\lambda})^\vee.$$

Proof. This follows immediately from the previous proposition, (4.3) and (6.5).

We add to the collection $\mathcal{C}_{\lambda,n}$ the hyperplanes $(z_k = z_l)$, $k < l$. This arrangement is denoted $\mathcal{C}^*_{\lambda,n}$. The projection $\pi : (V_{\lambda,n}, \mathcal{C}^*_{\lambda,n}) \to (V_n, \mathcal{C}_n)$ puts us in the situation studied in section 5. We use the notation of that section and so $A^\bullet_\pi$ will stand for the relative logarithmic differentials:

$$A^\bullet_\pi := A^\bullet(\mathcal{C}^*_{\lambda,n})/A^1(\mathcal{C}_n) \wedge A^{\bullet-1}(\mathcal{C}^*_{\lambda,n}) \otimes \mathbb{C}.$$

There is an evident chain map

$$(A^{|\lambda|-\bullet}(\mathcal{C}_{\lambda,n}) \otimes \mathbb{C}, \eta\wedge) \to (A^\bullet_\pi, \eta\wedge).$$

In view of (8.2), we therefore have a natural map

$$\psi : (H^p(\mathfrak{n}_+, M)_{\mu+\lambda})^\vee \to H^{|\lambda|-p}(A^\bullet_\pi, \wedge\eta)^{\mathfrak{S}_r}.$$

In (5.2) we found a KZ connection E_η on $(V_n, \mathcal{C}_n)$ with values in the right hand side and in (7.6) we found a KZ connection $\sum_{1\leq k<l\leq n} \omega_{k,l} \otimes B^\vee_{k,l}$ on $(V_n, \mathcal{C}_n)$ with values in $H^p(\mathfrak{n}_+, M)_{\mu+\lambda})$ and hence also one with values in its dual. We twist the former by adding to E_η the logarithmic form η' on $(V_n, \mathcal{C}_n)$ whose residue along $z_k = z_l$, $k < l$, is $B_0^\vee(\Lambda_k, \Lambda_l)$.

(8.3) Theorem. *The pull back of $E_\eta + \eta'$ under ψ is just the dual of the KZ connection $\sum_{1\le k<l\le n} \omega_{k,l} \otimes B^\vee_{k,l}$ defined in (7.6). In particular, the KZ connection on U_r with values in $H^{|\lambda|}(A^\bullet_\pi, \wedge\eta)^{\mathfrak{S}_r}$ pulls back under ψ to the KZ connection with values in* $\mathrm{Hom}(M, M(\mu - \lambda))$.

We omit the proof. The point of this theorem is that it gives the KZ connection $\sum_{1\le k<l\le n} \omega_{k,l} \otimes B^\vee_{k,l}$ with values in $\mathrm{Hom}(M, M(\mu - \lambda))$ a certain topological content.

References

[1] P. Deligne, *Equations differentielles à points réguliers singuliers*, Lecture Notes in Math., 163, Springer Verlag, Berlin and New York, 1970.

[2] P. Deligne, *Travaux de Griffiths*, Sém. Bourbaki Exp. 376, Lecture Notes in Math., 180, Springer Verlag, Berlin and New York, 1971.

[3] P. Deligne, Théorie de Hodge II, *Inst. Hautes Études Sci. Publ. Math.*, **40** (1971), 5–58; III, *Inst. Hautes Études Sci. Publ. Math.*, **44** (1975), 6–77.

[4] H. Esnault, V. Schechtman, E. Viehweg, Cohomology of local sytems on the complement of hyperplanes, *Invent. Math.*, **109** (1992), 557–661; Erratum, *Invent. Math.*, **112** (1993), 447.

[5] V. Kac, *Infinite dimensional Lie algebras*, Cambridge University Press, Cambridge.

[6] V. Knizhnik, A. Zamolodchikov, Current algebras and Wess-Zumino models in two dimensions, *Nucl. Phys.*, **B 247** (1984), 83–103.

[7] A. Matsuo, Integrable connections related to zonal spherical functions, *Invent. Math.*, **110** (1992), 95–121.

[8] V.V Schechtman, A.N. Varchenko, Arrangements of hyperplanes and Lie algebra homology, *Invent. Math.*, **106** (1991), 139–194.

Faculteit Wiskunde en Informatica,
Rijksuniversiteit Utrecht,
PO Box 80.010,
3508 TA Utrecht,
The Netherlands.

e-mail: looijenga@math.ruu.nl

The Signature of $f(x,y) + z^N$

András Némethi*

Dedicated to C. T. C. Wall on the occasion of his 60th birthday.

If the analytic germ $g : (\mathbf{C}^3, 0) \to (\mathbf{C}, 0)$ defines an isolated singularity, then its Milnor fiber F_g is a 4–dimensional manifold whose boundary ∂F_g is diffeomorphic to the real link K_g of $(g^{-1}(0), 0)$. The signature of g, by definition, is the signature of the manifold F_g, i.e. it is the signature of the symmetric (real) intersection form $H_2(F_g)^{\otimes 2} \to \mathbf{R}$. The basic object of this note is a surface singularity g of the form $g = f(x,y) + z^N$, where f is an isolated curve singularity. The note contains some results about its signature (identities, inequalities). We also present some of its connections with other mathematical objects. One of them is the link K_g of g. In the appendix, we give an algorithm for the computation of the plumbing diagram of K_g (or equivalently, of the resolution graph of $(\{g = 0\}, 0)$). A general reference for the terminology is [3].

1 The signature of $f + z^N$ in terms of f and N

As with some other invariants of hypersurface germs, we would like to compute the signature of $f + z^N$ in terms of invariants of f and the integer N. For example, the Milnor number satisfies $\mu(f + z^N) = \mu(f) \cdot (N - 1)$, the classical Sebastiani–Thom theorem [47] for the monodromy operators h gives $h(f + z^N) = h(f) \otimes h(z \to z^N)$, and the generalized Sebastiani–Thom type theorems provide similar relations for the Seifert form (or variation map) [44], or for the spectrum and spectral pairs [45, 53]. Unfortunately, the signature is a more complicated invariant, in the sense that there is no formula which provides the signature of $f + z^N$ in terms of the signature of f and a universal object depending only on N. The main reason for this is that the signature of manifolds with boundary is not additive with respect to cutting and glueing.

On the other hand, obviously, f has invariants which together with the integer N do provide the signature $\sigma(f + z^N)$, for example since the signature is computable from the Hodge numbers [49], the Sebastiani–Thom formula for these [45, 53] provides such a formula. In the following subsections we will present such relations. Each of them has its own flavour.

*Partially supported by NSF Grant No. DMS-9622724 and OSU Seed Grant

1.1. The signature $\sigma(f+z^N)$ in terms of the eta–invariant (differential topology, index theory viewpoint)

Let F be the Milnor fiber of f, $H = H_1(F, \mathbf{C})$; $b : H \to H^*$ corresponds to the intersection form $\langle\, , \rangle : H \otimes H \to \mathbf{C}$ via $b(x)(y) = \langle x, y\rangle$; $h : H \to H$ is the monodromy operator and $V : H^* \to H$ is the variation map (here $H_1(F, \partial F)$ is identified with the dual space H^* via the perfect pair $H_1(F) \otimes H_1(F, \partial F) \to \mathbf{C}$). For any natural number K, we define $V(K) := (I+h+\cdots+h^{K-1}) \circ V$. Then the system $(H; b, h^K, V(K))$ has a spectral decomposition $\oplus_\chi(H_\chi; b_\chi, (h^K)_\chi, V(K)_\chi)$ with respect to the automorphism h^K (i.e. H_χ is the χ–generalized eigenspace of h^K), and the spectral decomposition is compatible with the extra-structure $(b, V(K))$.

For any χ, we define (cf. [22, 23]):

$$\eta(f;K)_\chi = \begin{cases} (1-2c)\cdot \text{signature}(ib_\chi) & \text{if } \chi = e^{2\pi i c}, 0 < c < 1; \\ -\text{signature}[\Big(I + (h^K)_\chi^{-1}\Big)V(K)_\chi] & \text{if } \chi = 1. \end{cases}$$

Now, for any $K > 0$, the eta–invariant of f is defined by: $\eta(f;K) = \sum_\chi \eta(f;K)_\chi$. In [22, 23] it is proved that:

Theorem. *With the notation introduced above:*

$$\sigma(f+z^N) = \eta(f;N) - N\cdot\eta(f;1).$$

This relation has a flavour of index theory: in the case of f with non-degenerate Milnor lattice (i.e. when $H_1 = 0$), the eta–invariant defined above is exactly the eta–invariant, in the sense of Atiyah-Patodi-Singer [6], of the signature operator of the unit circle with the flat bundle provided by the monodromy of f. In the definition of $\eta(f;K)_1$, the author was substantially influenced by the work of W. Neumann, mainly by [33]. The connections of the above eta–invariant with "Meyer's cocycles" and "Wall's non–additivity formula for the signature" is explained in [22, 23] (see also Neumann's paper [33]).

1.2. $\sigma(f+z^N)$ in terms of the imbedded resolution graph of f (algebraic geometry viewpoint)

Let $f : (\mathbf{C}^2, 0) \to (\mathbf{C}, 0)$ be as above. Consider an embedded resolution $\phi : (\mathcal{Y}, D) \to (\mathbf{C}^2, f^{-1}(0))$ of $(f^{-1}(0), 0) \subset (\mathbf{C}^2, 0)$ (here $D = \phi^{-1}(f^{-1}(0))$). Let $E = \phi^{-1}(0)$ be the exceptional divisor and $E = \cup_{w\in\mathcal{W}} E_w$ be its decomposition to irreducible divisors. If $f = \prod_{a\in\mathcal{A}} f_a$ is the irreducible decomposition of f, then $D = E \cup \cup_{a\in\mathcal{A}} S_a$, where S_a is the strict transform of $f_a^{-1}(0)$. Let G_f be the resolution graph of f, i.e. its vertices $\mathcal{V} = \mathcal{W} \coprod \mathcal{A}$ consist of the non-arrowhead vertices $\mathcal{W}$ (corresponding to the irreducible exceptional divisors), and arrowhead vertices $\mathcal{A}$ (correponding to the strict transform divisors of D). We will assume that no irreducible exceptional divisor has a self intersection and $\mathcal{W} \neq \emptyset$. If two irreducible divisors corresponding to

$v_1, v_2 \in \mathcal{V}$ have an intersection point then (v_1, v_2) $(= (v_2, v_1))$ is an edge of G_f. The set of edges is denoted by $\mathcal{E}$.

For any $w \in \mathcal{W}$, we denote by $\mathcal{V}_w$ the set of vertices $v \in \mathcal{V}$ adjacent to w. Set $\delta_w = \#\mathcal{V}_w$ for any $w \in \mathcal{W}$. If $\delta_w > 2$, then $w \in \mathcal{W}$ is called a "rupture point". The set of rupture points is denoted by $\mathcal{R}$.

The graph G_f is decorated by the self intersection numbers $e_w := E_w \cdot E_w$ for any $w \in \mathcal{W}$. For any $v \in \mathcal{V}$ let m_v be the multiplicity of $f \circ \phi$ along the irreducible divisor corresponding to v. For any $e = (v_1, v_2) \in \mathcal{E}$, we define $m_e := \gcd(m_{v_1}, m_{v_2})$, and for any $w \in \mathcal{W}$, we define $M_w := \gcd(m_v | v \in \mathcal{V}_w \cup \{w\})$.

In [25], the eta-invariant of f is computed in terms of its embedded resolution graph G_f. This gives:

Theorem. *Let f be as above. Then: $\sigma(f+z^N) = \eta(f;N) - N\cdot\eta(f;1)$, where*

$$\eta(f;K) = \#\mathcal{A} - 1 + \sum_{e\in\mathcal{E}}\Big((K,m_e)-1\Big) - \sum_{w\in\mathcal{W}}\Big((K,M_w)-1\Big) +$$

$$+4\cdot\sum_{w\in\mathcal{R}}\sum_{v\in\mathcal{V}_w}\sum_{k=1}^{m_w}((\frac{km_v}{m_w}))\cdot((\frac{kK}{m_w})).$$

The above result, in the particular case when the link of $f+z^N$ is an integer homology sphere (equivalently, if f is irreducible, and N is relatively prime with all the integers which form the Newton pairs of f), was obtained by W. Neumann and J. Wahl (in a slightly different form) in [37].

1.3. $\sigma(f+z^N)$ in terms of generalized Dedekind sums (arithmetic viewpoint)

For arbitrary non–zero integers a, b and c, the generalized Dedekind sum [41, 59] is defined as follows:

$$s(b,c;a) = \sum_{k=1}^{a-1}((\frac{kb}{a}))((\frac{kc}{a})),$$

where $((x))$ is defined via the fractional part $\{x\}$ as: $((x)) = \{x\} - 1/2$ if $x \notin \mathbf{Z}$, and $=0$ otherwise. Therefore, the last expression in the formula (1.2) can be replaced by:

$$\sum_{w\in\mathcal{R}}\sum_{v\in\mathcal{V}_w}\sum_{k=1}^{m_w}((\frac{km_v}{m_w}))\cdot((\frac{kN}{m_w})) = \sum_{w\in\mathcal{R}}\sum_{v\in\mathcal{V}_w} s(m_v, N; m_w).$$

The formula in (1.2) shows that the function $N \mapsto \sigma(f+z^N)$ is a sum of a periodic function (namely $\eta(f;N)$) and of a linear function (namely $-N\cdot\eta(f;1)$). This fact was conjectured by E. Brieskorn, A. Durfee and D. Zagier, and first proved by W. Neumann [32] (cf. also [33]).

Rademacher's famous generalization of the reciprocity law of Dedekind [41] asserts the following. If a, b, c are strict positive, mutually coprime integers, then: $s(b,c;a)+s(c,a;b)+s(b,a;c)=-1/4+(a^2+b^2+c^2)/12abc$. This relation for arbitrary, strict positive integers reads as:

$$s(b,c;a)+s(c,a;b)+s(b,a;c)=-\frac{(a,b,c)}{4}+\frac{a^2(b,c)^2+b^2(a,c)^2+c^2(b,a)^2}{12abc}.$$

This applied to our case (after some computation) gives the following ([26]).

Theorem. *For any $a\in\mathcal{A}$, let $w_a\in\mathcal{W}$ be the unique vertex such that $(a,w_a)\in\mathcal{E}$. Then:*

$$\sigma(f+z^N)=(N-1)\cdot\#\mathcal{W}-\sum_{w\in\mathcal{W}}\Big((N,M_w)-1\Big)-4\sum_{(u,v)\in\mathcal{E}}s(m_u,m_v;N)$$

$$+\frac{1}{3N}\cdot\sum_{w\in\mathcal{W}}(-e_w)\Big[(N,m_w)^2-N^2\Big]+\frac{1-N^2}{3N}\sum_{a\in\mathcal{A}}m_{w_a}.$$

Example. Assume that $N=2$. Then:

$$\sigma(f+z^2)=\#\{w\in\mathcal{W}:2\not|M_w\}-\frac{1}{2}\Big[\sum_{\substack{w\in\mathcal{W}\\ 2\not|m_w}}(-e_w)+\sum_{a\in\mathcal{A}}m_{w_a}\Big].$$

1.4. $\sigma(f+z^N)$ in terms of the spectral pairs of f
(Hodge theory viewpoint)

The set of spectral pairs $Spp(f)\in\mathbf{Z}[\mathbf{Q}\times\mathbf{N}]$ codifies the equivariant Hodge numbers $h^{p,q}_\lambda$ of the mixed Hodge structure of the vanishing cohomology [49, 45], namely:

$$Spp(f)=\sum_{(\alpha,\omega)}h^{1+[-\alpha],\,\omega+s_\alpha-1-[-\alpha]}_{\exp(-2\pi i\alpha)}\ (\alpha,\omega)$$

where $s_\alpha=0$ if $\alpha\notin\mathbf{Z}$ and $s_\alpha=1$ if $\alpha\in\mathbf{Z}$. The spectrum of f (or the set of spectral numbers) is defined by $Sp(f)=\sum(\alpha)\in\mathbf{Z}[\mathbf{Q}]$, where the sum is over the spectral pairs (α,ω).

By a result of J. Steenbrink [49], the set of spectral pairs determines the signature (actually all the equivariant signatures). On the other hand, it is easier to handle the set of spectral numbers, but unfortunately this invariant does not determine the signature, as is explained in [22, 24]. In the next theorem, we will separate two expressions. One of them depends only on the spectral numbers of f, the other involves some equivariant Hodge numbers as well. The second expression will vanish if f has finite monodromy.

Following [22], we define $\Sigma p_{\lambda,\pm}(f)$ to be

$$\#\{c\mid c\text{ is a spectral number of }f\text{ with }e^{-2\pi ic}=\lambda,\ \text{and }(-1)^{[c]}=\pm1\}.$$

Then the general result (5.21) [22], in the plane singularity case reads as:

Theorem. $\sigma(f+z^N)=\eta(f;N)-N\cdot\eta(f;1)$, *where:*

$$\eta(f;K)=\sum_{\substack{\lambda^K=1\\ \lambda\neq 1}} h^{11}_\lambda(f)-\sum_{\substack{\lambda^K\neq 1 \text{ or } \lambda=1\\ \lambda=e^{-2\pi i c};\,0\leq c<1}}\Big(1-2\{Kc\}\Big)\Big(\Sigma p_{\lambda,-}(f)-\Sigma p_{\lambda,+}(f)\Big).$$

This gives $\eta(f;K)$ explicitly in terms of the spectral pairs of f. The connection with the resolution graph is given by [46], where the spectral pairs are computed in terms of the embedded resolution graph G_f.

Note that the expression

$$\sum_{e\in\mathcal{E}}\Big((N,m_e)-1\Big)-\sum_{w\in\mathcal{W}}\Big((N,M_w)-1\Big) \tag{1.4.1}$$

(used in 1.2-1.3) is exactly $\sum_{\lambda^N=1,\lambda\neq 1}h^{11}_\lambda(f)$. But for $\lambda\neq 1$, the Hodge number $h^{11}_\lambda(f)$ is the number of Jordan blocks of the monodromy operator h of f with eigenvalue λ and size two (in the sequel denoted by $\#_2 h_\lambda(f)$). If f is irreducible then this number is zero by a result of Lê [18]. The fact that the collection of numbers introduced in (1.4.1) (considered for all integers N) provide the Jordan block structure of the monodromy of f, was first proved by W. Neumann [35, 36] (see also [10]).

1.5. $\sigma(f+z^N)$ in terms of the equivariant signatures of f (topological viewpoint)

In the spirit of Sebastiani–Thom type theorems, the most natural result would be to express $\sigma(f+z^N)$ in terms of the signature of f. As we already noted, this is not possible, but it is possible to express it (mostly) in terms of the equivariant signatures $\sigma_\lambda(f)$ of f. Let r_f be the number of irreducible components of f $(=\#\mathcal{A})$.

Theorem. [29]

$$\sigma(f+z^N)=-(r_f-1)(N-1)+\sum_{\lambda^N=1,\lambda\neq 1}\#_2h_\lambda(f)-$$

$$-2\sum_{\substack{\lambda\neq 1\\ \lambda=e^{2\pi i\alpha}}}\sigma_\lambda(f)\cdot\Big(((N\alpha))-N\cdot((\alpha))\Big).$$

The connection between this theorem and the previous ones is realized by Neumann's formula of $\sigma_\lambda(f)$ in terms of G_f [35].

Example. Let $g=x^2+y^3$. Then $r_f=1$, and $\sigma_\lambda(f)=\pm 1$ if $\lambda=\exp(\pm 2\pi i/6)$. Therefore:

$$\sigma(x^2+y^3+z^N)=-4\cdot\Big(((N/6))-N((1/6))\Big).$$

1.6. The particular case of irreducible germs f

If f is irreducible then we can make some simplifications in the above relations: $r_f = \#\mathcal{A} = 1$, and $\#_2h_\lambda(f) = h_\lambda^{11}(f) = 0$ for any eigenvalue λ. But what is really surprising, is that the signature of $f+z^N$ is a sum of signatures of Brieskorn singularities.

Theorem. [25] *Assume that f is irreducible with Newton pairs $(p_i, q_i)_{i=1}^s$. Define the integers $\{a_i\}_{i=1}^s$ by:*

$$a_1 = q_1, \text{ and } a_{i+1} = q_{i+1} + p_{i+1}p_i a_i \text{ if } i \geq 1.$$

Additionally, set $d_i = (N, p_{i+1}\cdots p_s)$ for $1 \leq i < s$ and $d_s = 1$. Then:

$$\sigma(f+z^N) = \sum_{i=1}^{s} d_i \cdot \sigma(a_i, p_i, N/d_i),$$

where $\sigma(a,b,c)$ is the signature of $x^a + y^b + z^c$.

It would be nice, and very helpful, to have a similar expression in the reducible case as well, but the author knows of no such a relation in general.

In the case when the link of f is an integer homology sphere, the above theorem was proved by Neumann and Wahl in [37]. In that case $d_i = 1$, and $\sigma(f+z^N) = \sum_i \sigma(a_i, p_i, N)$.

1.7. The particular case of Brieskorn singularities

For germs $g = \sum_{i=1}^3 z_i^{a_i}$, E. Brieskorn's beautiful result [7] reads as:

Theorem.

$$\sigma(\sum_{i=1}^{3} z_i^{a_i}) = \sum_{k_1=1}^{a_1-1}\sum_{k_2=1}^{a_2-1}\sum_{k_3=1}^{a_3-1} \text{sign}\,\sin\Big(\pi\sum_{j=1}^{3} k_j/a_j\Big).$$

Hence σ is connected to the following lattice point problem. For $t = 0, 1, 2$ set:

$$S_t := \#\{\mathbf{k} \in \mathbf{Z}^3 : 1 \leq k_j \leq a_j - 1, (1 \leq j \leq 3), t < \sum_{j=1}^{3} k_j/a_j < t+1\}.$$

Then: $\sigma(\sum_j z_j^{a_j}) = \sum_t (-1)^t S_t$. Actually, this formula is much simpler. First notice that $\#\{(k_1,k_2,k_3) \in \mathbf{Z}^3;\ \sum_j k_j/a_j \in \mathbf{Z}\} = \mu_1$, where μ_1 is the dimension of the 1–eigenspace of $H_2(\sum z_j^{a_j})$, hence by [19],

$$\mu_1 = \frac{(a_1,a_2)(a_2,a_3)(a_3,a_1)}{(a_1,a_2,a_3)} - (a_1,a_2) - (a_2,a_3) - (a_3,a_1) + 2.$$

Since $S_0 = S_2$ by the symmetry of the lattice points, $S_0+S_1+S_2 = \mu_{\neq 1}(g) = (a_1-1)(a_2-1)(a_3-1) - \mu_1$, and $S_0 - S_1 + S_2 = \sigma(g)$, one has:

$$4 \cdot S_0 = \sigma(g) + (a_1-1)(a_2-1)(a_3-1) - \mu_1. \tag{1.7.1}$$

Therefore, the computation of the signature is equivalent to the counting of the number of lattice points in the open tetrahedron $T(a_1,a_2,a_3)$ with vertices $(0,0,0)$, $(a_1,0,0)$, $(0,a_2,0)$, and $(0,0,a_3)$. The latter was a famous classical problem solved by L. J. Mordell [20] in the case when the numbers a_j's are mutually coprime integers, and by Pommersheim [40] in general. Their formula provides S_0 in terms of some Dedekind sums involving the a_j's.

Now notice that the results presented in (1.2) or in (1.3), applied for $f = x^{a_1} + y^{a_2}$ and $N = a_3$, give a new proof of Mordell's and Pommersheim's lattice point formula.

1.8. Final remarks

Here we discuss briefly some further connections and remarks about the signature. Steenbrink computed the signature of any isolated hypersurface germ in terms of their Hodge numbers [49], and the signature of any weighted homogeneous isolated hypersurface singularity in terms of weights [50]. Brieskorn [7], and Hirzebruch-Meyer [15] present the signature as the crucial invariant which detects the exotic structures on the links of isolated singularities. We invite the reader to read Hirzebruch's beautiful paper [14] about a global picture of the equivariant index theorems, signatures, Dedekind sums and their geometrical-arithmetical symbiosis.

A different formula for $\sigma(f+z^N)$ is given by Ashikaga [5] who uses globalization methods, and an inductive procedure with respect to the graph G_f.

In the case $g = f + z^2$, the intersection pairing of the Milnor fiber of g is the same as the symmetrized Seifert matrix of the compound torus link $\{f=0\} \cap S^3_\epsilon \subset S^3_\epsilon$. In [48] Shinohara presents a formula for the signature of the symmetrized Seifert matrix of a compound link of one component, hence if f is irreducible, then this provides $\sigma(f+z^2)$ in terms of the Puiseux pairs of f. If f is not irreducible, then there still exists a (tedious) process for the signature of the symmetrized Seifert form given by Murasugi [21].

2 The signature $\sigma(f+z^N)$ and the zero set $f+z^N=0$

In this section we would like to connect the smoothing invariants $\sigma(f+z^N)$ and $\mu(f+z^N)$ (i.e. invariats of the Milnor fiber of $f+z^N$) to the analytic germ $(X_{f,N},0) = (\{f+z^N=0\},0)$. Basically, we are interested in the real link $K_{f,N} := X_{f,N} \cap S^3_\epsilon$ $(1 \gg \epsilon > 0)$ of $(X_{f,N},0)$ (as a topological invariant of $(X_{f,N},0)$), and in the geometric genus $p_g(X_{f,N})$ (as an analytic invariant). We present in (2.4) a formula for p_g. In order to give a more complete picture of germs g of type $g = f + z^N$, we provide the construction of the plumbing graph of $K_{f,N}$ (or the resolution graph of $X_{f,N}$) in the appendix.

First notice that if $g : (\mathbf{C}^3, 0) \to (\mathbf{C}, 0)$ is an isolated singularity, then its real link K_g, in general, does not determine the smoothing invariants $\sigma(g)$ and $\mu(g)$, nor the geometric genus $p_g(g)$. For example if $g_1 = x^2 + y^7 + z^{14}$ and $g_2 = x^3 + y^4 + z^{12}$, then their minimal resolution graphs are the same: they have only one vertex with self intersection -1 and genus 3 (see e.g. [16] or use §5). In particular their links are diffeomorphic. But, (μ, σ, p_g) in the first case is $(78, -48, 9)$, and in the second $(66, -40, 8)$.

On the other hand, these numerical invariants are not independent. Let $g : (\mathbf{C}^3, 0) \to (\mathbf{C}, 0)$ be an arbitrary germ with isolated singularity, and consider the resolution $\pi : (\mathcal{Y}, D) \to (g^{-1}(0), 0)$. Let $\mathcal{W}(g)$ be the set of irreducible exceptional divisors. Then the *canonical class* (associated with π) is an element $K = \sum r_w E_w \in H_2(\mathcal{Y}, \mathbf{Z})$ defined by the adjunction formula:

$$-K \cdot E_w = E_w^2 + 2 - 2 \cdot \text{genus}(E_w) \text{ for any } w \in \mathcal{W}(g).$$

Since the intersection matrix $(E_\alpha \cdot E_\beta)$ is negative definite, the above relations determine K. Notice that K depends on the choice of the resolution π, but the numerical invariant $K \cdot K + \#\mathcal{W}(g)$ is independent of π, and depends only on the link K_g.

2.1. Theorem. *Let $g : (\mathbf{C}^3, 0) \to (\mathbf{C}, 0)$ be an isolated singularity and define the numerical invariant $K \cdot K + \#\mathcal{W}(g)$ from one of the resolutions of $(\{g = 0\}, 0)$. Then:*

(a) [Durfee] [8] $3\sigma(g) + 2\mu(g) + 2 \cdot \dim H_1(K_g) + K \cdot K + \#\mathcal{W}(g) = 0.$

(b) [Laufer] [16] $\mu(g) = 12 \cdot p_g(g) - \dim H_1(K_g) + K \cdot K + \#\mathcal{W}(g).$

In particular, modulo the link K_g, the invariants $\mu(g)$, $\sigma(g)$ and $p_g(g)$ are equivalent.

The above theorem implies the following:

2.2. $$4p_g(g) = \mu(g) + \sigma(g) + \dim H_1(K_g).$$

This is equivalent to Durfee's formula: $2p_g(g) = \mu_0(g) + \mu_+(g)$ [8] because $\dim H_1(K_g) = \mu_0(g)$.

(In the sequel, $\mu_0(g)$ denotes the dimension of the kernel of the intersection form of g, $\mu_\pm(g)$ is the dimension of a maximal subspace where the intersection form is positive, respectively negative, definite. In particular $\mu = \mu_0 + \mu_+ + \mu_-$, and $\sigma = \mu_+ - \mu_-$.)

Now assume that $g(x,y,z) = f(x,y) + z^N$. Then $\mu(g) = \mu(f)(N-1)$, and $\sigma(g)$ is given in §1 in terms of the embedded resolution graph G_f of f. In the appendix, we present a construction of a resolution graph of $(X_{f,N}, 0)$ in terms of G_f and N. This will provide $\dim H_1(K_g)$ as well. But, even without the knowledge of this graph, we can compute $\dim H_1(K_g)$ as follows.

Fix a resolution π of $(X_{f,N},0)$ as above, and set $g'=\sum_{w\in\mathcal{W}(g)}\text{genus}(E_w)$.

2.3. Proposition. *Assume that $g=f+z^N$. Then:*

(a) $\dim H_1(K_g) = \dim H_1(\mathcal{Y}) = 2g'+c_\pi$, where c_π is the number of independent cycles of the resolution graph of π.

(b) In fact c_π does not depend on the choice of π: c_π is equal to the number of 2×2–Jordan blocks with eigenvalue 1 of the monodromy $h(z)$ of $z:(X_{f,N},0)\to(\mathbf{C},0)$; $(x,y,z)\mapsto z$. Since $h(z)=h(f)^N$, where $h(f)$ is the monodromy of f, the number in (1.4.1), given in terms of the graph G_f of f, is exactly c_π, i.e.:

$$c_\pi=\sum_{e\in\mathcal{E}}\Big((N,m_e)-1\Big)-\sum_{w\in\mathcal{W}}\Big((N,M_w)-1\Big).$$

(c) $\dim\ker(h(f)^N-1)=2g'+c_\pi+r_f-1$.

(d) the generalized 1–eigenspace of $h(f)^N$ has dimension $2g'+2c_\pi+r_f-1$.

For a proof, see [30]. Note also that the above identities yield another theorem of Durfee [9], that the monodromy of f is finite if and only if the resolution graphs of all the cyclic coverings $\{X_{f,N}\}_N$ contain no cycles.

The number in (d) can be computed from the zeta function of $f+z^N$ (which follows from results of Sebastiani–Thom [47] and of A'Campo's [1]). It is $1+\sum_{w\in\mathcal{W}}(N,m_w)(\delta_w-2)$. Since (b) gives c_π, by (a) one can determine $\dim H_1(K_g)$ in terms of the embedded resolution graph G_f of f. Finally, by (2.1), p_g can be determined from G_f. More precisely:

2.4. Theorem. *The geometric genus p_g of $f+z^N$, in terms of the embedded resolution graph G_f of f and the integer N, is given as follows:*

$$4p_g=N(2-r_f)+\sum_{w\in\mathcal{W}}\Big(m_w(N-1)+(N,m_w)\Big)(\delta_w-2)$$

$$+4\sum_{w\in\mathcal{R}}\sum_{v\in\mathcal{V}_w}\Big(\ s(m_v,N;m_w)-N\cdot s(m_v,1;m_w)\ \Big).$$

Example. If $f=x^2+y^3$, then $p_g(x^2+y^3+z^N)=[N/6]$ (where $[x]$ is the integer part of x).

The geometric genus $p_g(f+z^N)$ can be computed in the following way as well (but the proof is not simpler). By [43, 51], $p_g(g)$ =the number of spectral numbers of g situated in the interval $(-1,0]$. Now, compute this using the Sebastiani–Thom type result for the spectra [45, 53], and the formula of the spectral numbers in terms of G_f [46].

Notice that p_g and $\mu(g)$ depend only on the spectral numbers of f and the integer N, but $\sigma(g)$ and $\dim H_1(K_g)$ depend also on the weight filtration of the mixed Hodge structure of f. In both terms, the "guilty contribution" is c_π (cf. 1.4 and 2.4). Actually, c_π has other interpretations too: it is the equivariant signature $\sigma_1(g)$ and the equivariant Hodge number $h_1^{22}(g)$.

2.5. A fascinating (but difficult) problem is to find special families of germs, such that for each germ g of the family, the link K_g determines each of the numbers $\mu(g)$, $\sigma(g)$ and $p_g(g)$. A very interesting example is the following.

Casson in [2] defined an invariant $\lambda(\Sigma)$ for any integer homology 3–manifold Σ (generalized for rational homology spheres by Walker [55]).

2.6. Theorem. *Let g be as before, and assume that K_g is an integer homology sphere. Then $8\cdot\lambda(K_g)=\sigma(g)$ in the following cases:*

(a) if g is a Brieskorn singularity [Fintushel–Stern] [11];

(b) if $g(x,y,z)=f(x,y)+z^N$ [Neumann–Wahl] [37].

In particular, for these singularities, the analytic invariant $p_g(g)$, and the smoothing invariants $\mu(g)$ and $\sigma(g)$ are determined completely by the link K_g only!

Actually, Neumann and Wahl in [37] characterized those pairs (f,N) for which K_{f+z^N} is a $\mathbf{Z}$–homology sphere: f must be irreducible and $(N,l)=1$ for any $l\in\{p_1,\dots,p_s,a_1,\dots,a_s\}$, where the integers $\{a_i\}_i$ are defined in terms of the Newton pairs $(p_i,q_i)_{i=1}^s$ of f as in (1.6).

By the above theorem, the signature of g has a very subtle connection with the link. The starting point of the above result consisted of the following remark. The number of $SU(2)$ representations of $\pi_1(K(\sum_{i=1}^3 z_i^{a_i}))$ (where $(a_i,a_j)=1$ for $i\neq j$) is $N/4$, where N is the number of lattice points in the open tetrahedron with vertices $(a_1,0,0)$, $(0,a_2,0)$, $(0,0,a_3)$ and (a_1,a_2,a_3) [12]. But $N=\mu(g)-4S_0$ (see 1.7 for the notation), hence by (1.7.1) $N=-\sigma(\sum_{i=1}^3 z_i^{a_i})$.

A "*modulo* 16" connection between link and signature does not only work for germs $g=f+z^N$. We recall that the μ–invariant (or Rohlin invariant) of a $\mathbf{Z}_2$–homology sphere Σ (i.e. when $H_*(\Sigma,\mathbf{Z}_2)=H_*(S^3,\mathbf{Z}_2)$) is, by definition, the signature $\sigma(W^4)$ modulo 16 of any smooth compact parallelizable 4–manifold W^4 with boundary $\partial W^4=\Sigma$. For an isolated germ $g:(\mathbf{C}^3,0)\to(\mathbf{C},0)$, if K_g is a $\mathbf{Z}$–homology sphere, then the intersection lattice of g is even and unimodular, so $8|\sigma(g)$. Hence $\sigma(g)$ (*mod* 16) codifies exactly the μ–invariant of K_g in $8\mathbf{Z}/16\mathbf{Z}$ (take $W^4=$ the Milnor fiber of g). Moreover, it is well–known that for $\mathbf{Z}$–homology spheres Σ, the class of the Casson invariant $\lambda(\Sigma)$ in $\mathbf{Z}/2\mathbf{Z}$ is exactly the $\mu-$ (Rohlin) invariant of Σ.

But, even if K_g is only a $\mathbf{Z}_2$–homology sphere, it is still true that $2|\sigma(g)$ (see e.g. [34]), and the μ–invariant of K_g considered in $2\mathbf{Z}/16\mathbf{Z}$ is the class of $\sigma(g)$ in $2\mathbf{Z}/16\mathbf{Z}$ (see [34]).

3 Inequalities

3.1. In this section we present some inequalities satisfied by the numerical invariants σ, μ and p_g. We are interested basically in inequalities of type $\sigma\leq c_1\cdot\mu$ or $p_g\leq c_2\cdot\mu$, where $-c_1$ and c_2 are some (positive, rational) numbers.

First notice that using Durfee's formula $2p_g = \mu_0 + \mu_+$, the inequality $p_g \leq c_2 \cdot \mu$ is equivalent to $\sigma \leq (4c_2 - 1)\mu - \mu_0$. For example, if $c_2 = 1/6$, then $4c_2 - 1 = -1/3$, or if $c_2 = 1/8$ then $4c_2 - 1 = -1/2$. Our leading inequalities are formulated by two conjectures of Durfee [8], namely for any isolated singularity $g : (\mathbf{C}^3, 0) \to (\mathbf{C}, 0)$ (or even for any icis of dimension 2) conjecturally $\sigma(g) \leq 0$, or an even stronger estimate holds: $p_g \leq \mu/6$ (i.e. $\sigma \leq -\mu/3 - \mu_0$). Notice that, in general, the coefficient $c_2 = 1/6$ (or $c_1 = -1/3$) is the optimal one. Consider the example $g = \sum_{i=1}^3 z_i^d$, for which $-3\sigma = (d-1)(d^2+d-3)$, $\mu = (d-1)^3$, $\mu_0 = (d-1)(d-2)$. Unfortunately, we know very little about these inequalities in the general case.

On the other hand, for some particular cases, we do have some nice estimates. For example, if the germ g is simple (i.e. of type $A-D-E$), then $\sigma = -\mu$ and $p_g = 0$. Moreover, if g is at the beginning of the classification list, then even its Dynkin diagram is known, hence its signature (hence p_g too) is explicitely computed (for some examples, see Ebeling's book).

For weakly elliptic singularities, Xu and S. S.-T. Yau [56] proved that $12p_g \leq \mu + 4$. For $g = f(x,y) + z^2$, Tomari proved that $8p_g \leq \mu$ [52]. For weighted homogeneous singularities with multiplicity ν, Xu and S. S.-T.Yau [57] verified (by a rather long computation) Durfee's conjectures. Indeed they proved that $6p_g \leq \mu - \nu + 1$.

If $g = f(x,y) + z^N$ then a short proof of the negativity of the signature $\sigma(g)$ is given in [27].

3.2. Brieskorn singularities

Here we present in more detail some results about the signature $\sigma(a,b,c)$ of the germ $x^a + y^b + z^c$.

If $a = b = c$, then $-3\sigma(a,a,a) = (a-1)(a^2+a-3)$. But already for the cases $a = b \neq c$, the signature is not very simple:

$$\sigma(a,a,c) = 4a \cdot s(c,1;a) + a - 1 - c(a^2-1)/3.$$

For the proof use §1. The general inequality (for $a > 0$):

$$|s(c,1;a)| \leq (a-1)(a-2)/(12a) \tag{3.2.1}$$

gives the estimate:

$$|\sigma(a,a,c) + c(a^2-1)/3 - (a-1)| \leq (a-1)(a-2)/3.$$

If $a >> c$, then (3.2.1) is not very sharp, so we replace $s(c,1;a)$ with $s(a,1;c)$ via the Dedekind reciprocity law, and we apply (3.2.1) for the latter. This gives:

$$|\sigma(a,a,c) + 1 - (2c^2 + a^2 + (a,c)^2 - c^2a^2)/(3c)| \leq a(c-1)(c-2)/(3c).$$

Now, we will consider the general case of $x^a + y^b + z^c$.

Xu and S. S.-T. Yau in [58] proved that for $2 \leq a \leq b \leq c$ we have:

$$\sigma \leq -\mu/3 - \mu_0 - 2(a-1)/3$$

or equivalently:

$$6p_g \leq \mu - a + 1.$$

The author in [25] proved that for $2 \leq a \leq b \leq c$ we have:

$$\sigma \leq -\frac{\mu}{3} \cdot \frac{a+1}{a} - \frac{(c-1)(a-1)}{3a} \tag{3.2.2}$$

or equivalently:

$$p_g \leq \frac{1}{6}\left(1 - \frac{1}{2a}\right)\mu + \frac{1}{4}\mu_0 - \frac{(c-1)(a-1)}{12a}. \tag{3.2.3}$$

Notice that the coefficient of μ in (3.2.3) (or in 3.2.2) is better than the coefficient in Durfee's conjecture, and for $a = 2$ it is exactly $1/8$ as in Tomari's result $p_g \leq \mu/8$ [52]. It is not hard to see that all these relations imply Durfee's conjecture $p_g \leq \mu/6$.

3.3. The $g = f(x, y) + z^N$ case

It is convenient to use the following notation: if two plane curve singularities f_1 and f_2 have the same topological type, then we write $f_1 \sim f_2$.

Let ϵ_f be 1 if $f \sim x^2 + y^{n+1}$ $(n \geq 1)$, $f \sim y(x^2 + y^{2k+1})$ $(k \geq 1)$, $f \sim x^3 + y^4$, $f \sim (x^2 + y^{2k+1})(y^2 + x^{2l+1})$ $(k \geq 1, l \geq 1)$; otherwise $\epsilon_f = 0$.

Theorem. [26] *Assume that G_f is a minimal embedded resolution graph of f. Then the following inequalities hold. (For the notations, see (1.2).)*

(a) If f is irreducible then

$$p_g(f + z^N) \leq \mu(f + z^N)/6.$$

(b)

$$\sigma(f + z^2) \leq -\frac{1}{2}(\mu_f + \#\mathcal{A} - 1).$$

(c) For any f and N:

$$\sigma(f + z^N) \leq \frac{1-N}{N}\#\mathcal{E} + \frac{1-N^2}{3N}(\mu_f - \#\mathcal{W} - 2).$$

(d) There exists a constant $B(f)$, given in terms of G_f (for details see [26] (3.11.b)), such that for $N > B(f)$, one has:

$$\sigma(f + z^N) \leq \frac{1-N^2}{3N}(\mu_f - 3\epsilon_f).$$

(e) If $(N, m_w)=1$ for any $w \in \mathcal{W}(G_f)$, then:

$$\sigma(f+z^N) \le \frac{1-N}{N}(\#\mathcal{W}+\#\mathcal{E}) + \frac{1-N^2}{3N}(\mu_f - 3\epsilon_f);$$

and with the same assumption:

$$\sigma(f+z^N) \le -\mu(f+z^N)/3 \text{ or, equivalently: } p_g(f+z^N) \le \mu(f+z^N)/6.$$

In some cases it is helpful to compare the signature with the multiplicity ν_f of f. For example, in [26] it is established that (for arbitrary f and N):

3.4. $$\sigma(f+z^N) \le \frac{1-N}{N}\#\mathcal{E} + \frac{1-N^2}{3N}(\nu_f^2 - 2\nu_f - 2 - \#\mathcal{A}).$$

A similar inequality can be found in Ashikaga's paper [5].

Sometimes we only need the asymptotic behaviour of $N \mapsto \sigma(f+z^N)$.

3.5. Theorem. [26] *The limit* $\lim_{N\to\infty}(-\sigma_N)/N = \eta(f;1)$ *satisfies:*

$$\eta(f;1) \ge (\mu_f + \#\mathcal{W} + \#\mathcal{A} - 3 + B)/3,$$

where $B := \sum_{w\in\mathcal{W}}(-e_w) - 2\#\mathcal{W} \ge -1$; *and also:*

$$\eta(f;1) \ge (3\#\mathcal{W} + \#\mathcal{A} - 8)/3.$$

3.6. Final remarks

All these inequalities are subtly connected with the distribution of spectral numbers of singularities (see e.g. [42] and [27]). K. Saito in [42] presents a lot of very interesting conjectures about this distribution problem.

We also mention that Durfee's conjectures are not true for arbitrary smoothing of normal surface singularities [54], and that the distribution of the spectral numbers has "pathologies" [27].

4 Conjectures/Open problems

1. Generalize Theorem (1.6) for reducible germs f.

2. Generalize Theorem (2.6) for rational homology spheres.

For any f and N conjecturally the following inequalities hold:

3. $$\sigma(f+z^N) \le -\frac{N+1}{3N}\mu(f+z^N).$$

4. $$\eta(f;1) + \eta(f;N) - \eta(f;N+1) \ge 0$$

(which would imply the monotonity: $\sigma(f+z^{N+1}) \le \sigma(f+z^N)$).

5. $$\eta(f;1) + \eta(f;N) - \eta(f;N+1) \ge \mu_f/3,$$

(which would imply, for example, the inequality $\sigma \le -\mu/3$ for $f+z^N$).

5 Appendix:

The resolution graph $G(V_g)$ of $V_g := (\{g=0\},0)$ (where $g(x,y,z) = f(x,y)+z^N$, and f is an isolated plane curve singularity) in terms of the embedded resolution graph G_f of f and the integer N.

I guess that the result of this appendix is known for specialists, but I have never seen it written down in this form. The idea and the method is in the book of Laufer [17] (but is already in the "Jungian strategy"). Particular cases can be found in many places, e.g. [38, 39, 13, 4]. The proof, together with some applications, will be given in [31].

Actually, we will consider the germ $z : (V_g, 0) \to (\mathbf{C}, 0)$ (induced by the projection $(x,y,z) \mapsto z$), and we will determine the embedded resolution graph G_z of this germ z. Obviously, $G(V_g)$ is the graph G_z without arrowheads and multiplicities, so it is simpler, but computing G_z has an advantage. Sometimes, it is easier to compute multiplicities than self intersection numbers. So, computing G_z, first we compute all the multiplicities of the germ z along the exceptional divisors, then the self intersection numbers can be easily computed using the relation:

$$\text{(A0)} \qquad e_w m_w + \sum_{v \in \mathcal{V}_w(g)} m_v = 0 \quad \text{for any } w \in \mathcal{W}(g).$$

1. For any three positive integers u, v and N, with $(u,v,N)=1$, we consider the unique $0 \le x_1 < N/(u,N)$ and $m_1 \in \mathbf{N}$ with:

$$\text{(A1)} \qquad v + x_1 \cdot \frac{u}{(u,N)} = m_1 \cdot \frac{N}{(u,N)}.$$

If $x_1 \neq 0$, then consider the continuous fraction:

$$\frac{N/(u,N)}{x_1} = k_1 - \cfrac{1}{k_2 - \cfrac{1}{\ddots - \cfrac{1}{k_s}}}, \qquad k_1, \ldots, k_s \ge 2.$$

Consider the "string":

$$G(u,v,N): \quad (\tfrac{u}{(u,N)}) \longleftarrow \underset{(m_1)}{\overset{-k_1}{\bullet}} \text{---} \underset{(m_2)}{\overset{-k_2}{\bullet}} \text{---} \cdots \text{---} \underset{(m_s)}{\overset{-k_s}{\bullet}} \longrightarrow (\tfrac{v}{(v,N)})$$

where all vertices have genus $g=0$ (i.e. they represent rational irreducible exceptional divisors), their self intersection numbers are $-k_1, \ldots, -k_s$ respectively. The arrowheads have multiplicities $u/(u,N)$ and $v/(v,N)$, and the first vertex has multiplicity m_1 given by (A1). Therefore, $m_2, \ldots, m_s$ can be easily computed using (A0), namely

$$-k_1 m_1 + \frac{u}{(u,N)} + m_2 = 0, \text{ and } -k_i m_i + m_{i-1} + m_{i+1} = 0 \text{ for } i \ge 2.$$

If $x_1 = 0$, then the string $G(u,v,N)$ has no vertices, it is only an edge.

2. Now, consider the embedded resolution graph G_f of f. (For the notation, see 1.2.) The graph G_z can be considered as a "covering" $q : G_z \to G_f$.

(a) Above $w \in \mathcal{W}(G_f)$ there are (M_w, N) vertices of G_z, each with multiplicity $m_w/(m_w, N)$ and genus $\tilde{g}$, where:

$$2 - 2\tilde{g} = \frac{(2-\delta_w)(m_w, N) + \sum_{v \in \mathcal{V}_w}(m_w, m_v, N)}{(M_w, N)}.$$

The vertices in $q^{-1}(w)$ can be indexed by the group $\mathbf{Z}_{(M_w,N)}$.

(b) An edge $e = (w_1, w_2)$ of G_f

$$\overset{(m_{w_1})}{\bullet}\!\!-\!\!\!-\!\!\!-\!\!\!-\!\!\overset{(m_{w_2})}{\bullet}$$

is covered by (m_e, N) copies of strings in G_z, each of type

$$G(m_{w_1}/(m_e,N), m_{w_2}/(m_e,N), N/(m_e,N)).$$

These strings can be indexed by the group $\mathbf{Z}_{(m_e,N)}$. The arrowheads of the strings are identified with the vertices $q^{-1}(w_1)$, respectively $q^{-1}(w_2)$, via the natural morphisms $\mathbf{Z}_{(m_e,N)} \to \mathbf{Z}_{(M_{w_1},N)}$, respectively $\mathbf{Z}_{(m_e,N)} \to \mathbf{Z}_{(M_{w_2},N)}$.

(c) An arrowhead of G_f

$$\overset{(m_w)}{\bullet}\!\!\longrightarrow (1)$$

is covered by one string of type $G(m_w, 1, N)$, whose right arrowhead will remain an arrowhead of G_z with multiplicity 1, and its left arrowhead is identified with the unique vertex above w.

(d) In this way, we obtain all the vertices, edges and arrowheads of G_z, and all the multiplicities (of $z \circ \pi$ along the corresponding exceptional divisors and strict transforms). Moreover, by the description of the strings (cf. part 1), one has all the self intersection numbers of the vertices which are situated on the new strings. Now, the self intersection numbers of the vertices $q^{-1}(w)$ ($w \in \mathcal{W}(G_f)$) can be computed using (A0).

3. If we drop the arrowheads and multiplicities of G_z, we obtain $G(V_g)$. The graphs G_z and $G(V_g)$, in general, are not minimal. They can be simplified by blowing down the (-1)–rational curves with $\delta_w \leq 2$.

References

[1] A'Campo, N.: La fonction zeta d'une monodromie, *Comment. Math. Helv.*, **50** , 233-248 (1975).

[2] Akbulut, S. and McCarthy, J. D.: Casson's invariant for oriented homology 3–spheres, an exposition, Mathematical Notes **36**, Princeton University Press, Princeton, 1990.

[3] Arnold, V.I., Gusein-Zade, S.M. and Varchenko, A.N.: Singularities of Differentiable Mappings, Vol.2 , Birkhauser, Boston, 1988.

[4] Artal–Bartolo, E.: Forme de Seifert des singularités de surface, *C. R. Acad. Sci. Paris*, **t. 313**, Série I, 689-692 (1991).

[5] Ashikaga, T.: The Signature of the Milnor fiber of Complex Surface Singularities on Cyclic Coverings, preprint, (1995).

[6] Atiyah, M.F., Patodi, V.K. and Singer, I.M.: Spectral asymmetry and Riemannian geometry, I,II,III, *Math. Proc. Cambridge Philos. Soc.*, **77** 53-69 (1975), **78** 405-432 (1975), **79** 71-99 (1976).

[7] Brieskorn, E.: Beispiele zur Differentialtopologie von Singularitäten, *Invent. Math.*, **2**, 1-14 (1966).

[8] Durfee, A.: The Signature of Smoothings of Complex Surface Singularities, *Math. Ann.*, **232**, 85-98 (1978).

[9] Durfee, A. H.: The Monodromy of a Degenarating Family of Curves, *Inventiones Math.*, **28** 231-241 (1975).

[10] Eisenbud, D. and Neumann, W.: *Three-Dimensional Link Theory and Invariants of Plane Curve Singularities*, Ann. of Math. Studies **110**, Princeton University Press, 1985.

[11] Fintushel, R. and Stern, R.J.: Instanton homology of Seifert fibered homology 3–spheres, *Proc. London Math. Soc.*, (3) **61**, 109-137 (1991).

[12] Greenberg, L.: Homomorphisms of triangle groups into $PSL(2\mathbf{C})$, *Riemannian surfaces and related topics: Proc. of the 1978 Stony Brook Conference*, Annals of Math. Studies **97**, Princeton University Press 1981, 167-181.

[13] Harer, J., Kas, A. and Kirby, R.: Handlebody decomposition of complex surfaces, *Mem. Amer. Math. Soc.*, **62** (1986) no 350.

[14] Hirzebruch, F.: Reminiscenses and Recreation, *Annals of Math. Studies*, **70**, 3-31.

[15] Hirzebruch, F. and Meyer, K.: $O(n)$–Mannigfaltigkeiten, exotische Sphären and Singulariäten, *Lecture Notes in Math.*, **57**, Springer 1968.

[16] Laufer, H. B.: On μ for surface singularities, *Proc. of Symp. in Pure Math.*, **30**, 45-49 (1977).

[17] Laufer, H. B.: Normal two–dimensional singularities, *Annals of Math. Studies* **71**, Princeton University Press 1971.

[18] Lê Dũng Tráng: Sur les noeuds algébriques, *Compositio Math.*, **25**, 281-321 (1972).

[19] Milnor, J. and Orlik, P.: Isolated singularities defines by weighted homogeneous polynomials, *Topology*, **9**, 385-393 (1970).

[20] Mordell, L.J.: Lattice points in a tetrahedron and generalized Dedekind sums, *J. Indian Math.*, **15**, 41-46 (1951).

[21] Murasugi, K.: On a certain numerical invariant of link types, *Trans. Amer. Math. Soc.*, **117**, 387-422 (1965).

[22] Némethi, A.: The equivariant signature of hypersurface singularities and eta–invariant, *Topology*, **34**, 243-259 (1995).

[23] Némethi, A.: The eta–invariant of variation structures I, *Topology and its Applications*, **67**, 95-111 (1995).

[24] Némethi, A.: The real Seifert form and the spectral pairs of isolated hypersurface singularities, *Compositio Math.*, **98**, 23-41 (1995).

[25] Némethi, A.: Dedekind sums and the signature of $f(x,y)+z^N$, submitted.

[26] Némethi, A.: Dedekind sums and the signature of $f(x,y)+z^N$,II., submitted.

[27] Némethi, A.: On the spectrum of curve singularities, will appear in the Proc. of the Singularity Conf., Oberwolfach, 1996.

[28] Némethi, A.: Casson invariant of cyclic coverings via eta–invariant and Dedekind sums, submitted.

[29] Némethi, A.: Some topological invariants of isolated hypersurface singularities, will appear in the Proceedings of the EMS Summer School, Eger (Hungary), 1996.

[30] Némethi, A. and Steenbrink, J.: On the monodromy of curve singularities, *Math. Zeitschrift*, **223**, 587-593 (1996).

[31] Némethi, A. and Szilárd, Á.: The resolution graph of cyclic coverings, submitted.

[32] Neumann, W. D.: Cyclic suspensions of knots and periodicity of signature for singularities, *Bull. Amer. Math. Soc.*, Volume **80**, Number 5, 977-981 (1974).

[33] Neumann, W. D.: Signature related invariants of manifolds – I. Monodromy and γ–invariants, *Topology*, Vol. **18**, 147-172 (1979).

[34] Neumann, W. D.: An invariant of plumbed homology spheres, *Topology Symp. Siegen 1979, Lecture Notes in Math.* **788**, 125-144 (1980).

[35] Neumann, W. D.: Invariants of plane curve singularities, *Monographie No 31 de L'Enseignement Mathématique*, 223-232 (1983).

[36] Neumann, W. D.: Splicing Algebraic Links, *Advenced Studies in Pure Math.*, **8**, Complex Analytic singularities, 349-361 (1986).

[37] Neumann, W. and Wahl, J.: Casson invariant of links of singularities, *Comment. Math. Helv.* **65**, 58-78 (1991).

[38] Ono, I. and Watanabe, K.: On the singularity of $z^p + y^q + x^{pq} = 0$, *Sci. Rep. Tokyo Kyoika Daigaku Sect. A*, **12**, 123-128 (1974).

[39] Orlik, P. and Wagreich, P.: Isolated singularities of algebraic surfaces with $\mathbf{C}^*$ action, *Ann. of Math.* (2) **93**, 205-228 (1971).

[40] Pommersheim, J. E.: Toric varieties, lattice points and Dedekind sums, *Math. Ann*, **295**, 1-24 (1993).

[41] Rademacher, H.: Generalization of the Reciprocity formula for Dedekind sums, *Duke Math. Journal*, **21**, 391-397 (1954).

[42] Saito, K.: The Zeroes of Characteristic Function χ_f for the Exponents of a Hypersurface Isolated Singular Point, *Advanced Studies in Pure Math.* **1**, Algebraic Varieties and Analytic Varieties, 195-217 (1983).

[43] Saito, M.: On the exponents and the geometric genus of an isolated hypersurface singularity, *Proc. of Sympos. in Pure Math.*, **40**, Part 2, 465-472 (1983).

[44] Sakamoto, K.: The Seifert matrices of Milnor fiberings defined by holomorphic functions, *J. Math. Soc. Japan*, **26**, 714-721 (1974).

[45] Scherk, J. and Steenbrink, J. H. M.: On the Mixed Hodge Structure on the Cohomology of the Milnor Fiber, *Math. Ann.*, **271**, 641-665 (1985).

[46] Schrauwen, R., Steenbrink, J. and Stevens, J.: Spectral Pairs and Topology of Curve Singularities, *Proc. Sumpos. Pure Math.*, **53**, 305-328 (1991).

[47] Sebastiani, M. and Thom, R.: Un résultat sur la monodromie, *Inventiones Math.*, **13**, 90-96 (1971).

[48] Shinohara, Y.: On the signature of knots and links, *Trans. Amer. math. Soc.*, **156**, 273-285 (1971).

[49] Steenbrink, J.H.M.: Mixed Hodge structures on the vanishing cohomology, Nordic Summer School/NAVF, Symposium in Mathematics, Oslo, 1976.

[50] Steenbrink, J.: Intersection form for quasi-homogeneous singularities, *Compositio Math.*, vol. **34**, fasc. 2, 211-223 (1977).

[51] Steenbrink, J.H.M.: Mixed Hodge structures associated with isolated singularities, *Proc. Sumpos. Pure Math.*, **40**, Part 2, 513-536 (1983).

[52] Tomari, M.: The inequality $8p_g \leq \mu$ for hypersurface two-dimensional isolated double points, *Math. Nachr.*, **164**, 37-48 (1993).

[53] Varchenko, A. N.: Asymptotic Hodge structure in the vanishing cohomology, *Math. USSR Izv.*, **18**, 469-512 (1982).

[54] Wahl, J.: Smoothings of normal surface singularities, *Topology*, **20**, 219-246 (1981).

[55] Walker, K.: An extension of the Casson's invariant, Ann. of Math. Studies **126**, Princeton University Press, 1992.

[56] Xu, Y. and Yau, S.S.-T.: The inequality $\mu \geq 12p_g - 4$ for hypersurface weakly elliptic singularities, *Contemporary Math.*, **90**, 375-344 (1989).

[57] Xu, Y. and Yau, S.S.-T.: Durfee's conjecture and coordinate free characterization of homogeneous singularities, *J. Diff. Geometry* **37**, 375-396 (1993).

[58] Xu, Y. and Yau, S.S.-T.: A sharp estimate of the number of integral points in a tetrahedron, *J. reine angew. Math.*, **423**, 199-219 (1992).

[59] Zagier, D.: Higher dimensional Dedekind sums, *Math. Ann.*, **202**, 149-172 (1973).

The Ohio State University,
Dept. of Mathematics,
231 West 18th Avenue,
Columbus, OH 43210,
USA.

Spectra of $\mathcal{K}$–Unimodal Isolated Singularities of Complete Intersections

J.H.M. Steenbrink*

Abstract

We produce tables of spectra for all $\mathcal{K}$–unimodal isolated complete intersection singularities. Their computation requires slight generalisations of the known methods for hypersurface singularities.

1 Introduction

The spectrum of an isolated complete intersection singularity (abbreviated: ICIS) has been defined by W. Ebeling and the author [2]. Given an ICIS (X, x), one chooses an ICIS (X', x) with $\dim(X') = \dim(X) + 1$ and a holomorphic function germ $f : (X', x) \to (\mathbf{C}, 0)$ such that $X = f^{-1}(0)$ and the Milnor number $\mu(X', x)$ is as small as possible. Using the mixed Hodge structure on the vanishing cohomology of a holomorphic function f, one may define its spectrum $Sp(f) = \sum_{\alpha \in \mathbf{Q}} n_\alpha(\alpha) \in \mathbf{Z}[\mathbf{Q}]$. If $s_k(X')$ is the k–th Hodge number of any one-parameter smoothing of X' and $Sp(f)$ denotes the spectrum of f, then $Sp(X, x) := Sp(f) + \sum_k s_k(X')(k)$. Two properties of the spectrum have been proved in [2]: the symmetry and its semicontinuity under deformation. The spectrum is constant under deformations with both $\mu(X, x)$ and $\mu(X', x)$ constant.

In this paper we will give the spectra for all $\mathcal{K}$–unimodal isolated singularities of complete intersections; here $\mathcal{K}$ is the group of contact equivalences. The $\mathcal{K}$–unimodal isolated complete intersection singularities have been classified by Wall [10]. Let $\mathcal{R}$ stand for right equivalences, i.e. diffeomorphisms of the source of a map germ. Spectra for hypersurface singularities of $\mathcal{R}$–modality at most two have been calculated by Goryunov [4]; his list is reproduced in [1, p. 389]. All of these except the $J_{3,p}$–series are of $\mathcal{K}$–modality at most one. The zero–dimensional complete intersections can be handled easily so we will not list them all but give a general formula instead.

*The author thanks the School of Mathematics of the University of Minnesota for its hospitality and financial support.

2 Spectra of Isolated Complete Intersection Singularities

Spectra are associated with mixed Hodge structures on which an operator of finite order acts. If V is a mixed Hodge structure with Hodge and weight filtrations F, W and γ is an automorphism of (V, W, F) of finite order, one defines the equivariant Hodge numbers by

$$h_\lambda^{pq}(V,\gamma) = \dim\ker(\gamma - \lambda I : Gr^W_{p+q} Gr^p_F V)$$

and defines the spectrum $Sp(V,\gamma) \in \mathbf{Z}[\mathbf{Q}]$ by

$$Sp(V,\gamma) := \sum_{\alpha\in\mathbf{Q}} \sum_{w\in\mathbf{Z}} h_{e(\alpha)}^{[\alpha],w-[\alpha]}(\alpha).$$

Here, $[\alpha]$ denotes the integral part and $e(\alpha) = \exp(2\pi i\alpha)$.

Let (X,x) be an isolated complete intersection singularity of dimension n. The mixed Hodge structure which gives rise to $Sp(X,x)$ is defined in the following way. Choose a minimal smoothing pair $F = (f,g) : (\mathcal{X},x) \to (\mathbf{C}^2,(0,0))$ of (X,x), i.e. a two-parameter deformation such that $\mathcal{X}$ and $X' = g^{-1}(0)$ are isolated singularities and have the smallest possible Milnor numbers. Then on $\mathcal{X}$ one has the Hodge module $\mathbf{Q}^H_{\mathcal{X}} = (O_{\mathcal{X}}, \mathbf{Q}_{\mathcal{X}}[n+2])$. The category of mixed Hodge modules is stable by (perverse) vanishing and nearby cycle functors, denoted by ϕ_f and ψ_f respectively, and $\phi_f\psi_g\mathbf{Q}^H_{\mathcal{X}}$ has support at x, so it is a mixed Hodge structure, together with the finite order automorphism which is the semisimple part of the monodromy of f. The spectrum of this mixed Hodge structure is independent of the choice of the minimal smoothing pair F and is by definition the spectrum $Sp(X,x)$ of (X,x).

The $\mathcal{K}$–unimodal isolated complete intersection singularities are all given by one or two equations, so $\mathcal{X}$ is smooth for a minimal smoothing pair. Moreover, X' is either smooth (the hypersurface case) or a simple singularity of type A_1 or A_2 if $\dim X$ is positive. That means that in all cases, the invariants of X' are well-known.

Let $F = (f,g)$ be a minimal smoothing pair of (X,x), and let f' denote the restriction of f to X'. Then

$$Sp(X,x) = Sp(f') + [Sp(g)]$$

where [.] denotes again the integral part. So we only have to calculate $Sp(f')$.

We recall a suspension formula from [2]. Let us first fix some notation. For $p, n \in \mathbf{N}$ define

$$\alpha_{p,n} := \sum_{i=1}^{p}(\frac{n}{2} + \frac{i}{p+1}) \in \mathbf{Z}[\mathbf{Q}]$$

and

$$\delta_{p,n} := \sum_{i=1}^{p}(\frac{n}{2} + \frac{2i-1}{2p}).$$

Then the n–dimensional singularities of type A_k and D_k have spectrum $\alpha_{k,n}$ and $(\frac{n+1}{2}) + \delta_{k-1,n}$ respectively. For $p \in \mathbf{N}$ define a homomorphism

$$c'_p : \mathbf{Z}[\mathbf{Q}] \to \mathbf{Z}[\mathbf{Q}]$$

by

$$c'_p((\alpha)) := \sum_{m=0}^{p-1}([\alpha] + \{\alpha + \frac{m}{p}\}).$$

Also recall the degree functions $\deg_a : \sum_\alpha n_\alpha(\alpha) \mapsto \sum_{a \le \alpha < a+1} n_\alpha$ for each $a \in \mathbf{R}$. We let $*$ denote the multiplication in the group ring $\mathbf{Z}[\mathbf{Q}]$.

Theorem 1 *Let $F = (f, g) : (\mathbf{C}^{n+2}, 0) \to (\mathbf{C}^2, 0)$ be a minimal smoothing pair of the ICIS (X, x). Let $(\tilde{X}, \tilde{x}) \subset (\mathbf{C}^{n+3}, (0, 0))$ be defined by $f(y) + z^p = g(y) = 0$. Then*

$$Sp(\tilde{X}, \tilde{x}) = \alpha_{p-1,0} * Sp(f') + c'_p(Sp(g)) + [\alpha_{p-1,0} * Sp(g)].$$

This theorem can be applied in the case of the singularities of types $P_{k,l}$, G_{n+8}, $I_{1,0}$, J' and K' with $p = 3$ in the case $I_{1,0}$ and $p = 2$ in the other cases.

3 Resolution Methods

For a function on a normal surface singularity, the spectrum can often be calculated using an embedded resolution (see [7]). First consider the case of the singularity $FT_{k,l}$, given in [10] by $(xy + z^{l-1}, xz + y^{k-1} + yz^2)$.

Lemma 1 $(xz - 2z^3 + yz^2 + y^{k-1} + z^{l-1}, xy - z^3)$ *is a minimal smoothing pair for $FT_{k,l}$.*

Proof. Considering the quadratic terms (xz, xy) one sees that X is not a Cartier divisor on an ordinary double point, so the smoothing pair from the lemma is certainly minimal. It remains to be shown that it defines a singularity of type $FT_{k,l}$.

Recall from [10] that for maps

$$F(x, y_1, y_2) = (xy_1 + a(y_1, y_2), xy_2 + b(y_1, y_2))$$

one can put $\phi = y_1 b - y_2 a$ and one has a bijection between $\mathcal{K}$-classifications of F and ϕ. Wall's equations for $FT_{k,l}$ give $\phi(y, z) = y^k + y^2z^2 - z^l$ of type $T_{k,l}$ whereas the equations from the lemma give $\phi'(y, z) = z^2(z - y)^2 + y^k + yz^{l-1}$ which has the same $\mathcal{K}$-type.

To compute the spectrum of $FT_{k,l}$ we can use the Newton method of the next section if $k = 4$ or $l = 4$. If $k, l \geq 5$ we blow up the origin of the A_2–surface singularity. Two exceptional curves appear, both with self-intersection number -2. The first is cut transversely by a smooth branch,

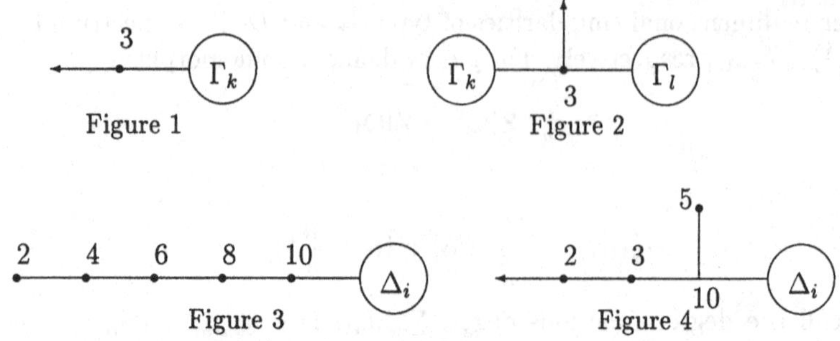

Figure 1
Figure 2
Figure 3
Figure 4

the second with multiplicity two by two branches with singularities of type A_{k-5} and A_{l-5} respectively. Let us decorate embedded resolution graphs of functions by the multiplicity of the function on each component. If the plane curve singularity of type D_k has resolution graph as in Figure 1, then $FT_{k,l}$ has resolution graph as in Figure 2. The formulas from [7] and comparison with the singularities of type D give

$$Sp(FT_{k,l}) = (\frac{1}{3}) + 3(1) + (\frac{5}{3}) + \delta_{k-1,1} + \delta_{l-1,1}.$$

Another case where this method is successful is the series $FW^{\#}_{1,i}$ given by the minimal smoothing pair $(xz + 2y^2z^2 - y^5 + z^2yl_i(z,y), xy + z^3)$ where $l_i(z,y) = zy^q$ if $i = 2q$ and y^{q+2} if $i = 2q+1$. If the resolution graph of a plane curve of type A_{i+9} is as in Figure 3, then the resolution graph of $FW^{\#}_{1,i}$ is as in Figure 4. As a consequence,

$$Sp(W^{\#}_{1,i}) = (\frac{3}{10}) + (\frac{9}{10}) + (1) + (\frac{11}{10}) + (\frac{17}{10}) + \alpha_{i+9,1}.$$

In a similar way, let $K^{\#}_{1,i}$ be given by the minimal smoothing pair $(x^2 + 2yz^2 + y^4 + zyl_i(z,y), xy - z^2)$, then it has resolution graph as in Figure 5.

The following applies to all of the curve singularities of type TZ. These are given by equations $(xy, xz + z^3 + yu(y,z))$ with $u \in (y,z)^3$.

Lemma 2 *Let C be a plane curve singularity given by the equation $z^m = yu(y,z)$ with $u(y,z) \in (y,z)^m$. Let X be the curve singularity in $\mathbf{C}^3$ with ideal $(xy, xz + z^m + yu(y,z))$ where $m \geq 3$. Then*
a) $(xy, xy + xz + z^m + yu(y,z))$ is a minimal smoothing pair for X and X' has type A_{m-1};
b) If C has resolution graph as in Figure 6, then X has resolution graph as in Figure 7.

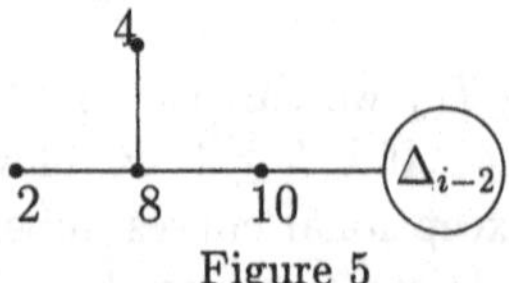

Figure 5

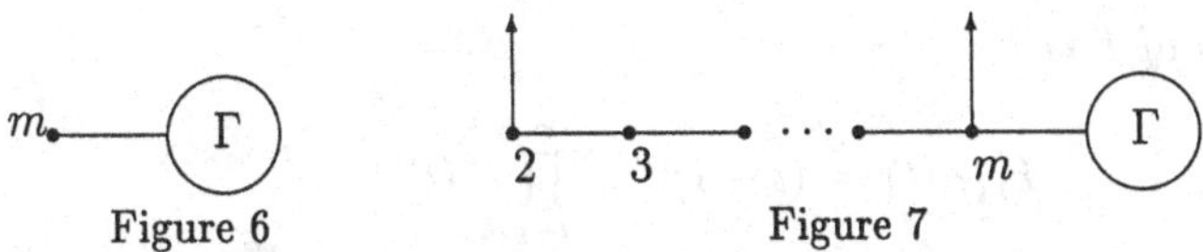

Figure 6 Figure 7

The proof is left to the reader. The lemma enables us to express the spectrum of X in terms of the spectrum of C and m:

$$Sp(X) = Sp(C) + \alpha_{m-1,0} + \alpha_{m-1,2} + (m-1)(1).$$

For all type FZ singularities, C is Newton nondegenerate, so the spectrum of C is easy to compute.

4 Newton Diagram Methods

The surface singularities of type A_m and the threefold singularity of type A_1 are toric; this makes it desirable to use toric methods for the computation of spectra of functions defined on them. Let us first fix some notation. We will closely follow [8].

Let $(Y,0)$ be a toric isolated singularity of dimension n: $Y = Spec\mathbf{C}[\sigma_{\mathbf{Z}}]$ where $\sigma \subset \mathbf{R}^n$ is a rational polyhedral cone which spans $\mathbf{R}^n$ and does not contain a proper linear subspace, and $\sigma_{\mathbf{Z}} := \sigma \cap \mathbf{Z}^n$. Let $f : Y \to \mathbf{C}$ be a polynomial map, say $f = \sum_{m\in\sigma_{\mathbf{Z}}} a_m z^m$. We define $\mathrm{supp}(f) = \{m \in \sigma_{\mathbf{Z}} | a_m \neq 0\}$, $\Gamma_+(f)$ the convex hull of $\cup_{m\in \mathrm{supp}(f)}(m+\sigma)$ and $\Gamma(f)$ the union of all compact faces of $\Gamma_+(f)$. This is called the *Newton diagram* of f.

Let γ be a closed face of $\Gamma(f)$. Denote the polynomial $\sum_{m\in\gamma} a_m z^m$ by f_γ. Then f is called *non-degenerate* if for any closed face γ of $\Gamma(f)$ the polynomials

$$x_1\frac{\partial f_\gamma}{\partial x_1}, \ldots, x_n\frac{\partial f_\gamma}{\partial x_n}$$

have no common zero in $(\mathbf{C}^*)^n$.

Let τ be an l-dimensional face of σ, with linear span L_τ. Let $\Gamma_i(\tau)$ ($i = 1, \ldots, j(\tau)$) be all $(l-1)$-dimensional faces of $L_\tau \cap \Gamma(f)$ and let $L_1, \ldots, L_{j(\tau)}$ be the $(l-1)$-dimensional affine subspaces of L_τ containing them. Let L_j be given by an equation $l_j(m) = m_j(\tau)$ with $l_j(\mathbf{Z}^n \cap L_\tau) = d_j\mathbf{Z}$ and $\gcd(m_j(\tau), d_j) = 1$.

Let V_l denote l-dimensional volume. Define $r_j(\tau) := (l-1)!V_{l-1}(\Gamma_j(\tau))$ and

$$P^l(t) := \prod_{\dim(\tau)=l} \prod_{j=1}^{j(\tau)} (t^{m_j(\tau)} - 1)^{r_j(\tau)}$$

for $l = 1, \ldots, n$.

Theorem 2 *Suppose that f has an isolated singularity at 0 and is non-degenerate with respect to $\Gamma(f)$. Then the characteristic polynomial of the* *monodromy of f is*

$$P_{\Gamma(f)}(t) := (t-1)^{(-1)^n} \prod_{l=1}^{n} P^l(t)^{(-1)^{n+l}}.$$

The proof of this theorem is exactly the same as that of [8, Theorem 4.1.3].

If $Y = \mathbf{C}^n/G$ is a quotient singularity, then one can consider the module $\tilde{\Omega}^n_Y/df \wedge \tilde{\Omega}^{n-1}_Y = Q^f$. Here for any toric variety Y, $\tilde{\Omega}^p_Y := j_*\Omega^p_{Y_{reg}}$ where $j : Y_{reg} \to Y$ is the inclusion map of the regular locus. If $\tilde{f}$ denotes the induced function on $\mathbf{C}^n$, then $Q^f \simeq (Q^{\tilde{f}})^G$. By an equivariant version of [5] or [9] one may compute the spectrum as $\sum_\alpha \dim Gr^\alpha_N Q^f(\alpha)$ where N denotes the Newton filtration. It seems plausible that this holds for any function with an isolated singularity defined on a toric singularity.

Instead of proving that statement, we will use the fact that the singularities we are dealing with have small dimension (in fact what follows will compute the spectrum if $\dim(X) \le 2$). We can compute the part of the spectrum belonging to $Gr^n_F(\phi_f \mathbf{C}^H_Y)$ as follows. Let $\Gamma_-(f)$ be the convex hull of $\{0\}$ and $\Gamma(f)$ and let $\nu_f : \Gamma_-(f) \to [0,1]$ be the piecewise linear function which has the value 1 at each point of $\Gamma(f)$. Let S denote the set of integral points in $\Gamma_-(f)$ which do not lie on any proper facet of σ. Then this part of the spectrum is given by

$$\sum_{m \in S} (n - \nu_f(m)).$$

To see this, consider for each m in the interior of $\sigma_{\mathbf{Z}}$ the differential form

$$\omega_m = z^m \frac{dz_1}{z_1} \wedge \cdots \wedge \frac{dz_n}{z_n}.$$

Lifting this to a toric resolution of f one sees that its geometrical weight with respect to the resolution is equal to $\nu_f(m)$. If this number is not greater than one, then $(n - \nu_f(m)$ is an effective contribution to the spectrum of f. If the cone σ is simplicial, this follows again from an equivariant version of the corresponding statement for Y smooth.

With this method, we can handle all unimodal ICIS with the exception of the cases dealt with in the previous section and the infinite series $I_{1,*}$, $L_{1,i}$, $L^\#_{1,i}$, $M_{1,i}$ and $M^\#_{1,i}$.

5 The Series Method

We will deal with the remaining cases using a result of M. Saito [6]. All of these are covered by the following situation. Let $Y \subset \mathbf{C}^n$ be an isolated singularity, defined by a weighted homogeneous polynomial $g(z_1, \ldots, z_n)$ of degree d (the variables z_i have weights $a_i \in \mathbf{N}$). Consider $f : Y \to \mathbf{C}$ with $f = f_0 + h$ where

- f_0 is weighted homogeneous of degree $d' > d$

- $f_0^{-1}(0)$ has a singular locus Σ of dimension one
- h is weighted homogeneous of degree $\geq d'$ and does not vanish identically on Σ.

Consider the map $\pi = (f_0, h) : Y \to \mathbf{C}^2$. Let C be a component of its critical locus, not contained in $\pi^{-1}(0)$. Then $C \setminus \{0\}$ is a $\mathbf{C}^*$–orbit $\{t * P | t \in \mathbf{C}^*\}$ for some $P \neq 0$, where $t * (z_1, \ldots, z_n) = (t^{a_1} z_1, \ldots, t^{a_n} z_n)$. Therefore $\pi(t * P) = (t^{d'} f_0(P), t^{\deg h} h(P))$. So $\pi(C)$ is not tangent to $\{0\} \times \mathbf{C}$.

Also note that π admits a globalization, as it is weighted homogeneous. Therefore we can apply [6, Thm 2.5] to π and obtain an expression for $Sp(f_0 + \epsilon h)$ for generic ϵ (if $\deg h > d'$ we can even take $\epsilon = 1$).

In the cases at hand Σ is irreducible and $f_0^{-1}(0)$ has transverse type A_1 along Σ. There are two possibilities: either the "vertical monodromy" along Σ is the identity, or it is minus the identity. Let k be the order at 0 of the restriction of h to Σ. In the former case we get the relation

$$Sp(f_0 + \epsilon h) = Sp(f_0) + (\frac{n-2}{2}) + \alpha_{k-1,n-2}$$

and in the latter case

$$Sp(f_0 + \epsilon h) = Sp(f_0) + \delta_{k,n-2}.$$

6 Tables

We follow the tables of Wall [10]. We use the following notation. As the spectrum is symmetric, we only need to list the spectrum numbers below the center of symmetry, which is 1 in the curve case and $\frac{3}{2}$ in the surface case. We will list the numerators of the spectrum numbers in increasing order, followed by their common denominator. E.g.

$$(1, 2, 2|2)$$

stands for $(\frac{1}{2}) + 2(1) + (\frac{3}{2})$. We preserve the notations $\alpha_{k,n}$ and $\delta_{k,n}$. The hypersurface singularities occurring are all of corank ≤ 3 and we will list the spectra for the functions in three variables as in [1].

We do not list the spectra for the surface singularities of type K' and J'. These are obtained from the curve singularities of type K and J respectively by suspension (with $p = 2$). Note that the Milnor numbers increase by 2 under suspension in these cases. The spectra are related by the formula

$$Sp(S') = (\frac{1}{2}) * Sp(S) + (1) + (2)$$

where $*$ stands for multiplication in the group ring $\mathbf{Z}[\mathbf{Q}]$ and the relation between S and S' is given in the following table:

S	K_n	$K_{1,i}$	$K^{\#}_{1,i}$	J_n	$J_{m+1,i}$
S'	K'_{n+2}	$K'_{1,i}$	$K^{\flat}_{1,i}$	J'_{n+2}	$J'_{m+1,i}$

The 0–modal hypersurfaces are the types A_k, D_k, E_6, E_7, E_8. The 0–modal space curves are the types $P_{k,2} (k \geq 2)$, $P_{3,3}$, $P_{4,3}$, $P_{5,3}$, J_7, J_8, J_9, K_8, K_9, G_9, G_{10} (see [3]). All 0–modal surface singularities are hypersurfaces (by [10]).

6.1 0–dimensional ICIS

If X has dimension 0, then X' is a reduced curve. Let δ denote its apparent number of double points and r its number of branches. Then $\mu(X') = 2\delta - r + 1$. Let $f : X' \to \mathbf{C}$ with $X \simeq f^{-1}(0)$. Let e_i denote the order of f on the i-th branch of X'. Then

$$Sp(X) = \delta(0) + \sum_{i=1}^{r} \alpha_{e_i-1,0} + \delta(1).$$

The proof is left to the reader.

6.2 Hypersurfaces

A_k	$\alpha_{k,2}$	$k \geq 1$
D_k	$\delta_{k-1,2} + (\frac{3}{2})$	$k \geq 4$
$T_{p,q,r}$	$(1) + (2) + \alpha_{p-1,2} + \alpha_{q-1,2} + \alpha_{r-1,2}$	$p \geq q \geq r \geq 2$
E_{6m+6}	$\alpha_{2,1} * \alpha_{3m+3,0}$	$m \geq 0$
E_{6m+7}	$(\frac{3}{2}) + \sum_{i=1}^{3} \sum_{j=1}^{2m+3} (\frac{(2m+3)i+2j}{6m+9} + \frac{1}{2})$	
E_{6m+8}	$\alpha_{2,1} * \alpha_{3m+4,0}$	$m \geq 0$
$E_{m+1,i}$	$\alpha_{2,1} * \alpha_{3m+2,0} + \alpha_{3m+2+i,2} - \alpha_{3m+2,2}$	$m \geq 1$ odd
	$\alpha_{2,1} * \alpha_{3m+2,0} + \delta_{3m+3+i,2} - \delta_{3m+3,2}$	$m \geq 0$ even
Z_{6m+5}	$(\frac{3}{2}) + \sum_{i=1}^{2} \sum_{j=1}^{3m+1} (\frac{(3m+1)i+3j}{9m+6} + \frac{1}{2})$	$m \geq 1$
Z_{6m+6}	$\sum_{i=1}^{3} \sum_{j=1}^{2m+2} (\frac{(2m+1)i+2j}{6m+5} + \frac{1}{2})$	$m \geq 1$
Z_{6m+7}	$(\frac{3}{2}) + \sum_{i=1}^{2} \sum_{j=1}^{3m+2} (\frac{(3m+2)i+3j}{9m+9} + \frac{1}{2})$	$m \geq 1$
$Z_{m-1,i}$	$(\frac{3}{2}) + \sum_{k=1}^{2} \sum_{j=1}^{m} (\frac{mk+j}{3m+1} + \frac{1}{2})$	$m \geq 1$ odd, $i \geq 0$
	$+ \alpha_{3m+i,2} - \alpha_{3m,2}$	
	$(\frac{3}{2}) + \sum_{k=1}^{2} \sum_{j=1}^{m} (\frac{mk+j}{3m+1} + \frac{1}{2})$	$m \geq 2$ even, $i \geq 0$
	$+ \delta_{3m+i+1,2} - \delta_{3m+1,2}$	
Q_{6m+4}	$\alpha_{2,0} * \delta_{3m+1,1} + (4\|3)$	$m \geq 1$
Q_{6m+5}	$\sum_{i=0}^{4m+1} (\frac{10m+7+4i}{12m+6}) + \sum_{i=0}^{2m} (\frac{14m+9+4i}{12m+6}) + (4\|3)$	$m \geq 1$
Q_{6m+6}	$\alpha_{2,0} * \delta_{3m+2,1} + (4\|3)$	$m \geq 1$
$Q_{m,i}$	$\alpha_{2,0} * \delta_{3m,1} + (4\|3) + \alpha_{3m+i-1,2} - \alpha_{3m-1,2}$	$m \geq 1$ odd, $i \geq 0$
	$\alpha_{2,0} * \delta_{3m,1} + (4\|3) + \delta_{3m+i,2} - \delta_{3m,2}$	$m \geq 2$ even, $i \geq 0$
W_{12}	$(19, 23, 24, 27, 28, 29\|20)$	
W_{13}	$(15, 18, 19, 21, 22, 23, 24\|16)$	
$W_{1,i}$	$(11, 14, 16, 17, 18\|12) + \delta_{6+i,2}$	$i \geq 0$
$W_{1,i}^{\#}$	$(11, 17\|12) + \alpha_{i+11,2}$	$i \geq 0$
W_{17}	$(18, 21, 23, 24, 26, 27, 28, 29, 30\|20)$	
W_{18}	$(25, 29, 32, 33, 36, 37, 39, 40, 41\|28)$	
S_{11}	$(15, 19, 20, 21, 24, 24\|18)$	
S_{12}	$(12, 15, 16, 17, 18, 19\|13)$	
$S_{1,i}$	$(9, 12, 13, 14\|10) + \delta_{i+5,2}$	$i \geq 0$
$S_{1,i}^{\#}$	$(9, 13\|10) + \alpha_{i+9,2}$	$i \geq 0$
S_{16}	$(15, 18, 20, 21, 22, 23, 24, 25\|17)$	
S_{17}	$(21, 25, 28, 29, 31, 32, 33, 35, 36\|24)$	
U_{12}	$(11, 14, 15, 15, 17, 18, 18\|12)$	
$U_{1,i}$	$(8, 11, 13\|9) + \alpha_{i+8,2}$	$i \geq 1$
U_{16}	$(13, 16, 18, 18, 19, 21, 21, 22\|15)$	

6.3 Space Curves

$P_{k,l}$	$(1,2,2\|2)+\alpha_{k-1,1}+\alpha_{l-1,1}$	$k\geq l\geq 2$
$FT_{k,l}$	$(1,3,3,3\|3)+\delta_{k-1,1}+\delta_{l-1,1}$	$k\geq l\geq 4$
FW_{13}	$(5,9,10,13,14,15\|16)+3(1)$	
FW_{14}	$(4,7,8,10,11,12\|13)+4(1)$	
$FW_{1,i}$	$(3,5,6,8,9\|10)+4(1)+\alpha_{i+4,1}$	$i\geq 0$
$FW_{1,i}^{\#}$	$(3,9\|10)+4(1)+\alpha_{i+9,1}$	$i\geq 0$
FW_{18}	$(5,8,10,11,13,14,15,16\|17)+4(1)$	
FW_{19}	$(7,11,14,15,18,19,21,22,23\|24)+3(1)$	
FZ_{6m+6}	$(1,2\|3)+4(1)+\sum_{i=1}^{2}\sum_{j=1}^{3m}(\frac{i}{3}+\frac{j}{3m+1})$	$m\geq 1$
FZ_{6m+7}	$(1,2\|3)+5(1)+\sum_{i=1}^{3}\sum_{j=1}^{2m}(\frac{i}{3}+\frac{2j}{6m+3})$	$m\geq 1$
FZ_{6m+8}	$(1,2\|3)+4(1)+\sum_{i=1}^{2}\sum_{j=1}^{3m+1}(\frac{i}{3}+\frac{j}{3m+2})$	$m\geq 1$
$FZ_{m-1,i}$	$(1,2\|3)+4(1)+\sum_{i=1}^{2}\sum_{j=1}^{3m-1}(\frac{mi+j}{3m})$ $+\alpha_{3m+i-1,1}-\alpha_{3m-1,1}$	$m\geq 2$ even, $i\geq 0$
	$(1,2\|3)+4(1)+\sum_{i=1}^{2}\sum_{j=1}^{3m-1}(\frac{mi+j}{3m})$ $+\delta_{3m+i,1}-\delta_{3m,1}$	$m\geq 1$ odd, $i\geq 0$
G_{n+8}	$(2,5\|6)+2(1)+\alpha_{n+4,1}$	$n\geq 1$
HA_{r+11}	$(2,4,5\|6)+4(1)+\delta_{r+3,1}$	$r\geq 0$
HB_{r+13}	$(3,6,8,9\|10)+3(1)+\delta_{r+4,1}$	$r\geq 0$
HC_{13}	$(4,7,8,10,10,11\|12)+3(1)$	
HC_{14}	$(4,5,7,8,8,10\|12)+4(1)$	
HC_{15}	$(2,4,5\|6)+(8,11,13,14\|15)+3(1)$	
HD_{13}	$(4,8,9,11,12,13\|14)+3(1)$	
HD_{14}	$(5,10,11,14,15,16,17\|18)+2(1)$	
J_{6m+7}	$\sum_{i=1}^{2}\sum_{j=0}^{3m+2}(\frac{(3m+4)i+3j}{9m+9})+2(1)$	$m\geq 0$
J_{6m+8}	$\sum_{i=1}^{3}\sum_{j=0}^{2m+1}(\frac{(2m+3)i+2j}{6m+7})+3(1)$	$m\geq 0$
J_{6m+9}	$\sum_{i=1}^{2}\sum_{j=0}^{3m+3}(\frac{(3m+5)i+3j}{9m+12})+2(1)$	$m\geq 0$
$J_{m+1,i}$	$\sum_{j=1}^{2}\sum_{k=0}^{3m+1}(\frac{(m+1)j+k}{3m+2})+2(1)$ $+\alpha_{3m+1+i,1}-\alpha_{3m+1,1}$	$m\geq 0$ even, $i\geq 0$
	$\sum_{j=1}^{2}\sum_{k=0}^{3m+1}(\frac{(m+1)j+k}{3m+2})+2(1)$ $+\delta_{3m+2+i,1}-\delta_{3m+2,1}$	$m\geq 1$ odd, $i\geq 0$
K_8	$(5,9,10,11\|12)+(1)$	
K_9	$(4,7,8,9\|10)+2(1)$	
$K_{1,i}$	$(3,6,7\|8)+\alpha_{i+3,1}+2(1)$	$i\geq 0$
$K_{1,i}^{\#}$	$(3,7\|8)+\alpha_{i+7,1}+(1)$	$i\geq 0$
K_{13}	$(5,8,10,11,12,13\|14)+2(1)$	
K_{14}	$(7,11,14,15,17,18,19\|20)+(1)$	

6.4 Surface Singularities

T_{pqrs}	$2(1) + \alpha_{p-1,2} + \alpha_{q-1,2} + \alpha_{r-1,2} + \alpha_{s-1,2} + 2(2)$	$p, q, r, s \geq 2$
$I_{1,i}$	$(5, 6, 7, 8, 8\|6) + \alpha_{i+5,2}$	$i \geq 0$
L_{10}	$(11, 12, 15, 16, 17, 18\|12)$	
L_{11}	$(9, 10, 12, 13, 14, 15, 15\|10)$	
$L_{1,i}$	$(7, 8, 10, 11, 12, 12\|8) + \delta_{i+4,2}$	$i \geq 0$
$L^{\#}_{1,i}$	$(7, 8, 11, 12\|8) + \alpha_{i+7,2}$	$i \geq 0$
L_{15}	$(12, 14, 15, 16, 17, 18, 19, 20, 21, 21\|14)$	
L_{16}	$(17, 20, 21, 24, 25, 28, 29, 30\|20)$	
M_{11}	$(8, 9, 11, 12, 12, 13\|9)$	
$M_{1,i}$	$(6, 7, 9, 10\|7) + \alpha_{i+6,2}$	$i \geq 0$
$M^{\#}_{1,i}$	$(6, 7, 9, 10\|7) + \alpha_{i+6,2}$	$i \geq 0$
M_{15}	$(10, 12, 13, 15, 15, 16, 17, 18, 18\|12)$	

6.5 Some Conclusions

For any ICIS X we let $w(X)$ denote the *spectral width* of X, i.e. the difference between the biggest and the smallest spectrum number. Then for ICIS X of positive dimension the following holds:

- $w(X) < 1 \Leftrightarrow X$ is a simple hypersurface singularity;
- if X is a unimodal ICIS then $w(X) \leq \frac{4}{3}$;
- if $w(X) < \frac{6}{5}$ then X is simple or unimodal.

The first two facts are proved by inspection of the tables. To prove the third fact, one needs the lists of the "first" bimodal singularities from [10], compute their spectral width and use the semicontinuity of the spectrum to show that the spectral width can only decrease under small deformations.

It would be very interesting to use the full power of the semicontinuity of the spectrum to get results about adjacencies between unimodal and simple ICIS.

References

[1] Arnold, V.I., S.M. Gusein–Zade, A.N. Varchenko, Singularities of differentiable maps Volume II. Birkhäuser 1988.

[2] Ebeling, W., and J.H.M. Steenbrink, Spectral pairs for isolated complete intersection singularities, submitted.

[3] Giusti, M., Classification des singularités isolées simples d'intersections complètes, Proc. Symp. Pure Math. 40 (1983) Part 1, 457–494.

[4] Goryunov, V.V., Adjacencies of spectra of certain singularities, Vestnik MGU ser. math. 1981, 4, 19–22, English translation: Moscow University Mathematics Bulletin 36 (1981), no. 4, 22–25.

[5] Saito, M., Exponents and Newton polyhedra of isolated hypersurface singularities, Math. Ann. 281 (1988), 411–417.

[6] Saito, M., On Steenbrink's conjecture, Math. Ann. 289 (1991), 703–716.

[7] Schrauwen, R., J.H.M. Steenbrink and J. Stevens, Spectral pairs and the topology of curve singularities, Proc. Symp. Pure Math. 53 (1991), 305–328.

[8] Varchenko, A.N., Zeta function of monodromy and Newton's diagram, Invent. math. 37 (1976), 253–262.

[9] Varchenko, A.N., and A.G. Khovanskii, Asymptotics of integrals over vanishing cycles and the Newton Polyhedron, Soviet Math. Dokl. 32 (1985) No, 1, 122–127.

[10] Wall, C.T.C., Classification of unimodal isolated singularities of complete intersections, Proc. Symp. Pure Math. 40 (1983) Part 2, 625–640.

Department of Mathematics,
University of Nijmegen,
Toernooiveld,
NL-6525 ED Nijmegen,
The Netherlands.

e-mail: steenbri@sci.kun.nl

Dynkin Graphs, Gabriélov Graphs and Triangle Singularities

Tohsuke Urabe

Abstract

We consider fourteen kinds of two-dimensional triangle hypersurface singularities, and which combinations of rational double points can appear on small deformations of these singularities. We show that the possible combinations can be described by Gabriélov graphs and Dynkin graphs.

1 Review of results by Russian mathematicians

In this article we assume that every variety is defined over the complex field.

First we briefly explain some results by Arnold and Gabriélov. In [1] Arnold introduced an invariant m (a non-negative integer) called *modality* or *modules number*, and made an extensive classification of hypersurface singularities. Though we find singularities of any dimension in Arnold's list, we consider singularities of dimension two in particular.

His class of singularities with $m = 0$ coincides with the class of rational double points. It is well known that each rational double point corresponds to a connected Dynkin graph of type A, D or E in the theory of Lie algebras. (Durfee [4].)

The class with $m = 1$ consists of three subclasses (λ is a parameter).

1. Three simple elliptic singularities: J_{10}, X_9, P_8.

2. Cusp singularities $T_{p,q,r}$ (where $\frac{1}{p} + \frac{1}{q} + \frac{1}{r} < 1$):

$$x^p + y^q + z^r + \lambda xyz = 0 \qquad (\lambda \neq 0).$$

3. Fourteen triangle singularities (also called exceptional singularities):

$$\begin{array}{llllllll} E_{12} & Z_{11} & Q_{10} & \quad & W_{12} & S_{11} & \quad & U_{12} \\ E_{13} & Z_{12} & Q_{11} & & W_{13} & S_{12} & & \\ E_{14} & Z_{13} & Q_{12} & & & & & \end{array}$$

where E_{12}, W_{12} and U_{12} are defined by

$$\begin{aligned} E_{12} &: x^7 + y^3 + z^2 + \lambda x^5 y = 0 \\ W_{12} &: x^5 + y^4 + z^2 + \lambda x^3 y^2 = 0 \\ U_{12} &: x^4 + y^3 + z^3 + \lambda x^2 yz = 0. \end{aligned}$$

For the polynomials defining the remaining singularities see Arnold [1].

We now consider Gabriélov's results (Gabriélov [5]). Let $f(x, y, z) = 0$ be one of the defining polynomials of the fourteen hypersurface triangle singularities. It defines a singularity at the origin. We consider the *Milnor fiber*, i.e.,

$$F = \{ (x, y, z) \in \mathbf{C}^3 \mid |x|^2 + |y|^2 + |z|^2 < \varepsilon^2,\ f(x, y, z) = t \}$$

where ϵ is a sufficiently small positive real number and t a non-zero complex number whose absolute value is sufficiently small compared with ϵ. The pair

$$(H_2(F, \mathbf{Z}),\ \text{the intersection form})$$

is called the *Milnor lattice*, and $\mu = \operatorname{rank} H_2(F, \mathbf{Z})$ is called the *Milnor number* of the singularity. Gabriélov computed the Milnor lattice for the fourteen hypersurface triangle singularities. According to him, there exists a basis $e_1, e_2, \ldots, e_\mu$ of $H_2(F, \mathbf{Z})$ such that each e_i is a vanishing cycle (thus in particular $e_i \cdot e_i = -2$) and the intersection form is represented by the dual graph shown in Figure 1. In this graph the basis $e_1, e_2, \ldots, e_\mu$ has a one-to-

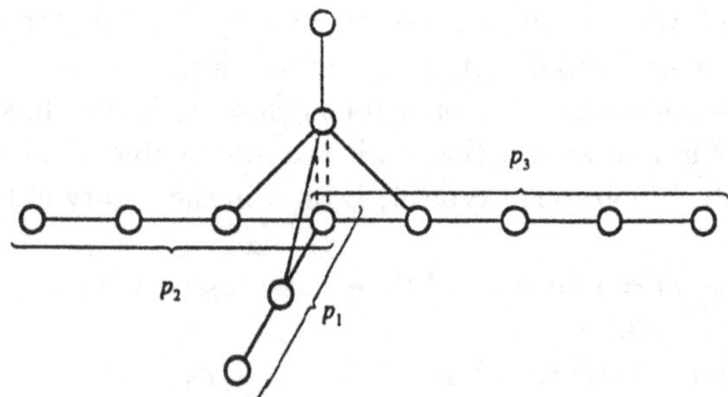

Figure 1.

one correspondence with vertices. Edges indicate intersection numbers. Two vertices corresponding to e_i and e_j are not connected if $e_i \cdot e_j = 0$. They are connected by a single solid edge if $e_i \cdot e_j = 1$, and a double dotted edge if $e_i \cdot e_j = -2$. The three integers p_1, p_2, p_3 are the numbers of vertices in the corresponding arms. They depend on the type of the triangle singularity. The triplets (p_1, p_2, p_3) corresponding to the above fourteen symbols are as follows (thus Figure 1 is the graph for S_{12}):

(2, 3, 7)	(2, 4, 5)	(3, 3, 4)	(2, 5, 5)	(3, 4, 4)	(4, 4, 4)
(2, 3, 8)	(2, 4, 6)	(3, 3, 5)	(2, 5, 6)	(3, 4, 5)	
(2, 3, 9)	(2, 4, 7)	(3, 3, 6)			

The main part of the graph is called the *Gabriélov graph*. This defines a lattice P^* with a basis $e_1, e_2, \ldots, e_{\mu-2}$ if we apply the above mentioned rule; see Figure 2. It is easy to check that P^* has signature $(1, \mu - 3)$, and $H_2(F, \mathbf{Z}) \cong P^* \oplus H$ as lattices. Here $H = \mathbf{Z}\,u + \mathbf{Z}\,v$ denotes the hyperbolic plane, i.e., a lattice of rank 2 with a basis u, v satisfying $u \cdot u = v \cdot v = 0$ and $u \cdot v = v \cdot u = 1$, and $\oplus$ denotes the orthogonal direct sum.

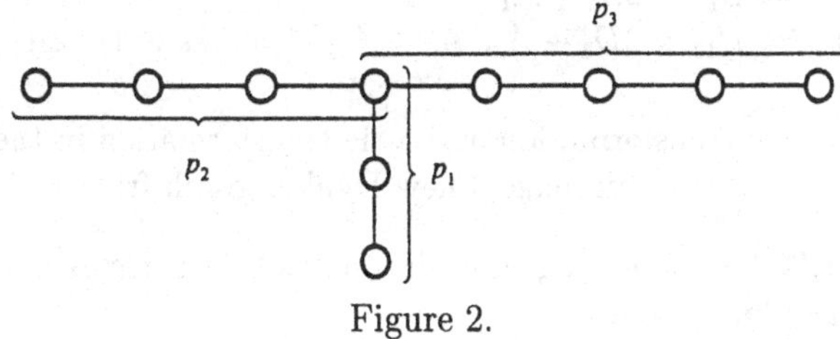

Figure 2.

2 Singularities on deformation fibers of triangle singularities

A finite disjoint union of connected Dynkin graphs is also called a Dynkin graph. Let T denote one of the above fourteen symbols of hypersurface triangle singularities. By $PC(T)$ we denote the set of Dynkin graphs G with several components such that there exists a small deformation fiber Y of a singularity of type T satisfying the following conditions:

1. Y has only rational double points as singularities.

2. The combination of rational double points on Y corresponds to the graph G exactly.

Here, the type of each component of G corresponds to the type of a rational double point on Y, and the number of components of each type corresponds to the number of rational double points of each type on Y. If G has a_k components of type A_k for each $k \geq 1$, d_ℓ components of type D_ℓ for each $\ell \geq 4$ and e_m components of type E_m for $m = 6,\ 7,\ 8$, we identify G with the formal sum $G = \sum a_k A_k + \sum d_\ell D_\ell + \sum e_m E_m$.

Mr. F.-J. Bilitewski informed me that he has a complete listing of Dynkin graphs of $PC(T)$ for every T of the above fourteen. (Bilitewski [2].)

Theorem 2.1. *Let T be one of the above fourteen symbols of hypersurface triangle singularities. Let G be a Dynkin graph with only components of type A, D or E. The following conditions* **(A)** *and* **(B)** *are equivalent:*

(A) $G \in PC(T)$.

(B) *Either* **(B-1)** *or* **(B-2)** *holds:*

(B-1) *G can be made by an elementary transformation or a tie transformation from a Dynkin subgraph of the Gabriélov graph of type T.*

(B-2) *G is one of the following exceptions:*

$$\begin{aligned} T &= Z_{13} : A_7 + A_4 \\ T &= S_{11} : 2A_4 + A_1 \\ T &= U_{12} : 2D_4 + A_2,\ A_6 + A_4,\ A_5 + A_4 + A_1,\ 2A_4 + A_1 \end{aligned}$$

An elementary transformation and a tie transformation in the above are operations by which we can make a new Dynkin graph from a given one.

Definition 2.2. The following procedure is called an *elementary transformation* of a Dynkin graph:

1. Replace each connected component by the corresponding extended Dynkin graph.

2. Choose in an arbitrary manner at least one vertex from each component (of the extended Dynkin graph) and then remove these vertices together with the edges incident with them.

An extended Dynkin graph is a graph obtained from a connected Dynkin graph by adding one vertex and one or two edges (Bourbaki [3]). Figure 3 shows extended Dynkin graphs. The numbers attached to the vertices are the coefficients of the maximal root (these appear in the definition of a tie transformation below). We can obtain the corresponding Dynkin graph if we erase one of the vertices which is labelled 1 together with the edges incident with it.

Definition 2.3. Assume that by applying the following procedure to a Dynkin graph G we have obtained the Dynkin graph $\bar{G}$. Then, we call the following procedure a *tie transformation* of a Dynkin graph:

1. Replace each component of G by the extended Dynkin graph of the same type. Attach the corresponding coefficient of the maximal root to each vertex of the resulting extended graph $\tilde{G}$.

2. Choose, in an arbitrary manner, subsets A, B of the set of vertices of the extended graph $\tilde{G}$ satisfying the following conditions:

   ⟨*a*⟩ $A \cap B = \emptyset$.

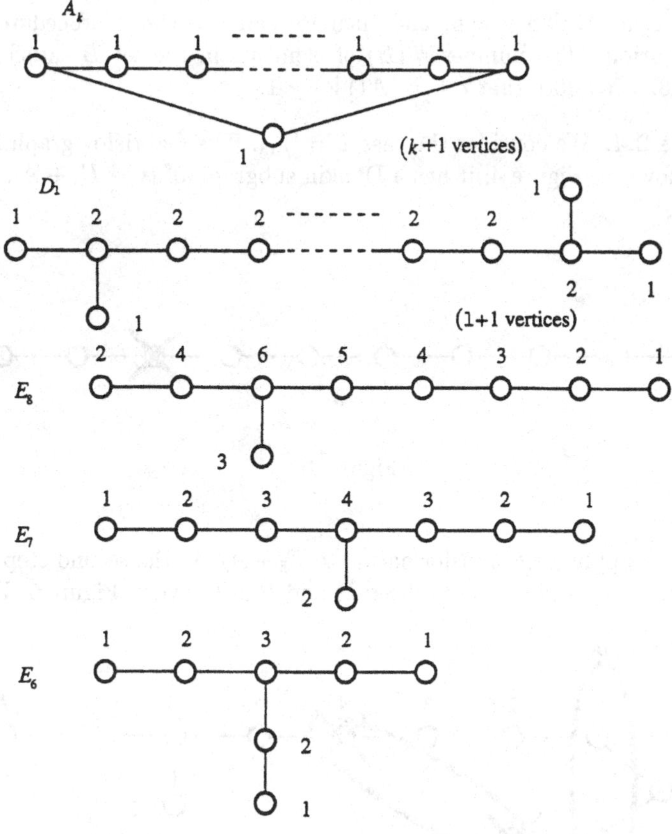

Figure 3.

⟨*b*⟩ Choose arbitrarily a component $\tilde{G}''$ of $\tilde{G}$ and let V be the set of vertices in $\tilde{G}''$. Let ℓ be the number of elements in $A \cap V$. Let $n_1, n_2, \ldots, n_\ell$ be the numbers attached to $A \cap V$. Also, let N be the sum of the numbers attached to elements in $B \cap V$ (if $B \cap V = \emptyset$ we set $N = 0$). Then, the greatest common divisor of the $\ell + 1$ numbers $N, n_1, n_2, \ldots, n_\ell$ is 1.

3. Erase all attached integers.

4. Remove vertices belonging to A together with the edges incident with them.

5. Draw another new vertex called θ. Connect θ and each vertex in B by a single edge.

Remark. After following the above procedure 1–5, the resulting graph $\bar{G}$ is often not a Dynkin graph. We consider only the cases where the resulting graph $\bar{G}$ is a Dynkin graph, and then we call the above procedure a tie transformation. The number $\#(B)$ of elements in the set B satisfies $0 \leq \#(B) \leq 3$. Also note that $\ell = \#(A \cap V) \geq 1$.

Example 2.4. We consider the case $T = W_{13}$. The Gabriélov graph in this case is shown in Figure 4; it has a Dynkin subgraph of type $E_8 + A_2$.

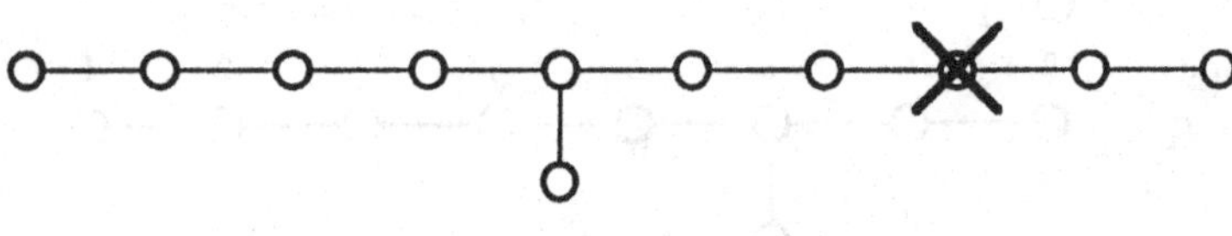

Figure 4.

First we apply a tie transformation to $E_8 + A_2$. In the second step of the transformation we can choose subsets A and B as shown in Figure 5. For the

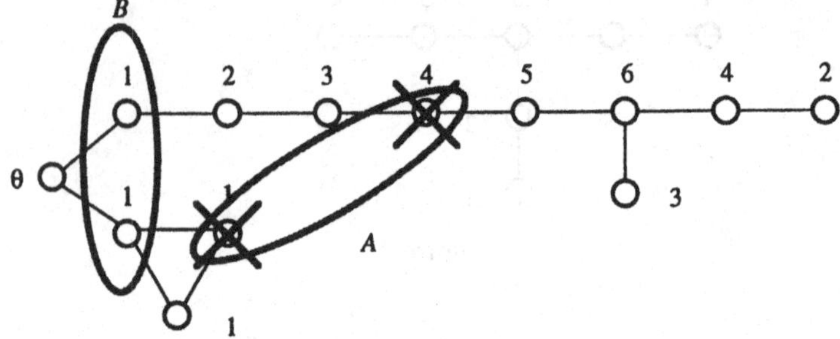

Figure 5.

component of type E_8: $\ell = 1$, $n_1 = 4$, $N = 1$ and thus $\gcd(n_1, N) = 1$. For the component A_2: $\ell = 1$, $n_1 = 1$, $N = 1$ and thus $\gcd(n_1, N) = 1$. One sees that the condition $\langle b \rangle$ is satisfied. As the result of the transformation one gets a graph of type $A_6 + D_5$. By our theorem one can conclude $A_6 + D_5 \in PC(W_{13})$.

Second we apply an elementary transformation to $E_8 + A_2$. As in Figure 6 we can get $E_6 + 2A_2$. Thus $E_6 + 2A_2 \in PC(W_{13})$.

3 K3 surfaces and lattice theory

It is known that the fourteen hypersurface triangle singularities enjoy an interesting property called strange duality. (Pinkham [7].) Let T be one of the above fourteen symbols of hypersurface triangle singularities. Associated

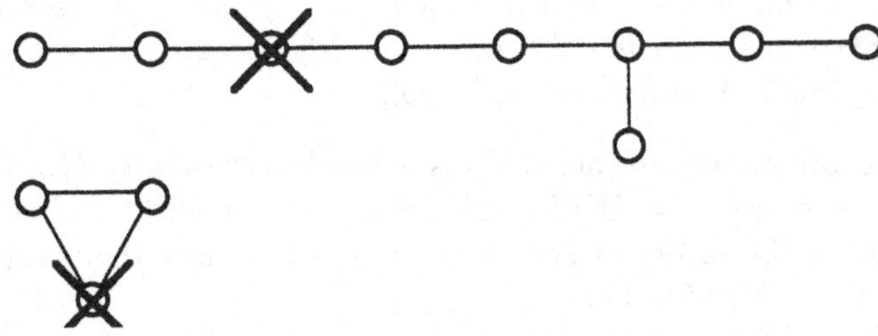

Figure 6.

with T, we have another symbol T^* also in the above fourteen symbols of hypersurface triangle singularities. This T^* is called the *dual* of T. The dual of the dual coincides with the original one, i.e., $(T^*)^* = T$.

For the following six singularities the dual coincides with itself (i.e., $T^* = T$): E_{12}, Z_{12}, Q_{12}, W_{12}, S_{12}, U_{12}. For the following four pairs the dual is another member of the pair: $\{E_{13}, Z_{11}\}$, $\{E_{14}, Q_{10}\}$, $\{Z_{13}, Q_{11}\}$, $\{W_{13}, S_{11}\}$.

Following Looijenga [6], we explain the relation between triangle singularities and K3 surfaces below. Let T be one of the above fourteen symbols of hypersurface triangle singularities. Let Γ^* be the Gabriélov graph of the dual T^*. We can define a reducible curve IF on a surface whose dual graph coincides with Γ^*. The curve $IF = IF(T)$ is called *the curve at infinity* of type T. The irreducible components are all smooth rational curves C with $C \cdot C = -2$ and have a one-to-one correspondence with the vertices of Γ^*. For two components C, C' of IF the intersection number $C \cdot C'$ is equal to one or zero, according as the corresponding vertices in Γ^* are connected in Γ^* or not.

Let G be a Dynkin graph with components of type A, D or E only. Assume that there exists a smooth K3 surface Z satisfying the following conditions **(a)** and **(b)**:

(a) Z contains the curve at infinity $IF = IF(T)$ of type T as a subvariety.

(b) Let E be the union of all smooth rational curves on Z disjoint from IF. The dual graph of the components of E coincides with the graph G.

(Note that an irreducible curve C on a K3 surface is a smooth rational curve if, and only if, $C \cdot C = -2$.)

Contracting every connected component of E to a rational double point and then removing the image of IF, we obtain an open variety $\tilde{Y}$.

Proposition 3.1 (Looijenga [6]). *1. Under the above assumption there exists a small deformation fiber Y of a singularity of type T homeomorphic to $\tilde{Y}$.*

2. Let Y be a small deformation fiber of a singularity of type T. Assume that Y has only rational double points as singularities, and the combi-

nation of rational double points on Y corresponds to a Dynkin graph G. Then, there exists a K3 surface satisfying **(a)** *and* **(b)**, *and the corresponding $\tilde{Y}$ is homeomorphic to Y.*

By the above proposition our study is reduced to the study of K3 surfaces containing the curve $IF = IF(T)$. K3 surfaces are complicated objects, but it is known, by the theory of periods, that we can reduce the study of K3 surfaces to the study of lattices.

Recall from (Urabe [10]) that a free module over $\mathbf{Z}$ of finite rank equipped with an integral symmetric bilinear form $(\ ,\ )$ is called a *lattice.* Further, if a free module L over $\mathbf{Z}$ of finite rank has a symmetric bilinear form $(\ ,\ )$ with values in the field of rational numbers, then L is called a *quasi-lattice.* For simplicity we write $x^2 = (x, x)$.

Let L be a quasi-lattice and M be a submodule. The submodule

$$\tilde{M} = \{\, x \in L \mid mx \in M \text{ for some non-zero integer } m \,\}$$

is called the *primitive hull* of M in L. We say that M is *primitive* if $M = \tilde{M}$, and an element $x \in L$ is *primitive* if $M = \mathbf{Z}\,x$ is primitive. We say that an embedding $M \to L$ of quasi-lattices is a *primitive embedding* if the image is primitive. If M is non-degenerate and primitive as a sub-quasi-lattice, we can define the canonical induced bilinear form on the quotient module L/M.

Let L be a quasi-lattice, and FL be a submodule of L such that the index $\#(L/FL)$ is finite. Set

$$\begin{aligned} R = \{\, \alpha \in FL \mid \alpha^2 = -2 \,\} \cup \{\, \beta \in L \mid \beta^2 = -1 \text{ or } \beta^2 = -2/3 \,\} \\ \cup \{\, \gamma \in L \mid \gamma^2 = -1/2,\ 2\gamma \in FL \,\}. \end{aligned}$$

The set $R = R(L, FL)$ is called the *root system* of (L, FL) and every element $\alpha \in R$ is called a *root.* If the pair (L, FL) satisfies the following conditions **(R1)** and **(R2)** then (L, FL) is called a *root module*:

(R1) $2\,(x, \alpha)\,/\alpha^2$ is an integer for every $x \in L$ and $\alpha \in R$.

(R2) $s_\alpha(FL) = FL$ for every $\alpha \in R$.

Under **(R1)**, for every $\alpha \in R$ we can define an isomorphism $s_\alpha : L \to L$ preserving the bilinear form, by setting for $x \in L$, $s_\alpha(x) = x - 2(x, \alpha)\alpha/\alpha^2$.

Let (L, FL) be a root module. If $L = FL$, we say that it is *regular* and simply write FL. Let M be a submodule of L. It is easy to check that the pair $(M, FL \cap M)$ is again a root module. Below we identify M with the pair $(M, FL \cap M)$. If the root system of M and the root system of $\tilde{M}$ coincide then we say that M is *full.* An embedding $M \to L$ of quasi-lattices is a *full embedding* if the image is full.

Let G be a Dynkin graph with several components of type A, D or E only. We can define a lattice and its basis such that the corresponding dual graph coincides with G. This lattice is called the *root lattice* of type G

and is denoted by $Q(G)$. Note $Q(G)$ is a regular root module with a basis $\alpha_1, \alpha_2, \ldots, \alpha_r$ with $\alpha_i^2 = -2$ for every i. Let Λ_N denote an even unimodular lattice with signature $(N, 16+N)$ for $N \geq 0$. The isomorphism class of Λ_N is unique if $N \geq 1$ and thus $\Lambda_N \cong \Lambda_{N-1} \oplus H$. For a K3 surface Z the second cohomology group $H^2(Z, \mathbf{Z})$ with the intersection form is a lattice isomorphic to Λ_3. Let $P = P(T)$ be the lattice whose dual graph is the Gabriélov graph Γ^* of the dual T^*. Assume that there exists a K3 surface Z satisfying the above condition **(a)**. The classes of the components of IF generate a primitive sublattice in $H^2(Z, \mathbf{Z})$, which is isomorphic to P.

Proposition 3.2. *1. If $N \geq 1$, there is a primitive embedding $P \to \Lambda_N$.*

2. *If $N \geq 2$, a primitive embedding $P \to \Lambda_N$ is unique up to automorphisms of Λ_N.*

3. *If $N \geq 1$, for any embedding $P \to \Lambda_N$, the pair $\left(\Lambda_N/\tilde{P}, F_N\right)$ is a root module, where F_N is the image of the orthogonal complement of P in Λ_N by the canonical surjective homomorphism $\Lambda_N \to \Lambda_N/\tilde{P}$.*

4. *For any primitive embedding $P = P(T) \to \Lambda_2$ the orthogonal complement F_2 of P in Λ_2 has a basis whose dual graph coincides with the Gabriélov graph of type T.*

With the aid of Looijenga's results in [6] we can show the following:

Proposition 3.3. *We fix a primitive embedding $P \to \Lambda_3$. There exists a K3 surface Z satisfying the above conditions* **(a)** *and* **(b)** *if, and only if, there is a full embedding $Q(G) \to \Lambda_3/P$.*

Corollary 3.4. *$G \in PC(T)$ if, and only if, there is a full embedding $Q(G) \to \Lambda_3/P(T)$.*

By Proposition 3.3 our study has been reduced to lattice theory. Next, we have to consider properties of the lattice $P = P(T)$ depending on T closely. Let T be one of fourteen symbols of hypersurface triangle singularities.

Proposition 3.5. *We fix $N \geq 1$.*

1. *For any T and any embedding $P(T) \to \Lambda_N$ the quasi-lattice $\Lambda_N/\tilde{P}(T)$ does not contain an element β with $\beta^2 = -1$.*

2. *The root module $\left(\Lambda_N/\tilde{P}(T), F_N\right)$ contains a root γ with $\gamma^2 = -1/2$ for some embedding $P(T) \to \Lambda_N$ if, and only if, $T = E_{13}$, Z_{12}, Q_{11}, W_{13} or U_{12}. It contains a root γ with $\gamma^2 = -1/2$ for some primitive embedding $P(T) \to \Lambda_N$ if, and only if, $T = E_{13}$, Z_{12} or Q_{11}.*

3. *The root module $\left(\Lambda_N/\tilde{P}(T), F_N\right)$ contains a root β with $\beta^2 = -2/3$ for some embedding $P(T) \to \Lambda_N$ if, and only if, $T = E_{14}$, Z_{13} or Q_{12}.*

Consider the case where (L, FL) is a root module such that the bilinear form on L has signature $(1, \operatorname{rank} L - 1)$. In this case we can apply hyperbolic geometry, and we can give the generalization of the theory in the negative definite case such as the Weyl chamber and the Dynkin graph. The generalized Dynkin graph in this case is called the *Coxeter-Vinberg graph* (Vinberg [12]).

We need to consider the Coxeter-Vinberg graph of $(\Lambda_2/P(T), F_2)$. By Proposition 3.2.4 we can expect that it is related to the associated Gabriélov graph. We fix a primitive embedding $P(T) \to \Lambda_2$.

Proposition 3.6. *Let $\tilde{\Gamma}$ denote the Coxeter-Vinberg graph of $(\Lambda_2/P(T), F_2)$.*

1. *We can draw $\tilde{\Gamma}$ in finite steps if, and only if, $T \neq S_{11}$, S_{12}.*
2. *If $T \neq W_{12}$, W_{13}, S_{11}, S_{12}, U_{12}, every vertex in $\tilde{\Gamma}$ corresponds to a root.*
3. *If $T = W_{12}$, every vertex in $\tilde{\Gamma}$ corresponds to either a root α with $\alpha^2 = -2$ or an element δ with $\delta^2 = -2/5$.*
4. *If $T = W_{13}$ or U_{12}, every vertex in $\tilde{\Gamma}$ corresponds to either a root α with $\alpha^2 = -2$ or an element δ with $\delta^2 = -1/2$ and $2\delta \notin F_2$.*
5. *If $T = E_{12}$, Z_{11} or Q_{10}, the Gabriélov graph coincides with $\tilde{\Gamma}$.*
6. *If $T = E_{13}$, E_{14}, Z_{12}, Z_{13}, Q_{11}, Q_{12}, W_{12} or U_{12}, the Gabriélov graph is the subgraph of $\tilde{\Gamma}$ consisting of all vertices corresponding to a root α with $\alpha^2 = -2$.*
7. *If $T = W_{13}$, the Gabriélov graph is the maximal subgraph of $\tilde{\Gamma}$ such that every vertex corresponds to a root α with $\alpha^2 = -2$, and if α, β are roots corresponding to two vertices, then $(\alpha, \beta) \neq -2$.*

We can explain the main ideas in the proof of Theorem 2.1 here. Let $\overline{PC}(T)$ denote the set of all Dynkin graphs made from a Dynkin subgraph of the Gabriélov graph of type T by an elementary transformation or a tie transformation. We assume that $G \in \overline{PC}(T)$ was made from a Dynkin subgraph G' of the Gabriélov graph, and fix a primitive embedding $P \to \Lambda_N$ for $N = 2, 3$. By Proposition 3.2.4 there is a primitive embedding $Q(G') \to F_2$. By the theory of elementary and tie transformations (Urabe [8], [9]) we can conclude that there is a full embedding $Q(G) \to F_3 \cong F_2 \oplus H$ into the regular root module F_3. Assume $T \neq E_{13}$, Z_{12}, Q_{11}, E_{14}, Z_{13}, Q_{12} here. By Proposition 3.5 the composition $Q(G) \to F_3 \subset \Lambda_3/P$ defines a full embedding into the root module $(\Lambda_3/P, F_3)$ in these cases. By Corollary 3.4 we have $G \in PC(T)$. Thus $\overline{PC}(T) \subset PC(T)$.

Next, we determine the difference $PC(T) - \overline{PC}(T)$. Let r be the number of vertices of a graph G. It is easy to see that if $G \in PC(T)$, then $r \leq \mu - 2$. In case $T \neq S_{11}$, S_{12}, using Proposition 3.6 we can show that conditions $G \in PC(T)$ and $G \in \overline{PC}(T)$ are equivalent if $r \leq \mu - 5$. Thus we can assume $r = \mu - 2$, $\mu - 3$ or $\mu - 4$. For triangle singularities the Milnor number μ is relatively small, and it is easy to check whether a Dynkin graph G belongs to $PC(T) - \overline{PC}(T)$ case-by-case. To tell the truth, we could not succeed in finding any effective method except case-by-case checking. This is a weak point of our theory. I regret this fact and hope that somebody can improve it. If $T = S_{11}$ or S_{12}, the checking becomes more complicated since we have no Coxeter-Vinberg graph.

Now, if $T \neq W_{12}, W_{13}, S_{11}, S_{12}, U_{12}$, then because of Proposition 3.6.2 we can formulate another theorem (Urabe [11]). Here we do not start from a Gabriélov graph but a Dynkin graph possibly with a component of type BC_1 or G_2, and the number of transformations is not one but two. No exception now appears, even in the case Z_{13}, since we can make $A_7 + A_4$ from $E_7 + G_2$ by two tie transformations. For $T = E_{13}, Z_{12}, Q_{11}, E_{14}, Z_{13}$ or Q_{12}, Theorem 2.1 follows from this alternative approach. Also for $T = W_{12}, W_{13}, S_{11}, S_{12}, U_{12}$ this alernative is also possible, but becomes very complicated, because Proposition 3.6.2 does not hold for these singularities. The approach here is not worth discussing. Details of the proof will appear elsewhere.

It is very strange that our Theorem 2.1 has a few exceptions in a few cases. Perhaps this is because our theory has a missing ingredient.

Problem. Find the missing part of the theory and give a simple characterization of the set $PC(T)$ without exceptions.

This may be very difficult, but I believe that there does exist a solution.

References

[1] V. I Arnold, Local normal forms of functions, *Invent. Math.*, 35 (1976), 87–109.

[2] F.-J. Bilitewski, *Deformationen von Dreiecks- und Viereckssingularitäten der Einbettungsdimension drei*, Bonner Mathematische Schriften **237**, Bonn, 1992.

[3] N. Bourbaki, *Groupes et algèbre de Lie*, Hermann, Paris, 1968.

[4] A. H. Durfee, Fifteen characterization of rational double points and simple critical points, *Enseign. Math. II*, 25 (1979), 131–163.

[5] A. M. Gabriélov, Dynkin diagrams for unimodular singularities, *Functional Anal. Appl.*, 8 (1974), 192–196.

[6] E. Looijenga, The smoothing components of a triangle singularity II, *Math. Ann.*, 269 (1984), 357–387.

[7] H. C. Pinkham, Singularités exceptionelles, la dualité étrange d'Arnold et les surfaces K-3, *C. R. Acad. Sci. Paris Sér. A*, 284 (1977), 615–618.

[8] T. Urabe, Elementary transformations of Dynkin graphs and singularities on quartic surfaces, *Invent. Math.*, 87 (1987), 549–572.

[9] T. Urabe, Tie transformations of Dynkin graphs and singularities on quartic surfaces, *Invent. Math.*, 100 (1990), 207–230.

[10] T. Urabe,*Dynkin graphs and quadrilateral singularities*, Springer-Verlag, Berlin Heidelberg New-York, 1993.

[11] T. Urabe, Dynkin graphs and triangle singularities, *Kodai Math.J.*, 17 (1994), 395–401.

[12] È. B. Vinberg, On groups of unit elements of certain quadratic forms, *Math. USSR Sbornik*, 16 (1972), 17–35.

Department of Mathematics
Tokyo Metropolitan University
Minami-Ohsawa 1-1, Hachioji-shi
Tokyo, 192-03 JAPAN

e-mail: urabe@comp.metro-u.ac.jp

Differential Forms on Singular Varieties and Cyclic Homology

J.P. Brasselet A. Legrand

To Terry Wall.

Abstract

A classic result of A. Connes asserts that the Frechet algebra of smooth functions on a smooth compact manifold X provides, by a purely algebraic procedure, the de Rham cohomology of X. The procedure uses the Hochschild and cyclic homology of this algebra.

For a Thom-Mather stratified variety, we construct a Frechet algebra of functions on the regular part and a module of poles along the singular part. We associate to these objects a complex of differential forms and a Hochschild mixed complex, on the regular part, both with poles along the singular part. The de Rham cohomology of the first complex and the periodic cyclic homology of the second one are the intersection homology of the variety, the corresponding perversity being determined by the orders of poles.

Detailed proofs will appear in a forthcoming publication.

1 Introduction

The aim of this paper is to extend to singular varieties the theorems of de Rham and Connes, using intersection homology, due to Goresky and MacPherson which is better adapted to singularities than classical homology. Let X be a C^∞ compact manifold, $\Omega^*(X)$ the de Rham complex and $H^*_{\mathrm{dR}}(X)$ its cohomology. A. Connes shows that the classical differential de Rham construction

$$X \Longrightarrow \Omega^*(X) \Longrightarrow H^*_{\mathrm{dR}}(X)$$

can be provided by a purely algebraic process

$$C^\infty(X) \Longrightarrow HH^*(C^\infty(X)) \Longrightarrow PHC^*(C^\infty(X))$$

where $HH^*(C^\infty(X))$ is the Hochschild homology of the Frechet algebra of C^∞-functions on X and $PHC^*(C^\infty(X))$ is the periodic cyclic homology [Co].

Now let X be a Thom-Mather stratified variety, with singular part $\Sigma \subset X$, the regular stratum $X - \Sigma$ is an open manifold. We want to construct an algebra $C^\infty_{\bar{\alpha}}(X)$ of differentiable functions on $X - \Sigma$ such that

$$C^\infty_c(X-\Sigma) \subset C^\infty_{\bar{\alpha}}(X) \subset C^\infty(X-\Sigma)$$

and which contains enough geometric information on the compactification of $X - \Sigma$ by the singular variety Σ to carry out Connes program in this singular framework. Evidently the two extremal algebras fail to have the desired property, we need to add asymptotic control to retain information "near Σ". Such control is defined in the following way. As X is a Thom-Mather stratified variety, for each singular stratum $S_i \subset \Sigma$, there is a distance function r_i defined on a neighbourhood of this stratum. Then we define $C^\infty_{\bar{\alpha}}(X)$ as the algebra of differentiable functions $f \in C^\infty(X - \Sigma)$ such that for each i, the ratio $f/r_i^{\alpha_i}$ is bounded in a neighbourhood of the singular stratum S_i, for some fixed power $\alpha_i > 0$ and with similar control for all derivatives.

Now we define two complexes: to obtain the first one, we consider the (non differential) graded algebra of differential forms such that the asymptotic behaviour of coefficients of k-forms belongs to the ideal $C^\infty_{\bar{\alpha}}(X)^k$. We multiply this graded algebra by a module of functions in the variables r_i admitting poles on Σ and whose maximum order β_i depends only on the strata S_i (and not the degree of the form). In fact, we also need to stabilize relative to the differential d. We obtain an "intersection complex", i.e. its cohomology is the Goresky-MacPherson intersection cohomology (Theorem 1). The perversity $\bar{p}$ is related to the power of controls $r_i^{\alpha_i}$ and the orders of poles $1/r^{\beta_i}$ by the formula

$$p_i = i - 2 - \left[\frac{\beta_i}{\alpha_i}\right].$$

So the perversity can be modified independently using either the orders of poles, or the control powers.

The second complex will be a mixed complex, which allows us to carry out Connes algebraic procedure: The Hochschild homology is the previous intersection complex and the periodic cyclic homology is the intersection cohomology (Theorem 3). In fact, at this level, we meet some obstacles, mainly that the control conditions are not compatible with the product and the exterior derivative. For simplicity, we make explicit this mixed complex, and give the result, in the case of the cone on a smooth compact manifold.

The authors thank the referee whose suggestions and remarks helped to improve the quality of the paper. The first author thanks the organizers of Terry Wall's 60th birthday meeting in Liverpool for the invitation to give a lecture during the conference.

2 Definition of Intersection Homology

We will consider a singular variety X endowed with a Thom-Mather C^∞ stratification i.e. a filtration of X by closed subsets

$$X = X_n \supset X_{n-1} \supset X_{n-2} \supset \cdots \supset X_1 \supset X_0 \supset X_{-1} = \emptyset \qquad (*)$$

where $\Sigma = X_{n-1}$ is the singular part. Each stratum $S_i = X_i - X_{i-1}$ is either empty or an i-dimensional C^∞-manifold and there are

an open neighborhood T_i of S_i in X,
a continuous retraction π_i of T_i on S_i,
a continuous function $\rho_i : T_i \to [0, 1[$,

such that $S_i = \{x \in T_i | \rho_i(x) = 0\}$ and the (T_i, π_i, ρ_i) satisfy the axioms of Mather [Ma].

These data imply the following local triviality condition:

$$\forall\, x \in S_{n-j}, \quad \exists\, U_x \subset X \text{ and a homeomorphism } \psi_x : U_x \to B^{n-j} \times cL_x$$

where B^{n-j} is the standard open $(n-j)$-dimensional ball and cL_x is the cone over the link L_x. The link is assumed to be stratified and independent of the point $x \in S_{n-j}$ and ψ_x preserves the stratifications of U_x (induced by the one of X) and the one of the product $B^{n-j} \times cL_x$. The parameter of the cone corresponds to the Mather distance function ρ_{n-j}. For a complete definition see for instance [GM2].

Now let us recall the definition of intersection homology due to M. Goresky and R. MacPherson [GM1]. Given a stratified singular variety, the idea of intersection homology is to consider chains and cycles whose intersections with the strata are "not too big". Allowable chains and cycles meet the strata with a controlled and fixed defect of transversality. This defect is given by a sequence of integers $\bar{p} = (p_0, p_1, p_2, \dots, p_j, \dots, p_n)$ (called the *perversity*), which satisfies:

$$p_0 = p_1 = p_2 = 0 \qquad \text{and} \qquad p_j \le p_{j+1} \le p_j + 1 \quad \text{for } j \ge 2$$

Let $C_i(X)$ be any "classical" chain complex on X with integer coefficients, we can define the complex of $\bar{p}$-*allowable* chains:

$$\begin{aligned} IC_i^{\bar{p}}(X) = \{\xi \in C_i(X) \ : \ & \dim(|\xi| \cap X_{n-j}) \le i - j + p_j \ \text{ and} \\ & \dim(|\partial\xi| \cap X_{n-j}) \le i - 1 - j + p_j\}\ , \end{aligned}$$

the *intersection homology groups* $IH_i^{\bar{p}}(X)$ are homology groups of this complex.

We shall be mainly interested in the axiomatic definition of intersection homology. Namely, if a complex of sheaves on X satisfies the so-called *perverse sheaves* axioms [GM2], then the hypercohomology of (X with values in) this perverse sheaf is the intersection homology of X. The main axioms of perverse sheaves originate from the following local computation property (cf [GM2]): let cL be the open cone over an $(n-1)$-dimensional (singular) variety L, and $\bar{p}$ be any perversity, then we have:

$$IH_i^{\bar{p}}(c(L)) \cong \begin{cases} IH_i^{\bar{p}}(L) & \text{if } i < n - p_n - 1, \\ 0 & \text{if } i \ge n - p_n - 1. \end{cases}$$

Intersection homology is the good theory for extending many classical results from manifolds to singular varieties such as Poincaré duality (intersection of cycles) [GM1], Morse theory [GM3], Lefschetz hyperplane theorem [GM3], hard Lefschetz theorem [BBD], Hodge decomposition [Sa] and de Rham theorem (see the following section).

3 de Rham Theorem for Stratified Varieties and Polar Forms

The constructions we will use are taken from those of Cheeger [Ch] and Cheeger-Goresky-MacPherson [CGM], who proved in particular situations the "standard" result concerning $\mathcal{L}^2$-cohomology of differential forms, i.e. the isomorphism:

$$H^*_{(2)}(X-\Sigma) \cong \mathrm{Hom}(IH^{\bar{p}}_*(X,\mathbf{R});\mathbf{R}) .$$

Many authors proved de Rham theorems for $\mathcal{L}^2$-forms, or $\mathcal{L}^p$-forms, in different situations but always in the framework of intersection homology (see [Bra] for a partial survey).

The constructions that we give are also related to the theory of shadow forms [BGM] which provides another way to extend the de Rham theorem (see the end of Section 3).

We shall begin by defining an intersection complex of differential forms on the cone $cL = [0,1[\times L/\{0\}\times L$. The base L is a smooth $(n-1)$-dimensional compact Riemannian manifold and the vertex is denoted by $\{s\}$. The coordinates are written (r,x).

For $\gamma \geq 0$, let us define $C^\infty_\gamma(cL)$ as the algebra of differentiable functions $f(r,x)$ on cL satisfying the following symbol condition

$$C^\infty_\gamma(cL) = \{f \in C^\infty(]0,1[\times L) \ : \ \forall\, n\geq 0, \forall\, D_L \text{ differential operator on } L, \sup_{(r,x)\in cL} |r^{-\gamma+n}(\partial_r^n D_L f)(r,x)| < \infty\}$$

This algebra can be also defined in a more geometric way. A function $f \in C^\infty_0(X-\Sigma)$ (i.e. $\lim_{r\to 0} f(r,x) = 0$) is in $C^\infty_\gamma(cL)$ if and only if its $r\partial_r$ and L derivatives (of arbitrary order) are pointwise bounded with respect to the metric

$$\nu_\gamma = r^{2\gamma}\left(\left(\frac{dr}{r}\right)^2 + g_L\right), \text{ with } g_L \text{ metric on } L.$$

Note that this metric is equivalent to the cone standard metric $\nu_1 = dr^2 + r^2 g_L$. [1]

Now we shall construct the intersection complex. Let us fix $\alpha > 0$ and $\beta > 0$. Denote by $\mathcal{B}^*(cL)$ the space of differential forms on $cL - \{s\}$ whose $r\partial_r$ and L derivatives of arbitrary order are pointwise bounded (for ν_α).

[1] This geometric definition was judiciously pointed out to us by the referee.

For a local description consider a locally finite covering $\{U_i\}_{i\in I}$ of L, and $x = (x_1, \dots, x_{n-1})$ a system of local coordinates in U_i, a differential form $\omega \in \Omega^k(]0,1[\times L)$ belongs to $\mathcal{B}^*(cL)$ if its restriction to each U_i is sum of terms

$$a dx^k + b\frac{dr}{r} \wedge dx^{k-1} \text{ with } a, b \in C^\infty_{k\alpha}(cL).$$

The space of β-controlled differential forms on the cone cL is defined by:

$$\mathcal{B}^*_\beta(cL) = r^{-\beta}\mathcal{B}^*(cL).$$

It is not a complex so we define the intersection complex by:

$$I\mathcal{B}^k_\beta(cL) = \{\omega \in \mathcal{B}^k_\beta(cL) \ : \ d\omega \in \mathcal{B}^{k+1}_\beta(cL)\}.$$

The number β controls the order of poles of the forms near the vertex $\{s\}$. With minor changes of control parameters, the following result is similar to [BL], Théorème 2.4.

Theorem 0 *Let cL be a cone over an $(n-1)$-dimensional smooth manifold L, and $\bar{p}$ any perversity such that $p_n = n - 2 - [\beta/\alpha]$. Suppose that β/α is not an integer, then*

$$H^k(I\mathcal{B}^*_\beta(cL)) \cong \mathrm{Hom}(IH^{\bar{p}}_k(cL, \mathbf{R}); \mathbf{R}) = \begin{cases} H^k_{\mathrm{dR}}(L) & \text{if } k \leq [\beta/\alpha], \\ 0 & \text{if } k > [\beta/\alpha]. \end{cases}$$

In order to generalize to stratified varieties the previous constructions of $C^\infty_\gamma(cL)$ and $\mathcal{B}_\beta(cL)$, we will use atlases whose charts are iterated cones.

Let x be a point in a stratum S_{n-j_1} *of the stratified space* X and $\psi_x : U_x \xrightarrow{\cong} B^{n-j_1} \times cL_x$ a distinguished open neighbourhood as previously defined. The link L_x is a singular variety and is covered by distinguished open sets of the same type. By iteration, we obtain a chart which defines an *iterated cone*:

$$W_{\bar{\jmath}} = \mathbf{B}^{n-j_1} \times c\left(\mathbf{B}^{j_1-1-j_2} \times c\left(\mathbf{B}^{j_2-1-j_3} \times \cdots \times c(\mathbf{B}^{j_\ell-1}) \cdots\right)\right)$$

where $\bar{\jmath} = \{n+1 = j_0 > j_1 > j_2 > \cdots > j_\ell > j_{\ell+1} = 0\}$ denotes a decreasing sequence of integers and $\mathbf{B}^{j_t-1-j_{t+1}}$ is an open ball in $\mathbf{R}^{j_t-1-j_{t+1}}$ (possibly a point).

Using the homeomorphism ψ_x, this chart corresponds to the following chain of elements of the filtration of X:

$$\emptyset = X_{n-j_0} \subset X_{n-j_1} \subset X_{n-j_2} \subset \cdots \subset X_{n-j_\ell} \subset X = X_n = X_{n-j_{\ell+1}} \, .$$

We will denote in the same way the iterated cone and its image in X.

We thus obtain a covering of U_x by charts which are iterated cones and correspond to different sequences $\bar{\jmath}$.

To define in a geometric way the controlled forms on a Thom-Mather stratified space X with such an atlas, we define a metric, called the $\bar{\alpha}$-metric on iterated cones, in an inductive way : consider a sequence of real numbers $\bar{\alpha} = (\alpha_0, \cdots, \alpha_n)$ associated to the filtration $(*)$, each α_j corresponding to the stratum S_{n-j}. On each open ball B^k, with coordinates $u = (u_1, \dots, u_k)$, the metric is the euclidean one: $du^2 = du_1^2 + \cdots + du_k^2$. On the regular part of each product $B^{j_t-1-j_{t+1}} \times c(L)$, the metric is the product metric $du^2 + r^{2\alpha_{j_{t+1}}}((dr/r)^2 + g_L)$ where r is the coordinate of the generatrix of the cone and g_L is the metric on the regular part of L.

We can now define controlled differential forms on a Thom-Mather stratified variety:

Let $\bar{\beta} = (\beta_1, \dots, \beta_n)$ be a fixed sequence of positive real numbers.

On a compact smooth manifold L, $\mathcal{B}^*_{\bar{\beta}}(L)$ is the de Rham complex of C^∞-differential forms. Suppose that for every Thom-Mather stratified space X of dimension $< n$, equipped with an $\bar{\alpha}$-metric, we have defined the graded sheaf of $\bar{\beta}$-controlled differential forms on X and let $\mathcal{B}^*_{\bar{\beta}}(X)$ be its space of global sections. This is naturally a graded Frechet space.

Let cL be a cone on a compact Thom-Mather stratified space L of dimension $n-1$ with an $\bar{\alpha}$-metric, then define

$$\mathcal{B}^k_{\bar{\beta}}(cL) = r^{k\alpha_n - \beta_n} \left[\left(A + A\frac{dr}{r} \right) \widehat{\otimes} \mathcal{B}^*_{\bar{\beta}}(L) \right]^k$$

where $\widehat{\otimes}$ is the completed projective tensor product [Gr], and A is the Frechet algebra of the (near $\{s\}$) bounded smooth functions relatively to the multiplicative group $\mathbf{R}^*_+$:

$$A = \{ a \in C^\infty(]0,1[) \ : \ \forall \ k \ \sup_{r \in]0,1[} |(r\partial_r)^k a(r)| < +\infty \}.$$

Note that, with the change $r = e^{-u}$, A is just the space of smooth functions in $u \in (0, \infty)$, all of whose derivatives are bounded. We have

$$C^\infty_\alpha(cL) \cong r^\alpha A \widehat{\otimes} C^\infty(L).$$

Now, let X be a Thom-Mather stratified space of dimension n covered by iterated cones W_{J}, equipped with an $\bar{\alpha}$-metric. For every W_{J}, the space $\mathcal{B}^*_{\bar{\beta}}(W_{\mathrm{J}})$ is well defined. The graded sheaf of $\bar{\beta}$-controlled differential forms on X, denoted by $\mathcal{IB}^*_{\bar{\beta}}$ is defined as the sheaf whose sections ω satisfy, on every open U, the following condition:

$$\omega \text{ and } d\omega \in \mathcal{B}^*_{\bar{\beta}}(W_{\mathrm{J}} \cap U) \text{ for all } W_{\mathrm{J}}.$$

The proof of the following theorem is similar to [BL], Théorème 3.5, up to minor changes of control parameters.

Theorem 1 *Let X be a Thom-Mather stratified space, endowed with a covering by iterated cones and an $\bar{\alpha}$-metric. Suppose $\bar{\beta}$ given, satisfying the following perversity condition:*

$$\left[\frac{\beta_j}{\alpha_j}\right] \leq \left[\frac{\beta_{j+1}}{\alpha_{j+1}}\right] \leq \left[\frac{\beta_j}{\alpha_j}\right] + 1 \qquad \forall j,$$

with β_j/α_j not an integer. Then, there is an isomorphism

$$H^k(I\mathcal{B}^*_{\bar{\beta}}(X)) \cong \mathrm{Hom}(IH^{\bar{p}}_k(X, \mathbf{R}); \mathbf{R})$$

with $\bar{p}_j = j - 2 - \left[\frac{\beta_j}{\alpha_j}\right]$.

This result is related to shadow forms whose framework is the following: On a polyedron (K) in euclidean space $\mathbf{R}^n$, with a given barycentric subdivision (K'), we associate, to each simplex σ in (K'), a differential form $\omega(\sigma)$ in the interior of simplices of maximal dimension, in a very explicit way. The shadow forms have poles over faces of (K): If the defect of transversality of σ with a face F of a (K)-simplex is p, then the maximum order of poles of $\omega(\sigma)$ on this face is p.

Proposition. *For $1 \leq q \leq \infty$ fixed, consider shadow forms whose maximum order of poles on each i-dimensional face F is $[(n-i)/q]$, and controlled forms such that $\alpha_i = 1$ and $[\beta_i] = [(n-i)/q]$. Then the following inclusions are quasi-isomorphisms of complexes (the first being defined only for polyedra):*

$$\{ \textit{ Shadow forms } \} \subset \{ \bar{\beta}\textit{-controlled forms } \} \subset \{\mathcal{L}^q\textit{- forms}\}.$$

4 Hochschild and Cyclic Homology of Controlled Functions

For the rest of the paper and for simplicity, we return to the case of a cone over a smooth manifold. Firstly we show that the controlled functions $C^\infty_{\bar{\alpha}}(cL)$ generate the intersection complex (Theorem 2), then we give relations with Hochschild and cyclic homology (Theorem 3). In this section all homological constructions are made in the framework of Frechet structures.

There are two methods to obtain a complex, starting with $\mathcal{B}^*_{\beta}$: in the first one we consider the intersection complex $I\mathcal{B}^*_{\beta}(cL)$ previously defined, in the second one we stabilize by the de Rham operator, i.e. we add coboundaries. It is not difficult to show directly that the two complexes are quasi-isomorphic

$$H^*(I\mathcal{B}^*_{\beta}(cL)) \cong H^*(\mathcal{B}^*_{\beta}(cL) + d\mathcal{B}^*_{\beta}(cL)).$$

The second complex has the important property of being generated by controlled functions.

Firstly, dealing with a complex (namely ω and $d\omega$ are $\bar{\beta}$-controlled) is translated in cyclic homology theory by the use of unitarized algebras. We define the A-unitarization of the A-algebra $r^\alpha A\widehat{\otimes}C^\infty(L)$ as the algebraic sum in $C^\infty(]0,1[\times L)$

$$IC^\infty_\alpha(cL) = r^\alpha A\widehat{\otimes}C^\infty(L) + A.$$

The Frechet A-algebra structure is provided when identifying $IC^\infty_\alpha(cL)$ with the quotient algebra $(r^\alpha A\widehat{\otimes}C^\infty(L)\oplus A)/I$ by the closed ideal

$$I = \{f + a \ : \ \forall\, g \in r^\alpha A\widehat{\otimes}C^\infty(L), (f+a)g = 0\}.$$

Every differential form $\omega \in \Omega^*(]0,1[\times L)$ can be written in an unique way $\omega = \eta + \frac{dr}{r}\wedge\varphi$ where φ and η are free of dr. Let us denote by $\Omega^*_P = \{\eta\}$, the space of differential forms relatively to the projection $P :]0,1[\times L \to]0,1[$.

Let $\Omega^*(IC^\infty_\alpha(cL))$ be the subcomplex of $\Omega^*(]0,1[\times L)$ generated by functions in $IC^\infty_\alpha(cL)$. With the previous notation, $\Omega^*_P(IC^\infty_\alpha(cL))$ is the A-complex generated by the elements η in $\Omega^*(IC^\infty_\alpha(cL))$ and with de Rham differential d_L of L. If the cone is regarded as a family of spaces, $L_r = L$ for $r \in]0,1[$ and $L_0 = \{s\}$, (this point of view appears implicitly in the A-module structure), then $\Omega^*_P(IC^\infty_\alpha(cL))$ is the complex of sections of the family of de Rham complexes on $]0,1[$ associated to the A-algebra $IC^\infty_\alpha(cL)$. The derivative in the variable r provides the underlying graded space of a flat connection

$$d_r : r^{-(\beta+\alpha)}\ \Omega^*_P(IC^\infty_\alpha(cL)) \to A\frac{dr}{r}\widehat{\otimes}_A\, r^{-(\beta+\alpha)}\ \Omega^*_P(IC^\infty_\alpha(cL)).$$

Consider the differential A-module of poles:

$$M^*_\beta = r^{-(\beta+\alpha)}A \oplus r^{-(\beta+\alpha)}A\frac{dr}{r}.$$

We define the complex

$$I\Omega^*_\beta(cL) = (M^*_\beta\widehat{\otimes}_A\Omega^*_P(IC^\infty_\alpha(cL)), d_L + d_r).$$

We observe that its differential is not the double complex differerential and that unitarization with $\mathbf{R}$ would not be sufficient to obtain a complex.

Theorem 2 *There is an isomorphism of complexes:*

$$I\Omega^*_\beta(cL) \cong \mathcal{B}^*_\beta + d\mathcal{B}^*_\beta.$$

In the assumptions of the Theorem 0, we have

$$H^k(I\Omega^*_\beta(cL)) = \mathrm{Hom}(IH^{\bar{p}}_k(cL,\mathbf{R});\mathbf{R}).$$

Before computing the Hochschild and periodic cyclic homology of controlled functions, recall some definitions, the references are [Co] and [Lo]. Let Λ be a ring and $\mathcal{A}$ be an Λ-algebra with unit, so $\Lambda \subset \mathcal{A}$. The Hochschild complex $(C_*(\mathcal{A}), b)$ is defined by $C_k(\mathcal{A}) = \mathcal{A} \otimes \mathcal{A}^{\otimes^k}$ and the Hochschild boundary is

$$b(a_0 \otimes \cdots \otimes a_k) =$$
$$\sum_{j=0}^{k-1} (-1)^j a_0 \otimes \cdots \otimes a_j a_{j+1} \otimes \cdots \otimes a_k + (-1)^k a_k a_0 \otimes a_1 \otimes \cdots \otimes a_{k-1}.$$

Its homology, called the Hochschild homology, is denoted by $HH_*(\mathcal{A})$. The reduced Hochschild complex $(C_*^{red}(\mathcal{A}), b)$ is the quotient of the Hochschild complex by the subcomplex generated by the elements $a_0 \otimes \cdots \otimes a_k$ where $a_i \in \Lambda$ for some $i > 0$. The reduced Hochschild complex is quasi-isomorphic to the Hochschild complex.

The Hochschild complex is a cyclic module, i.e. it admits a cyclic action

$$\tau(a_0 \otimes \cdots \otimes a_k) = (-1)^k a_k \otimes a_0 \otimes a_1 \otimes \cdots \otimes a_{k-1} .$$

We have $\tau^{k+1} = id$, so, for each k, τ defines an action of $\mathbf{Z}/(k+1)\mathbf{Z}$ on $C_k(\mathcal{A})$.

Connes cyclic homology is defined in the following way: Consider the situation where Λ is a field of characteristic 0. The cyclic homology of $\mathcal{A}$, denoted by $HC_*(\mathcal{A})$, is the homology of the complex $(C_*(\mathcal{A})/(1-\tau), b)$ where b is induced by the Hochschild boundary. The relation between Hochschild and Connes homology is given by the Connes exact sequence

$$\cdots \to HH_k(\mathcal{A}) \xrightarrow{I} HC_k(\mathcal{A}) \xrightarrow{S} HC_{k-2}(\mathcal{A}) \xrightarrow{B} HH_{k-1}(\mathcal{A}) \to \cdots .$$

Using the so called *periodicity operator* S we define the periodic cyclic homology

$$PHC_* = \lim_k \left[HC_k(\mathcal{A}) \xrightarrow{S} HC_{k-2}(\mathcal{A}) \right] .$$

The importance of the above definitions appears with the following result. Let X be a compact $\mathcal{C}^\infty$ manifold, $\mathcal{A} = C^\infty(X)$ the Frechet algebra of differentiable functions on X and $\Omega^*(X)$ the associated de Rham algebra. Replace $\otimes$ everywhere by the projective tensor product $\widehat{\otimes}$. Then the map $\pi : C_k(C^\infty(X)) \to \Omega^k(X)$ defined by

$$\pi(f_0 \otimes \cdots \otimes f_k) = f_0 df_1 \wedge \cdots \wedge df_k$$

induces the following isomorphims [Co]:

$$HH_*(C^\infty(X)) \cong \Omega^*(X); \quad PHC_{\substack{\text{odd} \\ \text{even}}}(C^\infty(X)) \cong \oplus_{\substack{\text{odd} \\ \text{even}}} H^*_{dR}(X).$$

If Λ is a ring, we need a more general setting, the notion of mixed complex that we briefly describe, as it appears in the singular framework.

A *mixed complex* is a triple (M_*, b, B), where M_* is a graded module and b resp. B is a differential on M_* of degree -1 resp. 1 such that $bB + Bb = 0$ (see [Ka]). It defines a bicomplex $M_*[u]$ with differentials $b(mu^k) = (bm)u^k$, $B(mu^k) = (Bm)u^{k-1}$ where degree$(u) = 2$. We define the *Hochschild homology* of (M_*, b, B) as $H_*(M_*, b)$ and the *cyclic homology* as $H_*(M_*[u], b+B)$. There is again a Connes exact sequence and we can define the periodic cyclic homology.

When Λ is a field of characteristic 0, the relation between the two previous definitions is the following. Replacing the quotient $C_k(\mathcal{A})/(1-\tau)$ by a $\mathbf{Z}/(k+1)\mathbf{Z}$-free resolution of $C_k(\mathcal{A})$, we obtain a bicomplex which is quasi-isomorphic to the bicomplex associated to a mixed complex. The mixed complex structure of $C_*(\mathcal{A})$ is given by the Hochschild boundary b and by the operator B defined by

$$\begin{aligned} B(a_0 \otimes \cdots \otimes a_k) = \sum_{j=0}^{k-1} (-1)^{kj} 1 \otimes a_j \otimes \cdots \otimes a_k \otimes a_0 \otimes \cdots \otimes a_{j-1} \\ - (-1)^{k(j-1)} a_{j-1} \otimes 1 \otimes a_j \otimes \cdots \otimes a_k \otimes a_0 \otimes \cdots \otimes a_{j-2}. \end{aligned} \tag{**}$$

Then the two definitions of cyclic homology agree.

Now, let us compute the Hochschild and periodic cyclic homology of controlled functions. We shall use a slight generalization of the previous cyclic construction.

Consider the A-Hochschild complex

$$C_k^A(IC_\alpha^\infty(cL)) = IC_\alpha^\infty(cL) \widehat{\otimes}_A \cdots \widehat{\otimes}_A IC_\alpha^\infty(cL)$$

($(k+1)$-terms) and denote by b_A its differential. Define the Hochschild intersection complex by

$$IC_k^\beta(cL) = M_\beta^* \widehat{\otimes}_A C_{k-*}^A(IC_\alpha^\infty(cL)).$$

The elements of degree k are sums of terms

$$r^{-(\beta+\alpha)} f_0 \otimes f_1 \otimes \cdots \otimes f_k + r^{-(\beta+\alpha)} \frac{dr}{r} g_0 \otimes g_1 \otimes \cdots \otimes g_{k-1}$$

such that $f_i, g_j \in IC_\alpha^\infty(cL)$. The total differential is given by

$$\begin{array}{ccc} r^{-(\beta+\alpha)}\, C_k^A(IC_\alpha^\infty(cL)) & \oplus & r^{-(\beta+\alpha)}\, A\frac{dr}{r}\, \widehat{\otimes}_A\, C_{k-1}^A(IC_\alpha^\infty(cL)) \\ b_A^{k+1} \uparrow\downarrow B_A^k & \searrow d_r & b_A^k \uparrow\downarrow B_A^{k-1} \\ r^{-(\beta+\alpha)}\, C_{k+1}^A(IC_\alpha^\infty(cL)) & \oplus & r^{-(\beta+\alpha)}\, A\frac{dr}{r}\, \widehat{\otimes}_A\, C_k^A(IC_\alpha^\infty(cL)) \end{array}$$

where the operator B_A is defined as in (**) but with cyclic action given by

$$\tau(r^{-(\beta+\alpha)}g_0 \otimes \cdots \otimes g_k) = (-1)^k r^{-(\beta+\alpha)} g_k \otimes g_0 \otimes \cdots \otimes g_{k-1}$$

and d_r corresponds to the r-derivation,

$$d_r(r^{-(\beta+\alpha)} f_0 \otimes f_1 \otimes \cdots \otimes f_k) =$$
$$(-1)^k r^{-(\beta+\alpha)} \frac{dr}{r} \wedge \left[(-(\beta+\alpha)) f_0 \otimes \cdots \otimes f_k + \sum_{i=0}^{k} f_0 \otimes \cdots \otimes \partial_r f_i \otimes \cdots \otimes f_k \right]$$

So $b_A^* \oplus b_A^{*-1}$ has degree -1 and $B_A^* \oplus B_A^{*-1} + d_r$ has degree 1.

Lemma. *The triple*

$$(IC_*^\beta(cL), b = b_A^* \oplus b_A^{*-1}, B = B_A^* \oplus B_A^{*-1} + d_r)$$

is a mixed complex (i.e. $b^2 = B^2 = bB + Bb = 0$).

Theorem 3.
i) The Hochschild homology of $IC_*^\beta(cL)$ (with differential $1 \otimes b_A^*$) is:

$$HH_k(IC_*^\beta(cL)) \cong I\Omega_\beta^k(cL).$$

ii) The periodic cyclic homology is:

$$PHC_k(IC_*^\beta(cL)) \cong IH_{\bar{q}}^k(cL) \oplus IH_{\bar{q}}^{k-2}(cL) \oplus \cdots$$

where the perversity $\bar{q}$ satisfies $q_n = [\frac{\beta}{\alpha}]$.

Sketch of the proof. i) The terms in $r^{-\beta}$ stay as common factor, so we omit them in the proof of part (i). Consider the reduced A-Hochschild complex

$$\begin{aligned} C_k^{\mathrm{red}} &= IC_\alpha^\infty(cL) \widehat{\otimes}_A C_\alpha^\infty(cL) \widehat{\otimes}_A \cdots \widehat{\otimes}_A C_\alpha^\infty(cL) \cong r^{k\alpha} IC_\alpha^\infty(c(L)) \widehat{\otimes} C^\infty(L^{\times k}) \\ &\cong r^{k\alpha} C^\infty(c(L^{\times(k+1)})) + r^{(k-1)\alpha} C^\infty(c(L^{\times k})). \end{aligned}$$

Using the lemma below, it suffices to prove that $H_k(C_*^{\mathrm{red}}) \cong \Omega_P^k(IC_\alpha^\infty(cL))$.

For every open $U \subset [0,1[$ we can define the Frechet module of controlled functions on $U \times L^{\times(k+1)}$ in the same way, as C_k^{red}. So we obtain a presheaf $U \mapsto C_k^{\mathrm{red}}(U)$ which define a fine sheaf $\mathcal{C}_k^{\mathrm{red}}$ using the same localization condition as $\mathcal{C}_{\bar{\gamma}}^\infty$. Its space of sections is the reduced A-Hochschild complex $C_k^{\mathrm{red}} = \mathcal{C}_k^{\mathrm{red}}([0,1[)$. We can also associate to $\Omega_P^k(IC_\alpha^\infty(cL))$ a fine sheaf which is denoted by $\mathcal{I}\Omega^k$. We define a sheaf morphism $\pi : \mathcal{C}_k^{\mathrm{red}} \to \mathcal{I}\Omega^k$ as above: for each U set $\pi(f_0 \otimes f_1 \otimes \cdots \otimes f_k) = f_0 df_1 \wedge \cdots \wedge df_k$ where f_i is controlled on U. For each $r \in]0,1[$, the fiber complex $(\mathcal{C}_*^{\mathrm{red}})_r$ is isomorphic to the standard Hochschild complex of $C_*(C^\infty(L))$ and the fiber complex $(\mathcal{I}\Omega^*)_r$ is isomorphic to $\Omega^*(L)$. These isomorphisms are induced by the following one in degree 0:

$$\begin{array}{ccc} IC^\infty_\alpha(cL)_r & \to & C^\infty(L) \\ r^\alpha a \otimes f + c & \mapsto & r^\alpha a f + c. \end{array}$$

By Connes's theorem ([Co] II, Théorème 46), for $r > 0$, the Hochschild homology of the fiber complex of C^{red}_* is isomorphic to $\Omega^*(L)$ so the morphism π induces a fiber isomorphism between the homology sheaf $\mathcal{H}_k(C^{\mathrm{red}}_*)$ and the sheaf $\mathcal{I}\Omega^*$. This can be extended for $r = 0$ and this implies that ([Go], 4.5):

$$H_k(\Gamma(cL, C^{\mathrm{red}}_*)) \cong \Gamma(cL, \mathcal{H}_k(C^{\mathrm{red}}_*)) .$$

ii) By the isomorphism $H_k(C^{\mathrm{red}}_*) \cong \Omega^k_P(IC^\infty_\alpha(cL))$, the differential B gives the de Rham differential. Using theorems 1 and 2, the proof of ii) is then similar to the non singular case.

Similar sheaf arguments establish the following lemma (used in the previous proof):

Lemma. *The A-Hochschild complex $C^A_*(IC^\infty_\alpha(cL))$ and the reduced A-Hochschild complex C^{red}_* are quasi-isomorphic.*

Remark. Although we often speak of A-objects, our construction does not agree with equivariant theories developed in [Bry], [BG]. We do not know if they can be adapted to this singular situation.

References

[BBD] A. Beilinson, J. Bernstein et P. Deligne, *Faisceaux pervers*, Asterisque n° 100, 1983.

[BG] J. Block and E. Getzler, Equivariant cyclic homology and equivariant differential forms, *Ann. Scient. Éc. Norm. Sup.* t. 27, (1994), 493-527.

[Bra] J.P. Brasselet, De Rham's theorems for singular varieties, *Contemporary Mathematics*, 161, (1994), 95-112.

[BGM] J.P. Brasselet, M. Goresky and R. MacPherson, Simplicial Differential Forms with Poles, *Amer. Journal of Maths.*, 113 (1991), 1019-1052.

[BL] J..P. Brasselet et A. Legrand, Un complexe de formes différentielles à croissance bornée sur une variété stratifiée, *Annali della Scuola Normale Superiore di Pisa Serie IV*, Vol XXI, Fasc.2 (1994), p. 213-234.

[Bry] J.L. Brylinski, Cyclic homology and equivariant theories, *Ann. Inst. Fourier*, vol. 37, 4 (1987), 15-28.

[Ch] J. Cheeger, On the Hodge theory of Riemannian pseudomanifolds, *Proc. of Symp. in Pure Math.*, vol 36 (1980), 91-146. Amer.Math.Soc., Providence R.I.

[CGM] J. Cheeger, M. Goresky and R. MacPherson, $\mathcal{L}^2$-cohomology and intersection cohomology for singular varieties, *Seminar on Differential Geometry, S.T.Yau, ed. Ann. of Math. Studies*, Princeton University Press, Princeton N.J., 102 (1982), 303-340.

[Co] A. Connes, Non commutative differential geometry, *Publ. Math. I.H.E.S.*, 62 (1986), 257-360.

[Go] R. Godement, *Théorie des faisceaux*, Actualités scientifiques et industrielles, 1252, Hermann 1964.

[GM1] M. Goresky and R. MacPherson, Intersection homology theory, *Topology*, 19 (1980), 135-162.

[GM2] M. Goresky and R.MacPherson, Intersection homology theory II, *Inv. Math.*, 71 (1983), 77-129.

[GM3] M. Goresky and R. MacPherson, *Stratified Morse theory*, Ergebnisse. Band 14, Springer-Verlag 1987.

[Gr] A. Grothendieck, *Produits tensoriels topologiques*, Mem. Amer.Math. Soc. 16, 1955.

[Ka] C. Kassel. Cyclic homology. Comodules and mixed complexes, *J. of Algebra*, 107 (1987), 195-216.

[Lo] J.L. Loday, *Cyclic Homology*, Grund. der math. Wiss. 301, Springer Verlag (1992).

[Ma] J. Mather, *Notes on topological stability*, Harvard University, 1970.

[Sa] M. Saito, *Modules de Hodge polarisables*, IAS, Princeton, 1986.

Jean-Paul Brasselet
IML - CNRS
Luminy Case 930
F-13288 Marseille Cedex 9
France

e-mail: jpb@iml.univ-mrs.fr

André Legrand
Laboratoire Emile Picard
Université Paul Sabatier
118 Route de Narbonne
F-31062 Toulouse Cedex
France

e-mail: legrand@picard.ups-tlse.fr

Continuous Controlled Vector Fields

A.A. du Plessis

Introduction

The idea that stratified vector fields controlled by tubular neighbourhoods of the strata are integrable is due to René Thom [11], more detail having been added by John Mather [3] and Klaus Wirthmüller [2, Chapter II]. The construction of such stratified vector fields by lifting a smooth vector field over a mapping submersive on strata is also treated in the articles cited; our contribution here is to show that the construction can be made to yield *continuous* controlled vector fields in the case where the stratification is C-regular in the sense of Karim Bekka [1], so in particular (see [1, p.52, Remarques 5]) when it is Whitney regular.

The result in the case of a Whitney regular stratification is announced by Masahiro Shiota in [8], where the result in the case where there are just two strata is proved. However, the extension to the general case, which is the most delicate construction in this article, is dismissed as trivial there. Shiota has, very recently, offered an easy construction in the general case, in his book [9, pp.10-11]; unfortunately, his construction is not sufficient to ensure the required continuity.

Continuous lifts of vector fields had previously been constructed for stratifications satisfying stronger regularity conditions. Jean-Louis Verdier [14] found "rugose" lifts for his notion of W-regular stratification, while Adam Parusinski [6] and Tadeusz Mostowski [4] discussed regularity conditions allowing construction of Lipschitz lifts.

The result given here is the best possible, in that *continuous* controlled vector fields cannot in general be constructed when C-regularity fails.

The result does not generalize to the relative case: a continuous controlled stratified vector field cannot in general be lifted over a stratified map to a *continuous* vector field controlled over the map; an example is given in [7, 2.5.10]. Thus finding *continuous* controlled stratified lifts in the context of Thom's second isotopy lemma, where the aim is to lift a smooth vector field first to the target of a stratified map over a map submersive on strata, and then over the stratified map itself, is not possible in general. An interesting question remains, however, as pointed out to me by Bekka: can such lifts be constructed after refining the stratification of the map?

The main construction given here was originally worked out in 1988, for the Whitney regular case, for the review section of the book [7] (see in particular pages 42 and 43), but in the event was not included there. Shortly thereafter David Trotman asked me whether I knew how to prove the result, and drew my attention to Shiota's paper, of which I had been unaware. My reply giving my construction drew an immediate response from Bekka, who pointed out that it worked more or less verbatim in the more general C-regular situation, and a more delayed response, from two other students of Trotman, Stephane Simon and Claudio Murolo, who used the result in their theses (see [10], [5]).

I thank Bekka, Murolo, Simon, Trotman, and Terry Wall for their interest and encouragement.

1 The results

The theorem we aim to prove is the following:

Theorem 1.1 *Let $f : N \to P$ be a smooth map between manifolds and let $(X, \mathcal{S})$ be a C-regular stratified subset of N such that f maps each stratum of $\mathcal{S}$ submersively into P. Then, given a smooth tube system for $\mathcal{S}$, any smooth vector field η on P can be lifted over f to a* continuous *vector field ξ on X, smooth on strata and controlled by the tube system.*

The notion of a tube system for a stratification is already touched on in [11], and has an elegant abstract formulation, see for example [15, 1.1, 1.2]. The existence of smooth tube systems for Whitney regular stratifications is also mentioned by Thom [11], but the first detailed statements and proofs are to be found in Mather's notes [3]. Since [3] is unpublished, we refer instead to [2, 2.6, 2.7], for the proof. Bekka proved the existence of smooth tube systems for C-regular stratifications in [1, 2.5].

The notion of a vector field controlled by a tube system also has an abstract formulation (see e.g. [15, 2.1], and the fact that smooth vector fields can be lifted over a map submersive on strata to controlled vector fields is the key to the proof of Thom's first isotopy lemma ([12, 2, 15]).

As in the articles cited, the key lemma from which all else follows by (very) careful inductive arguments is a two-stratum special case.

Lemma 1.2 *Let $X \subset M$ be a smooth submanifold, $\pi : T \to X$ the retraction of a smooth tubular neighbourhood of X, $\rho : T \to \mathbb{R}_+$ a smooth function such that $\rho^{-1}(0) = X$.*

Let $Y \subset T \setminus X$ be a smooth submanifold such that

(1) $\rho | Y$ has no critical points in a neighbourhood of X;

(2) *if* $\{y_i\} \subset Y$ *is a sequence converging to* $x \in X$, *such that* $\text{Ker } d(\rho|Y)_{y_i}$ *converges to* $\tau \subset TM_x$ *in the appropriate Grassmannian, then* $TX_x \subset \tau$.

Finally, let ξ_X *be a smooth vector field on* X. *Then there exists a neighbourhood* U *of* X *in* T *and a smooth vector field* ξ_Y *on* Y *lifting* $(\xi_X, 0)$ *on* $X \times \mathbb{R}$ *over* $(\pi, \rho) \mid Y \cap U$, *and such that*

$$\xi_X \cup \xi_Y : X \cup \{Y \cap U\} \to TM$$

is continuous.

Remark. The conditions (1) and (2) above are exactly Bekka's conditions [1, II, 1.1] for C-regularity of Y over X with respect to ρ.

Proof We claim that there is a neighbourhood U of X in M such that $(\pi, \rho) \mid Y \cap U$ is a submersion to $X \times (0, \infty)$. Indeed, the claim follows from Bekka's characterisation of C-regularity [1, II, 1.2], but for the reader's convenience we present a direct argument.

Suppose the claim is false, so that there exist $x \in X$ and a sequence $\{y_i\} \subset Y$ such that $d(\pi, \rho) \mid TY_{y_i} : TY_{y_i} \to TX_{\pi(y_i)} \oplus \mathbb{R}$ is not surjective. Since $d\rho_y \mid TY_y$ is surjective for y sufficiently near X, by hypothesis (1), the failure of surjectivity above is equivalent to $d\pi_{y_i} \mid \text{Ker } d(\rho|Y)_{y_i} : \text{Ker } d(\rho|Y)_{y_i} \to TX_{\pi(y_i)}$ not being surjective, at least for sufficiently large i.

Passing to a subsequence, we may suppose that $d\pi_{y_i}(\text{Ker } d(\rho|Y)_{y_i})$ has constant dimension, and converges in the appropriate Grassmannian to a *proper* subspace σ of TX_x. We may also suppose that $\text{Ker } d(\rho)_{y_i}$ converges to a subspace τ; and then we will have

$$\sigma = \lim_i d\pi_{y_i}(\text{Ker } d(\rho|Y)_{y_i}) \supset d\pi_x(\tau).$$

But by hypothesis (2), $TX_x \subset \tau$, so $d\pi_x(\tau) = TX_x$, giving a contradiction. Thus the supposition was false, and the claim is proved.

To prove the lemma, it is sufficient to show that there exists an open covering $\{U_\lambda\}_{\lambda \in \Lambda}$ of X in T such that $(\xi_X, 0) \mid (X \cap U_\lambda) \times \mathbb{R}$ can be lifted over $(\pi, \rho) \mid Y \cap U_\lambda$ to a smooth vector field $\xi_{Y,\lambda}$ defined on $Y \cap U_\lambda$ such that $(\xi_X \mid U_\lambda \cap X) \cup \xi_{Y,\lambda} : (X \cup Y) \cap U_\lambda \to TM$ is continuous. For we can take a smooth partition $\{a_\lambda\}$ of unity whose supports refine $\{U_\lambda\}$ and are locally finite, and then $\xi_Y = \sum_\lambda a_\lambda \xi_{Y,\lambda}$ clearly has the desired properties.

It is thus enough to work in the neighbourhood of a point $\mathbf{x}_0 \in X$. Taking local coordinates at $\mathbf{x}_0$ adapted to the submanifold X, we can thus reduce to the case

$$X = \mathbb{R}^n, M = T = \mathbb{R}^n \times \mathbb{R}^k = \mathbb{R}^{n+k}, \pi(\mathbf{x}, z) = \mathbf{x},$$

with $(\pi, \rho) \mid Y : Y \to X \times \mathbb{R}$ a submersion. Thus

$$d\pi_y \mid \text{Ker } d\rho_y \cap TY_y : \text{Ker } d\rho_y \cap TY_y \to \mathbb{R}^n$$

is surjective for all $\mathbf{y} \in Y$, so that if A_y denotes Ker $d\rho_y \cap TY_y$ and B_y denotes the orthogonal complement in A_y of $d\pi_y \,|\, A_y$, $d\pi_y$ induces an isomorphism $\theta_y : B_y \to \mathbb{R}^n$. We define

$$\xi_Y(\mathbf{y}) = \theta_y^{-1}(\xi_X(\pi(\mathbf{y}))).$$

It now follows at once that ξ_Y is a lift for $(\xi_X, 0)$ over $(\pi, \rho) \,|\, Y$. Moreover, $\theta_y^{-1} : Y \to \text{Inj}\ (\mathbb{R}^n, \mathbb{R}^{n+k})$ is clearly smooth, so ξ_Y is also smooth.

It thus remains to show that $\xi_X \cup \xi_Y : X \cup Y \to \mathbb{R}^{n+k}$ is continuous. This in turn will follow if we show that

$$\iota \cup \theta^{-1} : X \cup Y \to \text{Inj}\ (\mathbb{R}^n, \mathbb{R}^{n+k}),$$

where ι denotes the constant map to the inclusion, is continuous. Let $\{\mathbf{y}_k\}$ be a sequence in Y converging to $\mathbf{x} \in X$: we must show that $\theta_{y_k}^{-1}$ converges to ι. It will be enough to show that B_{y_k} converges to $\mathbb{R}^n$ in the appropriate Grassmannian, for if $\theta_y^{-1}(\mathbf{u}) = (\mathbf{u}, \psi_y(\mathbf{u}))$, B_y is the graph of ψ_y, so B_{y_k} converges if and only if ψ_{y_k} does, if and only if $\theta_{y_k}^{-1}$ does.

As a final reduction, note that it is enough to show that every sequence $\{\mathbf{y}_k\}$ as above has a subsequence such that the corresponding B_{y_k} converge to $\mathbb{R}^n$. Passing to a subsequence, we may thus suppose that A_{y_k} converges to a limit D, Ker $\pi \cap A_{y_k}$ converges to a limit P, and B_{y_k} converges to a limit Q, which we must show to be $\mathbb{R}^n$.

It is clear that

$$P = \lim_k \{\text{Ker}\,\pi \cap A_{y_k}\} \subset \text{Ker}\,\pi \cap \lim_k A_{y_k} = \text{Ker}\,\pi \cap D.$$

On the other hand, $\mathbb{R}^n \subset D$ by hypothesis, so that $\pi \,|\, D : D \to \mathbb{R}^n$ is surjective, and the spaces $\text{Ker}\,\pi \cap A_{y_k}$ and $\text{Ker}\,\pi \cap D$ have constant dimension. Thus P and $\text{Ker}\,\pi \cap D$ have the same dimension, and so are equal. It follows that $\{\text{Ker}\,\pi \cap A_{y_k}\}^\perp$ converges to $\{\text{Ker}\,\pi \cap D\}^\perp$; intersecting with the sequence A_{y_k} (which converges to D), we see that

$$Q = \lim_k B_{y_k} \subset D \cap \{\text{Ker}\,\pi \cap D\}^\perp.$$

Now, since $\mathbb{R}^n \subset D$, we may write $D = \mathbb{R}^n \oplus E$, with $E \subset \mathbb{R}^k$. So

$$D \cap \{\text{Ker}\,\pi \cap D\}^\perp = \{\mathbb{R}^n \oplus E\} \cap E^\perp = \mathbb{R}^n.$$

Thus $Q = \mathbb{R}^n$, completing the proof. □

Corollary 1.2.1 *Let $X \subset M$ be a smooth submanifold, T a smooth tubular neighbourhood of X in M, and $\pi : T \to X$, $\rho : T \to \mathbb{R}_+$ the associated retraction and Euclidean distance function. Let $Y \subset T \setminus X$ a smooth submanifold Whitney regular over X.*

Let ξ_X be a smooth vector field on X. Then there exists a smooth vector field ξ_Y on Y lifting $(\xi_X, 0)$ on $X \times \mathbb{R}$ over $(\pi, \rho)|Y$, and such that

$$\xi_X \cup \xi_Y : X \cup Y \to TM$$

is continuous.

Proof We need only see that Y is C-regular over X with respect to ρ. As Bekka [1] points out, this follows from [1], I, (1.2), $\beta \Rightarrow \alpha$, and a result of Thom's [11], see [13], β^1, p.581. However, it is easier to argue directly, so we do. It is sufficient to argue locally, so we may assume $X = \mathbb{R}^n \subset T = M = \mathbb{R}^n \oplus \mathbb{R}^k$, with π, ρ given by $\pi(\mathbf{x}, z) = \mathbf{x}$, $\rho(\mathbf{x}, z) = |z|^2$.

We claim first that 1.2, (1) holds, i.e. there is a neighbourhood U of X in M such that $\rho\,|\,Y \cap U$ has no critical points. Suppose not, so that there exist $(\mathbf{x}, 0) \in X$ and a sequence $\{(\mathbf{x}_i, z_i)\} \subset Y$ of critical points for $\rho\,|\,Y$ (so that $TY_{(bfx_i, z_i)}$ is perpendicular to $(0, z_i)$) which converge to $(\mathbf{x}, 0)$. Passing to a subsequence, we may suppose that the lines ℓ_i determined by the non-zero vectors $0 \oplus z_i$ converge to a line ℓ, and that the tangent spaces $TY_{(\mathbf{x}_i, z_i)}$ converge to a subspace $\tau \subset \mathbb{R}^n \oplus \mathbb{R}^k$. Since Y is Whitney regular over X, $\ell \subset \tau$ (this is Whitney's condition (b), see [16]). On the other hand, $\ell_i \subset TY_{y_i}^{\perp}$, by choice of the $(\mathbf{x}_i, z_i)$, so $l \subset \tau^{\perp}$. This contradiction shows the supposition false, and so the first claim is proved.

Now we show that 1.2, (2) holds, i.e., if $\{(\mathbf{x}_i, z_i)\} \subset Y$ is a sequence converging to $(\mathbf{x}, 0) \in X$ such that $\operatorname{Ker} d(\rho|Y)_{(\mathbf{x}_i, z_i)}$ converges to a subspace κ, then $\kappa \supset TX_{(\mathbf{x},0)}$. Passing to a subsequence, we may suppose that $TY_{(\mathbf{x}_i, z_i)} \to \tau$ and, if ℓ_i is the line determined by the non-zero vector $0 \oplus z_i$, then $\ell_i \to \ell$. We have

$$\operatorname{Ker} d(\rho|Y)_{(\mathbf{x}_i, z_i)} = \operatorname{Ker} d\rho_{(\mathbf{x}_i, z_i)} \cap TY_{(\mathbf{x}_i, z_i)} = \ell_i^{\perp} \cap TY_{(\mathbf{x}_i, z_i)},$$

so, taking limits, $\kappa = \ell^{\perp} \cap \tau$. Since ℓ_i is orthogonal to $TX_{(\mathbf{x}_i, 0)}$, we have $TX_{(\mathbf{x}_i,0)} \subset \ell_i^{\perp}$, and so, taking limits, $TX_{(\mathbf{x},0)} \subset \ell^{\perp}$. Also, since Y is Whitney regular over X, we have $TX_{(\mathbf{x},0)} \subset \tau$ (this is Whitney's condition (a), see [16]). Thus

$$TX_{(\mathbf{x},0)} \subset \ell^{\perp} \cap \tau = \kappa,$$

completing the proof. □

As a partial converse to Proposition 1.2, we have the very simple observation:

Lemma 1.3 *Let $X \subset M$ be a smooth submanifold, $\rho : U \to \mathbb{R}$ a smooth function defined on a neighbourhood U of X such that $\rho^{-1}(0) = X$. Let $Y \subset U \setminus X$ be another submanifold.*

Suppose that there exists a sequence $\{y_i\} \subset Y$ of non-critical points for $\rho\,|\,Y$ converging to $x \in X$, and such that $\operatorname{Ker} d(\rho\,|\,Y)$ converges in the appropriate Grassmannian to a subset $\kappa \subset TM_x$, with $TX_x \not\subset \kappa$.

Then there exists a smooth vector field ξ on X which cannot be extended to a continuous vector field $\xi \cup \eta$ on $X \cup Y$ with $\eta(y_i) \in \operatorname{Ker} d(\rho\,|\,Y)_{y_i}$.

Proof Let ψ be any smooth vector field whose value $\psi(x)$ at x is not contained in κ. If η is a vector field on Y with $\eta(y_i) \in \operatorname{Ker}\, d(\rho\,|\,Y)$, then $\psi(x)$ cannot be the limit of the $\eta(y_i)$, and $\psi \cup \eta$ is discontinuous at x. □

We now give

Proof of Theorem 1.1. We write

$$M^i = \cup\{S \in \mathcal{S} \mid \dim S = i\}, \quad X^i = \cup\{S \in \mathcal{S} \mid \dim S \leq i\}.$$

We will prove the theorem by induction on i, and so suppose that there is a continuous vector field ξ defined on X^{a-1}, smooth on each M^i, $i \leq a-1$, and controlled by the chosen smooth tube system; and we will show how to extend ξ to X^a (the induction starts trivially with $a = 0$ and $X^{-1} = \emptyset$). For this, we will only need the restriction of the tube system to X^a. We may take this to consist of, for each $i = 0, 1, \ldots, a-1$,

a neighbourhood T^i of M^i,

a smooth retraction $\pi^i : T^i \to M^i$,

a smooth function $\rho^i : T^i \to \mathbb{R}_+$ with $(\rho^i)^{-1}(0) = M^i$ with respect to which the $M^j (j > i)$ are C-regular over M^i,

so that, setting $Q^i = M^a \cap T^i$ for $0 \leq i < a$, we have the compatibility conditions:

1. $\pi^i \circ \pi^j = \pi^i$ on $Q^i \cap Q^j$, $0 \leq i < j < a$,
2. $\rho^i \circ \pi^j = \rho^i$ on $Q^i \cap Q^j$, $0 \leq i < j < a$,
3. $f \circ \pi^i = f$ on Q^i, $0 \leq i < a$,
4. $(\pi^i, \rho^i) : Q^i \to M^i \times \mathbb{R}$ is a submersion for $0 \leq i < a$.

That ξ is a controlled lift of η is expressed by the further compatibility conditions:

5. $T\pi^i \circ \xi = \xi \circ \pi^i$ on $X^{a-1} \cap T^i$, $0 \leq i \leq a-2$,
6. $T\rho^i \circ \xi(\mathbf{y}) = \mathbf{0}$ if $\mathbf{y} \in X^{a-1} \cap T^i$ and $f(\mathbf{y}) = f(\pi^i(\mathbf{y}))$.
7. $Tf \circ \xi = \eta \circ f$ on X^{a-1}.

By slightly shrinking the tubes T^i we obtain subsets $F^i \subset Q^i$ which are closed in M^a. All this is very much as in Wirthmüller's exposition in [2], see in particular p. 47; the only real difference is that the functions ρ^i are in our case not necessarily Euclidean distance functions.

By Lemma 1.2 there exist, for $1 \leq j < a$, smooth vector fields ξ_j on Q^j, with ξ_j a lift of $(\xi^j, 0)$ on $M^j \times \mathbb{R}$ over (π^j, ρ^j) and such that $\xi_j \cup \xi^j : Q^j \cup M^j \to TN$ is continuous (where $\xi^j = \xi \mid M^j$). Notice that, on Q^j,

$$Tf \circ \xi_j = Tf \circ T\pi^j \circ \xi_j = Tf \circ \xi^j \circ \pi^j = \eta \circ f \circ \pi^j = \eta \circ f$$

(using, in succession, 3), the fact that ξ_j lifts ξ^j, 7) and 3) again). As in [GWPL,§2], we call this equality (VFf). Similarly, on $Q^i \cap Q^j$ with $i < j$, we have

$$T\pi^i \circ \xi_j = T\pi^i \circ T\pi^j \circ \xi_j = T\pi^i \circ \xi^j \circ \pi^j = \xi^j \circ \pi^i \circ \pi^j = \xi^j \circ \pi^i$$

(by 1), the lifting property, 5) and 1)), and call this equality $(VF\pi)_i$. Also we have

$$T\rho^i \circ \xi_j = T\rho^i \circ T\pi^j \circ \xi_j = T\rho^i \circ \xi^j \circ \pi^j = 0$$

(by 2), the lifting property and 6)), and denote this $(VF\rho)_i$. Indeed, $(VF\pi)_j$ and $(VF\rho)_j$ also hold, since ξ_j is a lift of $(\xi^j, 0)$ over (π^j, ρ^j).

Now, for $0 \leq j < a$, define R^j to be the set of those $\mathbf{x} \in Q^j \cup M^j$ such that $d((\xi_j \cup \xi^j)(\mathbf{x}), (\xi_j \cup \xi^j)(\pi^j \mathbf{x})) < d(\mathbf{x}, X^{j-1})$ and $d(\mathbf{x}, \pi^j \mathbf{x}) < d(\mathbf{x}, X^{j-1})$, where d is the distance function on TN induced by a smooth metric. Since $\xi_j \cup \xi^j$ is continuous, R^j is an open neighbourhood of M^j in $Q^j \cup M^j$.

Let G^j be a closed neighbourhood of M^j in $M^a \cup M^j$ contained in R^j. We set $S^j = R^j \cup Q^j, H^j = G^j \cap F^j, U^a = M^a - \bigcup_{j=0}^{a-1} H^j$, and let ξ_a be a smooth vector field on U^a which is a lift for η over $f \mid U^a$. We set $U^j = S^j - \bigcup_{i=j+1}^{a-1} H^i$ for $0 \leq j < a$, let $\{\rho_j \mid 0 \leq j \leq a\}$ be a smooth partition of unity subordinate to $\{U^j \mid 0 \leq j \leq a\}$ and define $\xi^a = \sum_{j=0}^{a} \rho_j \xi_j$: then ξ^a is smooth.

Since, for $0 \leq j < a$, ξ_j satisfies (VFf) on Q^j and $(VF\pi)_i, (VF\rho)_i$ for $i \leq j$ on $Q^i \cap Q^j$, and hence on $H^i \cap H^j$, so does $\rho_j \xi_j$. Indeed, since $\rho_j \equiv 0$ on H^i for $i > j$, this holds also for $i > j$. It thus holds also for ξ^a. So $\xi \cup \xi^a$ is a controlled vector field on X^a which lifts η, and it remains only to show that it is continuous.

It suffices to show that for any $\mathbf{x} \in M^{i_0}$ with $i_0 < a$ and any sequence $\{\mathbf{y_k}\}$ in M^a converging to $\mathbf{x}$ we have a subsequence on which the $\xi^a(\mathbf{y}_k)$ converge to $\xi(\mathbf{x})$. Passing to a subsequence, we may suppose that each $\mathbf{y}_k$ belongs to $H^j - \bigcup_{i=j+1}^{a-1} H^i$, and hence to $\bigcup_{i=j}^{a-1} \operatorname{supp} \rho_i$, and indeed that for some fixed $I \subset \{j, ..., a-1\}$ we have $\mathbf{y}_k \in \operatorname{supp} \rho_i \iff i \in I$. Since the ρ_i are continuous and bounded, we may suppose that $\rho_i(\mathbf{y}_k)$ converges to a limit r_i; since $\sum \rho_i = 1$, we have $\sum r_i = 1$. For $i \in I$ we have $\mathbf{y}_k \in \operatorname{supp} \rho \subset R^i$, so

$$d(\xi_i(\mathbf{y}_k), \xi(\pi^i(\mathbf{y}_k))) < d(\mathbf{y}_k, X^{i-1}), \quad d(\mathbf{y}_k, \pi^i(\mathbf{y}_k)) < d(\mathbf{y}_k, X^{i-1}).$$

We claim that for $i \in I$, $\pi^i(\mathbf{y}_k)$ converges to $\mathbf{x}$. This is clear if $i = i_0$, for π^i is a continuous function on R^i and $\mathbf{x} \in R^i$. Otherwise $i > i_0$ and we have

$$d(\pi^i(\mathbf{y}_k), \mathbf{x}) \leq d(\pi^i(\mathbf{y}_k), \mathbf{y}_k) + d(\mathbf{y}_k, \mathbf{x}) < d(\mathbf{y}_k, X^{i-1}) + d(\mathbf{y}_k, \mathbf{x}) \leq 2d(\mathbf{y}_k, \mathbf{x})$$

(since $\mathbf{x} \in X^{i-1}$) and so converges to 0.

Next we show that for $i \in I$, $\xi_i(\mathbf{y}_k)$ converges to $\xi(\mathbf{x})$. This will establish the result, for then

$$\xi^a(\mathbf{y}_k) = \sum_{j=0}^{a} \rho_j(\mathbf{y}_k)\xi_j(\mathbf{y}_k) = \sum_{i \in I} \rho_i(\mathbf{y}_k)\xi_i(\mathbf{y}_k)$$

converges to $\sum_{i \in I} r_i \xi(\mathbf{x}) = \xi(\mathbf{x})$. Now from what we have just shown, with continuity of ξ on X^{a-1}, it follows that $\xi(\pi^i(\mathbf{y}_k))$ converges to $\xi(\mathbf{x})$. Thus if $i > i_0$, we have $d(\xi_i(\mathbf{y}_k), \xi(\mathbf{x})) \leq d(\xi(\mathbf{y}_k), \xi(\pi^i(\mathbf{y}_k))) + d(\xi(\pi^i(\mathbf{y}_k)), \xi(\mathbf{x}))$, and this in turn is $\leq d(\mathbf{y}_k, X^{i-1}) + d(\xi(\pi^i(\mathbf{y}_k)), \xi(\mathbf{x}))$, where both terms converge to 0; while if $i = i_0$ the assertion follows from continuity of $\xi_i \cup \xi^i$.

Thus the induction step, and hence the proof, is complete. □

References

[1] Bekka, K., *C-regularité et trivialité topologique.* In "Singularity theory and its applications, I", 42-62, Springer LNM 1462, Springer-Verlag (1991).

[2] Gibson, C. G., K. Wirthmüller, A. A. du Plessis, E. J. N. Looijenga, *Topological stability of smooth mappings.* Springer LNM 552, Springer-Verlag (1976).

[3] Mather, J. N., *Notes on topological stability.* Preprint, Harvard University (1970), 73 pages.

[4] Mostowski, T., *Lipschitz equisingularity.* Dissertationes Mathematicae **243** (1985) (whole volume).

[5] Murolo, C., *Semidifférentiabilité, transversalité, et homologie de stratifications régulières.* Thèse de Doctorat, Université de Provence, 1997.

[6] Parusinski, A., *Lipschitz properties of semianalytic sets.* Ann. Inst. Fourier, Grenoble **38** (1988), 189-213.

[7] du Plessis, A. A., C. T. C. Wall, *The geometry of topological stability.* LMS Monographs (new series) **9**, Oxford University Press (1995).

[8] Shiota, M., *Linearization of real analytic functions.* Publ. RIMS, Kyoto **20** (1984), 727-792.

[9] Shiota, M., *Geometry of subanalytic and semialgebraic sets.* Progress in Mathematics 150, Birkhäuser (1997).

[10] Simon, S., *Champs totalement radiaux sur une structure de Thom-Mather.* Ann. Inst. Fourier, Grenoble **45** (1995),1423-1447.

[11] Thom, R., *Local topological properties of differentiable mappings.* In "Differential analysis: Bombay colloquium", 191-202, Oxford University Press, 1964.

[12] Thom, R., *Emsembles et morphismes stratifiés.* Bull. AMS **75** (1969), 240-284.

[13] Trotman, D. J. A., *Comparing regularity conditions on stratifications.* In Proc. Symp. Math. **40** (II), 575-586, American Mathematical Society, 1983.

[14] Verdier, J.-L., *Stratifications de Whitney et théorème de Bertini-Sard.* Invent. Math. **36** (1976), 295-312.

[15] Verona, A., *Stratified mappings - structure and triangulability.* Springer LNM 1102, Springer-Verlag (1984).

[16] Whitney, H., *Tangents to an analytic variety.* Ann. of Math. **81** (1965), 469-549.

Matematisk Institut,
Universitetsparken,
Ny Munkegade,
8000 Aarhus C,
Denmark.

e-mail: matadp@mi.aau.dk

Finiteness of Mather's Canonical Stratification

A.A. du Plessis

Introduction

In [5, 6, 7], Mather showed how to construct a "canonical" Whitney-regular stratification of jet-space, with the property that a proper map whose jet-extension is multi-transverse to the stratification is topologically stable. (An alternative, rather simpler, presentation of this result is given in [2].)

Mather also claimed, in [7], that the stratification constructed has only finitely many connected strata. This is of considerable interest because the topological type of a map-germ transverse to a (connected component of a) canonical stratum is determined by the stratum, so the claim implies that there are only finitely many topological types of topologically stable map-germs in any given dimension-pair.

Unfortunately, Mather's proof [7, p.170] of this finiteness is not correct. The aim of this article is to give a correct proof.

We give this proof in §2. In §1 we give the necessary background; in particular we describe a construction for the canonical stratification rather simpler than both Mather's and that given in [2].

The arguments given here were originally worked out for the review section of the book [9], see in particular pp. 44-5, but in the event were not included there.

I thank Terry Wall for help with the preparation of this article.

1 The canonical stratification

Mather's original construction (see [7]) of the canonical stratification is rather complicated; we prefer therefore to use the approach of [2]. We begin by recalling the key results and definitions.

Let $f : N \to P$ be a smooth map such that $f|\Sigma(f)$ is finite-to-one. (Here $\Sigma(f)$ is the critical set of f.) A *critical value stratification* (CVS) $\mathcal{C}$ for f is a regular stratification of $f\Sigma(f) \subset P$ which satisfies, for any $U, V \in \mathcal{C}$ (possibly equal), the following conditions:

CVS 1 $f^{-1}U \cap \Sigma(f)$ is a smooth submanifold of N;

CVS 2 $f|f^{-1}U \cap \Sigma(f) \to U$ is a local diffeomorphism;

CVS 3 $f^{-1}V \cap \Sigma(f)$ is Whitney regular over $f^{-1}U \cap \Sigma(f)$;

CVS 4 $f^{-1}V \setminus \Sigma(f)$ is Whitney regular over $f^{-1}U \cap \Sigma(f)$.

(Such a stratification is called a *partial stratification* in [2], p.26; Per Holm suggested the above, more informative, name.)

If $f : N \to P$ has a critical value stratification $\mathcal{C}$, and if $f\Sigma(f)$ is closed in P, then f has a Thom stratification $(\mathcal{A}, \mathcal{A}')$ given by

$$\mathcal{A} = \{f^{-1}W \setminus \Sigma(f), f^{-1}W \cap \Sigma(f) | W \in \mathcal{C}\} \cup \{f^{-1}(P \setminus f\Sigma(f))\},$$

$$\mathcal{A}' = \mathcal{C} \cup \{P \setminus f\Sigma(f)\};$$

for a proof, see [2, I, 3.1].

This stratification is called the stratification *associated* with $\mathcal{C}$. Similarly, we say that its strata are the strata *associated* with $\mathcal{C}$.

A critical value stratification $\mathcal{C}$ for $f : N \to P$ is said to be *canonical* if its strata can be constructed by downward induction on dimension to be "as large as possible"; more precisely, if M_j is the union of the strata of $\mathcal{C}$ of dimension j, C_j the union of the strata of dimension $\leq j$, then $\mathcal{C}$ is canonical if, for each j, M_j is the largest subset of C_j satisfying the conditions:

M_j is either empty or a smooth submanifold of P of dimension j;

M_k is Whitney regular over M_j for $k > j$;

$f|f^{-1}M_j \cap \Sigma(f)$ is a local diffeomorphism onto M_j;

$f^{-1}M_k \cap \Sigma(f)$ is Whitney regular over $f^{-1}M_j \cap \Sigma(f)$ for $k > j$;

$f^{-1}M_k \setminus \Sigma(f)$ is Whitney regular over $f^{-1}M_j \cap \Sigma(f)$ for $k \geq j$.

It is easy to see that canonical critical value stratifications are preserved by smooth diffeomorphisms (see [2, I, 3.7]).

Let A, B be open semialgebraic subsets of $\mathbb{R}^n, \mathbb{R}^p$ respectively, and $f : A \to B$ a polynomial mapping such that $f|\Sigma(f)$ is finite-to-one. Then, by [2, I, 3.5], f has a canonical critical value stratification with finitely many strata, all semialgebraic. We may also suppose the strata to be connected, by [2, I, 2.2].

A representative $f : (U, \mathbf{x}) \to (V, \mathbf{y})$ of a germ $\hat{f} : (N, \mathbf{x}) \to (P, \mathbf{y})$ is *special* [2, IV, p. 134] if:

$\mathbf{x} = \Sigma(f) \cap f^{-1}(\mathbf{y})$;

$f \mid \Sigma(f)$ is proper and finite-to-one;

f admits a canonical critical value stratification $\mathcal{C}$ such that, if $(\mathcal{A}, \mathcal{A}')$ is the associated stratification, then the closure of any connected component of a stratum of $\mathcal{A}$ contains $\mathbf{x}$.

If $\hat{f}$ is a C^∞-stable germ, then it is finitely-$\mathcal{A}$- determined, so that in suitable local coordinates it admits a polynomial representative. Since in particular $\hat{f}$ is of finite singularity type, a restriction of this representative is finite to one on its critical set, and so, by the result [2, I, 3.5] quoted above, has a canonical critical value stratification. Indeed, by [2, IV, 2.1], $\hat{f}$ has a special representative.

It is asserted there that the stratifications of any two special representatives agree on their overlap. This is not quite true but (see [8, 3.3]) these stratifications do coincide on some neighbourhoods of $\mathbf{x}, \mathbf{y}$ respectively.

It should be observed that the existence of special representatives with the much stronger property of being locally conelike follows from the work of Fukuda [1]; but we do not need this in what follows.

Following [2, IV, pp. 134-5], we define $c(\hat{f})$, for any C^∞-stable map-germ $\hat{f} : (N, \mathbf{x}) \to (P, \mathbf{y})$, to be the codimension of the stratum containing $\mathbf{x}$ associated to the canonical CVS of a special representative. Any unfolding $\hat{F}$ of $\hat{f}$ is trivial, so $c(\hat{F}) = c(\hat{f})$. We can thus define $c(\hat{f})$ for any germ of finite singularity type to be $c(\hat{F})$ for some - hence any - stable unfolding $\hat{F}$ of $\hat{f}$. This is invariant under $\mathcal{K}$-equivalence. Moreover, if $z = j^l\hat{f} \notin W^l(n,p)$ the $\mathcal{K}$-class of $\hat{f}$, hence the $\mathcal{A}$-class of $\hat{F}$ is determined by z, so we can define $c(z) = c(\hat{F})$. (Here $W^l(n,p)$ is the set of l-jets whose representatives have $\mathcal{K}_e$-codimension $\geq l$.) Write $\mathcal{A}^l(n,p)$ for the partition of $J^l(n,p) - W^l(n,p)$ determined by the function c. Then $\mathcal{A}^l(n,p)$ is the canonical stratification of jet-space, and has the properties mentioned in the introduction (see [2, IV, 2.3, 3.1, 4.1]). The proofs are rather difficult; even to see that the elements of $\mathcal{A}^l(n,p)$ are smooth is very delicate.

An alternative, and rather simpler, construction can be made as follows. Write $G = J^l(n,p) - W^l(n,p)$, and write Φ for the evaluation map $\Phi : \mathbb{R}^n \times G \to \mathbb{R}^p \times G$ defined by $\Phi(\mathbf{x}, z) = (P^l(z)(\mathbf{x}), z)$, where $P^l(z)$ denotes the polynomial representative of degree $\leq l$ of the l-jet z. Write $\hat{\Phi}_z$ for the germ of Φ at $(0_n, z)$. Since $z \notin W^l(n,p)$, $\mathfrak{m}_n^l.\theta(P^l(z)_0) \subseteq T\mathcal{K}(P^l(z)_0)$, so $\hat{\Phi}_z$ versally unfolds $P^l(z)_0$ and so is C^∞-stable.

Let $\mathcal{C}$ be the canonical critical value stratification of a special representative $\tilde{\Phi}$ of $\hat{\Phi}_z$, S the stratum through $z \times \mathbf{0}$; then the germ S_z of S at $z \times \mathbf{0}$ does not depend on the choice of representative.

Proposition 1.1 *Let $A_z \in J^l(n,p)$ be the set-germ at z given by*

$$0_p \times A_z = \{0_p \times G\} \cap S_z.$$

Then A_z is the germ at z of the canonical stratum through z.

Proof. Let $T = \tilde{\Phi}^{-1}(S) \cap \Sigma(\tilde{\Phi})$, T_z its germ at $0_n \times z$. From the definition of CVS, T is a submanifold of $\mathbb{R}^n \times G$. Moreover, $0_n \times A_z = \{0_n \times G\} \cap T_z$. For, since $\Phi(0, z) = (0, z)$ for all $z \in G$, the right-hand side must contain $0_n \times A_z$; equality holds because, from the definition of a special representative, $\tilde{\Phi}$ is injective on T.

We claim that T_z is transverse to $0_n \times G$. In fact, we will prove:

(1) T_z contains the germ K_z at z of the set K of points $(x, z') \in \mathbb{R}^n \times G$ such that the germ of Φ at (x, z') has the same $\mathcal{K}$-class as $\hat{\Phi}_z$;

(2) K_z is transverse to $0_n \times G$.

To prove (1), we invoke Mather's theory of C^∞-stability [3], and observe that, shrinking $\tilde{\Phi}$ as necessary, we can suppose that it is locally C^∞-stable (i.e. all its multi-germs are C^∞-stable), that K is a connected smooth manifold, and that a smooth curve in K may be obtained as an integral curve of a lowerable vector field ψ. The flow of the corresponding liftable vector field preserves C, since it is canonical, so in particular preserves S. But then the flow for ψ must preserves T. So indeed $K \subset T$.

To prove (2), we remark that, if $f : \mathbb{R}^n \to \mathbb{R}^p$ is a polynomial map of degree $\leq l$ sending 0 to 0, then, for any $y \in \mathbb{R}^n$, the map $f_y : \mathbb{R}^n \to \mathbb{R}^p$ given by $f_y(x) = f(x - y) - f(-y)$ is also polynomial of degree $\leq l$, sends 0 to 0, and its germ at y is $\mathcal{A}$-equivalent to the germ of f at 0. It follows that the $\mathcal{A}$-class of the germ of Φ at points of the graph of $y \longmapsto j^l(P^l(z)_y)$ is constant, so that the germ of this graph at $(0, z)$ is contained in K_z. Since this graph is transverse to $0_n \times G$, (2) follows.

Since T is the stratum through $(0, z)$ associated to the canonical CVS of $\tilde{\Phi}$, it follows that A_z is a germ of submanifold of $J^l(n,p)$ of codimension $c(z)$. Since $\tilde{\Phi}$ is a special representative for the germ of $\tilde{\Phi}$ at any point of T, so in particular at any point of $T \cap \{0_n \times G\}$, we see that c is constant on this set, which is thus contained in a canonical stratum. But, by [2, IV, p. 137], the canonical stratum given by c-value $c(z)$ has codimension $c(z)$; and thus the germ of this stratum at $(0, z)$ is indeed A_z. □

Remark. It is claimed in [9, p. 44] that S is transverse to $0_p \times G$, but this is not true.

It will be seen that the canonical stratification can be defined, instead of just identified, via the construction just given. Indeed, doing so leads to rather simpler proofs of its key properties than those given in [2]. For example, it is easy to see that the canonical stratification is Whitney regular. To see this, let us first remark that if $\hat{f} = \hat{f}_1 \sqcup \ldots \sqcup \hat{f}_r : (N, \{\mathbf{x}_1, \ldots \mathbf{x}_r\}) \to (P, \mathbf{y})$ is C^∞-stable, then the canonical CVS of a locally C^∞-stable representative $f = f_1 \sqcup \ldots \sqcup f_r : U_1 \sqcup \ldots \sqcup U_r \to V$ with $f_i : U_i \to V$ a special representative of $\hat{f}_i, i = 1 \ldots r$, then the canonical CVS of f is the transverse intersection of the canonical CVS for $f_1, \ldots f_r$. This follows from [2], I, (3.10), the fact, proved above, that the submanifold of points of U_i where the germ of f_i has the same $\mathcal{K}$-class as $\hat{f}_i$ is contained in the associated stratum through $\mathbf{x}_i$, and the consequence of C^∞-stability, that the images of these submanifolds under f are in general position.

Now the source stratification associated to the canonical CVS of $\tilde{\Phi}$ is Whitney-regular, so that the germ at $(z, 0)$ of its intersection with $0_n \times G$ is too (by [2, I, 1.3]). This germ of stratification refines the germ of $0_n \times \{$the canonical stratification}, by the remark above, because (we can assume that) $\tilde{\Phi}$ is locally stable. Thus regularity follows from:

Lemma 1.2 *Let $\mathcal{A}$ be a smooth stratification of $X \subseteq \mathbb{R}^n$ such that for any $\mathbf{x} \in X$, there exist a neighbourhood U of $\mathbf{x}$ in $\mathbb{R}^n$ and a regular stratification $\mathcal{B}$ of $U \cap X$ refining $\mathcal{A}|(U \cap X)$ such that the strata of $\mathcal{B}$ and of $\mathcal{A}|(U \cap X)$ through $\mathbf{x}$ coincide. Then $\mathcal{A}$ is regular.*

Proof. Let S, T be strata of $\mathcal{A}$ and let $\{\mathbf{s}_i\} \subset S$, $\{\mathbf{t}_i\} \subset T$ be sequences converging to $\mathbf{x} \in T$ with $TS_{s_i} \to \tau$, $\frac{\mathbf{t}_i - \mathbf{s}_i}{\|\mathbf{t}_i - \mathbf{s}_i\|} \to \lambda$. We need to show $\lambda \in \tau$.

Let U, $\mathcal{B}$ be as in the hypothesis so that $U \cap T$ is a stratum of $\mathcal{B}$. Since $\mathcal{A} \mid (U \cap X)$ is refined by $\mathcal{B}$, $U \cap S$ is a finite disjoint union of strata of $\mathcal{B}$; passing to subsequences, we may thus suppose that all the $\mathbf{s}_i$ belong to a single stratum S' of $\mathcal{B}$. Passing to a further subsequence, we may suppose that TS'_{s_i} converges, to τ', say. By regularity of $\mathcal{B}$, $\lambda \in \tau'$. Since $TS'_{s_i} \subseteq TS_{s_i}$ for each i, $\tau' \subseteq \tau$. Thus $\lambda \in \tau$. □

2 The main theorem

Theorem 2.1 *The stratification $\mathcal{A}^l(n,p)$ has a finite number of strata.*

Proof. We recall the situation above. Set $G = J^l(n,p) - W^l(n,p)$ and define $\Phi : \mathbb{R}^n \times G \to \mathbb{R}^p \times G$ by $\Phi(\mathbf{x}, z) = (P^l(z)(\mathbf{x}), z)$. Then the germ $\hat{\Phi}_z$ of Φ at $(0_n, z)$ is stable. Choose a special representative $\hat{\Phi}$ and its canonical critical value stratification $\mathcal{C}$, with stratum S through $(0_n, z)$; then the germ of $S \cap \{0_p \times G\}$ coincides with $0_p \times A_z$, where A_z is the germ at $(0_n, z)$ of the stratum of $\mathcal{A}^l(n,p)$ through z.

The main part of the proof will consist in finding an open semi-algebraic neighbourhood U of $0_n \times G$ in $\mathbb{R}^n \times G$ such that $\Phi|U : U \to \mathbb{R}^p \times G$ is locally stable.

Once such U is found, the proof may be completed quickly as follows:

By [2], $\Phi|U$ has a canonical critical value stratification $\mathcal{C}$ and the strata of $\mathcal{C}$ are semialgebraic and finite in number. Note that since $\Phi|U$ is locally stable, for any open $U' \subseteq U$ - in particular the domain of $\tilde{\Phi}$ - the restriction of $\mathcal{C}$ to $\Phi(\Sigma(\Phi|U'))$ refines the canonical critical value stratification of $\Phi|U'$, by remarks made in §1.

Identifying $0_p \times G$ with G, we thus see that $\{\mathcal{C}, \mathbb{R}^p \times G \setminus \Phi\Sigma(\Phi)\} \cap 0_p \times G$ is identified with a finite partition by semi-algebraic sets of G which refines $\mathcal{A}^l(n,p)$. By semi-algebraicity, the collection of connected components of elements of this partition is also finite, and refines the collection of connected components of elements of $\mathcal{A}^l(n,p)$, which is thus finite, as desired.

Thus it remains to find U.

It is tempting to suppose that it is sufficient to find an open semi-algebraic neighbourhood V of $0_n \times G$ in $\mathbb{R}^n \times G$ such that $\Phi|V$ is finite-to-one, for such $\Phi|V$ also has a canonical critical value stratification. But it is not clear that this canonical critical value stratification would have the required refining property. The problem is that if X, Y are semi-algebraic sets canonically stratified by $\mathcal{A}$, $\mathcal{B}$ respectively, the canonical stratification of $X \cap Y$ need not refine the partition $\mathcal{A} \cap \mathcal{B}$. For example, let

$$X = \{(x, y, z) \in \mathbb{R}^3 \mid xy = z(x + y), x + y \geq 0\},$$

$$Y = \{(x, y, z) \in \mathbb{R}^3 \mid x + y = 0\}.$$

Then $\mathcal{A} = \{X - \{0\}, \{0\}\}$, $\mathcal{B} = \{Y\}$ so $\mathcal{A} \cap \mathcal{B} = \{z - \text{axis} - \{0\}, \{0\}\}$, while $X \cap Y = z - $ axis is a single stratum.

It is not sufficient to find an open semi-algebraic neighbourhood of $\mathbb{R}^n \times G$ such that the germ of Φ at any point of this neighbourhood is C^∞-stable, either, for the problem with refinements remains; this was Mather's error in [2, p.170].

We begin by showing the existence of an open neighbourhood U_1 of $0_n \times G$ in $\mathbb{R}^n \times G$ such that $\Phi|U_1$ is locally stable; the proof does not depend on this step, but it helps explain why the later arguments work. For each $z \in G$ we have a special representative with source U_z which we may suppose of the form $B(\epsilon_z) \times V_z$ where V_z is an open neighbourhood of z in G and $B(\epsilon)$ is the open ball with centre 0_n and radius ϵ in $\mathbb{R}^n$. Set $U_1 = \bigcup\{B(\epsilon_z) \times V_z \mid z \in G\}$. Then $\Phi|U_1$ is locally stable. For since Φ preserves G-levels, any finite set $\mathbf{S}$ of U_1 with a image under Φ a single point is contained in a finite union of open balls of form $B(\epsilon_{z'}) \times z''$, for some $z' \in G$, $z'' \in V_{z'}$, hence in the largest such ball, and so in some $U_{z'}$, so local stability at $\mathbf{S}$ follows. Of course, this set U_1 is far from semi-algebraic.

Now let V_1 be the set of $(\mathbf{x}, z) \in \mathbb{R}^n \times G$ at which the germ of Φ is stable. Since its complement is the preimage by $j^l\Phi$ of the set of unstable l-jets, which is algebraic, and $j^l\Phi$ is polynomial, V_1 is semi-algebraic and open.

We will inductively construct semi-algebraic open neighbourhoods V_i of $0_n \times G$ in V_{i-1} such that for any $\mathbf{S} \subseteq U_i$ with at most i points the germ of Φ at $\mathbf{S}$ is stable. Since any stable germ with each component singular has at most p points, all multigerms of $\Phi|V_{p+1}$ are stable, and we can take $V = V_{p+1}$.

The set B_i of unstable jets in ${}_iJ^l(V_{i-1}, \mathbb{R}^p \times G)$ is closed and semialgebraic, hence so is its preimage C_i by ${}_ij^l(\Phi|V_{i-1})$ in $V_{i-1}^{(i)}$. Since all germs of multiplicity less than i are stable, C_i consists entirely of i-tuples with a common image, hence a common level $z \in G$. The set

$$Q_i = \{((\mathbf{x}_1, z), \ldots, (\mathbf{x}_i, z)) \in (\mathbb{R}^n \times G)^i \mid \| \mathbf{x}_1 \|^2 \geq \| \mathbf{x}_j \|^2 \text{ for } 1 < j \leq i\}$$

is closed and semi-algebraic in $(\mathbb{R}^n \times G)^i$, hence so is $C_i \cap Q_i$. By the Tarski-Seidenberg theorem, its projection on the first factor is again semi-algebraic, and so is its closure. We set $V_i = V_{i-1} - \overline{\pi_1(C_i \cap Q_i)}$. This is clearly open and semi-algebraic. If $\mathbf{S} \subseteq V_i$ were any set of i points with common image under Φ such that the germ of Φ at $\mathbf{S}$ is not stable, then as noted above, all these points are at the same level z, and if we arrange them in decreasing order of distance from $(0_n, z)$ we obtain a point of $C_i \cap Q_i$. But then by construction, the first point does not belong to V_i. So no such $\mathbf{S}$ exists.

It remains to show that V_i is a neighbourhood of $0_n \times G$. Since, by inductive hypothesis, V_{i-1} is such a neighbourhood, this amounts to showing that $\overline{\pi_1(C_i \cap Q_i)}$ is disjoint from $0_n \times G$. Suppose not; then there exists a sequence of i-tuples

$$\{(\mathbf{x}_{1,k}, z_k), \ldots, (\mathbf{x}_{i,k}, z_k)\} \in C_i \cap Q_i$$

with $(\mathbf{x}_{1,k}, z_k)$ converging to $(\mathbf{0}, z)$, say. But, as in an earlier argument, $(\mathbf{0}, z)$ has a neighbourhood $B(\epsilon_z) \times V_z$ on which Φ is locally stable. Since for large k, $z_k \in V_z$ and $\epsilon_z > \| \mathbf{x}_{1,k} \| \geq \| \mathbf{x}_{j,k} \|$ for $1 < j \leq i$, we have found an unstable multigerm in this neighbourhood: a contradiction. So V_i has the required properties, the inductive step is completed, and with it the proof. □

References

[1] Fukuda, T., Local topological properties of differentiable mappings I, II, Invent. Math. **65** (1981) 227-250, Tokyo Jour. Math. **8** (1985) 501-520.

[2] Gibson, C.G., K. Wirthmüller, A.A. du Plessis and E.J.N. Looijenga, *Topological stability of smooth mappings*, Springer Lecture Notes in Math. **552** (1976).

[3] Mather, J.N., Stability of C^∞-mappings IV: Classification of stable germs by $\mathbb{R}$-algebras, Publ. Math. IHES **37** (1969) 223-248.

[4] Mather, J.N., Stability of C^∞-mappings V: Transversality, Advances in Math., **4** (1970) 301-335.

[5] Mather, J.N., Notes on topological stability, preprint, Harvard 1970.

[6] Mather, J.N., Stratifications and mappings, in *Dynamical Systems* (ed. M. M. Peixoto), Academic Press, pp. 195-232.

[7] Mather, J.N., How to stratify manifolds and jet spaces, Springer Lecture Notes in Math. **535** (1976) 128-176.

[8] du Plessis, A.A., On the genericity of topologically finitely determined map-germs, Topology **21** (1982) 131-156.

[9] du Plessis, A.A. and C.T.C.Wall, *The geometry of topological stability*, Oxford University Press, 1995.

Matematisk Institut,
Universitetsparken,
Ny Munkegade,
8000 Aarhus C,
Denmark.

e-mail: matadp@mi.aau.dk

Trends in Equisingularity Theory

Terence Gaffney* David Massey

Abstract

We survey results in the equisingularity theory of complex analytic sets and mappings in which the equisingularity condition is controlled by invariants depending only on the members of a family, not on the total space of the family.

Introduction

Describing the structure of a singular set or mapping remains a basic, but elusive, goal of complex-analytic geometry. A first approach to this problem is to consider families of sets or mappings. If a set varies in an analytic family, then the family puts strong constraints on the variation in the set's structure, and so it is often easier to tell when the set is similar to the general member of the family. The different notions of similarity are referred to as equisingularity conditions. The variation in the structure of the set is often controlled by invariants which come from some algebraic object associated with the singularity. Since the end result of this study is to gain a deeper understanding of the members of the family, it is desirable that the invariants and equisingularity conditions depend only on the members of the family and not on its total space.

In this paper, we survey equisingularity conditions of sets and mappings which are governed by invariants depending only on the members of the family. We will only consider the local case. The first section of the paper is concerned with families of hypersurfaces with isolated singularities because this is where many of the basic ideas of equisingularity theory first appear. The theorems of this section illustrate nicely the two different, but intertwined, approaches to equisingularity theory — the approach of topology, and that of commutative algebra/analytic geometry. The second and third sections develop in some detail the ideas underlying these approaches. The fourth section returns to the development of the subject. The fifth and sixth section describe recent progress in extending the model theorems for hypersurfaces, through the commutative algebra/analytic geometry approach.

*Supported in part by NSF grant 9403708-DMS.
It is a pleasure to thank Robert Gassler for reading over the manuscript carefully and making a number of helpful comments.

Section 5 is concerned with complete intersections with isolated singularities (ICIS germs); while Section 6 deals with non-isolated hypersurface singularities. The advances described in these sections are based on the theory of the integral closure of ideals and of modules. The new results here (Theorems 5.7, 5.8, 6.6) are included as ways of illustrating the power of some recent ideas and techniques. Section 7 shows the current circle of ideas that make up the topological approach at work in proving a far reaching extension of the Lê-Saito theorem.

It should be noted that there are many results in the theory of equisingularity not mentioned here because they focus on the total space of the family, not on individual members. Further, the theory of finite determinacy of sets and maps is certainly a branch of equisingularity theory in which the conditions controlling equisingularity depend only on the members of the family, not on the total space. However, to survey all the results in this area would take another paper of equal length.

For a marvellous survey of results in equisingularity theory up to 1975, see Teissier's paper [80].

The authors thank Bernard Teissier and Lê Dũng Tráng for helpful conversations over a period of years describing the development of their results, and of the theory of equisingularity.

1 Families of Hypersurfaces with Isolated Singularities

The first results of the type that we are interested in appeared in the early to middle seventies. Prior to this period, foundational work in the theory of equisingularity had been done by Zariski [90–94], Whitney [88, 89], Thom [85], Hironaka, and Mather [61–68]. Zariski began the study of equisingularity with a series of four papers which explored the properties a "good" theory of equisingularity should have, and which developed such a theory for families of plane curves. Zariski's point of view was based on the resolution of singularities. Whitney, Thom and Mather developed a theory of topological equisingularity based on stratified sets.

A stratified set is a set which is a union of submanifolds called strata; any point in the set is in the closure of only finitely many strata. In order to understand the geometry of a stratified set you must understand how the strata meet. To construct homeomorphisms of stratified sets by integrating vector fields, you need to understand the "infinitesimal structure" of a stratum in the limit at points in the closure of the stratum. The Whitney conditions (formulated by Whitney ([88], [89]), and described at the beginning of Section 2) provide one way to do this. Whitney showed that any analytic set has a stratification in which any pair of incident strata satisfy these conditions. Such stratifications are called Whitney stratifications.

The Whitney conditions provide sufficient control on the limiting tangent planes coming from strata incident to a point so that it is possible to lift vector fields from lower-dimensional strata to higher-dimensional ones. The resulting field can then be integrated giving homeomorphisms. This insight is due to Thom, and is the key idea in his two isotopy theorems ([85]). These theorems give criteria based on the Whitney conditions for a family of sets or mappings to be topologically trivial. In particular, the first isotopy lemma implies that a set with a Whitney stratification is locally topologically trivial along strata. The detailed development of the theory is due to Mather and appears in [67], [68]. A readable account can be found in [24]. The development of this theory for mappings requires the control of limiting tangent planes to the fibers of the map as well. The first condition for doing this was developed by Thom ([85], p. 281), and is called the A_f condition.

Because equisingularity conditions are concerned often with the behavior, in the limit, of tangent hyperplanes to the smooth part of an embedded analytic set $X \subset \mathbf{C}^N$, it is helpful to construct spaces associated to X which encode this information. Their geometry can then be used to describe whether or not the equisingularity condition holds. Define the *conormal space*, denoted $C(X)$, to be the closure of the set of pairs (x, H) in $\mathbf{C}^N \times \mathbf{P}^{N-1}$ where x is a smooth point of X and H is a tangent hyperplane to X at x. For example, if X is a hypersurface, defined by the vanishing of a function f, then $C(X)$ is just the blow-up of X with respect to the Jacobian ideal $J(f)$, which is the ideal on X generated by the partial derivatives of f.

It is with the work of Lê, Teissier, Ramanujam, Briançon, Saito and Speder that invariants enter as controls for various equisingularity conditions. The work of Lê and Teissier took, as one starting point, the work of Zariski and Hironaka for dealing with families of plane curves. The natural extension of the case of plane curves turned out to be hypersurfaces with isolated singularities. Our notation for this case is as follows.

Let X be a family of hypersurfaces defined by $G : (\mathbf{C} \times \mathbf{C}^{n+1}, \mathbf{C} \times 0) \to (\mathbf{C}, 0)$, $g_t(z) = G(t, z)$ defines X_t, a reduced hypersurface with isolated singularity at the origin. Denote $\mathbf{C} \times 0$ by T. Let $J_z(G)$ denote the ideal generated by the partials of G with respect to z.

The basic invariant for hypersurfaces with isolated singularities is the Milnor number. This invariant has two interpretations, one topological, the other algebraic. Given a germ $g : (\mathbf{C}^{n+1}, 0) \to (\mathbf{C}, 0)$ with an isolated singularity at the origin, the Milnor number is the rank of the degree n (middle) homology group of the Milnor fiber of g. It is also the multiplicity of the Jacobian ideal of g viewed as an ideal in $\mathcal{O}_{n+1}$. This double interpretation leads to two approaches to equisingularity theory–one based on topology, the other on a combination of analytic geometry and commutative algebra. We let $\mu(X_t)$ denote the Milnor number of X_t at the origin, and we let $\mu^{(*)}(X_t)$ denote the sequence of Milnor numbers of the germs obtained by intersecting X_t with generic linear spaces of dimension $0, \ldots, n+1$. (By convention, $\mu^0 = 1$.)

The following three theorems are models for the results we are interested in:

Theorem (1) (Lê-Ramanujam [51]) *In the above setup, suppose $\mu(X_t)$ is independent of t. Then, for all small t, if $n \neq 2$, the local, ambient, topological-type of X_t at the origin is independent of t.*

Theorem (2) (Lê-Saito [52]) *In the above setup, suppose that the Milnor number of X_t is independent of t. Then, the pair $(\mathbf{C}^{n+2} - T, T)$ satisfies Thom's A_G condition.*

Theorem (3) (Teissier, Briançon-Speder, [78], [5]) *Suppose X, G, g_t as above. Then the following are equivalent:*

(i) $X - T$ is smooth and Whitney regular over $T = \mathbf{C} \times 0$.

(ii) $\mu^{(*)}(X_t)$ is independent of t.

In each of the above theorems, an equisingularity condition is controlled by invariants depending only on members of the family, independent of how the member sits in the family. In each case, the constancy of the invariant(s) is necessary as well as sufficient. This is important, for it means that these invariants capture the essence of the equisingularity condition, motivating further study of their nature.

The theorems exemplify the two different approaches to equisingularity discussed above. Theorem (1) is proved by the h-cobordism theorem, (3) by the theory of integral closure of ideals. Theorem (2) can be proved by either method–Lê-Saito used Morse theory (Cerf Diagram), Teissier used the principle of specialization of integral dependence (PSID). The two different proofs of Theorem 2 each exploited the corresponding interpretation of μ; both approaches work to control the conormal space.

The history leading up to these results is interesting. Hironaka suggested to Lê and Teissier that μ constant families of plane curves should be topologically equisingular; Lê produced a proof of this in 1970 (Note to CRAS in January 1971). This result was over shadowed by his result with Ramanujam [51], which was the third part of his thesis defended in December 1971.

At the same time, Teissier was investigating Zariski's conjecture that the multiplicity of a hypersurface is a topological invariant [93]. Teissier realized that if μ constant passed to general hyperplane sections, then one could prove that it implies equimultiplicity. Then, the Lê-Ramanujam result influenced Teissier, and prompted him to prove, in the general case, that μ is a topological invariant [79]. The trend established here of the topological approach aiding the development of the geometric/algebraic approach still continues today. See the discussion later on the connection between the work of Massey and Gaffney-Gassler.

Theorems 1 and 3 show that the topological equisingularity of the generic plane sections of a family of isolated hypersurface singularities is equivalent to the Whitney conditions holding. This allows one to reduce the study of Whitney equisingularity back to topology at the cost of including the generic plane sections. This equivalence is a special case of a much more sweeping result, also proved by Lê and Teissier, which shows that the same result holds for general families of stratified sets [53].

To understand the methods used in the geometric/algebraic approach, we turn to the Whitney conditions and the theory of the integral closure of ideals.

2 The Whitney Conditions and the Theory of the Integral Closure of Ideals

The Whitney conditions are concerned with the relation between the tangent plane to a stratum and limiting tangent planes from higher-dimensional strata, so we need a way of measuring distance between linear spaces. Suppose A, B are linear subspaces at the origin in $\mathbf{C}^N$, then define the distance from A to B as:

$$\operatorname{dist}(A,B) = \sup_{\substack{u \in B^\perp - \{0\} \\ v \in A - \{0\}}} \frac{|(u,v)|}{\|u\|\,\|v\|}.$$

In the applications B is the "big" space and A the "small" space. (Note that $\operatorname{dist}(A,B)$ is not in general the same as $\operatorname{dist}(B,A)$.) This allows us to talk about the Whitney conditions holding with a certain exponent (a condition introduced by Hironaka in [38]). Suppose $Y \subset \overline{X}$, where X, Y are strata in a stratification of an analytic space, and $\operatorname{dist}(TY_0, TX_x) \leq C\operatorname{dist}(x,Y)^e$ for all x close to Y. Then the pair (X,Y) satisfies *Whitney A with exponent* e at $0 \in Y$.

To date, the cases of most interest are those when $e = 1$, which is known as *Verdier's condition W* (Cf. [87]), and when $e > 0$, but indeterminate (*Whitney A*). If one is working with the Whitney A condition, the value of e is usually not relevant.

Given X, Y as above, suppose that, for all sequences (x_i, y_i) where $x_i \in X$, $y_i \in Y$, and (x_i, y_i) converges to $(0,0)$, such that the sequences $\{TX_{x_i}\}$ and $\{sec(x_i, y_i)\}$ of tangent planes and secant lines converge to a plane H and a line l, we have that $l \subset H$. Then we say that the pair (X, Y) satisfies *Whitney's condition B* at $0 \in Y$.

Originally the relation between W and Whitney's conditions A and B was unclear. Since Teissier proved condition W is equivalent to these two Whitney conditions in the complex analytic case [82, V.1.2], we will use the two terms interchangeably.

If f is the germ of a complex analytic mapping defined on the closure of a stratum X, then it is useful to have a notion of condition W relative to

f; this is obtained from the above by replacing the tangent space of X by the tangent space of the fiber of f at points where f is a submersion onto its image. The W_f *condition* holds for the pair (X, Y) when the new condition holds with exponent 1; when it holds with some unspecified exponent, we say that *Thom's condition* A_f holds. This notion for hypersurfaces (although not explicitly named) appears in [78]. Teissier controlled this equisingularity condition using a condition, which he called *Condition C* ([80], p. 604). The formulation of this condition depends on the theory of the integral closure of ideals, and is discussed at the end of this section. See ([83], p.288) for a development of the relation between Condition C and W_f.

The terminology W_f and definition as stated here was introduced by Henry, Merle, and Sabbah, [36].

The *relative conormal space of* f, denoted $C(X, f)$ is used to study limiting tangent hyperplanes to the fibers of f. The construction is the relative form of the construction for $C(X)$: it is the closure of the set of pairs (x, H) in $\mathbf{C}^N \times \mathbf{P}^{N-1}$ where f is a submersion at $x \in X$ and H is a tangent hyperplane to the fiber of f at x.

The theory of the integral closure of ideals provides an algebraic description of these conditions in the hypersurface case. The basic source for the results quoted here is the work [54] of Monique Lejeune–Jalabert and Teissier. Another useful reference is Joseph Lipman's more algebraic survey [55].

Let $(X, 0) \subset (\mathbf{C}^N, 0)$ be a reduced analytic space germ. Let I be an ideal in the local ring $\mathcal{O}_{X,0}$ of X at 0, and f an element in this ring. Then f is integrally dependent on I if one of the following equivalent conditions obtain:

(i) There exists a positive integer k and elements a_j in I^j, so that f satisfies the relation $f^k + a_1 f^{k-1} + \cdots + a_{k-1} f + a_k = 0$ in $\mathcal{O}_{X,0}$.

(ii) There exists a neighborhood U of 0 in $\mathbf{C}^N$, a positive real number C, representatives of the space germ X, the function germ f, and generators $g_1, \ldots, g_m$ of I on U, which we identify with the corresponding germs, so that for all x in $X \cap U$ the following inequality obtains: $|f(x)| \leq C \max_i \{|g_i(x)|\}$.

(iii) For all analytic path germs $\phi : (\mathbf{C}, 0) \to (X, 0)$ the pull–back $\phi^* f$ lies in the ideal generated by $\phi^*(I)$ in the local ring, $\mathcal{O}_1$ of $\mathbf{C}$ at 0.

If we consider the normalization $\overline{B}$ of the blow-up B of X along the ideal I, we get another equivalent condition for integral dependence. Denote by $\overline{D}$ the pull–back to $\overline{B}$ of the exceptional divisor D of B, and the map from $\overline{B}$ to X by p.

(iv) For any component C of the underlying set of $\overline{D}$, the order of vanishing of the pull-back of f to $\overline{B}$ along C is no smaller than the order of the divisor $\overline{D}$ along C.

There is a condition that is related to integral dependence which is also important. If we strengthen (iii) so that we require $\phi^* f$ be in the product of m_1 and the ideal generated by $\phi^*(I)$, where m_1 is the maximal ideal of $\mathcal{O}_1$, then we say that f is *strictly dependent* on I. If I has finite colength, this condition is equivalent to asking that the order of vanishing of f on the components of the exceptional divisor $\overline{D}$ be strictly greater than the order of vanishing of the pull-back of I. In this form, the condition was called $\overline{\nu}(f) > 1$, and appears in [78].

The elements f in $\mathcal{O}_{X,0}$ that are integrally dependent on I form an ideal called the integral closure of I, and denoted $\overline{I}$. Often we are only interested in the properties of the integral closure of an ideal I; so we may replace I by an ideal J contained in I with the same integral closure as I. Such an ideal J is called a *reduction* of I.

It is easy to see that J is a reduction of I if and only if there exists a finite map $\mathrm{Bl}_I X \to \mathrm{Bl}_J X$. In particular, the fibers of the two blow-ups over 0 have the same dimension. Samuel proved that any ideal I has a reduction generated by at most n elements where n is the dimension of X at 0. In fact, n generic linear combinations of given generators of I generate a reduction.

We denote the set of elements strictly dependent on I by $I^\dagger$.

Some of the implications linking the above four conditions are easy to see. Assume that (i) holds; then pulling this relation back to $\overline{B}$ and dividing by $(g_i \circ p)^k$ we have that

$$(f \circ p/g_i \circ p)^k + \tilde{a}_1(f \circ p/g_i \circ p)^{k-1} + \cdots + \tilde{a}_{k-1}(f \circ p/g_i \circ p) + \tilde{a}_k = 0$$

holds locally, where $g_i \circ p$ is a local generator of the pull-back of I. Because $\overline{B}$ is normal and the pull-back of I is principal, it follows that $f \circ p$ is in the pull-back of I. Condition (iv) then follows. Given (iv), (ii) holds in $\overline{B}$, so we can push it downstairs. Condition (iii) follows immediately from (ii).

In practice, (iii) is used to reduce the kind of tangent space calculations that have played such an important role in singularities of mappings since Mather, to calculations in modules over $\mathcal{O}_1$, the ring of analytic functions in one variable. It is also useful in performing calculations for specific examples, and sometimes for establishing equisingularity conditions directly. Condition (ii) is the link between the theory of integral closure and the inequalities defining the various equisingularity conditions. Before discussing the role that condition (iv) plays we give an example, which shows how the parts work together; this is a result of Teissier ([78], Remark 3.10).

Theorem 2.1 *In the hypersurface set-up, the pair* $(\mathbf{C}^{n+2} - \Sigma(G), T)$ *satisfies* a_G *if and only if* $\frac{\partial G}{\partial t}$ *is strictly dependent on* $J_z(G)$.

Proof. The set of limiting tangent planes at the origin to the fibers of G restricted to $\mathbf{C}^{n+2} - \Sigma(G)$ is exactly the fiber of $\mathrm{Bl}_{J(G)}(\mathbf{C}^{n+2})$ over the origin. This means that for each limiting tangent plane H, you can find a curve $c(s)$, such that the image of c lies in $\mathbf{C}^{n+2} - \Sigma(G)$ except for $c(0)$, and the limit of the tangent planes to the fiber of G at $c(s)$ is H. The curve c exists because a curve with analogous properties exists on $\mathrm{Bl}_{J(G)}(\mathbf{C}^{n+2})$, and we can push it down to $\mathbf{C}^{n+2}$. This means that

$$\lim_{s\to 0}(1/s^k)(DG(c(s))) = (a_1,\ldots,a_{n+2})$$

where $a_1z_1+\cdots+a_{n+1}z_{n+1}+a_{n+2}t=0$ defines H, and k is the minimum of the orders of the first non-vanishing term in $\partial G/\partial t\circ\phi,\ldots,\partial G/\partial z_{n+1}\circ\phi$. The condition that H contains the t-axis is just that $a_{n+2}=0$. This is equivalent to asking that $\partial G/\partial t\circ c\in m_1c^*(\partial G/\partial z_1,\ldots,\partial G/\partial z_n)$. Thus, the condition that every limiting tangent plane at the origin contain the t-axis is equivalent to $\partial G/\partial t$ being strictly dependent on $(\partial G/\partial z_1,\ldots,\partial G/\partial z_{n+1})$. By the remark after the definition of strict dependence, and the argument used to show that (iv) implies (ii), it follows that the inequality used to define A_G holds with some exponent. (The exponent would depend on the ratio of the order of vanishing of $\partial G/\partial t$ and $J_z(G)$ on the components of $\overline{D}$.) This completes the proof.

The fourth formulation of integral closure is the one that lends itself to control by algebraic invariants. If I is an ideal of finite colength in $\mathcal{O}_{X^d,x}$, then the *multiplicity of* I, denoted $e(I)$, is the intersection number of the exceptional divisor of the blow-up with $d-1$ hyperplanes in $\mathbf{P}^{n-1}$ where n is the number of generators of I. Two basic results due to Rees are that if $I\subset J$, then $e(I)\geq e(J)$ with equality holding if $\overline{I}=\overline{J}$, and if (X,x) is equidimensional, then $e(I)=e(J)$ if and only if $\overline{I}=\overline{J}$. You can think of the multiplicity as describing how the exceptional divisor "sits" in projective space. The importance of the multiplicity stems from the following fact. Given a family of ideals, we can consider the exceptional divisor of the blow-up of the total space by this ideal sheaf. If the behavior of this exceptional divisor changes drastically over some $x\in X$, then the multiplicity of that member of the family at x must jump. This is made precise by the next theorem which is due to Teissier.

To state this theorem, we introduce a piece of notation often used in what follows. If we have a family of ideals and the fiber of the support of $\mathcal{O}_X/\mathcal{I}$ over t is finite and is $x_{t,1},\ldots,x_{t,k(t)}$, then the sum of the multiplicities of the ideals induced by $\mathcal{I}$ in the local rings $\mathcal{O}_{X(t),x_{t,i}}$ is denoted $e(\mathcal{I}\mathcal{O}_{X(t)})$.

Theorem 2.2 (Teissier [78], Principle of Specialization of Integral Dependence of Ideals.) *Consider a reduced equidimensional family $X\to Y$ of analytic spaces, and an ideal sheaf $\mathcal{I}$ on X with finite co-support over Y. Suppose h is a section of $\mathcal{O}_X$ so that for all t in a Zariski-open dense subset of Y the induced section of $\mathcal{O}_{X(t)}$ on the fiber over t is integrally dependent on the induced ideal sheaf $\mathcal{I}\mathcal{O}_{X(t)}$. If the multiplicity $e(\mathcal{I}\mathcal{O}_{X(t)})$ is independent of t in Y, then h is integrally dependent on $\mathcal{I}$.*

Proof. We give a brief sketch of the proof the principle as contained in [81, Appendix 1]. For simplicity, assume that $\mathcal{O}_X/\mathcal{I}$ is supported on the image of a section $t \mapsto x_t$ of X over Y. Denote the fiber dimension of $X \to Y$ by n. By Samuel's theorem, we can find n elements $g_1, \dots, g_n$ in the stalk of $\mathcal{I}$ at x_0 the image of which in $\mathcal{O}_{X(0),x_0}$ generate a reduction of the ideal induced by $\mathcal{I}$. We identify all germs with representatives in a small neighborhood of x_0 in X. Denote the ideal sheaf generated by $g_1, \dots, g_n$ by $\mathcal{J}$, its stalk at 0 by J. Now, we use the upper-semicontinuity of the multiplicity:

$$e(\mathcal{I}\mathcal{O}_{X(0),x_0}) = e(\mathcal{J}\mathcal{O}_{X(0),x_0}) \geq e(\mathcal{J}\mathcal{O}_{X(t),x_t}) \geq e(\mathcal{I}\mathcal{O}_{X(t),x_t}) = e(\mathcal{I}\mathcal{O}_{X(0),x_0}).$$

Hence, all inequalities are equalities and, by Rees' theorem, the ideal induced by $\mathcal{J}$ in $\mathcal{O}_{X(t),x_t}$ is a reduction of the ideal induced by $\mathcal{I}$ for all t in Y close to 0.

Now, note that the underlying set of the exceptional divisor D of the blow-up B of X along $\mathcal{J}$ equals $Y \times \mathbf{P}^{n-1}$. Also, as the normalization $\overline{B} \to B$ is finite, all components of the pull-back $\overline{D}$ of D to $\overline{B}$ are equidimensional over Y. Thus, it is not hard to see, using (iv), that an element induced by a section h of $\mathcal{O}_X$ in $\mathcal{O}_{X,x_0}$ is integrally dependent on $\mathcal{J}$ if, for all t in a Zariski-open subset of Y, the element induced by h in $\mathcal{O}_{X(t),x_t}$ is integrally dependent on the ideal induced by $\mathcal{J}$. In fact, the order of vanishing of the pull-back of h to $\overline{B}$ along a component C of the underlying set of $\overline{D}$ can be computed in the fiber over some generic t in Y. By assumption, this is no smaller than the order of vanishing of $\overline{D}$ along C. This proves the theorem.

With this result we can describe the steps in the proof that $\mu^{(*)}$ constant implies the Whitney conditions.

The first step is to show that the Whitney conditions hold if and only if

$$\partial G/\partial t \in \overline{m_{n+1}J_z(G)},$$

where $m_{n+1}J_z(G) \subset \mathcal{O}_X$. This amounts to using the formulation of integral closure in terms of inequalities to convert the inequality defining the Whitney conditions into a statement about integral closure. The next step is to observe that the Whitney conditions hold generically, so we can assume that the above inclusion holds on a Z-open subset of T. Now use the following formula proved by Teissier ([78], p. 322) which relates the multiplicity of $m_{n+1}J(g_t)$ to the $\mu^{(*)}$ invariants:

$$e(m_{n+1}J(g_t)) = \sum_{i=0}^{n} \binom{n}{i} (\mu^{i+1}(X(t), 0) + \mu^i(X(t), 0)).$$

This formula shows that the $\mu^{(*)}$ invariants are independent of t if and only if the multiplicity of $m_{n+1}J(g_t)$ is. These two steps establish the hypotheses needed to apply the principle of specialization of integral dependence.

If we consider the integral closure of $m_{n+1}J_z(G)$ in $\mathcal{O}_{n+2}$, then we can define Condition C by the inclusion

$$\partial G/\partial t \in \overline{m_{n+1}J_z(G)} \qquad \text{(Condition C)}.$$

As mentioned before, this condition is equivalent to W_G. A proof similar to the one given above shows that $\mu^{(*)}$ constant implies W_G as well as W.

At this point, we pass to a discussion of the ideas underlying the topological approach to equisingularity theory.

3 Topological Data and Non-Isolated Singularities

Now we describe how the ideas used to study hypersurfaces with isolated singularities from the topological point of view evolved to deal with non-isolated singularities. The topological approach to hypersurfaces begins with Milnor, who defined the fibration that now bears his name in [69] in 1968. Let us recall the definition of the Milnor fibration.

Let $\mathcal{U}$ be an open subset of $\mathbf{C}^{n+1}$ and let $f : \mathcal{U} \to \mathbf{C}$ be a complex analytic function. For each point $p \in f^{-1}(0)$, the *Milnor fibration of f at p* is defined up to (stratified) homeomorphism by the following.
For all $\epsilon > 0$, let $B^\circ_\epsilon(p)$ denote the open ball of radius ϵ in $\mathcal{U}$ centered at p. For all $\eta > 0$, let D_η denote the closed disc centered at the origin in $\mathbf{C}$, and let ∂D_η denote its boundary, which is a circle of radius η. Then, there exists $\epsilon_0 > 0$ such that, for all ϵ such that $0 < \epsilon \leq \epsilon_0$, there exists $\eta_\epsilon > 0$ such that, for all η such that $0 < \eta \leq \eta_\epsilon$, the restriction of f to a map $B^\circ_\epsilon(p) \cap X \cap f^{-1}(\partial D_\eta) \to \partial D_\eta$ is a locally trivial fibration whose (stratified) homeomorphism-type is independent of the choice of ϵ and η. This fibration is called the *Milnor fibration* of f at p and the fibre is the *Milnor fibre* of f at p, which we denote by $F_{f,p}$. (In this generality, the existence of the Milnor fibration was proved by Lê in [46]).

In the case where f has an isolated critical point at p, Milnor proved that $F_{f,p}$ has the homotopy-type of a finite bouquet (wedge, one-point union) of spheres of real dimension n (which is half of the real dimension of $F_{f,p}$ as a manifold – hence, the usual term is "middle" dimension); the number of spheres in this bouquet is the *Milnor number*, $\mu_p(f)$, of f at p. Algebraically, $\mu_p(f)$ can be calculated as the complex dimension of the Jacobian algebra of f at p.

We would now like to give a sketch of the topological proof of the result of Lê and Saito on the A_f condition. We do not use the exact phrasing of Lê and Saito, rather we state the proof in a way that generalizes nicely (see Section 7).

Let $\mathcal{U}$ be a open neighborhood of the origin in $\mathbf{C}^{n+1}$, let D° be an open disk centered at the origin in $\mathbf{C}$, and let $f : (D^\circ \times \mathcal{U}, D^\circ \times \{0\}) \to (\mathbf{C}, 0)$ be an analytic function. Define the family $f_t(z_0, \ldots, z_n) := f(t, z_0, \ldots, z_n)$. Assume that each f_t has an isolated critical point at the origin, and that the Milnor number of f_t at the origin is independent of $t \in D^\circ$.

Before we can begin our sketch of the Lê-Saito result, we must introduce another object of fundamental importance in the topological study of singularities: the *relative polar curve*, $\Gamma^1_{f,t}$. This is defined by $\Gamma^1_{f,t} := \overline{\mathrm{crit}(f,t) - \mathrm{crit}(f)}$. The relative polar curve seems to have first been studied by Hamm and Lê in [31] and, independently, by Teissier in [78]. The first discussion of $\Gamma^1_{f,t}$ as a relative version of the absolute polar varieties appearing in [83] (see, also, Section 4) was in Teissier's note [84].

Now that we have the notion of the relative polar curve, the first step of the proof is to show that the Milnor number of f_t being independent of t implies that $\Gamma^1_{f,t} = \emptyset$. Let B_ϵ denote a small closed ball of radius ϵ centered at the origin inside the hyperplane $t = 0$ (a *Milnor ball* for f_0). If we have $0 < \eta \ll \epsilon$ and D_η denotes a complex disk of radius η centered at 0, then $B_\epsilon \cap f_0^{-1}(\eta)$ is the Milnor fibre of f_0 at the origin and $B_\epsilon \cap f_0^{-1}(D_\eta)$ is contractible (a *Milnor tube*). The pair $(B_\epsilon \cap f_0^{-1}(D_\eta), B_\epsilon \cap f_0^{-1}(\eta))$ has relative cohomology only in degree $n+1$, and there it is free Abelian of rank equal to $\mu(f_0)$. The pair lifts to $t = t_0$, where t_0 is small and unequal to 0, to yield a homeomorphic pair $(B_\epsilon \cap f_{t_0}^{-1}(D_\eta), B_\epsilon \cap f_{t_0}^{-1}(\eta))$. Note, though, that both ϵ and η may be too large for $B_\epsilon \cap f_{t_0}^{-1}(\eta)$ to be the Milnor fibre of f_{t_0} at the origin in the hyperplane $t = t_0$. However, Morse theory implies that the cohomology of this pair in degree $n+1$ is the sum of the Milnor numbers of $f_{t_0} - f_{t_0}(p)$ at each point p in $B_\epsilon \cap f_{t_0}^{-1}(D_\eta) \cap \left(\mathrm{crit} f_{t_0} \cup (\Gamma^1_{f,t} \cap V(t - t_0))\right)$. As we are assuming that the Milnor number at the origin is constant in the family f_t, it follows that $\Gamma^1_{f,t}$ must be empty (and that $\mathrm{crit} f$ equals the t-axis).

But, it is easy to show that an empty relative polar curve implies that the exceptional divisor of the blow-up of $D^\circ \times \mathcal{U}$ along the Jacobian ideal of f does not have a component lying over the origin (see the first example of Section 6). As the exceptional divisor is purely $(n+1)$-dimensional and as the A_f condition holds generically, this shows that the only component of this exceptional divisor is the projectivized conormal to the t-axis. This is easily seen to imply the A_f condition.

While Milnor himself proved a large number of interesting results for **isolated** critical points of an analytic function, f, a nice treatment in the case of non-isolated critical points was hampered by a number of problems – not the least of which was how to encode the data provided by all the Milnor fibrations at all of the critical points.

Sheaf theory immediately comes to mind. But what sort of sheaf theory should one use? Essentially, one wants a sheaf of Milnor fibres; a sheaf which has as stalks the various Milnor fibres and in which the restriction maps tell us how the Milnor fibres at different points are related. Actually, though, we will only try to encode this information on the level of cohomology.

The problem is: at each point of the critical locus of f, we wish to have the information about the cohomology of the Milnor fibre at that point. This means that **after we look at the stalk at a point in the critical locus**, we still wish to have cohomology groups in all dimensions at our disposal.

It does not take long to realize that what you need is a complex of sheaves. But, really, one frequently only cares about this complex of sheaves up to cohomology (or chain-homotopy equivalence). Very loosely speaking, this is what the *derived category* gives you.

Hence, in 1969 in [13], Deligne defined the sheaf of vanishing cycles as an object in the derived category. While we shall not give a rigorous definition of the vanishing cycles here, we do wish to describe the stalk cohomology.

Let X denote a complex analytic space and $f : X \to \mathbf{C}$ a complex analytic function. For convenience, we assume that X is contained in some open subset $\mathcal{U}$ of some affine space $\mathbf{C}^{n+1}$. The *vanishing cycles along* f encode cohomological data about all of the Milnor fibres of f at all points of $f^{-1}(0)$, and also contain information on how nearby Milnor fibres relate to a given one. The vanishing cycles are actually a bounded, constructible complex of sheaves (an object in the derived category).

The coefficients that one uses for the vanishing cycles – when calculating the cohomology of the Milnor fibres – are extremely important; if we were only interested in the case where X is a local complete intersection, then we could get by using only the constant coefficients $\mathbf{C}$. However, since we want to deal with completely general analytic spaces, we find that we must allow coefficients which themselves are bounded, constructible complexes of sheaves.

Let $F^\bullet$ be a bounded constructible complex of sheaves of $\mathbf{C}$-vector spaces on X; we write $F^\bullet \in D_c^b(X)$. The *vanishing cycles of* $F^\bullet$ *along* f are denoted by $\phi_f F^\bullet$; they are an object in $D_c^b(f^{-1}(0))$ with stalk cohomology at any point $p \in f^{-1}(0)$ given by the relative (hyper-)cohomology of a ball in X centered at p modulo the Milnor fibre of f at p with coefficients in $F^\bullet$, i.e. for all small $\epsilon > 0$ and all $\xi \in \mathbf{C}^*$ with $|\xi| \ll \epsilon$,

$$H^i(\phi_f F^\bullet)_p \cong H^{i+1}(B_\epsilon^\circ(p) \cap X, B_\epsilon^\circ(p) \cap X \cap f^{-1}(\xi); F^\bullet).$$

In particular, in the case where $F^\bullet$ equals the constant sheaf $\mathbf{C}_X^\bullet$, we obtain

$$H^i(\phi_f \mathbf{C}_X^\bullet)_p \cong H^{i+1}(B_\epsilon^\circ(p) \cap X, B_\epsilon^\circ(p) \cap X \cap f^{-1}(\xi); \mathbf{C}_X^\bullet) \cong \widetilde{H}^i(F_{f,p}\, ; \mathbf{C}),$$

where $\widetilde{H}$ denotes reduced singular cohomology.

An important fact about the vanishing cycles is that their support is contained in the *stratified critical locus* of f; that is, if $\mathcal{S}$ is a Whitney stratification of X with respect to which $F^\bullet$ is constructible, then the vanishing cycles, $\phi_f F^\bullet$, have non-zero stalk cohomology only at points in $\bigcup_{S \in \mathcal{S}} \operatorname{crit}(f_{|S})$.

While the formalism of the derived category enables one to prove many interesting results about $\phi_f F^\bullet$ – and so, provides interesting topological information about hypersurface singularities – the vanishing cycles themselves contain far too much information for us to calculate them algebraically. Hence, we wish to extract some of the information from the vanishing cycles;

we want enough topological data to control Thom's A_f condition, but we want the data to be more amenable to algebraic calculation. For this purpose, we need the notion of the *characteristic cycle* of a complex of sheaves. The characteristic cycle was introduced by Sato, Kashiwara, and Kawaï in 1973 [77] in the context of $\mathcal{D}$-modules.

If $\mathcal{S} = \{S_\alpha\}$ is a Whitney stratification (with connected strata) of $X \subseteq \mathcal{U}$ with respect to which $F^\bullet$ is constructible, then the characteristic cycle of $F^\bullet$ measures – in a fairly weak way – how the cohomology of X (with coefficients in $F^\bullet$) changes as one moves in normal directions to the strata. In order to describe the characteristic cycle, we first need to recall the definition of the *link of a stratum.*

The *(complex) link of a stratum* $S_\alpha \in \mathcal{S}$ is defined up to homeomorphism by

$$B^\circ_\epsilon(p) \cap X \cap N \cap L^{-1}(\eta)$$

where p is any point in the stratum S_α, with normal slice N to S_α at p, $L : (\mathcal{U}, p) \to (\mathbf{C}, 0)$ is such that d_pL is a non-degenerate covector at p, $L_{|S_\alpha}$ has a Morse singularity at p, $\epsilon > 0$ is small, and $0 < |\eta| \ll \epsilon$ (see [27]). The homeomorphism-type of the link is independent of all the choices.

The characteristic cycle of a general complex of sheaves $F^\bullet$ is an analytic cycle in the cotangent space $T^*\mathcal{U}$; the coefficients of this cycle are obtained – up to sign – by taking the difference between the Euler characteristic of the link of the stratum and the Euler characteristic of a point on the stratum **with coefficients in the complex** $F^\bullet$.

The *characteristic cycle,* $\mathrm{Ch}(F^\bullet)$, *of* $F^\bullet$ in $T^*\mathcal{U}$ is the linear combination $\sum_\alpha m_\alpha \overline{T^*_{S_\alpha}\mathcal{U}}$, where $\overline{T^*_{S_\alpha}\mathcal{U}}$ denotes the closure of the conormal bundle to S_α in $T^*\mathcal{U}$ and the m_α are integers given by

$$(3.1) \qquad m_\alpha := (-1)^{\dim X - 1}\chi(\phi_{L_{|X}} F^\bullet)_p = (-1)^{\dim X - d - 1}\chi(\phi_{L_{|N}} F^\bullet{}_{|N})_p$$

for any point p in the d-dimensional stratum S_α, with normal slice N to S_α at p, and any $L : (\mathcal{U}, p) \to (\mathbf{C}, 0)$ such that d_pL is a non-degenerate covector at p (with respect to our fixed stratification; see [27]) and $L_{|S_\alpha}$ has a Morse singularity at p. This cycle is independent of all the choices made.

The fact that d_pL is non-degenerate is equivalent to saying that if S_α is the stratum containing p, then (p, d_pL) is not in $\overline{T^*_{S_\beta}\mathcal{U}}$ for $S_\alpha \neq S_\beta$; thus, (p, d_pL) is a smooth point of $\mathrm{Ch}(F^\bullet)$.

Hence, for such a p and L, we have an equality of intersection numbers

$$(\mathrm{Ch}(F^\bullet) \cdot \mathrm{Im}\, dL)_{(p,d_pL)} = \left(m_\beta \overline{T^*_{S_\beta}\mathcal{U}} \cdot \mathrm{Im}\, dL\right)_{(p,d_pL)} = m_\beta\ ,$$

since

$$\left(\overline{T^*_{S_\beta}\mathcal{U}} \cdot \mathrm{Im}\, dL\right)_{(p,d_pL)} = \left(T^*_{S_\beta}\mathcal{U} \cdot \mathrm{Im}\, dL\right)_{(p,d_pL)} = 1$$

as $L_{|S_\beta}$ is Morse.

Therefore, at a smooth point $(p, \xi) \in \mathrm{Ch}(F^\bullet)$, if L is Morse with respect to $\{S_\alpha\}$, with $L(p) = 0$ and $d_pL = \xi$, then

$$(-1)^{s-1}\chi(\phi_L F^\bullet)_p = (\mathrm{Ch}(F^\bullet) \cdot \mathrm{Im}\, dL)_{(p,\xi)} \cdot$$

The result of Ginsburg [25], Sabbah [76], and Lê [48] is that this equality holds at any point $(p, \xi) \in \mathrm{Ch}(F^\bullet)$ and for any L such that $L(p) = 0, d_pL = \xi$, and (p, ξ) is an isolated point in the intersection $|\mathrm{Ch}(F^\bullet)| \cap \mathrm{Im}\ dL$.

Example. We wish to give one easy, but important example of both vanishing cycles and the characteristic cycle.

Let $M = \mathbf{C}^3$ and consider $f : M \to \mathbf{C}$ given by $f = y^2 - x^a - tx^b$, where $a > b > 1$. The critical locus of f consists of $\mathcal{T} :=$ the t-axis.

As a family parameterized by t, f defines a family of isolated singularities with Milnor number $\mu(f_t)$ equal to $b-1$ if $t \neq 0$, and equal to $a-1$ if $t = 0$. Looking at the Milnor fibres of the total function f at points $p \in \mathcal{T}$, we see that $F_{f,p}$ is homotopy-equivalent to a wedge of $b-1$ one-spheres if $p \neq 0$, and homotopy-equivalent to a single two-sphere if $p = 0$.

The vanishing cycles $\phi_f\mathbf{C}^\bullet_M$ is a complex of sheaves on $X := f^{-1}(0)$ which is supported on the line $\mathcal{T}$. A Whitney stratification of X with respect to which $\phi_f\mathbf{C}^\bullet_M$ is constructible is given by $\{X - \mathcal{T}, \mathcal{T} - \{0\}, \{0\}\}$. We wish to calculate the characteristic cycle of $\phi_f\mathbf{C}^\bullet_M$ in $\mathbf{C}^3$.

Using Formula 3.1, the coefficient of $\overline{T^*_{X-\mathcal{T}}\mathbf{C}^3}$ in $\mathrm{Ch}\,(\phi_f\mathbf{C}^\bullet_M)$ is easily seen to be equal to 0, since near any point $p \in X - \mathcal{T}$ the vanishing cycles are zero.

To calculate the coefficient $m_\mathcal{T}$ of $\overline{T^*_{\mathcal{T}-\{0\}}\mathbf{C}^3}$ in $\mathrm{Ch}\,(\phi_f\mathbf{C}^\bullet_M)$, we take a normal slice N to $\mathcal{T} - \{0\}$ at some point $p = (0, t_0)$ where $t_0 \neq 0$. We may use the slice $N = V(t - t_0)$. Now, viewing N as a copy of $\mathbf{C}^2$, we pick a generic linear form $L : N \to \mathbf{C}$. According to Formula 3.1, the coefficient $m_\mathcal{T}$ is given by

$$\begin{aligned} m_\mathcal{T} &= (-1)^{2-1-1}\chi\left(\phi_L\left(\phi_f\mathbf{C}^\bullet_M\right)_{|N}\right)_p \\ &= \chi\left(H^{*+1}\left(B^\circ_\epsilon(p) \cap N,\ B^\circ_\epsilon(p) \cap N \cap L^{-1}(\eta)\ ;\ \phi_f\mathbf{C}^\bullet_M\right)\right), \end{aligned}$$

where $0 < |\eta| \ll \epsilon \ll 1$. Note that the restricted complex $\left(\phi_f\mathbf{C}^\bullet_M\right)_{|N}$ has cohomology supported solely at the point p and there only in degree 1. Therefore, the complex link has zero cohomology and

$$\begin{aligned} m_\mathcal{T} &= \chi\left(H^{*+1}\left(B^\circ_\epsilon(p) \cap N,\ B^\circ_\epsilon(p) \cap N \cap L^{-1}(\eta)\ ;\ \phi_f\mathbf{C}^\bullet_M\right)\right) \\ &= \dim H^1(\phi_f\mathbf{C}^\bullet)_p \\ &= b-1. \end{aligned}$$

For the stratum $\{0\}$, a normal slice consists of the entire ambient space. To calculate the coefficient m_0 of $T_{\{0\}}\mathbf{C}^3$ in $\mathrm{Ch}\,(\phi_f\mathbf{C}^\bullet_M)$, we pick a generic linear form $L : \mathbf{C}^3 \to \mathbf{C}$. According to Formula 3.1, the coefficient m_0 is given by

$$\begin{aligned} m_0 &= (-1)^{2-0-1}\chi\left(\phi_L\left(\phi_f\mathbf{C}^\bullet_M\right)\right)_0 \\ &= -\chi\left(H^{*+1}\left(B^\circ_\epsilon(0),\ B^\circ_\epsilon(0) \cap L^{-1}(\eta)\ ;\ \phi_f\mathbf{C}^\bullet_M\right)\right), \end{aligned}$$

where $0 < |\eta| \ll \epsilon \ll 1$. Now, the only point in the support of $\phi_f\mathbf{C}^\bullet_M$ which lies in $B^\circ_\epsilon(0) \cap L^{-1}(\eta)$ is a single point on $\mathcal{T}$, where we know that the Milnor fibre has the homotopy-type of a wedge of $b-1$ one-spheres. Recall, also, that the Milnor fibre at the origin has the homotopy-type of a single two-sphere. Thus,

$$\begin{aligned} m_0 &= -\chi\left(H^{*+1}\left(B^\circ_\epsilon(0),\, B^\circ_\epsilon(0)\cap L^{-1}(\eta)\ ;\ \phi_f\mathbf{C}^\bullet_M\right)\right) \\ &= -\chi\left(H^{*+1}\left(B^\circ_\epsilon(0)\ ;\ \phi_f\mathbf{C}^\bullet_M\right)\right)+\chi\left(H^{*+1}\left(B^\circ_\epsilon(0)\cap L^{-1}(\eta)\ ;\ \phi_f\mathbf{C}^\bullet_M\right)\right) \\ &= -(-1)+(b-1) \\ &= b. \end{aligned}$$

Therefore, $\mathrm{Ch}(\phi_f\mathbf{C}^\bullet_M)=(b-1)\ \overline{T^*_{T-\{0\}}\mathbf{C}^3}\ +\ b\,T_{\{0\}}\mathbf{C}^3$.

Note that the exponent a does not appear in the characteristic cycle – nor should it; after an analytic change of coordinates given by $u=x^{a-b}+t$, we have that $f=y^2-x^a-tx^b=y^2-ux^b$. However, the fact that $T_{\{0\}}\mathbf{C}^3$ appears with a non-zero coefficient in the characteristic cycle is a reflection of the fact that the Milnor numbers in the family f_t jump from $b-1$ to $a-1$ at $t=0$.

Having described the vanishing cycles and the characteristic cycle, we would like to revisit the proof of the result of Lê and Saito on the A_f condition. We wish to sketch a proof of their result which generalizes nicely to arbitrary underlying spaces (see Section 7). The above example will be useful to keep in mind.

Assume we have a family f_t as before, where each f_t has an isolated critical point at the origin and in which the Milnor number of f_t at the origin is independent of $t\in D^\circ$. The proof of Lê and Saito's result that we will outline breaks up into two pieces: one piece which says that if the Milnor numbers, $\mu(f_t)$, (at the origin) of the members of the family f_t are independent of t, then $\mathrm{Ch}(\phi_f\mathbf{C}^\bullet_M)$ has a special form, and a second piece which says that $\mathrm{Ch}(\phi_f\mathbf{C}^\bullet_M)$ controls Thom's A_f condition. The proof goes as follows.

As before, the first step of the proof is to show that the Milnor number of f_t being independent of t implies that $\Gamma^1_{f,t}=\emptyset$. The proof of this is the same as at the beginning of the section.

Now, by Lê's attaching result [42], which consists of more Morse theory, $\Gamma^1_{f,t}=\emptyset$ implies that the canonical inclusion of the Milnor fibre of f_0 into the Milnor fibre of f induces a homotopy-equivalence $F_{f_0,0}\simeq F_{f,0}$; the same argument applies at all small t_0, and so we conclude that, for all small t_0, $F_{f_{t_0},0}\simeq F_{f,(t_0,0)}$. As we are assuming that $\mu(f_t)$ is independent of t, we conclude that $F_{f,0}\simeq F_{f,(t_0,0)}$. This tells us that the Euler characteristic of the Milnor fibre of f does not change along the t-axis; thus, when we calculate the coefficient of $T^*_{\{0\}}\mathcal{U}$ in the characteristic cycle of the sheaf of vanishing cycles (as in the above example), we get a 0. Hence, $T^*_{\{0\}}\mathcal{U}$ does not appear in $\mathrm{Ch}(\phi_f\mathbf{C}^\bullet_M)$, and so the only conormal variety which does appear in $\mathrm{Ch}(\phi_f\mathbf{C}^\bullet_M)$ is $T^*_{D^\circ\times\{0\}}\mathcal{U}$.

At this point, one uses a result of Kashiwara [40] and Lê and Mebhkout [50] which says that the projectivization of $\mathrm{Ch}(\phi_f \mathbf{C}^\bullet_M)$ is isomorphic to the exceptional divisor of the Jacobian blow-up of f in $\mathcal{U}$. Finally, one shows easily that if $T^*_{D^\circ\times\{0\}}\mathcal{U}$ is the only component of the exceptional divisor of the Jacobian blow-up of f, then $(\mathcal{U} - (D^\circ \times \{0\}), D^\circ \times \{0\})$ satisfies A_f.

The proof of Lê and Saito in [52] contains pieces which correspond to all of the argument sketched out above, **except** one will find nothing remotely resembling the fact that the projectivization of $\mathrm{Ch}(\phi_f \mathbf{C}^\bullet_M)$ is isomorphic to the Jacobian blow-up of f in $\mathcal{U}$; the reason for this is that – in the affine setting of [52] – this isomorphism is trivial. However, proving in the general case that the projectivization of $\mathrm{Ch}(\phi_f \mathbf{C}^\bullet_M)$ can be realized as a blow-up is the major step in a more general proof of the Lê and Saito result (see Section 7).

Having completed a first round of discussions on some of the key ideas in the topological and geometric approaches to equisingularity, we return to the development of the theory in the seventies and eighties.

4 Progress Since Cargese

The three model theorems of Section 1 had all been published by the time the proceedings of the Cargese conference appeared in 1973 (with the exception of [5]). Since then, one main direction of work has been to find invariants that describe the Whitney equisingularity of general complex analytic spaces. The invariants used to do this were the polar multiplicities of an analytic space [83] which we now define.

Suppose $(X^n, 0) \subset \left(\mathbf{C}^N, 0\right)$ is an equidimensional analytic set of dimension d. Suppose $D_k \subset \mathbf{C}^N$ is a linear subspace of codimension k. Let p_k denote a linear submersion with kernel D_k. The polar variety of $(X^n, 0)$ with respect to D_k is $\left(\overline{\Sigma\,(p_k|_{X_{reg}})}, 0\right)$ where Σ denotes the critical set of p_k. The polar variety is denoted $P(X^d, D_k)$ and its multiplicity at the origin is denoted $m(X^d, D_k)$. Teissier showed that the polar multiplicity was independent of the choice of D_k for D_k in a Z-open set in Grassman space; if D_k is such a generic element, then we suppress it from the notation. Teissier also proved many other foundational results about the polar multiplicities: they are upper-semicontinuous, analytic invariants, and behave well with respect to taking generic plane sections. (For many such results, see [82].) However, the polar varieties only exist in positive dimension–there is no polar variety of dimension 0. This means, for example, that the Milnor number of a hypersurface singularity cannot be computed through polar varieties. So, in the case of a family of hypersurfaces with isolated singularities, if you restrict yourself to the polar multiplicities, you must work at the level of the total space. There, the polar curve will detect some of the same problems that the multiplicity of the Jacobian ideal detected in the case of isolated hypersurface singularities. This lack of a good zero-dimensional invariant prevented the extension of the three model theorems of Section 1 to more general spaces.

An important development was the realization that, in the general case, the Whitney conditions were better studied using limits of tangent hyperplanes instead of tangent planes. If $X^n \subset \mathbf{C}^N$ the conormal modification sits in $X \times \mathbf{P}^{N-1}$; given a codimension k plane H in $\mathbf{P}^{N-1}$, we can intersect $C(X)$ with $X \times H$. For H chosen generically, the projection of this set is the polar variety of codimension $d - n + k + 1$ in X. To see this, think of points of $\mathbf{P}^{N-1}$ as hyperplanes in $\mathbf{C}^N$. Then H consists of all hyperplanes containing a fixed plane K of codimension k in $\mathbf{C}^N$. Generically the points of $(X \times H) \cap C(X)$ consist of pairs (x, H) where H contains the tangent plane to X at x and which also contains K. This means that x is in the polar variety defined by the projection with kernel K. Since the discriminant of this projection has dimension $N - k - 1$, the polar variety has codimension $n - N + k + 1$. Thus, using all of the cohomology classes of $\mathbf{P}^{n-1}$ to pin down the position of $C(X)$ in $X \times \mathbf{P}^{n-1}$ amounts to computing the polar varieties. If we look at the Nash modification which is formed from the closure of the tangent planes to the smooth part of X, and which sits in the product of X with a Grassman space, computing the polar varieties only uses part of the cohomology ring of the Grassman space, if X is not a hypersurface. This means that the polar multiplicities might be constant along a stratum Y, yet the fibers of the blow-up of X by $m_Y J(G)$ might have different dimensions over different points of Y. This makes the proof which we sketched for Theorem 3 of Section 1 impossible. The realization of the necessity of working with the conormal space was due to V. Navarro [71], and J.P.G. Henry and M. Merle [33]. With the incorporation of this insight, Teissier showed that the independence of the polar multiplicities of the total space along a stratum Y was equivalent to the Whitney conditions holding between the open stratum of the space and Y was equivalent to the dimension of the fiber of $\mathrm{Bl}_{m_Y}(\mathrm{C}(X))$ being constant over Y ([82, V.1.2]).

Now we turn to results on particular classes of spaces, beginning with complete intersections. One of the techniques for extending results from the hypersurface case to the complete intersection case is to chose a generic set of generators of the family $(F_1, \ldots, F_p)$, and define the chain of complete intersections $X_0 = \mathbf{C}^n$, X_1 with ideal (F_1), X_2 with ideal (F_1, F_2) etc. You can then consider $F_{i|X_{i-1}}$; this puts you back into the hypersurface case, but with a singular domain. Parameswaran [72] used this approach to give an extension of the Lê-Ramanujam theorem to complete intersections. To an ICIS germ $(X, 0)$, he associated [72, Def. 1, p. 324] the sequence of numbers,

$$\mu_* := \mu_0, \mu_1, \ldots, \mu_k,$$

where k is the embedding codimension and where μ_i is the smallest Milnor number of any ICIS germ that serves as the total space of an i-parameter (flat) deformation of $(X, 0)$. Given a chain (or nested sequence) of deformations $(X_i, 0) \to (Y_i, 0)$ for $1 \leq i \leq k$ such that each Y_i is of dimension

i, call the chain μ_*-minimal if the Milnor number of $(X_i, 0)$ is equal to μ_i. Parameswaran proved [72, Lem. 3, p. 325] that there exists a μ_*-minimal chain where each Y_i is smooth. He said [72, Def. 6, p. 331] that two ICISs have the same *topological type* if each has a μ_*-minimal chain such that the two chains are embedded homeomorphic, and he proved that this notion does not depend on the choice of chains. Finally, he proved [72, Thm. 2, p. 332] that, in a family of ICISs of dimension other than 2, the topological types of the members are the same if their μ_*-sequences are the same. The methods of proof are similar to those used earlier by Lê. It should be noted that Parameswaran's definition of topological type is sometimes more rigid than Whitney equisingularity, since it requires the embedding codimension of X_t to be constant.

The case of curves was extensively studied by Buchweitz and Greuel [8] and by Briançon, Galligo and Granger [6]. Following Bassein [1], Buchweitz and Greuel gave a formula for the "Milnor number" μ of an arbitrary reduced curve singularity (X_0, x_0) by means of the dualizing module of Grothendieck. If (X_0, x_0) is a complete intersection, then their formula gives the Milnor number of X_0. They show that in a 1-parameter family X_t of curves that $\mu(X_0) - \mu(X_t)$ is the rank of $H^1(X_t, \mathbf{C})$, so μ measures the vanishing cohomology in the family (Theorem 4.2.2). (If X_t is smooth for small t, then $\mu(X_t) = 0$.) If it is assumed that the total space of the family is singular only along the parameter axis, then they show that the constancy of $\mu(X_t)$ is equivalent to the constancy of the topology of the family. Their results are some of the few dealing with families of sets which are not complete intersections. Here is how the dualizing sheaf helps: given an isolated hypersurface singularity, then the Jacobian ideal has finite colength inside $\mathcal{O}_{\mathbf{C}^{n+1}}$, but if the curve X is not a complete intersection then neither the Jacobian ideal nor Jacobian module has finite colength in either the local ring of the ambient space or of X. However, the map from $\mathcal{O}_X$ to the differential forms on X given by derivation extends to a map to the dualizing module, and this map has a cokernel of finite length. This length is the generalized Milnor number.

Briançon, Galligo, and Granger [6] described the relations between many different equisingularity conditions including Whitney equisingularity, for complete intersection curves; they showed, for example, that a family of such curves is Whitney equisingular if and only if the Milnor number and the multiplicity are constant.

Briançon and Speder [5] also found numerical invariants which describe Zariski equisingularity for a family X of hypersurfaces in $\mathbf{C}^3$ with isolated singularities. In this equisingularity condition, you ask that the restriction of any suitably generic family of projections to $\mathbf{C}^2$ of X be a *Whitney equisingular family*, that is, there exist Whitney stratifications of the source and target of the family such that the family of projections preserve strata, that the mapping between strata is a submersion, pairs of incident strata in the source satisfy the A_f condition, and that the parameter spaces in source and target be strata.

Zariski proposed this equisingularity condition in the hope that it would be invariant under blowing-up along the parameter stratum.

The two new invariants were the number of vanishing double folds and the number of vanishing cusps. To define them, start with an isolated hypersurface singularity $X \subset \mathbf{C}^3$, and find its miniversal unfolding $F : \mathbf{C}^3 \times \mathbf{C}^k \to \mathbf{C} \times \mathbf{C}^k$. The fibers of F are representatives of all nearby hypersurfaces to X, most of which are smooth; project them to $\mathbf{C}^2$ using a linear projection π whose kernel is "generic". Denote the restriction of the projection to $F^{-1}(y)$ by π_y. Most of the maps π_y will be stable, that is the only singularities of the maps will be folds, cusps, and double folds. (A double fold is a point in the target where two fold curves cross.) The number of cusps and double folds which come together as π_y tends to π_0 is called the number of vanishing cusps and vanishing double folds. These numbers can be calculated by computing the local degree of the projection of the closure of the relevant singularity of $F \times \pi$ to the parameter stratum. More information about these invariants, and this equisingularity condition can be found in [3], [34].

The work of Briançon and Speder, besides controlling Zariski equisingularity, is also part of the beginning of the study of the equisingularity of mappings on singular spaces. Not surprisingly, when the mappings are as simple as possible (i.e., the restriction of linear submersions), their geometry is controlled by their sources.

Similar techniques were used by Gaffney [16] to describe the Whitney equisingularity type of families of finitely-determined map germs of discrete stable type. A map germ $f : (\mathbf{C}^n, 0) \to (\mathbf{C}^p, 0)$ is finitely-determined of order k if whenever the k-th Taylor polynomial of f and g agree then f and g are equivalent by smooth coordinate changes of source and target. Finitely-determined germs also have versal unfoldings in which representatives (up to smooth coordinate change) of all nearby germs appear. (In fact, their properties are often analogous to those of ICIS singularities.) A germ f has discrete stable type if (up to smooth coordinate changes) only a finite number of types of stable germs appear in a versal unfolding of f. The existence of a versal unfolding makes it possible to define invariants counting the number of vanishing points of a stable type which appear with dimension 0 in nearby germs. (This was first noticed by Mond in [70]). Even more can be done. Denote the versal unfolding of f by F, and the projection to the parameter space by π_s. Suppose X is the closure of the set of points of a given stable type S in either the source or target of F; choose a generic linear map π from $(\mathbf{C}^n, 0) \to \mathbf{C}$, and take the closure of the critical set of $(\pi \times \pi_s)$ restricted to the smooth part of X. The degree of this set over the parameter space plays the part of a zero-dimensional polar multiplicity in ensuring that the stratum of points of type S satisfies the Whitney conditions over the parameter stratum.

Call a 1-parameter unfolding F of f *good* if the there exist neighborhoods U and W of the origin in $\mathbf{C}^n \times \mathbf{C}$ and $\mathbf{C}^p \times \mathbf{C}$, respectively, such that $F^{-1}(W) = U$, F maps $(U \cap \Sigma(F)) - (0 \times \mathbf{C})$ to $W - (0 \times \mathbf{C})$ and if $(y_0, t_0) \in$

$W - (0 \times \mathbf{C})$ with $S = F^{-1}(y_0, t_0) \cap \Sigma(F)$, then the germ $f_{t_0} : (\mathbf{C}^n, S) \to (\mathbf{C}^p, y_0)$ is a stable germ. If $n = p$, then we ask the degree of f_t at 0 to be constant as well. Then it is shown in [16] that if the polar multiplicities of all the stable types which occur in the discriminant, $\Delta(f_0)$, singular set or inverse image of $\Delta(f_0)$ are constant, then the family is Whitney equisingular. The converse also holds if the strata in the Whitney stratifications are by stable type, as they must be in the nice dimensions. Gaffney uses this result to analyze mappings from the plane to the plane and from the plane to space in detail, showing for example that a 1 parameter unfolding F of a finitely-determined germ f_0 is Whitney equisingular if and only if $\mu(\Delta(f_t))$ is constant. More results in this framework can be found in [22], [23].

Now we turn back to the extensions of our model theorems. There are two different directions in which to proceed; the first is to try to understand singularities of codimension greater than 1, the second is to study non-isolated hypersurface singularities.

5 Moving Beyond Codimension 1

In moving beyond hypersurfaces, it seems necessary to shift from working with Jacobian ideals to Jacobian modules. If X is defined by $G = 0$, where $G : \mathbf{C}^n \to \mathbf{C}^p$, let $JM(X) \subset \mathcal{O}_X^p$ denote the Jacobian module of G, which is the module generated by the partial derivatives of G.

The previous work in equisingularity gives two good reasons for this. In the eighties, it became clear that the conormal space offered a superior approach to studying the Whitney conditions than the Nash blow-up; at points on X where G has maximal rank, the projectivization of the row space of the Jacobian matrix of G is exactly the fiber of $C(X)$. This indicates a link between the Jacobian module and the conormal space.

Back in the sixties, Mather used the Jacobian module to prove results about the analytic triviality of families of analytic sets. In fact, the Jacobian module played the same role for general analytic spaces that the Jacobian ideal did in Mather's work on hypersurfaces. Given the role that the integral closure of the Jacobian ideal played in Teissier's work, it was natural for any disciple of Mather to think about the integral closure of the Jacobian module.

The notion of the integral closure of a module goes back to Rees [73]. Rees considered a module M which was a submodule of a module L; he used the discrete integer- valued valuations on the quotient field of the ground ring to say when an element of L was integrally dependent on M. His definition is related to that of a complete module as given by Zariski in Appendix 4 of [95].

Because of the ease of application to problems in stratification theory, we follow the development of this concept as it appears in [17]. Suppose (X, x) is a complex analytic germ, M a submodule of $\mathcal{O}_{X,x}^p$. Then $h \in \mathcal{O}_{X,x}^p$ is in the *integral closure of* M, denoted $\overline{M}$, if and only if for all $\phi : (\mathbf{C}, 0) \to (X, x)$,

$h \circ \phi \in (\phi^* M)\mathcal{O}_1$.

One link back to the theory for ideals is provided by the fitting ideals of M. We denote the $(p-k)$-th fitting ideal of $\mathcal{O}^p_{X,x}/M$ by $J_k(M)$; if h is an element of $\mathcal{O}^p_{X,x}$, we denote by (h, M) the module generated by M and h.

Proposition 5.1 ([17], Prop 1.7) *Suppose M is a submodule of $\mathcal{O}^p_{X,x}$, $h \in \mathcal{O}^p_{X,x}$ and the rank of (h, M) is k on each component of (X, x). Then $h \in \overline{M}$ if and only if $J_k(h, M) \subset \overline{J_k(M)}$.*

The next proposition is the analogue of (ii) in the list of properties relating to integral closure that appeared in Section 2.

Proposition 5.2 ([17], Prop 1.11) *Suppose $h \in \mathcal{O}^p_{X,x}$, M a submodule of $\mathcal{O}^p_{X,x}$. Then $h \in \overline{M}$ if and only if for each choice of generators $\{s_i\}$ of M, there exists a constant $C > 0$ and a neighborhood U of x such that for all $\psi \in \Gamma(Hom(\mathbf{C}^p, \mathbf{C}))$,*

$$||\psi(z) \cdot h(z)|| \leq C||\psi(z) \cdot s_i(z)||$$

for all $z \in U$.

This proposition can then be used to convert analytic inequalities defining conditions like W and W_F into integral closure terms.

Proposition 5.3 ([17], Theorem 2.5) *Suppose $(X^{d+k}, 0) \subset (\mathbf{C}^{n+k}, 0)$ is an equidimensional complex analytic set, $X = G^{-1}(0)$, Y a smooth subset of X, $G : \mathbf{C}^{n+k} \to \mathbf{C}^p$, coordinates chosen so that $\mathbf{C}^k \times 0 = Y$, $m_Y = (z_1, ..., z_n)$ denoting the ideal defining Y, G defines X with reduced structure. Then, $\frac{\partial G}{\partial s} \in \overline{m_n JM(G)}$ for all tangent vectors $\frac{\partial}{\partial s}$ to Y if and only if W holds for (X_0, Y).*

In order to state the result for W_F, fix a function germ $F : (X, 0) \to (\mathbf{C}, 0)$, and let $X_{\text{sm}(F)}$ denote the set of simple points of X where F is a submersion. Denote the zero set of F in X by Z. Let H be the map with components (G, F).

Proposition 5.4 *In the setup above, $(X_{\text{sm}(F)}, Y)$ satisfies W_F at 0 if and only if*

$$\frac{\partial H}{\partial s} \in \overline{\mathbf{m}_Y JM_z(H)}$$

for all tangent vectors $\frac{\partial}{\partial s}$ to Y.

Proof. Cf. [21], Lemma 6.1. However, we indicate the proof in the case that Y is one-dimensional and embedded as the t-axis. Applying the definition for W_F for the pair $(X_{\text{sm}(F)}, Y)$, we see that we can take $v = < 0, \ldots, 0, 1 >$, while u is the complex conjugate of $\psi(z,t) \cdot D(H)(z,t)$, where ψ is any element of $\Gamma(Hom(\mathbf{C}^{p+1}, \mathbf{C}))$, since at a point of $X_{\mathrm{sm}(F)}$, the rows of the Jacobian matrix of H are a basis for the vectors orthogonal to the fibers of F on X. This gives the inequality:

$$\frac{||\psi(z,t)\cdot\frac{\partial H}{\partial t}(z,t)||}{||\psi(z,t)\cdot D(H)(z,t)||}\leq C||(z_1,\ldots,z_n|| \qquad \forall\psi.$$

This is equivalent to:

$$\left|\left|\psi(z,t)\cdot\frac{\partial H}{\partial t}(z,t)\right|\right|\leq C||(z_1,\ldots,z_n||||\psi(z,t)\cdot D(H)(z,t)|| \qquad \forall\psi.$$

By Proposition 5.2, it is easy to see that this is equivalent to:

$$\frac{\partial H}{\partial t}\subset\overline{\mathbf{m}_Y JM_z(H)}.$$

This finishes the proof.

In order to describe the multiplicity of a module, and the proof of the principle of specialization for modules, some constructions need to be made. (The approach to the multiplicity described here is due to Kleiman and Thorup and appears in [41].) Let $\mathcal{N}$ be a coherent submodule of $\mathcal{M}$, which in turn is a submodule of $\mathcal{E}$, a free $\mathcal{O}_X$ module of rank p. Let $\mathcal{SE}$ denote the symmetric algebra, and $\mathcal{RM}$ and $\mathcal{RN}$ the subalgebras generated by $\mathcal{M}$ and $\mathcal{N}$. The subalgebra $\mathcal{RM}$ is called the Rees algebra of M. Form the analytic homogeneous spectra,

$$P:=\mathrm{Projan}(\mathcal{SE}),\ P':=\mathrm{Projan}(\mathcal{RM})\text{ and }P'':=\mathrm{Projan}(\mathcal{RN}).$$

The space $P:=\mathrm{Projan}(\mathcal{SE})$ is $X\times\mathbf{P}^{p-1}$, while P', P'' are realized by picking a set of generators for each module, using the inclusion in $\mathcal{E}$ to form matrices with the generators as columns, considering the projectivized row space at all points where the matrices have maximal rank, then taking the closure of this set in $X\times\mathbf{P}^{k-1}$, where k is the number of generators of the module. The part of $\mathcal{RM}$ of positive degree generates an ideal in $\mathcal{SE}$, and this in turn induces an ideal sheaf on P, which we denote by $\mathcal{I}(\mathcal{M})$. Form a new space B by considering the blow-up of P by $\mathcal{I}(\mathcal{M})$.

We then have the diagram:

$$\begin{array}{ccc}\mathrm{Bl}_{\mathrm{IM}}(\mathrm{P}) & \xrightarrow{p} & \mathbf{P}\\ \downarrow & & \downarrow\\ \mathbf{P}' & \xrightarrow{p'} & \mathbf{X}\end{array}$$

In the ideal case, B is P', while P is X. The multiplicity is the intersection number gotten by intersecting the exceptional divisor with c_1^{n-1} where c_1 is the first Chern class of the tautological sheaf $\mathcal{O}_{P'}(1)$ on P'. In the general case, if the module has finite colength, the exceptional divisor sits in a product of projective spaces, so there are two Chern classes whose products can be used to specify the position of the exceptional divisor. Denote the first Chern classes of $\mathcal{O}_{P'}(1)$ and $\mathcal{O}_P(1)$ by ℓ' and ℓ. Denote the exceptional divisor of B by D. Finally, define the Segre numbers $s^i(M)$ of M to be:

$$s^i(M) := \int \ell'^{i-1}\ell^{r-i}[D] \qquad \text{for } i = 1, \ldots, r, \qquad r = \dim X + p - 1.$$

Kleiman and Thorup showed that the the Buchsbaum-Rim multiplicity of the module M, denoted $e(M)$, is the sum of the Segre numbers. Given a family of modules of finite colength, on an equidimensional space X, parameterized by a smooth space Y, Gaffney and Kleiman were able to show that the constancy of the Segre numbers is equivalent to the fibers of D over all points $y \in Y$ being equidimensional or empty (Cf. Theorem 2.2 [21]). The arguments are very similar to those used by Teissier in his proof of the principle of specialization in the ideal case, and yield a version of the principle in this context.

However, a better version of the theorem is possible. As before, $e(y)$ denotes the sum of the multiplicities of $\mathcal{M}_x$ at all points x over y.

Theorem 5.5 (Specialization of integral dependence.) *Assume that X is equidimensional, and that $y \mapsto e(y)$ is constant on Y. Let h be a section of $\mathcal{E}$ whose image in $\mathcal{E}(y)$ is integrally dependent on the image of $\mathcal{M}$ for all y in a dense Zariski open subset of Y. Then h is integrally dependent on $\mathcal{M}$.*

Proof. Cf. Theorem 1.8 [21]. The idea of the proof is that the constancy of the multiplicity means that $\mathcal{M}$ has a reduction $\mathcal{N}$ which is generated by $\dim(X(y)) + p - 1$ generators. Now replace M by the submodule generated by N and h. The dimension of P'' over our base point $x_0 \in X$ is $\dim(X(y)) + p - 2$, which is one less than the number of generators, and one can show that P' is finite over P'' over some Z-open subset of Y. (This is where we use the hypothesis on h.) The theorems of [41], culminating in Proposition 10.7, then show that the small fiber dimension of P'' over x_0 forces P' to be finite over P'' at all points proving that $\overline{N} = \overline{M}$.

Given the principle of specialization, it is now possible to prove versions of Theorems 2 and 3 from Section 1 for families of complete intersections with isolated singularities. Here is the result for W_F.

Theorem 5.6 *Suppose $(X^{d+k}, 0) \subset (\mathbf{C}^{n+k}, 0)$ is a complete intersection, $X = G^{-1}(0)$, $G : \mathbf{C}^{n+k} \to \mathbf{C}^p$, Y a smooth subset of X, coordinates chosen so that $\mathbf{C}^k \times 0 = Y$. Suppose $F : (X, Y) \to (\mathbf{C}, 0)$ and that $Z = F^{-1}(0)$ is also a complete intersection, which is nowhere dense in X. The following three conditions are equivalent:*

(i) the pair $(X - Y, Y)$ satisfies W_F at 0;

(ii) *the two pairs* $(X-Y, Y)$ *and* $(Z-Y, Y)$ *satisfy both Whitney conditions at* 0;

(iii) *The sets* $X(y)$ *and* $Z(y)$ *are complete intersections with isolated singularities and the* $\mu^{(*)}$ *sequence of* $X(y)$ *and* $Z(y)$ *are independent of* y *for all* $y \in Y$ *near* 0.

Proof. Cf. Theorem 6.4 and Lemma 6.3 of [21]. It is clear that either i) or ii) imply iii). By means of intersection theory, there exists an expansion formula for $e(m_Y JM(H_y))$ in terms of the $\mu^{(*)}$ sequence of $X(y)$ and $Z(y)$. In fact we have:

$$\begin{aligned} e(m_Y JM(H_y)) &= \binom{n-1}{n-p} + \sum_{i=0}^{n-p} \binom{n-1}{i} \mu_i(X(y),0) \\ &\quad + \sum_{i=0}^{n-p-1} \binom{n-1}{i} \mu_i(Z(y),0). \end{aligned}$$

(Here $\mu_i(X)$ is the Milnor number of the intersection of X with a generic plane of codimension i.) Thus the constancy of the $\mu^{(*)}$ sequences implies that $e(m_Y JM(H_y))$ is constant. Since it is known that W_F holds generically, the inclusion of Proposition 5.4 holds generically. The principle of specialization then applies and the inclusion of Proposition 5.4 holds at all y values. Note that, in the above theorem, the expansion theorem plays a key role, relating the invariant which controls the equisingularity condition, which is $e(m_Y JM(H_y))$, with invariants more directly connected to the geometry. Without this relation, it would be difficult to prove that the constancy of $e(m_Y JM(H_y))$ was necessary as well as sufficient.

The generalization of the Lê-Saito theorem depends on a lemma which shows that Projan$(\mathcal{RM})$ shares an important property with the exceptional divisor of a blow-up. Recall that the cosupport of a coherent submodule M of a free module, denoted $C(M)$, is the set of points where the two are unequal. If M is an ideal, this is exactly $V(M)$.

Lemma 5.7 *Let* M *be a submodule of a free* $\mathcal{O}_{X^n,x}$ *module of rank* p; *assume that the cosupport of* M *is a nowhere dense subset of* X^n, *and that* X^n *is equidimensional. Then, each component of* Projan $\mathcal{R}(M)|_{C(M)}$ *has dimension* $n+p-2$ *and, hence, has codimension* 1 *in* Projan $\mathcal{R}(M)$.

Proof. The idea of the proof is to show that locally Projan $\mathcal{R}(M)|_{C(M)}$ is the exceptional divisor of a blow-up. We first fix some notation, then give the framework in which we will be blowing-up. Fix a matrix of generators of M, $[M]$. Let M_i be the submodule of $\mathcal{O}^{p-1}_{X,x}$ whose matrix of generators $[M_i]$ is obtained by deleting the i-th row from $[M]$. We know that the Rees space of M_i has a canonical line bundle on it, since Projan $\mathcal{R}(M_i) \subset X \times \mathbf{P}^{n-1}$, where n is the number of generators of M. The line bundle itself then is a subset of $X \times \mathbf{P}^{n-1} \times \mathbf{C}^n$. We can project the line bundle to $X \times \mathbf{C}^n$. Call the image $L'(M_i)$. Note that off the cosupport of M_i on X, the fiber of $L'(M_i)$ is a $(p-1)$-dimensional subspace of $\mathbf{C}^n$, so the dimension of $L'(M_i)$ is $n+p-1$. The i-th row of $[M]$ defines a map from X to $\mathbf{C}^n$. Consider the graph of this map and the blow-up $L'(M_i)$ using the graph as center. Call the exceptional divisor of this blow-up E_i. When we project to X, the image of E_i lies in the cosupport of M, and when we project E_i to $X \times \mathbf{P}^{n-1}$, the map is an isomorphism. Now let $x \in C(M)$. Then, we can find some M_i whose matrix of generators has the same rank at x as $[M]$. We claim that the fiber of $\operatorname{Projan} \mathcal{R}(M)$ over x is the fiber of the image of E_i in $\mathbf{P}^{n-1}$. Denote the projection to $\mathbf{P}^{n-1}$ by π. We claim that $\pi(E_i)_x$ is contained in $\operatorname{Projan} \mathcal{R}(M)$.

Suppose $(0, l)$ is a point of $\pi(E_i)_x$; then we can find curves ϕ, ψ such that $\phi(0) = x$,

$$\lim_{t \to 0} \sum_i \psi_i(t) \cdot [M_i \circ \phi(t)] - [m_i \circ \phi(t)] = 0,$$

$$\lim_{t \to 0} < \sum_i \psi_i(t) \cdot [M_i \circ \phi(t)] - [m_i \circ \phi(t)] >= l.$$

Here $[m_i]$ denotes the i-th row of $[M]$. It is clear that

$$< \sum_i \psi_i(t) \cdot [M_i \circ \phi(t)] - [m_i \circ \phi(t)] >$$

is a point of $\operatorname{Projan} \mathcal{R}(M)$ for $t \neq 0$, so the same is true for l and $\operatorname{Projan} \mathcal{R}(M)$.

Now suppose $(x, H) \in (\operatorname{Projan} \mathcal{R}(M))_x$. Then, there exist curves ϕ, ψ and an exponent r such that

$$\lim_{t \to 0} < \frac{1}{t^r}(\psi_1(t) \cdot [M_i \circ \phi(t)] - \psi_2(t) \cdot [m_i \circ \phi(t)]) >= H.$$

By assumption, we know that $[m_i(x)]$ is in the span of the rows of $[M]$; by choosing a linear isomorphism of $\mathbf{C}^p$ if necessary, we can replace M by an isomorphic submodule M' of $\mathcal{O}^p_{X,x}$ such that $[m_i(x)] = 0$, and $\operatorname{Projan} \mathcal{R}(M) = \operatorname{Projan} \mathcal{R}(M')$ because the row spaces are the same.

If $o(\psi_2) < o(\psi_1 \cdot [M_i \circ \phi(t)])$ then $\frac{\psi_1(t)}{\psi_2(t)} \cdot [M_i \circ \phi(t)]$ gives a lift of ϕ to $L'(M_i)$ and

$$< \frac{\psi_1(t)}{\psi_2(t)} \cdot [M_i \circ \phi(t)] - [m_i \circ \phi(t)] >$$

tends to H, showing that H lies in $\pi(E_i)_x$.

If $o(\psi_2) \geq o(\psi_1 \cdot [M_i \circ \phi(t)])$, then

$$o(\psi_2 \cdot [m_i \circ \phi(t)]) > o(\psi_1 \cdot [M_i \circ \phi(t)]),$$

so H is the limit as t tends to zero of $< \psi_1 \cdot [M_i \circ \phi(t)] >$. If the order of the components of $[m_i \circ \phi(t)]$ is greater than 1, then we use the same ϕ

but replace ψ_1 by $\psi(t)_1/t^k$ where k is chosen so that the order of $(\psi(t)_1/t^k)\cdot[M_i(\phi(t))]$ is greater than 0, but less than the order of $[m_i \circ \phi(t)]$. Then again $(\psi(t)_1/t^k)\cdot[M_i(\phi(t))]$ provides a lift of ϕ to $L'(M_i)$, and the limit of $< (\psi(t)_1/t^k)\cdot[M_i(\phi(t))] - [m_i \circ \phi(t)] >$ is again H. If the order of the components of $[m_i \circ \phi(t)]$ is 1, then we re-parameterize ϕ so that the order of $[m_i \circ \phi(t)]$ is again greater than 1. This finishes the proof.

It would be interesting to find an algebraic proof of this fact.

Theorem 5.8 *Suppose $(X, 0)$, G, Y, F, Z as in Theorem 5.6.*

A) Suppose X_y and Z_y are complete intersections with isolated singularities with Milnor numbers independent of y. Then the union of the singular points of F_y is Y or is empty, and the pair of strata $(X - Y, Y)$ satisfies Thom's A_F condition.

B) Suppose $\Sigma(F)$ is Y or is empty, and the pair $(X - Y, Y)$ satisfies Thom's A_F condition. Then the Milnor numbers of X_y and Z_y are independent of y.

Proof. The proof of part B) is already contained in [G-K], so it suffices to prove A). The condition that the Milnor numbers of X_y and Z_y are independent of y is equivalent to the condition that the multiplicity of the Jacobian module $JM(H_y)$ is independent of y. (Recall that $H = (G, F)$.) This condition implies that the cosupport of $JM_z(H)$ does not split, which implies that union of the singular points of F_y is Y or is empty. This shows that $(X - Y, Y)$ are a pair of strata. The A_F condition is equivalent to requiring that the relative conormal $C(X, F)|_Y$ be a subset of $C(Y)$. Since A_F holds generically on Y (because the target of F is $\mathbf{C}$), any component of $C(X, f)|_Y$ which surjects onto Y must lie in $C(Y)$. Again, because the A_F condition holds generically on Y, we have that $\partial H/\partial y_k$ for all k, is in the integral closure of $JM_z(H)_y$, for all y in the Z-open subset of Y where A_F holds. By the principle of specialization of integral dependence for modules, it follows that $\partial F/\partial y_k$ for all k, is in the integral closure of $JM_z(H)_y$ for all $y \in Y$. This implies that $JM_z(H)$ is a reduction of $JM(H)$. In turn, this implies that the map from $C(X, f) = \text{Projan}(\mathcal{R}(JM(H)))$ to $\text{Projan}(\mathcal{R}(JM_z(H)))$ is finite. Now, $\text{Projan}(\mathcal{R}(JM_z(H)))$ is a subset of $X \times \mathbf{P}^{n-1}$, so if V is a component of $C(X, f)|_Y$, its image in $\text{Projan}(\mathcal{R}(JM_z(H)))$ must be an analytic set in $Y \times \mathbf{P}^{n-1}$ of the same dimension as V. By Lemma 4.7, V has dimension $n + k - 1$, so its image must be $Y \times \mathbf{P}^{n-1}$; hence, V must surject onto Y. This completes the proof.

This proof is very similar to the proof that Teissier gave of the Lê-Saito theorem in [82]. The new ingredient which allows it to go through is the lower bound on the dimension of the components of the relative conormal over Y. It would be interesting to see a topological proof in the same spirit as the proof of Lê and Saito.

6 Hypersurfaces with Non-Isolated Singularities

To extend the three model theorems to the case of families of hypersurfaces with non-isolated singularities, the topological approach needs invariants connected with the Milnor fiber, computable from the defining equation, while the commutative algebra-analytic geometry approach needs invariants which control the integral closure of an ideal for ideals of non-finite colength.

Massey made the first significant progress in the topological direction. Shortly after his thesis work, he began to study a formula of Lê and Iomdin [39], [45]. Given a function $h : \mathbf{C}^{n+1} \to \mathbf{C}$ with a one-dimensional singular set, if z_0 is a coordinate with the property that $h_{|V(z_0)}$ has an isolated singularity, then the Lê-Iomdin formula relates the Milnor number of $h + z_0^j$ for j large to the Euler characteristic of the Milnor fiber F of h and the transverse Milnor number of h. (Define the transverse Milnor number T of h to be the sum of the Milnor numbers of $h|_{l=\epsilon}$ where ϵ is small and $l = 0$ defines a generic hyperplane.) The technique of adding on a high power of a generic variable is a valuable one in the study of non-isolated singularities, as it reduces the dimension of the singular locus of the map by 1, enabling inductive proofs. The formula is:

$$\mu(h + z_0^j) = b_n(F) - b_{n-1}(F) + jT.$$

To calculate the Milnor number of $h + z_0^j$, it is necessary to calculate the intersection number $[(\frac{\partial h}{\partial z_0} + j z_0^{j-1})] \cdot [(\frac{\partial h}{\partial z_1}, \ldots, \frac{\partial h}{\partial z_n})]$. As a cycle $[(\frac{\partial h}{\partial z_1}, \ldots, \frac{\partial h}{\partial z_n})]$ has two parts; one is the relative polar curve $\Gamma^1(h)$, the other is $\Lambda^1(h)$, the singular set of h, with the correct cycle structure. Because $\frac{\partial h}{\partial z_0}$ is zero on the singular set of h,

$$\begin{aligned} \left[(\frac{\partial h}{\partial z_0} + j z_0^{j-1}) \right] \cdot \Lambda^1(h) &= (j-1)[z_0] \cdot \Lambda^1(h) \\ &= (j-1)T. \end{aligned}$$

For j large,

$$\left[(\frac{\partial h}{\partial z_0} + j z_0^{j-1}) \right] \cdot \Gamma^1(h) = \left[(\frac{\partial h}{\partial z_0}) \right] \cdot \Gamma^1(h).$$

It follows from the Lê-Iomdin formula that the difference of $[(\frac{\partial h}{\partial z_0})] \cdot \Gamma^1(h)$ and $[z_0] \cdot \Lambda^1(h)$ is the reduced Euler characteristic of the Milnor fiber of h up to a sign.

This showed clearly the significance of the quantity $[(\frac{\partial h}{\partial z_0})] \cdot \Gamma^1(h)$, which Massey defined to be the zero-dimensional Lê number $\lambda^0(h)$ of h, while T became the Lê number of highest dimension. The other Lê numbers are defined in a similar way in general; use the relative polar variety of dimension $k+1$ to pick out a cycle of dimension k in $\Sigma(h)$; take the multiplicity of the cycle to get $\lambda^k(h)$. More precisely, if coordinates $\mathbf{z}$ are chosen appropriately at a point p, and $\Gamma^{k+1}_{h,z}$ denotes the relative polar variety of h of dimension $k+1$, then the Lê cycle of dimension k with respect to $\mathbf{z}$, $\Lambda^k_{h,z}$ is defined to be:

$$\left[\Gamma^{k+1}_{h,z} \cap V(\frac{\partial h}{\partial z_k})\right] - [\Gamma^k_{h,z}].$$

Massey defined the Lê numbers to be the intersection numbers:

$$\lambda^k_{h,z}(p) = \Lambda^k_{h,z} \cdot V(z_0 - p_0, \ldots, z_{k-1} - p_{k-1}).$$

He proved many results with the Lê numbers showing that their alternating sum was the reduced Euler characteristic of the Milnor fiber ([58] Theorem 3.3), that their constancy implied that the Milnor fibrations in a one-parameter family X were of the same diffeomorphism type if the codimension of the critical set was at least 3 in X_0, and of the same homotopy-type if the codimension was 2 ([58] Theorem 9.4), and that if X was a k-parameter family of hypersurfaces parametrized by a smooth $Y \subset X$ then the A_f condition held for $((\mathbf{C}^{n+1} \times \mathbf{C}^k) - \Sigma(f), Y)$ ([58] Theorem 6.9).

Gaffney (among others) realized that the Lê cycles of f were gotten by intersecting the exceptional divisor of $\mathrm{Bl}_{J(f)}\mathbf{C}^{n+1}$ with the pull-backs of generic planes in $\mathbf{P}^n$. In particular, that the intersection formula for λ^0 was the intersection formula for the multiplicity of the Jacobian ideal in the case of an isolated singularity. He conjectured that since the Lê numbers had some of the properties that you would expect a generalization of the multiplicity to have, and you could prove results with them similar to what you could prove with the generalization of the principle of specialization of integral dependence, that both generalizations should exist. These ideas were developed in [20], written with Gassler, developments which we now describe.

In what follows, assume I is an ideal which does not vanish on a component of X^n. Suppose I has k generators. The blow-up of X by I is embedded in $X \times \mathbf{P}^{k-1}$, so a hyperplane h in $\mathbf{P}^{k-1}$ induces a Cartier divisor H on B via the pull-back p^*, provided the blow-up is not contained in the product of X and the hyperplane. We call the pull-back a *hyperplane on* $\mathrm{Bl_I X}$. Then, the *polar varieties* $P_j(I)$ and *Segre cycles* $\Lambda_j(I)$ of I on X are defined as follows:

$$P_0(I) := X, \quad P_j(I) := b(H_1 \cap \cdots \cap H_j \cap \mathrm{Bl_I X}),$$

$$\Lambda_j(I) := b_*(H_1 \cdots H_{j-1} \cdot E \cdot \mathrm{Bl_I X}),$$

for generic H_i. If $[V] = \Sigma s_j[V_j]$ is a cycle with $[V_j]$ reduced, then the multiplicity of $[V]$ is $\Sigma s_j m(V_j)$. The Segre numbers $e_j(I, X)$ and the polar multiplicities $m_j(I, X)$ of I are the multiplicities of the corresponding cycles. (Notice that, as usual, subscripts indicate codimension.) Both of these notions appear in a more general setting in [41].

Example. If $I{=}J(f)$, $X = \mathbf{C}^{n+1}$ then $e_i(J(f))$ are the Lê numbers of Massey, while the polar multiplicities are the relative polar multiplicities of f. If coordinates are chosen generically, then

$$b(H_1 \cap \cdots \cap H_j \cap \mathrm{Bl}_{J(f)}\mathbf{C}^{n+1}) \subset V\left(\frac{\partial f}{\partial z_{n+1-j}}, \ldots, \frac{\partial f}{\partial z_n}\right)$$

with equality holding at smooth points of f. From this description, it is clear that the relative polar curve is empty if and only if the intersection $H_1 \cap \cdots \cap H_j \cap \mathrm{Bl}_{J(f)}\mathbf{C}^{n+1}$ is empty if and only if the fiber over the origin is not $\mathbf{P}^n$, i.e., is not a component.

The Segre numbers generalize the notion of the multiplicity of an ideal; if I has finite colength, then only $e_n(I)$ is non-zero, and it is equal to the multiplicity of I. The polar varieties also play an important role. Their multiplicities enter into the formulas relating the Segre numbers of I to the Segre numbers of the restriction of I to a generic slice of X. If we restrict I to the polar variety P^l, then $\Lambda^j(I,P^l) = \Lambda^j(I,X)$ for $j < l$. This supports proofs based on induction on the dimension of X.

If L_{n-k} is a generic $(n-k)$-codimensional linear subspace of $\mathbf{C}^N$, the following restriction formulas hold:

$$\begin{aligned} e_i(I, X \cap L_{n-k}) &= e_i(I,X) \quad (\text{for } i = 1, \ldots, k-1) \\ e_k(I, X \cap L_{n-k}) &= m_k(I,X) + e_k(I,X). \end{aligned}$$

These formulas offer another route for proofs by induction on the dimension of the underlying space. In addition, there is an expansion formula which relates the invariants of mI to the polar numbers and Segre numbers of I.

Theorem 6.1 (The expansion formulas, [20] Theorem 3.5.) *Suppose I an ideal in $\mathcal{O}_{X^n,x}$, m the maximal ideal in $\mathcal{O}_X$.*

$$\begin{aligned} e_n(mI, X) &= \sum_{i=0}^{n-1} \binom{n}{i} m_i(I,X) + \sum_{i=1}^{n} \binom{n-1}{i-1} e_i(I,X), \\ e_k(mI, X) &= \sum_{i=1}^{k} \binom{n-1}{i-1} e_i(I,X). \end{aligned}$$

The proof of the above theorem and the restriction formulas are done through intersection theory. Form the blow-up of X by mI; call this space B and the exceptional divisor D. The map to X factors through the blow-up of X by I, denoted B_2 with exceptional divisor D_2, call the map $\tilde{b}_2$. Denote the blow-up of m in X by B_1 with exceptional divisor D_1 and the analogous map to B_1 by $\tilde{b}_1$. Denote the blow-up maps to X from B_1 and B_2 by b_1 and b_2 respectively. These spaces and maps give the commutative diagram:

$$\begin{array}{ccc} \mathbf{B} & \xrightarrow{\tilde{b}_2} & \mathbf{B_2} \\ \downarrow \tilde{b}_1 & & \downarrow b_2 \\ \mathbf{B_1} & \xrightarrow{b_1} & \mathbf{X} \end{array}$$

Formulas for the Segre numbers and polar multiplicities can be obtained by using intersection products of the first Chern classes of the tautological line bundles on the above blow-ups and their exceptional divisors. Working in the fiber of B over 0, which is compact, allows the use of rational equivalence in deriving the formulas.

Semicontinuity is an essential property of an invariant if it is to be useful in proofs. The upper-semicontinuity of the Segre numbers is a little subtle, and indicates something of the role they play in the geometry of $V(I)$.

Example. ([58]) Let $f_t(x, y, z) = z^2 - y^3 - txy^2$. Then for $t \neq 0$, f_t defines a Whitney umbrella, while f_0 defines the product of a cusp with a line. Let I_t be the Jacobian ideal of f_t, with $X_t = \mathbf{C}^3$. A priori, $e_3(I_t)$ and $e_2(I_t)$ may be non-zero since f defines a family of hypersurfaces with a 1-dimensional singular set. Since $e_2(I_t)$ is the transverse Milnor number, it is 1 for $t \neq 0$ and 2 for $t = 0$, while $e_3(I_0) = 0$, since the zero set of f_0 is a product. For the polar surface of the total space, we can use the plane with equations $z = 0$ and $3y - 2tx = 0$. These equations are gotten by setting f_z and f_y equal to zero and removing the components in the singular locus of f. The fibers of this polar surface over the t-axis are the polar curves of I_t at the origin for $t \neq 0$; at $t = 0$ the fiber has fallen into $V(I_0)$. To calculate $e_3(I_t)$, calculate the multiplicity of I_t in the local ring of the polar curve at the origin; this gives a value of 2 for $t \neq 0$ and 0 again for $t = 0$, since there is no polar curve. The example shows that if e_2 jumps up at the origin, then e_3 is free to jump down and the fiber of the polar surface may fail to specialize.

However, the e_i are lexicographically upper-semicontinuous. To be precise:

Proposition 6.2 *The n-tuple $(e_1(t), \ldots, e_n(t))$ is lexicographically upper-semicontinuous: If for some k the map $t \to (e_1(t), \ldots, e_k(t))$ is constant on Y, then $t \to e_{k+1}(t)$ is upper-semicontinuous. In particular, $e_1(t)$ is always upper-semicontinuous.*

Proof. Cf. Corollary 4.5 [20]. It is convenient to discuss the proof using the previous example. Take a generic 3-plane containing the t-axis and move it off the t-axis a little in a parallel way–call the new plane P. Then the restriction of I to P defines a family of ideals of finite colength, and the multiplicity of $I_{t|P_t}$ is still $e_2(I_t)$, where as usual P_t denotes the fiber of P over t of the projection to the t-axis. The theory of the multiplicity shows that $e_2(I_t)$ is upper-semicontinuous. Further, if $e_2(I_t)$ were independent of t, then the theory for ideals of finite colength says that no component of the blow-up of P by $I_{|P}$ projects to a point, which means that in the example

$I_{|P}$ would have no polar curve, so the polar surface of I could not dive into $V(I_0)$ at $t = 0$. This implies that the fiber of the polar surface of I specializes to the polar curve of I_0 over $t = 0$; so we can restrict I to its polar surface, again get a family of ideals of finite colength, where the multiplicity will now be $e_3(I_t)$. This means that $e_3(I_t)$ is upper-semicontinuous.

The above discussion shows that the constancy of the Segre numbers of codimension less than or equal to 2 prevents a codimension 1 or 2 subset of X_0 from being the image of a component of D_2. Using the ideas in the above discussion, it is not hard to show:

Theorem 6.3 *Suppose X^{n+k} is a family of equidimensional spaces over Y^k, Y a smooth subset of X. Then the map $t \to (e_1(t), \dots, e_n(t))$ is constant on Y if and only if all components of D_2^Y and $\tilde{D}_2^{X-Y} \cdot \tilde{D}_1$ are equidimensional over Y.*

Proof. For a more precise statement and the proof, see Proposition 4.6 of [20]. In the above statement, D_2^Y is the set of components of the exceptional divisor of D_2, whose image lies in Y, while $\tilde{D}_2^{X-Y} \cdot \tilde{D}_1$ is gotten by taking the components of D_2, whose image lies outside of Y, taking their strict transforms after blowing-up B_2 by the pull-back of m_Y, and taking the part of the strict transform over Y.

The control that this gives over the components of the exceptional divisor of D_2 then makes it possible to prove:

Theorem 6.4 (The Principle of Specialization.) *Suppose in the set-up of 4.3, $h \in \mathcal{O}_{X,0}$ is a function such that $h\mathcal{O}_{X(t),0}$ is integrally dependent on I_t, for all t in a Zariski-open subset of Y. If the numbers $e_1(t), \dots, e_n(t)$ are independent of t, then h is integrally dependent on I.*

Proof. Cf. Theorem 4.7 of [20]. The idea of the proof is similar to the last part of the argument in Teissier's original proof. Since all of the components of D_2 contain Y in their image, the same is true for the normalization of B_2. Therefore, since the pull-back of h to the normalized blow-up vanishes on the fiber of every component of the exceptional divisor to the desired degree generically, it vanishes on the component to the desired degree generically and, hence, vanishes to the desired degree.

Using this result, it is possible to recover Massey's extension of the Lê-Saito theorem and to extend the $\mu^{(*)}$ constant theorem to the case of families of non-isolated hypersurface singularities. We now describe this extension.

As mentioned earlier, a family of sets is Whitney equisingular if and only if the family and all its plane sections are topologically equisingular. In order to have invariants which are both necessary and sufficient for Whitney equisingularity, it is useful to link your controlling invariants to topological invariants of the family and its plane sections. The formulation of the $\mu^{(*)}$

theorem is done with this in mind. Let X be a family of hypersurfaces defined by $F : \left(\mathbf{C}^k \times \mathbf{C}^{n+1}, \mathbf{C}^k \times 0\right) \to (\mathbf{C}, 0)$, $f_y(z) = F(y,z)$ defines X_t, a reduced hypersurface which contains the origin. Denote $\mathbf{C}^k \times 0$ by Y.

Fix y in Y. For $L^i \subset \mathbf{C}^{n+1}$ a generic i-dimensional linear subspace, we denote the Euler characteristic of the Milnor fiber of $f_y|_{L^i}$ at 0 by $\chi^{(i)}(y)$. We define

$$\chi^*(y) := (\chi^{(n+1)}(y), \ldots, \chi^{(2)}(y)).$$

Note that it is obvious that $\chi^*(y)$ will be independent of y for any Whitney equisingular family of hypersurfaces.

Theorem 6.5 *Suppose $(m_1(f_y), \ldots, m_n(f_y), \chi^*(y))$ is independent of y on Y. Then, the pair $(\mathbf{C}^{n+k} - \Sigma(F), Y)$ satisfies W_f at 0. In particular, the smooth part of X is Whitney regular along Y at 0. Also, any stratum of X whose closure is the image of a component of the exceptional divisor of $\mathrm{Bl}_{\mathrm{J(F)}}\mathrm{C}^{\mathrm{n+k}}$ satisfies the Whitney conditions along Y.*

Proof. Cf. Theorem 6.2 of [20]. The constancy of the relative polar varieties of f_y and the $\chi^*(y)$ imply that the Lê numbers of f_y are constant as well, because of the restriction formula relating the Lê numbers of the restriction of f_y to L^i to the Lê numbers of f_y and the relative polar multiplicities. By the expansion formula, this implies that the Segre numbers of $m_y J(f_y)$ are constant. Since W_f holds generically, the principle of specialization then applies showing that W_f holds for the pair $(\mathbf{C}^{n+k} - \Sigma(F), Y)$.

But much more is true. If the closure of a stratum S of V(F) is the image of a component of the exceptional divisor of $\mathrm{Bl}_{\mathrm{J(F)}}\mathrm{C}^{\mathrm{n+k}}$, then for dimensional reasons, and because A_F holds generically, that component is the conormal space of $\overline{S}$. Essentially by Proposition 6.3, the blow-up of this component by the pull-back of m_Y has an exceptional divisor which is equidimensional over Y. Hence, by the theorem of Teissier's ([82, V.1.2]) mentioned earlier, the pair (S, Y) satisfies condition W.

If X is a family of hypersurfaces with isolated singularities, then this result does become the $\mu^{(*)}$ constant result, for then χ^* constant becomes $\mu^{(*)}$ constant, and the relative polar multiplicities constant also becomes $\mu^{(*)}$ constant. In [20], it is also shown (Theorem 6.3) that if X has a Whitney stratification with Y as a stratum, then $(m_1(f_y), \ldots, m_n(f_y), \chi^*(y))$ is independent of y, so the constancy of the invariants is necessary for Whitney equisingularity.

Theorem 6.5 underscores the importance of understanding which singularities correspond to the different components of the exceptional divisor of $J(f)$. As was mentioned earlier in Section 3, this is equivalent to understanding which singularities correspond to components of the characteristic cycle of the sheaf of vanishing cycles of f, so this is an important point of intersection of the topological and geometric approaches. Here is one class of examples for which we have enough machinery to show that all of the strata correspond to components of the exceptional divisor.

Theorem 6.6 *Suppose $F : (\mathbf{C}^n \times \mathbf{C}^k, 0) \to (\mathbf{C}^p \times \mathbf{C}^k, 0)$ is a k-parameter family of finitely-determined map-germs, whose locus of instability is $Y = 0 \times \mathbf{C}^k \subset \mathbf{C}^p \times \mathbf{C}^k$. Suppose (n,p) is in Mather's nice range of dimensions or on the boundary of the nice dimensions with $p \leq n$. Then, the family of discriminants, $\Delta(f_y)$, are Whitney equisingular if and only if*

$$(m_1(f_y), \ldots, m_{p-1}(f_y), \chi^*(y))$$

is independent of y.

Proof. The hypotheses imply that $\Delta(F)$, the total space of the discriminant family, has a stratification by stable types and the parameter space Y such that W holds for every pair of strata, except possibly over Y. By the previous results, it suffices to show that each stable type is the image of a component of the exceptional divisor of the blow-up of $\mathbf{C}^p \times \mathbf{C}^k$ by the Jacobian ideal of $\Delta(F)$. We first consider those stable types presented by a germ at a single point.

It suffices to show that if f_S is the minimal stable unfolding which represents a stable type S which appears in F with positive dimension, then the relative polar curve of $\Delta(f_S)$ at the origin is non-empty. This will imply that the origin is the image of a component of the exceptional divisor. Since the discriminant is analytically a product along strata of $\Delta(F)$ different from Y, this will imply that the stratum presenting the stable type S is also the image of a component of the exceptional divisor. Recall that the nice dimensions are the pairs (n,p) of source and target dimensions for which the stable mappings are dense, and for which all finitely-determined germs are of discrete stable type. Since (n,p) is on the boundary of the nice dimensions, the source and target dimensions of f_S, (n',p') is inside the nice dimensions, since both n and p must decrease by at least 1, since S appears in F with positive dimension. Now Massey observed in ([59], p365) that the multiplicity of the polar curve is just the number of spheres in the homotopy-type of the complex link of the singularity, regardless of the dimension of the critical locus (in the isolated case, this fact can be found in [53] and [78]). Damon [12] found a way to calculate this quantity in the case at hand. Here is Damon's result. Take a generic hyperplane H that passes through the origin in $\mathbf{C}^{p'}$, and consider the induced map germ $f_{S,0} : f_S^{-1}(H) \to H$. The map germ $f_{S,0}$ will be contact equivalent to f_S, but will not be stable. Form the stabilization of $f_{S,0}$ by moving H off the origin in a family H_t, so that the family of induced germs will be stable for $t \neq 0$. Now, the number of spheres in the homotopy-type of the complex link is just the number of spheres in the homotopy-type of the discriminant of a stabilization $f_{S,t}$. Damon showed that the number of spheres of this last type was equal to the A_e-codimension of $f_{S,0}$ which, therefore, must be greater than zero.

Now we consider stable types presented by multigerms. If two smooth sheets of $\Delta(F)$ intersect, then this has codimension 0 in $\Sigma(\Delta(F))$, so this stratum must be the image of a component of the exceptional divisor. Otherwise, we can assume that the discriminant is locally the union of two hypersurfaces X_1 and X_2 embedded in $\mathbf{C}^k \times \mathbf{C}^{n-k}$, with equations $g_1(z_1, \ldots, z_k) = 0$, $g_2(z_{k+1}, \ldots, z_n) = 0$. There are two cases: $k = n-1$, with $g_2 = z_n$, and g_1, g_2 both defining singular hypersurfaces.

Suppose first that $g_2 = z_n$. Then, we can assume that Ψ_1 parametrizes a branch of the polar curve of g_1, and that $g_{1,z_i} \circ \Psi_1$ is zero for $1 \leq i \leq n-2$. We know that g_1 is in the integral closure of $J(g_1)$, so the ratio $(g_1/g_{1,z_{n-1}}) \circ \Psi_1$ is an analytic germ ψ_n. Define $\Psi = (\Psi_1, \psi_n)$; then, if $f = z_n g_1$ $f_{z_i} \circ \Psi = 0$ for $1 \leq i \leq n-2$ and $(f_{z_n} - f_{z_{n-1}}) \circ \Psi = (g_1 - z_n g_{1,z_{n-1}}) \circ \Psi = 0$. So, Ψ parametrizes a branch of the polar curve of f.

Now suppose that $f = g_1 g_2$, where we can assume that Ψ_1 and Ψ_2 parameterize branches of the polar curves of g_1 and g_2. Then $\Psi = (\Psi_1, \Psi_2)$ parametrizes a polar surface for f, for it is easy to check that $f_{z_i} \circ \Psi = 0$ for $1 \leq i < k$ and for $k+1 \leq i < n$. It is easy to see that for almost any choice of A, B that $(Af_{z_k} + Bf_{z_n}) \circ \Psi$ will define a curve singularity; parametrize a branch by ϕ, then $\Psi \circ \phi$ will parametrize a branch of the polar curve of f. This finishes the proof.

The results of Damon show that for stable germs in the nice dimensions presented at a single point with dimension 0, then the number of spheres in the homotopy-type of the complex link of the discriminant at the origin is 1, so each of these singularities contributes a component with coefficient 1 to the characteristic cycle of the sheaf of vanishing cycles of the discriminant.

This kind of result indicates that singularities of mappings should provide a place for the convergence of the two points of view being discussed here.

In the next section, we look more closely at the connection between the characteristic cycle and the Lê-Saito theorem.

7 A Further Extension of the Lê-Saito Theorem

In Section 3, we discussed the topological machinery of the sheaf of vanishing cycles and the characteristic cycle. In this closing section, we will briefly describe how this machinery can be used to generalize the result of Lê and Saito on Thom's A_f condition. The generalization that we present here is one in which the underlying space is allowed to be arbitrary, but in which we are still looking at families of generalized isolated singularities.

The generalization of a critical point that we consider here is the notion of a *topological critical point*; we define the topological critical locus, $\Sigma_{top} f$, of f to be the closure of the set of points, x, at which the Milnor fibre of f does not have the complex cohomology of a point, i.e.,

$$\Sigma_{top} f := \overline{\{x \mid \widetilde{H}^*(F_{f,x}; \mathbf{C}) \neq 0\}}.$$

As we mentioned in Section 3, $\Sigma_{top} f$ is contained in the stratified critical locus of f, regardless of what Whitney stratification one uses; thus, $\Sigma_{top} f$ is a more subtle, and intrinsic, notion of "generalized" critical locus. (Note that the "Milnor fibre" of f is defined for functions with arbitrary analytic domains by the results of Lê in [49].)

In this section, we will outline how to generalize the result of Lê and Saito to a family f_t, where the domain X of the total function f is arbitrary, but $\Sigma_{top}(f_t)$ has the origin as an isolated point for all small t. However, just as one can pass from a number of results for isolated affine singularities to non-isolated singularities by using the Lê cycles, the results that we present here generalize in a Lê-cycle way to results for generalized non-isolated singularities on arbitrary analytic spaces; this "super generalization" of the result of Lê and Saito is contained in the thesis of M. Green [28]. Unfortunately, even if we merely state the necessary preliminary results without proof, the presentation becomes too technical and takes us too far afield. Therefore, we will simply discuss our results and (hopefully) give the reader the flavor of the approach, and we will give a complete statement of our final result. The full statements and details of all of these results can be found in [57].

A quick look at our modified proof of the Lê-Saito theorem at the end of Section 3 shows that the proof breaks up into three pieces:

- constant Milnor number throughout a family implies the global polar curve is empty;
- the global polar curve being empty implies that the characteristic cycle of the vanishing cycles (of the constant sheaf) has only one component — a component corresponding to the conormal variety to the parameter axis;
- because the projectivization of the characteristic cycle of the vanishing cycles is isomorphic to the exceptional divisor of the Jacobian blow-up, if this characteristic cycle has only a component corresponding to the conormal variety to the parameter axis, then the pair (ambient space minus the parameter axis, parameter axis) satisfies Thom's A_f condition.

If we wish to deal with the case where our domain is an arbitrary analytic space, X, rather than just an open subset of affine space, U, what changes must we make in our argument? That is, how must each of the three pieces of the above argument be altered?

First of all, everywhere that we used the constant sheaf on U, we must now use the (shifted) perverse cohomology (or, perverse projection) of the constant sheaf on X (see [2]). If X is a local complete intersection, this perverse cohomology simply returns the constant sheaf; however, in general, one obtains "the perverse sheaf closest to the constant sheaf".

Secondly, in keeping with the previous paragraph, "Milnor number" must now mean the middle Betti number of the cohomology of the Milnor fibre **with the perverse cohomology of the constant sheaf as coefficients.**

Once these changes are made, it is once again true that constant Milnor number implies that the global polar curve is empty, and that empty polar curve implies that the vanishing cycles (of the perverse projection of the shifted constant sheaf) has a single component, corresponding to the conormal to the parameter axis.

Finally, one must prove something analogous to "the projectivization of the characteristic cycle of the vanishing cycles of f is isomorphic to the exceptional divisor of the Jacobian blow-up". What we prove is that, in the general case, the projectivization of the characteristic cycle of the vanishing cycles of any bounded, constructible complex, $A^\bullet$, is isomorphic to the exceptional divisor of the blow-up along the image of df of the characteristic cycle of $A^\bullet$. We apply this result to the case where $A^\bullet$ is the perverse cohomology of the constant sheaf on X.

The reader may be thinking: "Who cares if you can prove results using the machinery of perverse sheaves, if one cannot understand the statements of the results without the language of the derived category and perverse sheaves?" The point is that we can, in fact, state our main results without the language of the derived category (though the statements tend to be lengthy without the concise formulations in the derived category).

Theorem. *Let X be a complex analytic subset of an open subset of affine space $\mathcal{U}$. Suppose that $0 \in X$ and $C := (\mathbf{C} \times \{0\}) \cap \mathcal{U} \subseteq X$, i.e., the z_0-axis, near 0, is contained in X. Let $f : (X, C) \to (\mathbf{C}, 0)$ be an analytic function and, for all $(a, 0) \in C$, let f_a denote the restriction of f to $X_a := X \cap V(z_0 - a)$.*

Suppose that f is in the square of the maximal ideal of X at 0. Suppose that, $0 \notin \Sigma_{top}(z_{0|X})$. Suppose, further, that 0 is an isolated point in $\Sigma_{top}(f_0)$.

Fix an integer k. If the reduced Betti number $\tilde{b}_{k-1}(F_{f_a,(a,0)})$ is independent of a for all small a, then, near 0, for every $(k+1)$-dimensional component X^{k+1} of X, $\Sigma(f_{|_{X^{k+1}_{reg}}}) \subseteq C$ and the pair $(X^{k+1}_{reg} - \Sigma(f_{|_{X^{k+1}_{reg}}}),\ C)$ satisfies Thom's A_f condition at 0.

References

[1] R. Bassein, On smoothable curve singularities: Local methods, *Math. Ann.*, 230 (1977), 273-277.

[2] A. Beilinson, J. Berstein, and P. Deligne, *Faisceaux Pervers,* Astérisque, 100, Soc. Math de France (1983).

[3] J. Briançon and J.P.G. Henry, Equisingularite generique des familles de surfaces a singularite isolee, *Bull. Soc. Math. Fr.*, 108, 2 (1980), 259-281.

[4] J. Briançon, P. Maisonobe and M. Merle, Localisation de systèmes différentiels, stratifications de Whitney et condition de Thom, *Invent. Math.*, 117 (1994), 531-550.

[5] J. Briançon and J. P. Speder, Les conditions de Whitney impliquent μ^* constant, *Ann. Inst. Fourier*, 26 (1976), 153-163.

[6] J. Briançon, A. Galligo, M. Granger, *Deformations equisingulieres des germes de courbes gauches reduits,* Memoire de la Societe Mathematique de France, Nouvelle serie, #1, (1980).

[7] D. A. Buchsbaum and D. S. Rim, A generalized Koszul complex. II. Depth and multiplicity, *Trans. Amer. Math. Soc.*, 111 (1963), 197-224.

[8] R.-O. Buchweitz and G.-M. Greuel, The Milnor number and deformations of complex curve singularities, *Invent. Math.*, 58 (1980), 241-281.

[9] J.N. Damon, Finite Determinancy and Topological Triviality I, *Invent. Math.*, 62 (1980), 299-324.

[10] J.N. Damon, Sufficient conditions and topological stability, Compositio Math., 47 (1982), 101-132.

[11] J.N. Damon, *Topological triviality and Versality for Subgroups of A and K,* Memoir of the American Mathematical Society, 389, (1988).

[12] J.N. Damon, *Higher multiplicities and almost free divisors and complete intersections,* Memoir of the American Mathematical Society, 589, (1996).

[13] P. Deligne, Comparaison avec la théorie transcendante, *Séminaire de géométrie algébrique du Bois-Marie*, SGA 7 II, Springer Lect. Notes 340 (1973).

[14] W. Fulton, *Intersection Theory,* Ergebnisse der Mathematik und ihrer Grenzgebiete, 3. Folge · Band 2, Springer–Verlag, Berlin, 1984.

[15] T. Gaffney, Aureoles and integral closure of modules, *Stratifications, Singularities and Differential Equations II*, (1997) Herman, Travaux en Cours #55. p55-62.

[16] T. Gaffney, Polar Multiplicities and Equisingularity of Map Germs, *Topology*, 32 (1993), 185-223.

[17] T. Gaffney, Integral closure of modules and Whitney equisingularity, *Invent. Math.*, 107 (1993), 301-322.

[18] T. Gaffney, Equisingularity of plane sections, t_1 condition, and the integral closure of modules, in *Real and Complex Singularities, Proceedings of the Third International Workshop on Real and Complex Singularities at Sao Carlos, Brasil 1994*, W. L. Marar (ed.) Pitman Research Notes in Mathematics 333 (1995), 95-111.

[19] T. Gaffney, Multiplicities and equisingularity of ICIS germs, *Invent. Math.*, 123 (1996), 209-220.

[20] T. Gaffney and R. Gassler, Segre numbers and hypersurface singularities, to appear in *Journal of Algebraic Geometry.*

[21] T. Gaffney and S. Kleiman, Specialization of integral dependence for modules, to appear in *Invent. Math.*.

[22] T. Gaffney and D. Mond, Cusps and Double Folds of Germs of Analytic Maps $\mathbf{C}^2 \to \mathbf{C}^2$, *Journal of the London Math. Society, second series*, 139, Vol. 43 (1991),185-192.

[23] T. Gaffney and D. Mond, Weighted Homogeneous Maps from the Plane to the Plane, *Math. Proc. Camb. Phil. Soc.*, 109 (1991), 451-470.

[24] C.G. Gibson, E. Looijenga, A. du Plessis, K. Wirthmuller, *Topological stability of smooth maps*, SLN #555, Springer-Verlag, N.Y. (1976).

[25] V. Ginsburg, Characteristic Varieties and Vanishing Cycles, *Inv. Math.*, 84 (1986), 327-403.

[26] M. Goresky and R. MacPherson, Intersection homology II, *Inv. Math.*, 71 (1983), 77-129.

[27] M. Goresky and R. MacPherson, *Stratified Morse Theory,* Ergebnisse der Math. 14, Springer-Verlag, Berlin (1988).

[28] M. Green, *Dissertation, Northeastern University*, 1997.

[29] M. D. Green and D. B. Massey, Vanishing cycles and Thom's a_f conditions, Preprint 1996.

[30] G. M. Greuel, *Der Gauss–Manin Zusammenhang isolierter Singularitäten von vollständigen Durchschnitten,* Dissertation, Göttingen (1973), also *Math. Ann.*, 214 (1975), 235-266.

[31] H. Hamm and Lê D. T., Un Théorème de Zariski du type de Lefschetz, *Ann. Sci. L'Ecole Norm. Sup.*, 6 (1973), 317-366.

[32] J.P.G. Henry and M. Merle, Limites d'espaces tangents et transversalité de variétés polaires, in *Proc. La Rábida, 1981.* J. M. Aroca, R. Buchweitz, M. Giusti and M. Merle (eds.), *Springer Lecture Notes in Math.*961 (1982), 189-199.

[33] J.P.G. Henry and M. Merle, Limites de normales, conditions de Whitney et éclatement d'Hironaka, *Singularities, Proc. Symposia Pure Math.*, vol. **40**, part 1, Amer. Math. Soc. (1983), 575-84.

[34] J.P.G. Henry and M. Merle, *Fronces et doubles plis,* Thesis de Doctorat d'Etat, M. Merle, Universite de Paris VII, 1990.

[35] J.P.G. Henry and M. Merle, Conormal Space and Jacobian module. A short dictionary, in *Proceedings of the Lille Congress of Singularities, J.-P. Brasselet (ed.)*, London Math. Soc. Lecture Notes **201** (1994), 147-74.

[36] J.P.G. Henry, M. Merle and C. Sabbah, Sur la condition de Thom stricte pour un morphisme analytique complexe, *Ann. Scient. Éc. Norm. Sup.*, 17 (1984), 227-68.

[37] H. Hironaka, Stratifications and flatness, in *Real and complex singularities, Nordic Summer School, Oslo*, (1976), Sijthoff and Noordhoff, 1977.

[38] H. Hironaka, Normal cones in analytic Whitney stratifications, *Inst. Hautes Etudes Sci. Publ. Math.*, 36 (1969), 127-138.

[39] I.N. Iomdin, Local topological properties of complex algebraic sets, *Sibirsk. mat. Z.*, 15(4) (1974), 1061-1082.

[40] M. Kashiwara, *Systèmes d'équations micro-différentielles,* Dépt. de Math., Univ. Paris-Nord 8 (Notes by T. M. Fernandes), 1978.

[41] S. Kleiman and A. Thorup, A geometric theory of the Buchsbaum–Rim multiplicity, *J. Algebra*, 167 (1994), 168-231.

[42] D.T. Lê, Calcul du Nombre de Cycles Evanouissants d'une Hypersurface Complexe, *Ann. Inst. Fourier, Grenoble*, 23 (1973), 261-270.

[43] D.T. Lê, Topological Use of Polar Curves, *Proc. Symp. Pure Math.*, 29 (1975), 507-512.

[44] D. T. Lê, Calculation of Milnor number of isolated singularity of complete intersection, *Funct. Anal. Appl.*, 8 (1974), 127-131.

[45] D. T. Lê, Ensembles analytiques complexes avec lieu singulier de dimension un (d'apres I. N. Iomdin), *Seminaire sur les Singularites (Paris, 1976-1977)*, Publ. Math. Univ. Paris VII (1980), 87-95.

[46] D. T. Lê, Le théorème de la monodromie singulier, *C.R.A.S., Paris Ser. A*, 288 (1979), 985-988.

[47] D. T. Lê, Le concept de singularité isolée de fonction analytique, *Advanced Studies in Pure Math.*, 8 (1986), 215-227.

[48] D. T. Lê, Morsification of *d*-modules, Preprint (1988).

[49] D. T. Lê, Sur les cycles évanouissants des espaces analytiques, *C.R. Acad. Sci. Paris, Ser. A*, 288 (1979), 283-285.

[50] D. T. Lê and Z. Mebkhout, Variétés caractéristiques et variétés polaires, *C.R. Acad. Sci. Paris Se. A*, 296 (1983), 129-132.

[51] D. T. Lê and C. P. Ramanujam, The invariance of Milnor's number implies the invariance of the topological type, *Amer. J. Math.*, 98 (1976), 67-78.

[52] D. T. Lê and K. Saito, La constance du nombre de Milnor donne des bonnes stratifications, *C. R. Acad. Sci. Paris*, 277 (1973), 793-795.

[53] D. T. Lê and B. Teissier Cycles evanescents, sections planes et conditions de Whitney, II, *Proc. Symposia Pure Math.*, 40, part 2, Amer. Math. Soc. (1983), 65-104.

[54] M. Lejeune-Jalabert and B. Teissier, *Clôture integrale des ideaux et equisingularité, chapitre 1* Publ. Inst. Fourier (1974).

[55] J. Lipman, Equimultiplicity, reduction and blowing-up, in *Commutative algebra: analytic methods, Dekker Lecture Notes in Pure and Applied Math.*, 68 (1982), 111-148.

[56] E. J. N. Looijenga, *Isolated Singular Points on Complete Intersections,* London Math. Soc. Lect. Note Series 77, Cambridge Univ. Press (1984),

[57] D. Massey, Critical Points of Functions on Singular Spaces, to appear in *Topology and its Applications.*

[58] D. Massey, *Lê Cycles and Hypersurface Singularities,* Springer Lecture Notes in Mathematics 1615, (1995).

[59] D. Massey, Numerical invariants of perverse sheaves, *Duke Mathematical Journal*, 73 (2) (1994), 307-369.

[60] D. Massey, Hypercohomology of Milnor Fibres, *Topology*, 35 (1996), 969-1003.

[61] J. N. Mather, Stability of C^{∞}-mappings: I The division theorem, *Ann. of Math*, 87 (1968), 87-104.

[62] J. N. Mather, II Infinitesimal stability implies stability, *Ann. of Math*, 89 (1969), 259-291.

[63] J. N. Mather, III finitely determined map-germs, *Pub. Math., IHES*, 35 (1969), 127-156.

[64] J. N. Mather, IV Classification of stable germs by R algebras, *Publ. Math. IHES* (1970), 223-248.

[65] J. N. Mather, V Transversality, *Advances in Math.*, 4 (1970), 301-335.

[66] J. N. Mather, VI The nice dimensions, *Proceedings of the Liverpool Singularities Symposium I*, Springer Lecture Notes in Mathematics, 192 (1971), 207-255.

[67] J. N. Mather, Stratifications and Mappings, in *Dynamical Systems*, Academic Press (1973), 195-232.

[68] J. N. Mather, How to stratify manifolds and jet spaces, Springer Lecture Notes in Mathematics, 535 (1976), 128-76.

[69] J. Milnor, *Singular Points of Complex Hypersurfaces,* Annals of Math. Studies, no. 77 (1968).

[70] D. Mond, Some remarks on the geometry and classification of germs of maps from surfaces to 3-space, *Topology*, 26 (1987), 207-255.

[71] V. Navarro Aznar, *Stratificationes regulieres et varietes polaires locales,* manuscript (1981).

[72] A. J. Parameswaran, Topological equisingularity for isolated complete intersection singularities, *Compositio Math.*, 80 (1991), 323-336.

[73] D. Rees, Reduction of modules, *Math. Proc. Camb. Phil. Soc.*, 101 (1987), 431-449.

[74] D. Rees, *Gaffney's problem,* manuscript dated Feb.22 1989.

[75] D. Rees and D. Kirby, Multiplicities in graded rings I: The general theory, in *Commutative algebra: syzygies, multiplicities, and birational algebra,* W. J. Heinzer, C. L. Huneke, J. D. Sally (eds.), *Contemp. Math.*, 159 (1994), 209-267.

[76] C. Sabbah, Quelques remarques sur la géométrie des espaces conormaux, *Astérisque*, 130 (1985), 161-192.

[77] M. Sato, T. Kawaï, M. Kashiwara, Hyperfunctions and pseudo-differential equations, in *Hyperfunctions and Pseudo-differential equations, Proc. Katata 1971*, Springer Lect. Notes 287 (1973), 265-529.

[78] B. Teissier, Cycles évanescents sections planes et conditions de Whitney, in *Singularités á Cargèse, Astérisque*, 7-8 (1973).

[79] B. Teissier, Déformations à type topologique constant II, in *Séminaire Douady-Verdier, Secrétariat E.N.S.*, Paris, 1972.

[80] B. Teissier, Introduction to equisingularity problems, *Proc. Symposia Pure Math.*, 29, Amer. Math. Soc. (1975), 575-584.

[81] B. Teissier, Résolution simultanée et cycles évanescents, in *Sém. sur les singularités des surfaces. Proc. 1976-77.* M. Demazure, H. Pinkham and B. Teissier (eds.), *Springer Lecture Notes in Math.*777 1980 82-146.

[82] B. Teissier, Multiplicités polaires, sections planes, et conditions de Whitney, in *Proc. La Rábida, 1981.* J. M. Aroca, R. Buchweitz, M. Giusti and M. Merle (eds.), *Springer Lecture Notes in Math.*961 (1982), 314-491.

[83] B. Teissier, Variétés polaires I: Invariants polaires des singularités d'hypersurfaces, *Invent. Math.*, 40 (1977), 267-292.

[84] B. Teissier, Variétés polaires locales et conditions de Whitney, *C. R. Acad. Sci. Paris Sér. A*, 290 (1980), 799.

[85] R. Thom, Ensembles et morphismes stratifies, *Bull. Amer. Math. Soc.*, 75 (1969), 240-284.

[86] D. Trotman, On the canonical Whitney stratification of algebraic hypersurfaces, *Sem. sur la géométrie algébrique realle, dirigé par J.-J. Risler, Publ. math. de l'université Paris VII*, Tome I (1986), 123-152.

[87] J. L. Verdier, Stratifiactions de Whitney et théorème de Bertini–Sard, *Invent. Math.*, 36 (1976), 295-312.

[88] H. Whitney, Local properties of analytic varieties, *In Differential and combinatorial topology,* S.S. Cairns (ed.), Princeton University Press, 205-244.

[89] H. Whitney, Tangents to an analytic variety, *Ann. of Math.*, 81 (1965), 496-549.

[90] O.Zariski, Studies in equisingularity I, *Amer. J. Math.*, 87 (1965), 507-536.

[91] O.Zariski, Studies in equisingularity II, *Amer. J. Math.*, 87 (1965), 972-1006.

[92] O.Zariski, Studies in equisingularity III, *Amer. J. Math.*, 90 (1968), 961-1023.

[93] O.Zariski, Some open questions in the theory of singularities, *Bull. AMS*, 77 (1971), 481-491.

[94] O.Zariski, The elusive concept of equisingularity and related questions, *Algebraic geometry, The Johns Hopkins centennial lectures,* J.I. Igusa (ed.), Johns Hopkins Univeristy Press (1977).

[95] O.Zariski and P. Samuel, *Commutative Algebra,* vol 2, Van Nostrand, 1960.

Department of Mathematics,
Northeastern University,
Boston, MA 02115,
USA.

e-mail: gaff@neu.edu
e-mail: dmassey@neu.edu

Regularity at Infinity of Real and Complex Polynomial Functions

Mihai Tibăr

Dedicated to Professor C.T.C. Wall.

1 Introduction

Let $f : \mathbb{K}^n \to \mathbb{K}$ be a polynomial function, where $\mathbb{K}$ is $\mathbb{C}$ or $\mathbb{R}$. A value $a \in \mathbb{K}$ is called *typical* for f if the fibre $f^{-1}(a)$ is nonsingular and the function f is a locally trivial topological fibration at a. Otherwise a is called *atypical*. The set of atypical values is known to be finite [Th] (for a proof, see e.g. [Ph, Appendix], which uses resolution of singularities, or [V], [ST] and Corollary 2.12 in this paper, where stratification theory is used).

One of the most natural problems is to identify the set of noncritical atypical values of f and to describe how the topology of fibres changes at such a value. There is, as yet, no solution to this problem in full generality. One can solve it only in the case when the singularities occuring "at infinity" are in some sense isolated and $\mathbb{K} = \mathbb{C}$ (see [Pa-1], [ST], [Ti-2]), including the case $n = 2$ [HL] and the case of "no singularity at infinity" [Br-2], [NZ].

To prove topological triviality at infinity (i.e. on $\mathbb{K}^n \setminus K$, where K is some large compact set) one could try to produce a foliation which is transversal to the fibres of f. A natural attempt is to integrate the vector field $\operatorname{grad} f$ (or $\operatorname{grad} f / \| \operatorname{grad} f \|^2$), which is a lift by f of the vector field $\partial/\partial t$ on $\mathbb{K}$. The resulting foliation may have leaves that "disappear" at infinity (since $\operatorname{grad} f$ may tend to 0 along some non-bounded sequence of points), and it would then not be of the kind we want. (Nevertheless such a foliation was used by Fourrier [Fo-1], [Fo-2] to characterise, in two complex variables, topological right-equivalence at infinity of polynomials.)

To construct a "good" foliation, one needs some regularity conditions on the asymptotic behaviour of the fibres of f. Keeping in mind the idea of using *controlled* vector fields (rather than just $\operatorname{grad} f$), we introduce two regularity conditions "at infinity": *t-regularity* and *ρ-regularity*. The former depends on the compactification of f, but allows one to apply algebro-geometric tools, more effectively in the complex case. It already appeared with a different definition in [ST] and implicitly in [Ti-2], where it was used for localizing the change of topology at infinity, in suitable circumstances. The second condition does not depend on any extension, but on the choice of a proper non negative C^1-function ρ which defines a codimension one foliation.

These regularity conditions correspond to the two main strategies used, to date, in the study of affine functions, which were regarded as parallel methods, which now roughly describe. One method is to "compactify" in some way the function (i.e. to extend it to a proper one). This has the advantage of having "infinity" as a subspace of the total space Y, but at the same time creates the problem of getting rid of it in the end. The space Y has singularities (usually nonisolated ones) exactly "at infinity"; one may either resolve those singularities or endow Y with a stratification. The second main strategy is to stay within the affine situation and use, for instance, Milnor-type methods.

We prove in this paper that t-regularity implies ρ-regularity, which fact provides a link between the two aforementioned approaches. We treat not only the complex case, but also the real case, which seems to have been less explored up tp now (see Durfee's papers [Du], [D&all] for polynomials in two variables). Our approach relies on the study of the limits at infinity of tangents to the levels of f and their behaviour with respect to the divisor at infinity. We have already used this point of view in investigating the topology "at infinity" of complex polynomial functions (see [ST], [Ti-2]).

We further explain, in the complex case, the relation between t-regularity and affine polar curves associated to f. Local and affine polar curves were used in [ST] and [Ti-2] to get information on the topology of the fibres of f. In two variables, there are even stronger relations among the investigated notions; we discuss this in Section 3.

We finally apply our study to prove a Lê-Ramanujam type result for the *global monodromy fibration* within families of complex polynomial functions having singularities at infinity.

2 Regularity conditions at infinity

We introduce two regularity conditions which ensure topological triviality at infinity. Let $f : \mathbb{K}^n \to \mathbb{K}$ be a polynomial function and let:

$$\mathbf{X}_{\mathbb{K}} := \{\tilde{f}(x, x_0) - t x_0^d = 0\} \subset \mathbb{P}^n_{\mathbb{K}} \times \mathbb{K},$$

be the closure in $\mathbb{P}^n_{\mathbb{K}} \times \mathbb{K}$ of its graph, where $\tilde{f}$ denotes the homogenization of f by the new variable x_0 and d is the degree of the polynomial f.

Let us denote by $t : \mathbb{P}^n_{\mathbb{K}} \times \mathbb{K} \to \mathbb{K}$ the second projection. Then the restriction $\hat{f} := t_{|\mathbf{X}}$ is a proper function and if we identify $\mathbb{K}^n$ with the graph of f, then the restriction $t_{|\mathbb{K}^n}$ is just f. Let us denote by $\mathbf{X}^\infty_{\mathbb{K}} := \mathbf{X}_{\mathbb{K}} \cap \{x_0 = 0\}$ the divisor at infinity. For simplicity, we shall often suppress the lowercase $\mathbb{K}$.

This proper extension of f has been used several times in the study of polynomial functions at infinity [Br-1], [Ph], [Pa-1], [ST], [Ti-2], etc.

2.1 Definition We say that f *is topologically trivial at infinity at* t_0 if there is a neighbourhood D of $t_0 \in \mathbb{K}$ and a compact set $K \subset \mathbb{K}^n$ such that the restriction $f_| : (X \setminus K) \cap f^{-1}(D) \to D$ is a topologically trivial fibration. We say that f is *locally trivial at* $y \in \mathbf{X}^\infty$ if there is a fundamental system of neighbourhoods U_i of y in $\mathbf{X}$ and, for each i, some small enough neighbourhood D_i of $\hat{f}(y)$, such that the restriction $f_| : U_i \cap (\mathbf{X} \setminus \mathbf{X}^\infty) \cap f^{-1}(D_i) \to D_i$ is a topologically trivial fibration.

2.2 Remark Local triviality at all points $y \in \mathbf{X}^\infty \cap \hat{f}^{-1}(t_0)$ *does not imply* topological triviality at infinity at t_0. To be able to glue together a finite number of locally trivial nonproper fibrations, one needs a global control over these, which is not available in general. See Example 2.13. This should be contrasted to the regularity conditions we define next.

2.3 Definition (**ρ-regularity**) Let $\rho : \mathbb{K}^n \setminus K \to \mathbb{R}_{\geq 0}$ be a proper C^1-submersion, where $K \subset \mathbb{K}^n$ is some compact set. We say that f is ρ-regular at $y \in \mathbf{X}^\infty$ if there is a neighbourhood U of y in $\mathbf{X} \setminus K$ such that f is transversal to ρ at all points of $U \cap \mathbb{K}^n$.

We say that the fibre $f^{-1}(t_0)$ is *ρ-regular at infinity* if f is ρ-regular at all points $y \in \mathbf{X}^\infty \cap \hat{f}^{-1}(t_0)$.

2.4 Remark The definition of ρ-regularity at infinity of a fibre $f^{-1}(t_0)$ *does not depend on any proper extension of* f, since it is equivalent to the following: for any sequence $(x_k)_{k\in\mathbb{N}} \subset \mathbb{K}^n$, $|x_k| \to \infty$, $f(x_k) \to t_0$, there exists some $k_0 = k_0((x_k)_{k\in\mathbb{N}})$ such that, if $k \geq k_0$ then f is transversal to ρ at x_k. It also follows from the definition that if $f^{-1}(t_0)$ is ρ-regular at infinity then this fibre has at most isolated singularities.

The transversality of the fibres of f to the levels of ρ is a "Milnor type" condition. In case ρ is the Euclidean norm, denoted in this paper by ρ_E, this condition has been used by John Milnor in the local study of singular functions [Mi, §4,5]. For complex polynomial functions, transversality to large spheres (i.e. ρ_E-regularity, in our definition) was used in [Br-2, pag. 229] and later in [NZ], where it is called *M-tameness*.

2.5 Example $\rho : \mathbb{K}^n \to \mathbb{R}_{\geq 0}$, $\rho(x) = (\sum_{i=1}^n |x_i|^{2p_i})^{1/2p}$, where $(w_1, \dots, w_n) \in \mathbb{N}^n$, $p = \operatorname{lcm}\{w_1, \dots, w_n\}$ and $w_i p_i = p$, $\forall i$. This function is "adapted" to polynomials which are quasihomogeneous of type $(w_1, \dots, w_n)$. By using it one can show that a value $c \in \mathbb{K}$ is atypical for such a polynomial if and only if c is a critical value of f (hence only the value 0 can be atypical). Namely, let $E_r := \{x \in \mathbb{K}^n \mid \rho(x) < r\}$ for some $r > 0$. Then the local Milnor fibre of f at $0 \in \mathbb{K}^n$ (i.e. $f^{-1}(c) \cap E_\varepsilon$, for some small enough ε and $0 < |c| \ll \varepsilon$) is diffeomorphic to the global fibre $f^{-1}(c)$, since $f^{-1}(c)$ is transversal to $\partial\overline{E_r}$, $\forall r \geq \varepsilon$.

The first pleasant property of ρ-regularity is that it implies topological triviality, more precisely we prove the following:

2.6 Proposition *If the fibre $f^{-1}(t_0)$ is ρ-regular at infinity then f is topologically trivial at infinity at t_0.*

Proof. If f is ρ-regular at $y \in \hat{f}^{-1}(t_0) \cap \mathbf{X}^\infty$ then one can lift the (real or complex) vector field $\partial/\partial t$ defined in a neighbourhood D of $t_0 \in \mathbb{K}$ to a (real or complex) vector field ξ on $U \cap \mathbb{K}^n$ tangent to the levels ρ =constant. If $f^{-1}(t_0)$ is ρ-regular at infinity then we may glue these local vector fields by a partition of unity and get a vector field defined in some neighbourhood of $\mathbf{X}^\infty$ without $\mathbf{X}^\infty$. This is a controlled vector field which can be integrated to yield a topologically trivial fibration $f_| : (\mathbb{K}^n \setminus K) \cap f^{-1}(D) \to D$, as shown by Verdier in his proof [V, Theorem 4.14] of the Thom-Mather isotopy theorem [Th], [Ma]. (Note however that the Thom-Mather theorem does not directly apply since f is not proper.) We use here ρ as "fonction tapissante", a notion introduced by Thom in [Th]. In the case $n = 2$, a similar procedure was used by Hà H.V. and Lê D.T. [HL]. □

To introduce the second regularity condition we need some preliminaries. First, let us define the relative conormal, following [Te], [HMS] then state some technical results which we need.

2.7 Let $\mathcal{X} \subset \mathbb{K}^N$ be a $\mathbb{K}$-analytic variety. In the real case, assume that $\mathcal{X}$ contains at least a regular point. Let $U \subset \mathbb{K}^N$ be an open set and let $g : \mathcal{X} \cap U \to \mathbb{K}$ be $\mathbb{K}$-analytic and nonconstant. The *relative conormal* $T^*_{g|\mathcal{X}\cap U}$ is a subspace of $T^*(\mathbb{K}^N)_{|\mathcal{X}\cap U}$ defined as follows:

$$T^*_{g|\mathcal{X}\cap U} := \text{closure}\{(y,\xi) \in T^*(\mathbb{K}^N) \mid y \in \mathcal{X}^0 \cap U, \xi(T_y(g^{-1}(g(y)))) = 0\} \subset T^*(\mathbb{K}^N)_{|\mathcal{X}\cap U},$$

where $\mathcal{X}^0 \subset \mathcal{X}$ is the open dense subset of regular points of $\mathcal{X}$ where g is a submersion. The relative conormal is conical (i.e. $(y,\xi) \in T^*_{g|\mathcal{X}\cap U} \Rightarrow (y,\lambda\xi) \in T^*_{g|\mathcal{X}\cap U}$, $\forall \lambda \in \mathbb{K}^*$). The canonical projection $T^*_{g|\mathcal{X}\cap U} \to \mathcal{X} \cap U$ will be denoted by π.

2.8 Lemma [Ti-2] *Let $(\mathcal{X}, x) \subset (\mathbb{K}^N, x)$ be a germ of an analytic space and let $g : (\mathcal{X}, x) \to (\mathbb{K}, 0)$ be a nonconstant analytic function. Let $\gamma : \mathcal{X} \to \mathbb{K}$ be analytic such that $\gamma(x) \neq 0$ and denote by W a neighbourhood of x in $\mathbb{K}^N$. Then $(T^*_{g|\mathcal{X}\cap W})_x = (T^*_{\gamma g|\mathcal{X}\cap W})_x$.* □

2.9 Definition Let $U_i = \{x_i \neq 0\}$, for $0 < i \leq n$, be an affine chart of $\mathbb{P}^n$ and let $y \in (U_i \times \mathbb{K}) \cap \mathbf{X} \cap \{x_0 = 0\}$. The relative conormal $T^*_{x_0|\mathbf{X}\cap U_i \times \mathbb{K}}$ is well defined. We denote by $(\mathcal{C}^\infty_{\mathbb{K}})_y$ the fibre $\pi^{-1}(y)$ and call it the *space of characteristic covectors at infinity, at y*. It follows from Lemma 2.8 that $(\mathcal{C}^\infty)_y$ does not depend on the choice of affine chart U_i.

2.10 Definition (*t*-**regularity**) We say that f (or that the fibre $f^{-1}(t_0)$) is *t-regular* at $y \in \mathbf{X}^\infty \cap \hat{f}^{-1}(t_0)$ if $(y, \mathrm{d}t) \notin (\mathcal{C}^\infty_{\mathbb{K}})_y$.

We also say that $f^{-1}(t_0)$ is *t-regular at infinity* if this fibre is t-regular at all its points at infinity.

The definition of t-regularity was first formulated (somewhat differently) by D. Siersma and the author [ST], for a complex polynomial function $f : \mathbb{C}^n \to \mathbb{C}$.

2.11 Proposition *If f is t-regular at $y \in \mathbf{X}^\infty$ then f is ρ_E-regular at y, where ρ_E is the Euclidean norm.*

Proof. The mapping $d^\infty : \mathbf{X}_\mathbb{K} \to \mathbb{R}$, defined by:

$$\begin{cases} d^\infty(x, f(x)) = 1/\rho_E^2(x), & \text{for } x \in \mathbb{K}^n \\ d^\infty(y) = 0, & \text{for } y \in \mathbf{X}^\infty \end{cases}$$

is analytic and defines $\mathbf{X}^\infty$. In the real case we have $(T^*_{d^\infty|\mathbf{X}})_y = (T^*_{|x_0|^2|\mathbf{X}})_y$, by Lemma 2.8, and the latter is in turn equal to $(T^*_{x_0|\mathbf{X}})_y = (\mathcal{C}^\infty_\mathbb{R})_y$. The ρ_E-regularity at $y \in \mathbf{X}^\infty$ is certainly implied by $(y, \mathrm{d}t) \notin (T^*_{d^\infty|\mathbf{X}})_y$, which is just t-regularity, by the above equalities. This finishes the proof in the real case.

Now the complex case. Let us first introduce the map $\iota : \mathbb{P}T^*(\mathbb{R}^{2n}) \to \mathbb{P}T^*(\mathbb{C}^n)$ between the real and the complex projectivised conormal bundles (where $\mathbb{R}^{2n}$ is the real underlying space of $\mathbb{C}^n$) defined as follows: if ξ is conormal to a hyperplane $H \subset \mathbb{R}^{2n}$ then $\iota([\xi])$ is the conormal to the unique complex hyperplane included in H. This is clearly a continuous map. We then have the following equality:

$$\mathbb{P}(\mathcal{C}^\infty_\mathbb{C})_y = \iota(\mathbb{P}(T^*_{|x_0|^2|\mathbf{X}})_y),$$

since the complex tangent space $T_x\{x_0 = \text{constant}\}$ is exactly the unique complex hyperplane contained in the real tangent space $T_x\{|x_0|^2 = \text{constant}\}$. Equality follows from the fact that ι commutes with taking limits. Now $(y, \mathrm{d}t) \notin (\mathcal{C}^\infty_\mathbb{C})_y$ implies $(y, \iota^{-1}([\mathrm{d}t]) \notin \mathbb{P}(T^*_{|x_0|^2|\mathbf{X}})_y$, which in turn implies ρ_E-regularity at y since, as above in the real case, we still have $(T^*_{d^\infty|\mathbf{X}})_y = (T^*_{|x_0|^2|\mathbf{X}})_y$. □

2.12 Corollary *Let $f : \mathbb{C}^n \to \mathbb{C}$ be a complex polynomial. Then the set of values t_0 such that $f^{-1}(t_0)$ is not ρ_E-regular at infinity is finite. (In particular, the set of atypical values of f is finite.)*

Proof. Take a Whitney stratification $\mathcal{W} = \{\mathcal{W}_i\}_i$ of $\mathbf{X}_\mathbb{C}$ with a finite number of strata and with $\mathbb{C}^n$ as a stratum. It turns out from [ST, Lemma 4.2] or [Ti-2, Theorem 2.9] that any pair of strata $(\mathbb{C}^n, \mathcal{W}_i)$ with $\mathcal{W}_i \in \mathbf{X}^\infty$ has the Thom property with respect to the function x_0, in any local chart. If $\hat{f}^{-1}(t_0)$ is transversal to a stratum $\mathcal{W}_i \subset \mathbf{X}^\infty$, then $f^{-1}(t_0)$ is t-regular at infinity. Now the restriction of the projection $\hat{f} : \mathbf{X}_\mathbb{C} \to \mathbb{C}$ to a stratum contained in $\mathbf{X}^\infty$ has a finite number of critical values. This implies that the values t_0 such that $f^{-1}(t_0)$ is not t-regular at infinity are finitely many. The conclusion follows by Proposition 2.11. □

Nevertheless, the ρ_E-regularity is really weaker than t-regularity. We show by the next example that the converse of Proposition 2.11 is not true. This makes ρ-regularity interesting and raises questions concerning real methods and their interplay with the complex ones.

2.13 Example Let f be the following complex polynomial function $f := x+x^2y : \mathbb{C}^3 \to \mathbb{C}$ in 3 variables x, y, z. We show below that f is not t-regular at a whole line $L := \{x_0 = x = t = 0\}$ within $\mathbf{X}^\infty_\mathbb{C}$, hence has a *nonisolated* t-singularity. We can also prove (see [Ti-2]) that, at any point $p \in L$, f is locally trivial at p, according to Definition 2.1, by using [ST, Proposition 5.4] and the fact that the affine polar locus $\Gamma(l, f)$ is empty for a general linear $l : \mathbb{C}^3 \to \mathbb{C}$ (see Definition 3.2 and Proposition 3.5).

On the other hand, there is a single point in $\mathbf{X}^\infty_\mathbb{C}$ at which f is not ρ_E-regular. Therefore the singularity is *isolated* in this sense. Here are the computations.

Let $F := xx_0^2+x^2y-tx_0^3$. The t-regularity in the chart $y \neq 0$ is equivalent to $|y| \cdot \|\frac{\partial f}{\partial x}\| \not\to 0$, as $\|x,y\| \to \infty$, cf. [ST, p. 780]. But in our example $|y| \cdot \|1+2xy\|$ tends to 0, for instance if $y \to \infty$, $x = 1/y^3 - 1/(2y)$, $z = ay$. The limit points in $\mathbf{X}^\infty$ form the 1-dimensional set L.

We now find the set of points $(p,t) \in \mathbf{X}^\infty$ where f is not ρ_E-regular. This amounts to finding the solutions of the equation $\overline{\operatorname{grad} f} = (\lambda x, \lambda y, \lambda z)$. One can assume $\lambda \neq 0$. It follows $\|x\|^2\bar{x} = y + 2\bar{x}\|y\|^2$, $z = 0$. The solution is a 2-dimensional real algebraic set A and the set we are looking for is $\bar{A} \cap \mathbf{X}^\infty$. If we work in two variables instead, i.e. with $g := x + x^2y : \mathbb{C}^2 \to \mathbb{C}$, then the single point where g is not ρ_E-regular is $x_0 = x = t = 0$. In our case (3 variables), one intersects with $z = 0$, hence $\bar{A} \cap \mathbf{X}^\infty$ consists of a single point.

2.14 The relation with the Malgrange condition. In the complex case, F. Pham formulated a regularity condition which had been found by B. Malgrange [Ph, 2.1]. We give below a definition which also works in the real case, together with a localised version at infinity.

Definition We consider sequences of points $x_i \in \mathbb{K}^n$ and the following properties:

(L_1) $\|x_i\| \to \infty$ and $f(x_i) \to t_0$, as $i \to \infty$.

(L_2) $x_i \to y \in \mathbf{X}^\infty_\mathbb{K}$, as $i \to \infty$.

One says that the fibre $f^{-1}(t_0)$ satisfies *Malgrange condition* if there is $\delta > 0$ such that, for any sequence of points with property (L_1), one has

(M) $\|x_i\| \cdot \|\operatorname{grad} f(x_i)\| > \delta$.

We say that f satisfies *Malgrange condition at* $y \in \mathbf{X}^\infty_\mathbb{K}$ if there is $\delta > 0$ such that one has (M), for any sequence of points with property (L_2).

2.15 Note It clearly follows from the definition that $f^{-1}(t_0)$ satisfies the Malgrange condition if and only if f satisfies the Malgrange condition at any $y = (z, t_0) \in \mathbf{X}^{\infty}_{\mathbb{K}}$. We have proved in [ST, Proposition 5.5] that t-regularity implies the Malgrange condition, both at a point or on a fibre (the proof works in the complex case as well as the real one). Conversely the Malgrange condition implies t-regularity, as has more recently been proven in the complex case by Parusiński [Pa-2, Theorem 1.3]. In fact the same proof works over the reals. Briefly, we have the following relations:

(1) Malgrange condition $\Longleftrightarrow$ t-regularity $\Longrightarrow$ ρ_E-regularity.

The Malgrange condition at $y \in \mathbf{X}^{\infty}_{\mathbb{K}}$ is equivalent to saying that the *Lojasiewicz number $L_y(f)$ at y* is ≥ -1, where $L_y(f)$ is defined as the smallest exponent $\theta \in \mathbb{R}$ such that, for some neighbourhood U of y and some constant $C > 0$ one has:

$$|\operatorname{grad} f(x)| \geq C|x|^{\theta}, \quad \forall x \in U \cap \mathbb{C}^n.$$

There are two other regularity conditions used in the literature which are similar to Malgrange condition but clearly stronger: Fedoryuk's condition (or *tameness*, see Proposition 4.1) [Fe], [Br-1], [Br-2] and Parusiński's condition [Pa-1].

3 Polar curves and regularity conditions

We define the affine polar curves attached to f and show how they are related to the t-regularity condition.

Given a polynomial $f : \mathbb{K}^n \to \mathbb{K}$ and a linear function $l : \mathbb{K}^n \to \mathbb{K}$, one denotes by $\Gamma(l, f)$ the closure in $\mathbb{K}^N$ of the set $\operatorname{Crt}(l, f) \setminus \operatorname{Crt} f$, where $\operatorname{Crt}(l, f)$ is the critical locus of the map $(l, f) : \mathbb{K}^N \to \mathbb{K}^2$. A basic and useful result is that $\Gamma(l, f)$ is a curve (or void) if l is general enough. Below we give the precise statement (a particular case of [Ti-2, Lemma 1.4]). For some hyperplane $H \in \check{\mathbb{P}}^{n-1}$, one denotes by $l_H : \mathbb{K}^n \to \mathbb{K}$ the unique linear form (up to multiplication by a constant) which defines H.

3.1 Polar Curve Theorem [Ti-2] *There exists an open dense set $\Omega_f \subset \check{\mathbb{P}}^{n-1}$ (Zariski-open in the complex case) such that, for any $H \in \Omega_f$, the critical set $\Gamma(l_H, f)$ is a curve or it is void.* □

3.2 Definition For $H \in \Omega_f$, we call $\Gamma(l_H, f)$ the *affine polar curve* of f with respect to l_H. A system of coordinates $(x_1, \ldots, x_n)$ in $\mathbb{K}^n$ is called *generic* with respect to f iff $\{x_i = 0\} \in \Omega_f$, $\forall i$.

It follows from 3.1 that such systems are generic among the systems of coordinates.

We first relate the affine polar curves to the local polar curves at infinity, as follows. Let us consider the map germ:

$$(t, x_0) : (\mathbf{X}, p) \to (\mathbb{K}^2, 0),$$

for some $p \in \mathbf{X}^\infty$, with critical locus denoted by $\mathrm{Crt}_p(t, x_0)$. The local (nongeneric) *polar locus* $\Gamma_p(t, x_0)$ is defined as the closure of $\mathrm{Crt}_p(t, x_0) \setminus \mathbf{X}^\infty$ in $\mathbf{X}$.

Let $p = ([0 : p_1 : \cdots : p_n], \alpha)$, where $p_i \neq 0$, for some fixed i. In the chart $U_i \times \mathbb{K}$, the polar locus $\Gamma_p(t, x_0)$ is the germ at p of the analytic set $\overline{\mathcal{G}_i} \subset \mathbf{X}$, where

$$\mathcal{G}_i := \left\{ ([x_0 : x], t) \in \mathbf{X} \setminus \mathbf{X}^\infty \mid \frac{\partial f^{(i)}}{\partial x_1} = \cdots = \frac{\partial \hat{f}^{(i)}}{\partial x_i} = \cdots = \frac{\partial f^{(i)}}{\partial x_n} = 0 \right\},$$

and $f^{(i)} := \tilde{f}(x_0, x_1, \ldots, x_{i-1}, 1, x_{i+1}, \ldots, x_n)$.

On the intersection of charts $(U_0 \cap U_i) \times \mathbb{K}$, the function $\frac{\partial f^{(i)}}{\partial x_j}$ differs from $\frac{\partial f^{(0)}}{\partial x_j}$, by a nowhere zero factor, for $j \neq 0, i$.

Thus the germ of $\overline{\mathcal{G}_i}$ at p is the germ at p of the following algebraic subset of $\mathbb{P}^n \times \mathbb{K}$:

$$\text{closure} \left\{ (x, f(x)) \in \mathbb{K}^n \times \mathbb{K} \mid \frac{\partial f}{\partial x_1} = \cdots = \frac{\partial f}{\partial x_{i-1}} = \frac{\partial f}{\partial x_{i+1}} \cdots = \frac{\partial f}{\partial x_n} = 0 \right\},$$

which is equal to $\overline{\Gamma}(x_i, f)$, where:

$$\overline{\Gamma}(l, f) := \text{closure}\{(x, t) \in \mathbb{K}^n \times \mathbb{K} \mid x \in \Gamma(l, f),\ \ t = f(x)\} \subset \mathbb{P}^n \times \mathbb{K}.$$

One can now easily prove the following finiteness result:

3.3 Lemma *If the system of coordinates $(x_1, \ldots, x_n)$ is generic with respect to f, then there is a finite number of points $p \in \mathbf{X}^\infty$ for which the polar locus $\Gamma_p(t, x_0)$ is nonvoid and at such a point $\Gamma_p(t, x_0)$ is a curve.*

Proof. Fix a generic system of coordinates. By Theorem 3.1, the set $\Gamma(x_i, f)$ is a curve (or void) and therefore $\overline{\Gamma}(x_i, f) \subset \mathbf{X}$ is a curve too (or empty), $\forall i \in \{1, \ldots, n\}$. We have shown above that, for a point $p \in \mathbf{X}^\infty$ with $p_i \neq 0$, one has equality of germs $\Gamma_p(t, x_0) = \overline{\Gamma}(x_i, f)_p$. Then the assertion follows from the fact that the intersection $\mathbf{X}^\infty \cap (\cup_{i=1}^n \overline{\Gamma}(x_i, f)$ is a finite set. □

In the complex case, the affine polar curves are closely related to the *isolated t-singularities*, defined as follows.

3.4 Definition We say that f has isolated t-singularities at the fibre $f^{-1}(t_0)$ if this fibre has isolated singularities and the set $\{p \in \mathbf{X}^\infty \mid f^{-1}(t_0)$ is not t-regular at $p\}$ is a finite set.

3.5 Proposition *Let $f : \mathbb{K}^n \to \mathbb{K}$ be a polynomial function and let $p \in \mathbf{X}^\infty \cap \hat{f}^{-1}(t_0)$. Then the following are equivalent:*

(a) $\Gamma_p(t, x_0) \neq \emptyset$.

(b) $\exists H \in \check{\mathbb{P}}^{n-1}$ *such that* $\overline{\Gamma}(l_H, f) \ni p$.

(c) $\overline{\Gamma}(l_H, f) \ni p$, $\forall H \in \check{\mathbb{P}}^{n-1}$ *with* $H \not\ni y$, *where* $p = (y, t_0)$.

If $\mathbb{K} = \mathbb{C}$ and f has isolated t-singularities at $f^{-1}(t_0)$ then these conditions are moreover equivalent to the following ones:

(d) *p is a t-singularity.*

(e) *$\lambda_p \neq 0$, where λ_p is the* Milnor number at infinity *at p, as defined in* [ST, Definition 3.4].

Proof. The equivalence (a) $\iff$ (b) follows from the equality $\Gamma_p(t, x_0) = \overline{\Gamma}(x_i, f)_p$ already established, where we take an i such that $p_i \neq 0$. Let us show (a) $\iff$ (c). For a given $H \in \check{\mathbb{P}}^{n-1}$ with $y \notin H$, we consider a linear change of variables $(x'_1, \ldots, x'_n)$ such that $l_H = x'_1$. In the new coordinates, f, t, p become respectively f', t', p', and x_0 is invariant. Then $\Gamma_{p'}(t', x_0) = \overline{\Gamma}(x'_i, f')_{p'}$ as above and we have $p' \in \overline{\Gamma}(x'_1, f') \iff p \in \overline{\Gamma}(l_H, f)$.

The equivalence (a) $\iff$ (d) is Proposition 5.3 in [ST] and (d) $\iff$ (e) is a consequence of [ST, Proposition 4.5] and [Ti-2], see Remarks 3.6 below. □

In the case of a nonisolated t-singularity at $p \in \mathbf{X}^\infty$, the general affine polar curve might be empty at p. See Example 2.13 for a discussion of such a case.

3.6 Remarks on the Milnor number at infinity. In case of a complex polynomial with isolated t-singularities, we prove in [ST, Corollary 3.5 (b)] the following formula, which shows in particular that the fibre $F_a := f^{-1}(a)$ is atypical if and only if its Euler characteristic is different from the one of a general fibre $F_u := f^{-1}(u)$:

$$\chi(F_a) - \chi(F_u) = (-1)^n(\mu_{F_a} + \lambda_{F_a}), \tag{2}$$

where μ_{F_a} is the sum of the Milnor numbers at the isolated singularities on the fibre F_a and λ_{F_a} is the sum of the so-called Milnor numbers at infinity at the t-singularities on $\mathbf{X}^\infty \cap t^{-1}(a)$.

In the more general case of a polynomial with isolated singularities in the affine space, the t-singularities at infinity may be non-isolated and one does not have numbers λ_p anymore. However, one can still give a meaning to the number λ_{F_a} by taking the relation (2) as its definition:

3.7 Definition Let $f : \mathbb{C}^n \to \mathbb{C}$ be a polynomial with isolated singularities. We call the following number the *Euler-Milnor number at infinity of the fibre F_a*:

$$\lambda_{F_a} := (-1)^{n-1}(\chi(F_u) - \chi(F_a)) - \mu_{F_a}.$$

If the t-singularities at infinity are not isolated then we have $\lambda_{F_a} = 0$ if F_a is a typical fibre, whereas the converse is not true in general. However, one has the following interesting statement, which gives an extension, at the Euler characteristic level, of [ST, Corollary 3.5(a)]:

3.8 Proposition *If $f : \mathbb{C}^n \to \mathbb{C}$ is a polynomial with isolated singularities then*

$$\chi(F_u) = 1 + (-1)^{n-1} \sum_{a \in \Lambda} (\lambda_{F_a} + \mu_{F_a}),$$

where Λ is the set of atypical values of f.

Proof. Let δ_a be a small disc centered at $a \in \mathbb{C}$, which does not contain any other atypical value. Let γ_a, $a \in \Lambda$, be suitable paths (non self-intersecting, etc.) from a to a typical value u. Then $1 = \chi(\mathbb{C}^n) = \chi(\bigcup_{a\in\Lambda}(f^{-1}(\delta_a) \cup f^{-1}(\gamma_a)) = \chi(F_u) + \sum_{a\in\Lambda}[\chi(F_a) - \chi(F_u)]$, by retraction and by a standard Mayer-Vietoris argument. □

3.9 Remark The general fibre of a complex polynomial with isolated t-singularities at all its fibres is homotopy equivalent to a bouquet of spheres of real dimension $n-1$, by [ST], [Pa-2], [Ti-2]. This property is also shared by all complex polynomials of two variables with irreducible generic fibre and by complex polynomials with isolated singularities which are ρ-regular at all points $y \in \mathbf{X}^\infty$. The latter fact follows from the definition of ρ-regularity and its proof is left to the reader.

4 The case $n = 2$

Let us first remark that any polynomial function $f : \mathbb{C}^2 \to \mathbb{C}$ has isolated t-singularities at its reduced fibres. Not only that Proposition 3.5 applies in this case, but one has further relations. See [ST], [Pa-1], [Ti-2] for other results.

4.1 Proposition *For $t_0 \in \mathbb{C}$, the following are equivalent:*

(a) *f is topologically trivial at infinity at t_0.*

(b) *$f^{-1}(t_0)$ is ρ_E-regular at infinity.*

(c) *$f^{-1}(t_0)$ is t-regular at infinity.*

(d) *f is tame at t_0 (i.e. there is no sequence of points $x_i \in \mathbb{C}^2$ with property (L_1) and such that $|\operatorname{grad} f(x_i)| \to 0$ as $i \to \infty$.)*

Proof. The proof of (a) $\Rightarrow$ (c) runs as follows. By contradiction suppose that $f^{-1}(t_0)$ is not t-regular at infinity. Its t-singularities are isolated, as remarked above, and we may apply Proposition 3.5 together with [ST, Cor. 3.5] to deduce that there is a nonzero Milnor number at infinity $\lambda_{f^{-1}(t_0)}$ which gives a jump "at infinity" in the Euler characteristic of the fibres at $f^{-1}(t_0)$. Therefore f is not topologically trivial at infinity at t_0 (cf. Definition 2.1). We get a contradiction. Now (c) $\Rightarrow$ (b) is Proposition 2.11 and (b) $\Rightarrow$ (a) is Proposition 2.6 (both valid for $n \geq 2$).

The equivalence (a) $\Leftrightarrow$ (d) was proved by Hà H.V. [Hà] , see also Durfee's preprint [Du]. Notice however that (d) $\Rightarrow$ (a) is obvious. □

4.2 Note The above result shows that, given a polynomial function $f : \mathbb{C}^2 \to \mathbb{C}$ and some point $p \in \mathbf{X}^\infty$, if the Łojasiewicz number $L_p(f)$ is ≥ -1 then $L_p(f) \geq 0$. This is no longer true for $n \geq 3$ (see Example 2.13). Moreover, in the following example $f(x,y,z) = x + x^2y + x^4yz$ taken from [Pa-2], f is not tame at t_0, for any $t_0 \in \mathbb{C}$, whereas all except a finite number of fibres are ρ_E-regular at infinity (cf. Corollary 2.12).

4.3 Remark If we consider "tameness" in the sense of Broughton [Br-2] (i.e. $\exists K > 0$ such that, for any sequence $|x_i| \to \infty$, one has $|\operatorname{grad} f(x_i)| \geq K$) then "$f$ is tame" obviously implies "f is tame at t, for any $t \in \mathbb{C}$", whereas the converse is not true, see [NZ] for examples.

4.4 Remarks on polynomials with at most one atypical value.
If a polynomial function $f : \mathbb{K}^2 \to \mathbb{K}$ has no critical points and all its fibres are t-regular at infinity then there are neither critical nor atypical values. In the complex case, it follows from the well known Abhyankar-Moh Lemma [AM] that f is equivalent (modulo an automorphism of $\mathbb{C}^2$) to a linear form.

Assume now that f is a complex polynomial and has a single atypical value, say $0 \in \mathbb{C}$. In what follows we set $F_0 := f^{-1}(0)$. Take a small disc $D_0 \subset \mathbb{C}$ centered at 0. Then the restriction $f_| : \mathbb{C}^2 \setminus f^{-1}(D_0) \to \mathbb{C} \setminus D_0$ is a locally trivial fibration, hence $\mathbb{C}^2$ retracts to $f^{-1}(D_0)$. In turn, $f^{-1}(D_0)$ retracts to a tubular neighbourhood of F_0. Therefore we get $\chi(F_0) = \chi(\mathbb{C}^2) = 1$.

Proposition *Let $f : \mathbb{C}^2 \to \mathbb{C}$ be a polynomial function with at most one atypical value, say $0 \in \mathbb{C}$, and such that F_0 is reduced.*

(a) *If F_0 is connected and smooth then f is equivalent (modulo* $\operatorname{Aut} \mathbb{C}^2$*) to a linear form.*

(b) [ZL] *If F_0 is connected and irreducible then f is equivalent (modulo* $\operatorname{Aut} \mathbb{C}^2$*) to $x^p + y^q$, for some relatively prime integers $p, q \geq 1$.*

(c) *If F_0 is smooth but not connected then f is equivalent (modulo* $\operatorname{Aut} \mathbb{C}^2$*) to $x(1 + xh(x,y))$, for some $h \in \mathbb{C}[x,y]$.*

Proof. (b) is the Zaïdenberg-Lin result [ZL, Theorem A]. We refer the reader to [ZL, Theorem B] for the case when F_0 is only assumed to be connected. If F_0 is connected and smooth, then one gets (a) by applying the same theorem, or by using the fact that F_0 is isomorphic to $\mathbb{C}$ and Abhyankar-Moh lemma. One argument for $F_0 \simeq \mathbb{C}$ is the following. There is a compactification of F_0 to a Riemann surface M which is also smooth and connected. Then $1 = \chi(F_0) = \chi(M) - \chi(M \setminus F_0) = 2 - 2g - s$, where g is the genus of M and s is the number of points in $M \setminus F_0$. This implies $g = 0$, $s = 1$ and therefore $M \simeq \mathbb{P}^1$ and $F_0 \simeq \mathbb{C}$. This finishes (a).

To prove (c), we first note that there is at least a connected component F_{0i} of F_0 such that $\chi(F_0) = 1$ (since $1 = \chi(F_0) = \sum_{i \in I} \chi(F_{0i})$, but $\chi(F_{0i}) \leq 1$, $\forall i \in I$). Since F_{0i} is smooth, we may apply (a) to it and obtain that f is equivalent (modulo $\operatorname{Aut} \mathbb{C}^2$) to $x \cdot v(x,y)$ and further that the polynomial v is of the form $\alpha + xh(x,y)$, with α a constant $\neq 0$ and some $h \in \mathbb{C}[x,y]$. The assertion (c) was also proved in [Ass] in an algebraic way. □

The study of singularities at infinity might be useful for proving (or disproving) the well known Jacobian Conjecture, which has the following equivalent formulation, as we have shown in [ST, pag. 781].

Conjecture Let $f : \mathbb{C}^2 \to \mathbb{C}$ be a polynomial function with smooth fibres, which has singularities at infinity (i.e. there is at least a fibre of f which is not t-regular at infinity). Then for any $g \in \mathbb{C}[x,y]$, the Jacobian $\operatorname{Jac}(f,g)$ cannot be a nonzero constant.

4.5 Remark One can prove that a polynomial f with at most one atypical value, except in case (a) of Proposition 4.4, satisfies the above Conjecture; the nontrivial case is (c), but, as communicated to us by M. Oka and A. Assi, this follows by elementary results proved in [Oka] using the Newton polygon.

5 Families of complex polynomials with singularities at infinity

Let $f_s : \mathbb{C}^n \to \mathbb{C}$, $s \in [-\varepsilon, \varepsilon]$, $\varepsilon > 0$ be a family of polynomial functions. If s is fixed, then there is a (large enough) open disc $D_s \subset \mathbb{C}$ such that all the atypical values of f_s are inside D_s. For such a disc, the following locally trivial fibration:

$$f_{s|} : f_s^{-1}(\partial D_s) \to \partial D_s \tag{3}$$

is called the *global monodromy fibration* of f_s.

5.1 Lemma *For any polynomial function $f_s : \mathbb{C}^n \to \mathbb{C}$ there exists $R_s \in \mathbb{R}_+$ such that the fibration* (3) *is diffeomorphic to the following one:*

$$f_{s|} : f_s^{-1}(\partial D_s) \cap B_R \to \partial D_s, \tag{4}$$

for all $R \geq R_s$, where $B_R \subset \mathbb{C}^n$ is the ball centered at 0, of radius R.

Proof. This follows from Corollary 2.12. □

We consider the following rather large class of polynomial functions $f : \mathbb{C}^n \to \mathbb{C}$ (see Remark 3.9 for examples):

(V) The typical fibre of f is homotopy equivalent to a bouquet of spheres $\bigvee_\gamma S^{n-1}$.

For some polynomial function f_s which satisfies condition (V), we denote by γ_s the number of spheres in the bouquet.

5.2 Theorem *Let f_s be a family of polynomial functions with the following properties for some $\varepsilon > 0$:*

(a) *for any $s \in [-\varepsilon, \varepsilon]$, f_s verifies property* (V).

(b) *γ_s is constant, for $s \in [-\varepsilon, \varepsilon]$.*

(c) *the atypical values of f_s are inside some fixed disc D, $\forall s \in [-\varepsilon, \varepsilon]$.*

Then the global monodromy fibrations in the family are fibre homotopy equivalent. If $n \neq 3$ then the monodromy fibrations are actually diffeomorphic to each other.

Proof. One applies the original method of proof of Lê D.T. and C.P. Ramanujam [LR] to the monodromy fibrations (4); it actually works with minor modifications under the assumed conditions (a), (b), (c). We leave to the reader the task of step-by-step checking, following the lines of the (local case) proof in [LR]. Finally use Lemma 5.1 to conclude. □

5.3 Example Let $\{f_s\}_{s \in [-\varepsilon,\varepsilon]}$ be a family of polynomials such that f_s is ρ_s-regular at infinity, for some ρ_s as in Definition 2.3. Then f_s satisfies property (a), by Remark 3.9. If property (b) holds, then property (c) is also satisfied, hence the conclusion of Theorem 5.2 would be true for this family. To see this, one has to prove that the set of atypical values of f_s (which are just the critical values in this case, by Proposition 2.6) is bounded as $s \to 0$. But this is a consequence of the constancy of γ_s, which, under the assumed conditions, is the sum of the local Milnor numbers at all critical points of f_s (= total Milnor number).

These observations apply to families of ρ_E-regular polynomial functions with constant total Milnor number. It is a particular case of Theorem 5.2, previously proved in [HZ]; notice that there are no singularities at infinity in this situation.

5.4 Example Consider the following families of polynomial functions in two variables, with $\mu + \lambda =$ constant and $\lambda \neq 0$ (from the classification lists of

Siersma and Smeltink [SS]):

$$\begin{aligned}
\mathbf{A^2BC}_{5,\lambda}: \quad & x^4 - x^2y^2 + Qxy + Kx^3 + Px^2 + A, \\
& Q \neq 0; \quad \lambda = 1,2,3,4; \quad \mu + \lambda = 5. \\
\mathbf{A^2B^2}_{4,\lambda}: \quad & x^2y^2 + y^3 + Qxy + Ry^2 + By, \\
& Q \neq 0; \quad \lambda = 1,2,3; \quad \mu + \lambda = 4. \\
\mathbf{A^2B^{2+}}_{3,\lambda}: \quad & x^2y^2 + Qxy + Ry^2 + By, \\
& QR \neq 0; \quad \lambda = 1,2; \quad \mu + \lambda = 3.
\end{aligned}$$

The parameters are A, B, K, P, Q, R and they depend on the variable s. Since $n = 2$ and since all listed polynomials have isolated critical points, the conditions (a) and (b) hold within each family. Namely these polynomials have isolated $\mathcal{W}$-singularities at infinity (in the sense of [ST]) and $\gamma = \mu + \lambda$, by [ST, Corollary 3.5]. In all the cases, there is a singularity at infinity which occurs at the point $p = ([0:1:0], \frac{Q^2}{4})$ in the first family, resp. at the point $p = ([0:1:0], -\frac{Q^2}{4})$ in the two last families, with Milnor number at infinity denoted by λ (cf. [ST, Definition 3.5]). There are several critical points in affine space, with total Milnor number denoted by μ. In each case, an atypical value is either a critical value, or the value $Q^2/4$ (resp. $-Q^2/4$). It follows that condition (c) is also satisfied for any of the listed families and therefore Theorem 5.2 applies.

References

[AM] S.S. Abhyankar, T.T. Moh, *Embeddings of the line in the plane*, J. Reine Angew. Math. **276** (1975), 148–166.

[Ass] A. Assi, *Familles de corbes planes ayant une seule valeur irrégulière*, C.R. Acad. Sci. Paris t. **322**, I (1996), 1203–1207.

[Br-1] S.A. Broughton, *On the topology of polynomial hypersurfaces*, Proceedings A.M.S. Symp. in Pure. Math., vol. 40, I (1983), 165–178.

[Br-2] S.A. Broughton, *Milnor number and the topology of polynomial hypersurfaces*, Inventiones Math. **92** (1988), 217–241.

[Du] A. Durfee, *Critical points at infinity of polynomials in two variables*, manuscript.

[D&all] A. Durfee, N. Kronenfeld, H. Munson, J. Roy, J. Westby, *Counting critical points of real polynomials in two variables*, Amer. Math. Monthly **100**, 3 (1993), 255–271.

[Fe] M.V. Fedoryuk, *The asymptotics of the Fourier transform of the exponential function of a polynomial*, Docl. Acad. Nauk **227** (1976), 580–583; Soviet Math. Dokl. (2) **17** (1976), 486–490.

[Fo-1] L. Fourrier, *Classification topologique à l'infini des polynômes de deux variables complexes*, CRAS Paris, t. **318**, I (1994), 461–466.

[Fo-2] L. Fourrier, *Topologie d'un polynôme de deux variables complexes au voisinage de l'infini*, Ann. Inst. Fourier, Grenoble, **46**, 3 (1996), 645–687.

[Hà] Hà H.V., *Nombre de Łojasiewicz et singularités à l'infini des polynômes de deux variables complexes*, CRAS t. **311**, I (1990), 429–432.

[HL] Hà H.V., Lê D.T., *Sur la topologie des polynômes complexes*, Acta Math. Vietnamica, **9** (1984), pp. 21–32.

[HZ] Hà H.V., A. Zaharia, *Families of polynomials with total Milnor number constant*, Math. Ann., **304** (1996), pp. 481–488.

[HMS] J.P. Henry, M. Merle, C. Sabbah, *Sur la condition de Thom stricte pour un morphisme analytique complexe*, Ann. Scient. Ec. Norm. Sup. **4**^{e} série, t. 17 (1984), 227–268.

[LR] Lê D.T., C.P. Ramanujam, *The Invariance of Milnor number implies the invariance of the topological type*, American J. of Math. **98**, 1 (1976), 67–78.

[Ma] J. Mather, *Notes on topological stability*, Harvard University (1979).

[Mi] J. Milnor, *Singular points of complex hypersurfaces*, Ann. of Math. Studies **61**, Princeton 1968.

[NZ] A. Némethi, A. Zaharia, *On the bifurcation set of a polynomial and Newton boundary*, Publ. RIMS **26** (1990), 681–689.

[Oka] M. Oka, *On the bifurcation of the multiplicity and topology of the Newton boundary*, J. Math. Soc. Japan, **31** (1979), 435–450.

[Pa-1] A. Parusiński, *On the bifurcation set of a complex polynomial with isolated singularities at infinity*, Compositio Math. **97** (1995), 369–384.

[Pa-2] A. Parusiński, *A note on singularities at infinity of complex polynomials*, in: "Simplectic singularities and geometry of gauge fields", Banach Center Publ. vol. **39** (1997), 131–141.

[Ph] F. Pham, *Vanishing homologies and the n variable saddlepoint method*, Arcata Proc. of Symp. in Pure Math., vol. **40**, II (1983), 319–333.

[SS] D. Siersma, J. Smeltink, *Classification of singularities at infinity of polynomials of degree 4 in two variables*, preprint **945** (1996), University of Utrecht.

[ST] D. Siersma, M. Tibăr, *Singularities at infinity and their vanishing cycles*, Duke Math. Journal **80**:3 (1995), 771–783.

[Te] B. Teissier, *Varietés polaires 2: Multiplicités polaires, sections planes et conditions de Whitney*, Géométrie Algèbrique à la Rabida, Springer L.N.M. **961**, pp. 314–491.

[Th] R. Thom, *Ensembles et morphismes stratifiés*, Bull. Amer. Math. Soc. **75** (1969), 249–312.

[Ti-1] M. Tibăr, *Bouquet decomposition of the Milnor fibre*, Topology **35**, no.1 (1996), 227–242.

[Ti-2] M. Tibăr, *Topology at infinity of polynomial maps and Thom regularity condition*, Compositio Math., to appear.

[V] J.-L. Verdier, *Stratifications de Whitney et théorème de Bertini-Sard*, Inventiones Math. **36** (1976), 295–312.

[ZL] M.G. Zaïdenberg, V. Ya. Lin, *An irreducible simply connected algebraic curve in $\mathbb{C}^2$ is equivalent to a quasihomogeneous curve*, Soviet Math. Dokl., **28**, no. 1 (1983), 200–204.

U.F.R. de Mathématiques
Université de Lille 1
59655 Villeneuve d'Ascq cedex
France

e-mail: tibar@gat.univ-lille1.fr

and: Institute of Mathematics of the Romanian Academy.

A Bennequin Number Estimate for Transverse Knots

V.V. Goryunov J.W. Hill

Abstract

We show that the Bennequin number of a transverse knot in the standard contact 3-space or solid torus is bounded by the negative of the lowest degree of the framing variable in its HOMFLY polynomial. For $\mathbf{R}^3$, this fact was established earlier by Fuchs and Tabachnikov [7] by comparison of the results of [1] and [5, 11]. We develop a different, direct approach based on the lowering of the polynomial to transversally framed regular planar curves and the results of [4]. We show, by providing explicit examples, that for knots in $\mathbf{R}^3$ with at most 8 double points in their diagrams the estimate is exact.

It is very well known (see, for example, [8]) that any topological knot type in a contact 3-manifold has a Legendrian representative. If the contact structure is coorientable a Legendrian knot gets a natural framing. The question of Legendrian representability of an arbitrary framed knot type has in general a negative answer: according to a classical result by Bennequin [1], the self-linking numbers (called in this case also *Bennequin numbers*) of all canonically framed Legendrian representatives of a fixed unframed knot type in $\mathbf{R}^3$ are bounded from one side.

Paper [1] gives an estimate for the Bennequin number in the standard contact 3-space in terms of the genus of a knot. This has a disadvantage of being insensitive to passing from a knot to its mirror image. One feels this immediately, considering the two trefoils: the Bennequin estimate is exact for one of them and is far from being such for the other.

In recent papers [7, 4, 2, 3] there were obtained a series of estimates for the Bennequin number of a Legendrian knot in standard 3-space (and its analog for the solid torus introduced by Tabachnikov in [13]) which do respect the change of orientation of the ambient space and are more efficient than the original by Bennequin. According to them, the self-linking number of a Legendrian knot K is at least the negative of the lowest power of the framing variable in (the unframed versions of) the HOMFLY and Kauffman polynomials of K (all the signs here follow our further choice of the orientations). The estimate coming from the Kauffman polynomial is usually better: for example, it copes with the Legendrian trefoils immediately, while the HOMFLY information falls a little short of being sufficient.

Another class of knots which is natural to consider in a contact 3-manifold are *transverse knots*, that is those everywhere transverse to the contact distribution. Their theory is parallel to the Legendrian one: any knot type has a transverse representative, they possess the canonical framing if the contact structure is parallelisable, the self-linking (Bennequin) number of a canonically framed transverse knot in standard contact 3-space is at least the Euler characteristic of any Seifert surface of the knot [1]. In a similar vein to the Legendrian case, the Bennequin number of a transverse knot in $\mathbf{R}^3$ cannot be less than the lowest degree of the framing variable in its HOMFLY polynomial [7] (it is not very reasonable to consider the Kauffman polynomial in the transverse setting, since the set of transverse knots is not closed under the main skein relation of that polynomial). In fact, it did not take too much effort to obtain the latter estimate in [7]: it came immediately from the comparison of some intermediate results of [1] with the results of [5, 11]. As in the Legendrian case, this easier estimate has proved to be more effective than that by Bennequin (once again, the example of the trefoils already demonstrates this).

The main goal of the present note is to obtain an estimate on the Bennequin number of a transverse knot in the standard solid torus $ST^*\mathbf{R}^2$ which is analogous to that of [7] for 3-space. We do this by direct methods, rather different from those of [7]. Namely, we follow the approach of [4, 2, 3] to the study of Legendrian knot invariants in 3-space and solid torus via invariants of their planar fronts, and study transverse knots in $ST^*\mathbf{R}^2$ via their projections to the plane. The latter are regular curves equipped with a natural transverse framing. We lower the framed version of Turaev's HOMFLY polynomial for the solid torus [14] to our planar curves. After this, a straightforward application of the results of [4] immediately implies the desired estimate. Passing to the universal cover of our solid torus, we get another proof of the Fuchs-Tabachnikov estimate.

We also give experimental results which rather surprisingly show that the Fuchs-Tabachnikov estimate for transverse representation is sharp for all the knots in $\mathbf{R}^3$ with at most 8 double points in their diagram.

Acknowledgements. The authors are grateful to H. Morton and S. Tabachnikov for useful discussions.

1 Links in the solid torus and 3-space, and framed planar curves

1.1 The two standard contact spaces

A *cooriented contact element* at a point of $\mathbf{R}^2$ is a line in the tangent space at this point along with a choice of one of the two half-planes into which it divides the tangent plane. Any such element can be represented by the unit normal n to the line pointing into the chosen half-plane (Fig.1). Thus

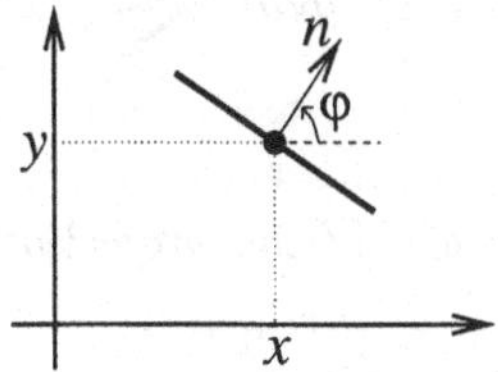

Figure 1: *A cooriented contact element on the plane and its coordinates in the solid torus* $ST^*\mathbf{R}^2$.

the variety of all cooriented contact elements of the plane is the spherisation $ST^*\mathbf{R}^2$ of its cotangent bundle. This solid torus possesses the standard contact structure, that is maximally non-integrable field of tangent planes: at each point of $ST^*\mathbf{R}^2$ one takes the tangent plane which is mapped by the canonical projection $p : ST^*\mathbf{R}^2 \to \mathbf{R}^2$ onto the contact element represented by that point. In the coordinates of Fig.1 this is the field of kernels of the 1-form $\alpha = (\cos\varphi)dx + (\sin\varphi)dy$. The standard contact structure is naturally cooriented: the coorientation of a contact element lifts via p to a coorientation of the tangent plane.

In what follows we will need an orientation on $ST^*\mathbf{R}^2$. We fix it to be the orientation of $\mathbf{R}^2$ followed by the direction of positive (counter-clockwise) rotation in the plane. This is opposite to the orientation $\alpha \wedge d\alpha$ traditionally taken in contact geometry.

Along with $ST^*\mathbf{R}^2$ we will be considering its universal cover $\mathbf{R}^3$ with its standard contact structure (induced via the covering mapping). The orientation of 3-space will be that inherited from the solid torus.

1.2 Links as framed planar curves

Any link L, that is an embedded collection of circles, in $ST^*\mathbf{R}^2$ can be put, by an arbitrary small perturbation, in generic position with respect to the projection p. Then its p-image F_L in the plane is an immersed collection of circles whose only singularities are transverse double points. At every point $p(a) \in F_L$ consider the normal n_a coorienting the contact element $a \in L$. We equip F_L with the unit framing e, $e(p(a)) \in a$, such that the pair $\{e(p(a)), n_a)\}$ orients the plane positively (see Fig.2). For our further considerations, the framing e is more visual than n_a itself.

For a Legendrian link, for example, this framing is everywhere tangent to the planar curve. But this is an infinitely-degenerate case. In general position, e is transverse to F_L except for a finite number of isolated points of their tangency. The latter are projections of the points of tangency of the link L to the contact structure.

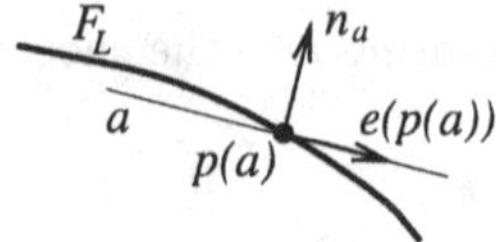

Figure 2: *Canonical framing e of the projection of a generic curve from $ST^*\mathbf{R}^2$ to the plane.*

For a generic link L, the two framing vectors at any double point of F_L are distinct. On the other hand, making an elementary change of the link topology via a generic homotopy in $ST^*\mathbf{R}^2$ involving a double point, we instantaneously observe a regular planar curve with a double point at which the two vectors coincide (Fig.3). Such a double point will be called *an elementary framing degeneracy point.*

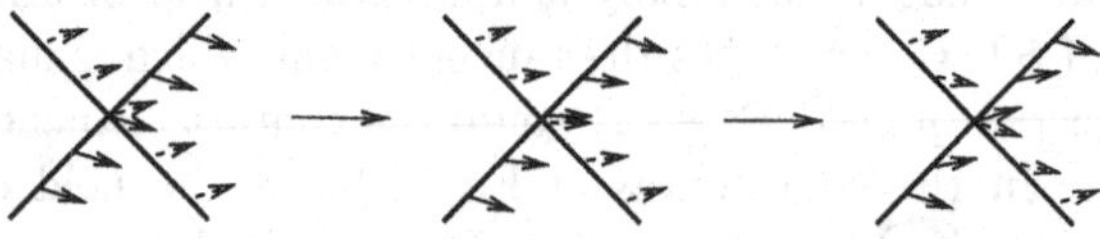

Figure 3: *An elementary framing degeneracy at a double point of a framed planar curve.*

Reversing our construction we can start with a generic framed planar curve F. Then the direction of the framing lifts F to a link L_F in $ST^*\mathbf{R}^2$.

To represent a link in $\mathbf{R}^3$ in a similar way, we need each of the components of a framed immersed planar collection F of circles to have a rotation (or Whitney winding) number zero. If F has more than one component, to make its lifting to the universal cover of $ST^*\mathbf{R}^2$ well-defined we must choose a point on each of the components and specify one of countably many possibilities to assign a phase φ to the framing at that point. The elementary framing degeneracy at a double point now requires coincidence of the corresponding real phases (not reduced modulo 2π).

1.3 Transverse links and transverse invariants

Definition 1.1 A link in a contact space is called *transverse* if it is everywhere transverse to the contact structure.

Definition 1.2 A transverse link L in $ST^*\mathbf{R}^2$ is *positive* if it is oriented and the framing e, followed by the orientation of F_L, orients the plane positively.

For a generic transverse link $L \subset ST^*\mathbf{R}^2$, the framing e is everywhere transverse to F_L. We can consider e up to homotopy which does not change the topology of L. This means that we may assume e to be normal to F_L

except for small neighbourhoods of some double points. Such double points will be called *abnormal* and marked by a dot aside the figure (see Fig.4). Such a specification of the double points, along with the coorientation of F_L by e, is obviously sufficient to restore the link type of $L \subset ST^*\mathbf{R}^2$. All the other information about the framing may be suppressed.

For a one-component or positive transverse link L, even the coorientation of F_L can be omitted. Thus any such link in the solid torus can be identified with an oriented planar curve with some double points marked. The $\mathbf{R}^3$-version of this approach is obvious.

Example 1.3 The move of Fig.3 is a homotopy from a normal to abnormal double point.

Figure 4: *The convention to depict normal and abnormal double points of transversally framed planar curves.*

We will be searching for invariants of transverse links in terms of transversally framed regular planar curves. This means that we will be interested only in such invariants of those curves which change in generic 1-parameter families that involve curves with elementary framing degeneracy double points. Such invariants will be called *transverse* invariants. Thus we are ignoring generic degenerations of transversally framed regular planar curves which have no relation to the framing, that is self-tangencies and triple points. Only families of immersed planar curves will be considered. Their framing is not allowed to be tangent at any time. At a double point, any interaction between a framing vector and the tangent to the other branch is ignored.

Remark 1.4 The above shows that, in fact, we will be working within the space of immersions of a finite number of circles into $ST^*\mathbf{R}^2$ which are transversal to the contact structure. Let us call such immersions *transverse curves.*

The description of the set of connected components of the space of 1-component transverse curves is as follows.

A generic oriented transverse knot K in $ST^*\mathbf{R}^2$ has two obvious invariants read from its projection F_K to the plane: the rotation number $w(F_K) \in \mathbf{Z}$, and the orientation $\varepsilon \in \mathbf{Z}_2$ of the frame made up by the framing e and the velocity of F_K. The Whitney-Graustein theorem [15] implies

Proposition 1.5 *Connected components of the space of 1-component oriented transverse curves in the standard solid torus are enumerated by the numbers* $(w, \varepsilon) \in \mathbf{Z} \times \mathbf{Z}_2$.

1.4 Self-linking of transverse links

Consider any oriented framed link L in $ST^*\mathbf{R}^2$. Following Tabachnikov [13], we define its self-linking number as the index of intersection of a small shift of L along the framing with a film that realises homology between L and the multiple of the fibre of $p : ST^*\mathbf{R}^2 \to \mathbf{R}^2$ over a sufficiently distant point of the plane. This is a natural generalisation of the traditional definition for $\mathbf{R}^3$.

For a generic transversally framed oriented planar curve, consider a counter-clockwise rotation of its framing by a small angle. Lifting this to $ST^*\mathbf{R}^2$ or, if possible, to its universal cover we obtain a canonical framing of the corresponding transverse link.

Definition 1.6 The self-linking number β of a canonically framed transverse oriented link in the solid torus or 3-space is called the *Bennequin number* of the link.

In the solid torus, the Bennequin number of a link L is read from the link diagram as usual, that is as half the sum of the signs of crossings of L and its shift along the framing (see Fig.5).

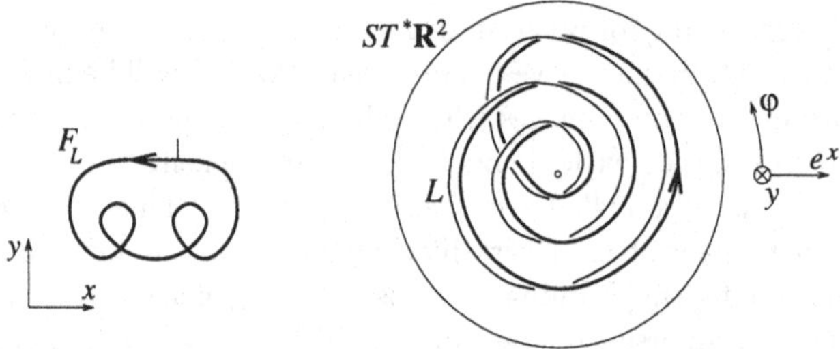

Figure 5: *Lifting of a planar curve to a positive transverse knot in $ST^*\mathbf{R}^2$ with the canonical framing. In absence of abnormal double points, φ is the direction of the velocity. The Bennequin number is 5.*

To calculate β for a transverse link L in $\mathbf{R}^3$, given as an oriented and cooriented regular curve $F_L \subset \mathbf{R}^2$ with marked double points (see section 1.3), one may resolve all the double points of F_L according to the phases $\varphi \in \mathbf{R}$ and thus obtain a link diagram of L. Then the canonical framing is blackboard with respect to the projection back to $\mathbf{R}^2$. Thus, $\beta(L)$ is the sum of the signs of all the double points.

1.5 Transverse representation of topological knot types

Proposition 1.7 *Any oriented topological link type in $\mathbf{R}^3$ and $ST^*\mathbf{R}^2$ has a positive transverse representative.*

This follows, for example, from the validity of the same statement for Legendrian representatives whose fronts are regular curves (the constructive proof of this fact for knots given in [4] can be easily generalised for links). Indeed, a shift of such a Legendrian representative by a small non-zero constant along the fibres of the fibration $p : ST^*\mathbf{R}^2 \to \mathbf{R}^2$ provides a transverse link.

According to [1], the Bennequin numbers of canonically framed transverse knots of a fixed topological type are bounded from below (for our orientation of the solid torus). This was proved by Bennequin for knots in $\mathbf{R}^3$, but implies the result for the solid torus as well via a contact embedding of the solid torus into the 3-space. Thus, the analogue of Proposition 1.7 for framed knots and links is not true.

In the next section we derive a new restriction on the Bennequin number of a transverse knot in the standard solid torus. This is the main result of our paper.

2 The HOMFLY polynomials

From now on we are considering only positive transverse links and corresponding transversally framed planar curves.

2.1 The solid torus

Following Turaev [14], we introduce the framed version of the HOMFLY polynomial of an oriented link in a solid torus [4].

Theorem 2.1 *For any framed link L in a solid torus, there exists an element $P(L) \in \mathbf{Z}[x^{\pm 1}, y^{\pm 1}, \xi_{\pm 1}, \xi_{\pm 2}, \ldots]$ which is uniquely defined by the skein relations and initial data of Fig.6.*

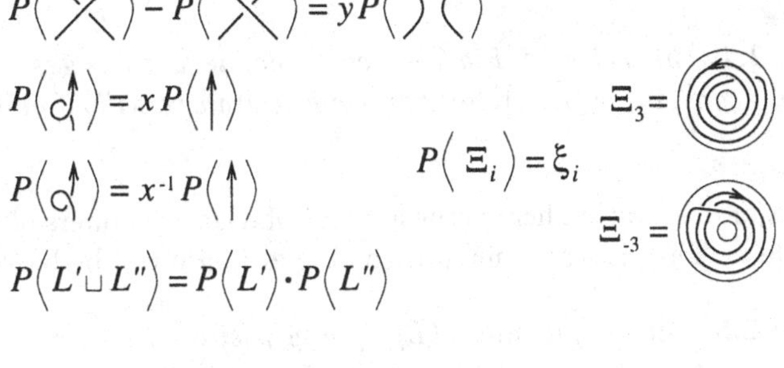

$$P(L' \sqcup L'') = P(L') \cdot P(L'') \qquad P(\Xi_i) = \xi_i$$

Figure 6: *Definition of the framed version of the HOMFLY polynomial for oriented links with the blackboard framing in a solid torus. In the last line, the links L' and L'' are mutually unlinked.*

Example 2.2 *For an unknot with trivial framing* $P = \frac{x-x^{-1}}{y}$.

Now we assume a link $L \subset ST^*\mathbf{R}^2$ to be transverse. We set $P(F_L) = P(L)$ and would like to calculate $P(L)$ entirely in terms of the planar curves. To do this we use the following obvious observation concerning the representation of Section 1.3 (see Fig.3).

Lemma 2.3 *A generic homotopy of positive transverse curves in* $ST^*\mathbf{R}^2$ *passing from a negative crossing to a positive one is seen, in terms of the underlying framed planar curves, as a homotopy from a normal to abnormal double point via an elementary framing degeneracy.*

Now, for positive transverse links represented as oriented regular planar curves with some of the double points marked (as in Section 1.3), the rules of Fig.6 imply the set of rules shown in Fig.7. The collections F' and F'' of its last line are lying in disjoint half-planes. According to the second rule of Fig.6, the relation between the transverse generators z_i we are using now and the blackboard generators ξ_i is $z_i = x^{|i|}\xi_i$: it is easy to show that the lift of Z_i has the unframed knot type of Ξ_i, and its canonical framing differs from the blackboard one of Ξ_i by $2|i|$ positive half-twists (see Fig.5, where $i = 3$).

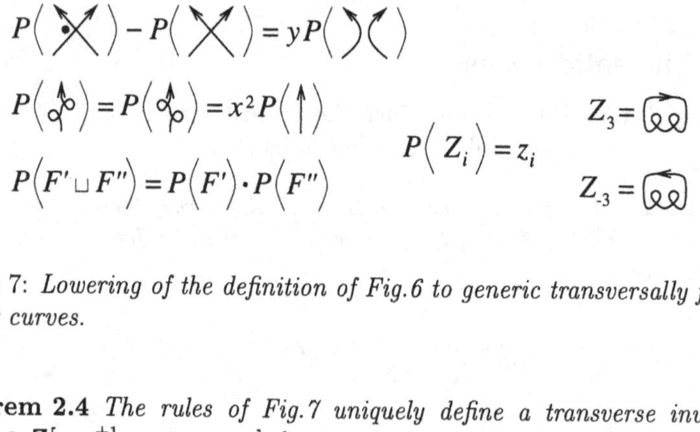

$$P(F' \sqcup F'') = P(F') \cdot P(F'') \qquad P(Z_i) = z_i$$

Figure 7: *Lowering of the definition of Fig.6 to generic transversally framed planar curves.*

Theorem 2.4 *The rules of Fig.7 uniquely define a transverse invariant* $P(F) \in \mathbf{Z}[x, y^{\pm 1}, z_{\pm 1}, z_{\pm 2}, \ldots]$ *for any generic transversally framed planar curve* F.

The major observation here is the absence of negative powers of the framing variable x. That is what our forthcoming estimate will be based on.

Corollary 2.5 *The polynomial* $P(L)$ *of any positive transverse link* L *in* $ST^*\mathbf{R}^2$ *is a genuine polynomial in* x, *not a Laurent one.*

Remark 2.6 Allowing components with both positive and negative orientations we would not be able to stay within the set of transverse links after application of the main skein relation.

To prove Theorem 2.4 we need the following

Lemma 2.7

$$P(\;\;) - P(\;\;) = yP(\;\;)$$

Proof.

$$P(\;\;) = P(\;\;) - yP(\;\;) = P(\;\;) - yP(\;\;)$$ □

Proof of Theorem 2.4. If a transversally framed planar curve has no abnormal double points, we calculate its polynomial P using the relation of the above lemma instead of the main skein relation of Fig.7. The renewed set of rules is exactly that which uniquely defined the framed version of the HOMFLY polynomial of a regular plane curve in [4]. According to [4], the element so defined is unique and lies in $\mathbf{Z}[x, y^{\pm 1}, z_{\pm 1}, z_{\pm 2}, \ldots]$.

If a planar curve has some abnormal double points, we start with applying the main skein relation of Fig.7 to reduce their number. Now the assertion of the theorem follows by induction. □

2.2 The standard contact 3-space

The set of rules which uniquely defines the framed version of the HOMFLY polynomial $P_0(L) \in \mathbf{Z}[x^{\pm 1}, y^{\pm 1}]$ for a link L with the blackboard framing in $\mathbf{R}^3$ is that of Fig.6 with all the information about the variables ξ_i omitted. According to what was said about the representation of transverse links in the universal cover of $ST^*\mathbf{R}^2$ in Section 1.2, the restriction of P_0 to positive transverse links in the standard $\mathbf{R}^3$, in terms of transversally framed planar curves, is that defined by Fig.7 without mentioning the z_i. The main skein relation is now applicable only when the homotopy between the normal and abnormal framings in its left-hand side passes through a double point at which the phases $\varphi \in \mathbf{R}$ of the planar framing coincide. In all the other cases the change normal-abnormal does not affect the polynomial (along with the link type in the 3-space). We refer to the rules obtained as *modified.*

As in the previous subsection, we have

Theorem 2.8 *For any transverse link L_0 in standard 3-space, the modified rules of Fig.7 uniquely define an element $P_0(L_0) \in \mathbf{Z}[x, y^{\pm 1}]$.*

2.3 The estimate

Let ℓ be the self-linking number of an oriented framed knot L or L_0 in either the solid torus or 3-space respectively. Then the polynomials

$$P^u(L) = x^{\ell(L)}P(L) \qquad \text{and} \qquad P_0^u(L_0) = x^{\ell(L_0)}P(L_0)$$

are invariants of unframed knot types [14, 6].

Corollary 2.5 and Theorem 2.8 imply

Theorem 2.9 *Let k be the lowest power of the framing variable x in the unframed version of the HOMFLY polynomial of an oriented knot K in the solid torus or 3-space. Equip the ambient space with the standard contact structure. Then the Bennequin number β of any transverse knot, whose topological type is that of K, is at least $-k$.*

Remark 2.10 The Bennequin number β of a transverse knot is odd. Indeed, according to Proposition 1.5, one can pass from any transverse knot to the lift of either a basic curve of Fig.7 or of the curve ∞ by a series of ordinary change-crossings which change β by ± 2. Now the lift of any basic curve, as well as that of the curve ∞, has odd β.

Example 2.11 Due to Example 2.2, for an unknot $\beta \geq 1$ [1]. The lift of the curve ∞ is minimal, with $\beta = 1$.

Example 2.12 The transverse knots represented by the curves Z_i of Fig.7 have minimal possible Bennequin numbers $2|i| - 1$ allowed for their topological type. On the other hand, the original Bennequin estimate provides, via a contact embedding of the solid torus into the 3-space which unknots all such knots, a weaker lower bound 1 independent of i.

Example 2.13 For $(2, q)$-torus knots in $\mathbf{R}^3$, our lower bounds are $2 - q$ and $2 + q$ for the left- and right-handed cases respectively. Those bounds are exact as Fig.8 shows. For better illustration the double points there are resolved according to the phase φ.

Figure 8: *Transverse representatives of left- and right-handed $(2, q)$-torus knots in the standard 3-space with minimal Bennequin numbers $2 - q$ and $2 + q$.*

3 Minimal representatives of knots in 3-space with few double points in a knot diagram

Proposition 3.1 *The estimate of Theorem 2.9 is exact for all knots in the standard contact 3-space with at most 8 double points in a knot diagram.*

This rather surprising fact follows from the explicit examples given in the three pages below. There we show the minimal realisations of the table knots [12, 9] and, when their topology is different, of their mirror images. The lowest degrees of the HOMFLY polynomials P_0^u were taken from [10] (the entry for the knot 8_{13} there required correction).

Remark 3.2 We do not know any examples of more complicated knots for which the estimate is not exact.

Remark 3.3 Taking a similar approach to that used in this note, it looks very convenient to study links in the variety of all cooriented contact elements of any Riemannian surface in terms of framed curves on the surface.

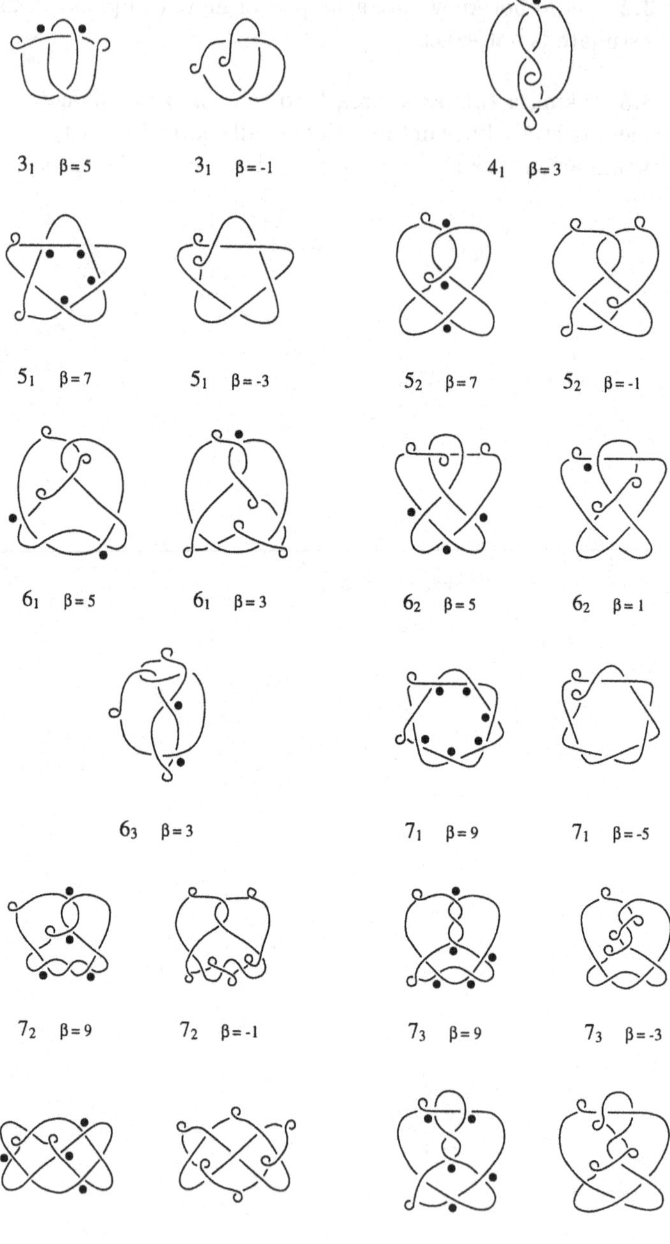

3_1 β=5
3_1 β=-1
4_1 β=3
5_1 β=7
5_1 β=-3
5_2 β=7
5_2 β=-1
6_1 β=5
6_1 β=3
6_2 β=5
6_2 β=1
6_3 β=3
7_1 β=9
7_1 β=-5
7_2 β=9
7_2 β=-1
7_3 β=9
7_3 β=-3
7_4 β=9
7_4 β=-1
7_5 β=9
7_5 β=-3

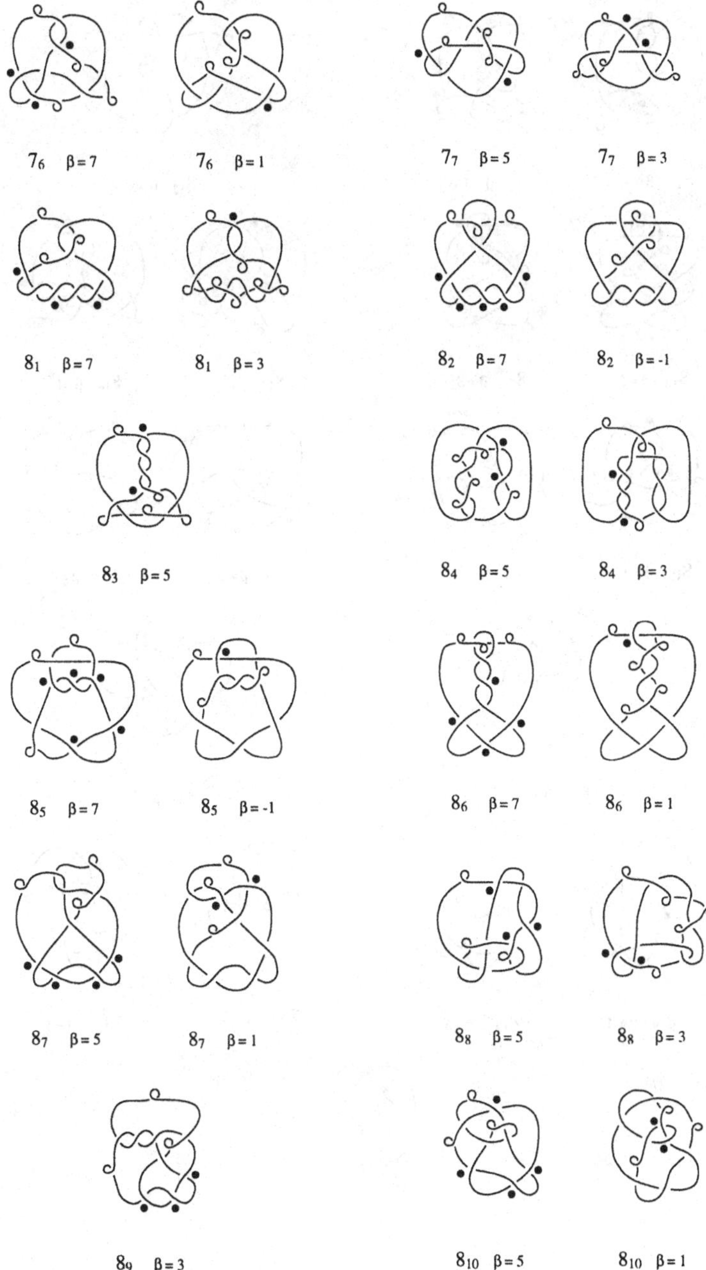
7_6 $\beta=7$
7_6 $\beta=1$
7_7 $\beta=5$
7_7 $\beta=3$
8_1 $\beta=7$
8_1 $\beta=3$
8_2 $\beta=7$
8_2 $\beta=-1$
8_3 $\beta=5$
8_4 $\beta=5$
8_4 $\beta=3$
8_5 $\beta=7$
8_5 $\beta=-1$
8_6 $\beta=7$
8_6 $\beta=1$
8_7 $\beta=5$
8_7 $\beta=1$
8_8 $\beta=5$
8_8 $\beta=3$
8_9 $\beta=3$
8_{10} $\beta=5$
8_{10} $\beta=1$

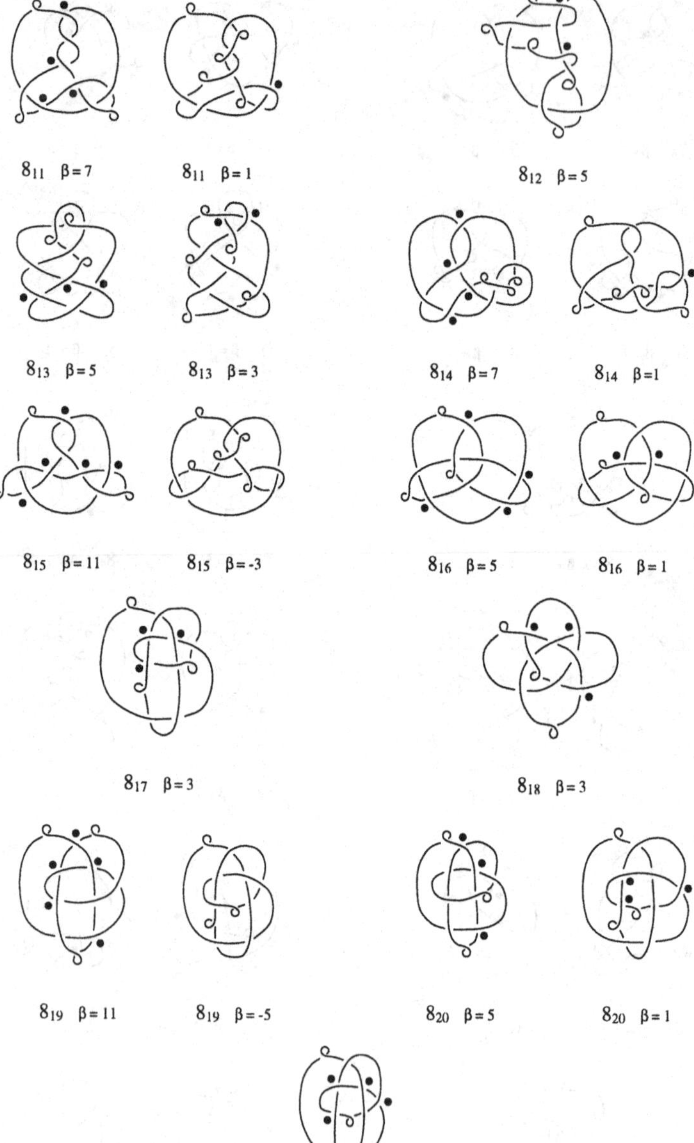
8_{11} $\beta = 7$
8_{11} $\beta = 1$
8_{12} $\beta = 5$
8_{13} $\beta = 5$
8_{13} $\beta = 3$
8_{14} $\beta = 7$
8_{14} $\beta = 1$
8_{15} $\beta = 11$
8_{15} $\beta = -3$
8_{16} $\beta = 5$
8_{16} $\beta = 1$
8_{17} $\beta = 3$
8_{18} $\beta = 3$
8_{19} $\beta = 11$
8_{19} $\beta = -5$
8_{20} $\beta = 5$
8_{20} $\beta = 1$

8_{21} $\beta = 7$

References

[1] D. Bennequin, *Enterlacements et équation de Pfaff*, Astérisque **107-108** (1983) 87–162.

[2] S. Chmutov and V. V. Goryunov, *Polynomial invariants of Legendrian links and plane fronts*, Preprint, University of Liverpool, 1996.

[3] S. Chmutov and V. V. Goryunov, *Polynomial invariants of Legendrian links and their fronts*, Preprint, University of Liverpool, 1996. To appear in the Proceedings of the Conference on Knot Theory, Tokyo, July 1996.

[4] S. Chmutov, V. V. Goryunov and H. Murakami, *Regular Legendrian knots and the HOMFLY polynomial of immersed plane curves*, Preprint, University of Liverpool, 1996.

[5] J. Franks and R. Williams, *Braids and the Jones polynolmial*, Trans. AMS **303** (1987) 97–108.

[6] P. Freyd, D. Yetter, J. Hoste, W. B. R. Lickorish, K. C. Millet and A. Ocneanu, *A new polynomial invariant of knots and links*, Bull. Amer. Math. Soc. (N.S.) **12** (1985) 239–246.

[7] D. Fuchs and S. Tabachnikov, *Invariants of Legendrian and transverse knots in the standard contact space*, to appear in Topology.

[8] M. Gromov, *Partial differential relations*, Springer-Verlag, Berlin and New York 1986

[9] L. Kauffman: On knots. Annals of Mathematics Studies **115** (1987), Princeton University Press, Princeton, NJ, 480 pp.

[10] W. B. R. Lickorish and K. C. Millet, *A polynomial invariant of oriented links*, Topology **26** (1987) 107–141.

[11] H. R. Morton, *Seifert circles and knot polynomials*, Math. Proc. Camb. Phil. Soc. **99** (1986) 107–109.

[12] D. Rolfsen, Knots and links. Mathematics Lecture Series **7** (1976), Publish or Perish, Wilmington, DE, 439 pp.

[13] S. Tabachnikov, *Computation of the Bennequin invariant of a Legendrian curve from the geometry of its front*, Func. Anal. Appl. **22** (1988) 246–248.

[14] V. G. Turaev, *The Conway and Kauffman modules of the solid torus with an appendix on the operator invariants of a tangle*, LOMI preprint E-6-88, Leningrad, 1988.

[15] H. Whitney, *On regular closed curves in the plane*, Comp. Math. **4** (1937) 276–284.

Department of Mathematical Sciences
Division of Pure Mathematics
The University of Liverpool
Liverpool, L69 3BX,
UK.

e-mail: goryunov@liv.ac.uk
j.w.hill@liv.ac.uk

Abelian Branched Covers of Projective Plane

A. Libgober*

To Professor C.T.C. Wall.

Abstract

This note outlines a relationship between the fundamental groups of the complements to reducible plane curves and certain geometric invariants (polytopes of quasiadjuntion) depending on the local type and *configuration* in the plane of singularities of the curves. This generalizes the relationship between the Alexander polynomial of plane curves and the position of singularities (cf. [Z],[L1],[LV]). Complete details will appear elsewhere.

A relationship between the homology of cyclic covers of $\mathbf{CP}^2$ branched over a singular plane curve and the position of singularities was discovered by Zariski. In [L1] we described how a certain portion of the fundamental group of an *irreducible* curve depends on the degree, the local type and the position of singularities. One associates to a curve C the module

$$\mathcal{A}(C) = H_1(\mathcal{M}_f, \mathbf{Q}) \tag{1}$$

where $\mathcal{M}_f$ is the Milnor fiber of the (non-isolated if C is singular) singularity $f(x,y,z) = 0$ at the origin where f is the defining equation of C. $\mathcal{A}(C)$ has the structure of a $\mathbf{Q}[t,t^{-1}]$-module where t acts on $H_1(\mathcal{M}_f, \mathbf{Q})$ as the monodromy operator of the singularity. Alternatively

$$\mathcal{A}(C) = G'/G'' \otimes \mathbf{Q} \tag{2}$$

where $G = \pi_1(\mathbf{CP}^2 - C), G' = [G,G]$ is the commutator of $G, G'' = [G',G']$ is the second commutator, and t acts as the generator of $\mathbf{Z}/(\deg C)\mathbf{Z} = G/G'$ with G/G' acting on G'/G'' in the standard way. It turns out that $\mathcal{A}(C) \otimes \mathbf{C}$ also can be described as follows. There exist a collection of rational numbers (constants of quasiadjunction) $\kappa_1, \kappa_2, \ldots, \kappa_N$ depending on the local type of the singularities of C such that inequality $\mathcal{A}(C) \neq 0$ is possible only if

*Supported by NSF.

$\deg C \cdot \kappa_i \in \mathbf{Z}$ for all i (these constants of quasiadjunction κ_i were related to the Arnold-Steenbrink spectrum [A],[St] of singularities in [LV]). Moreover to the sequence κ_i is associated a collection of sheafs of ideals $\mathcal{J}_{\kappa_1}, \ldots, \mathcal{J}_{\kappa_N}$ so that for each κ_i the quotient $\mathcal{O}_{\mathbf{CP}^2}/\mathcal{J}_{\kappa_i}$ is supported at the union of singular points of C and

$$\mathcal{A}(C) \otimes \mathbf{C} = \oplus_{\kappa_i} (\mathbf{C}[t, t^{-1}]/(t - e^{2\pi i \kappa_i})(t - e^{-2\pi i \kappa_i}))^{dim H^1(\mathbf{CP^2}, \mathcal{J}(d-3-d\cdot\kappa_i))}. \quad (3)$$

For example if $d = \deg C$, and C has nodes and ordinary cusps as its only singularities, then $\mathcal{A}(C) = 0$ if 6 does not divide d and otherwise $\mathcal{A}(C) = \oplus_s \mathbf{C}[t, t^{-1}]/(t^2 - t + 1)$, where s is the difference between the dimension of the space of curves of degree $d - 3 - \frac{d}{6}$ passing through the cusps of C and the expected dimension of this space (i.e. $\frac{1}{2}(d - 2 - \frac{d}{6})(d - 1 - \frac{d}{6})$). (In particular if $\deg C = 6$ and the number of cusps is 6, then $\mathcal{A}(C)$ is either zero or $\mathbf{Q}[t, t^{-1}]/(t^2 - t + 1)$ depending on whether the cusps belong to a curve of degree $6 - 3 - \frac{6}{6} = 2$ or not.)

In the case when C is reducible, one still can define a $\mathbf{Q}[t, t^{-1}]$-module $\mathcal{A}_\ell(C)$ similarly to (1) or to (2) (and $\mathcal{A}_\ell(C)$ can be related to the position of singularities ([LV])), which depends on the complement and on the homomorphism $\ell : G = \pi_1(\mathbf{CP}^2 - C) \to \mathbf{Z}/(\deg)\mathbf{Z}) \cdot \mathbf{Z}$ given by the linking number with C. This linking number of a loop $\alpha \in \pi_1(\mathbf{CP}^2 - C)$ is equal to the intersection number of a 2-chain σ such that $\partial\sigma = \alpha$ with C, and is well defined as an integer modulo d: if σ^1 is another cochain with $\partial\sigma^1 = \alpha$ then $(\sigma - \sigma^1, C) \equiv 0 \pmod{d}$. In this case $\mathbf{Z}/d\mathbf{Z}$ acts on K/K' where $K = \ker \ell : G \to \mathbf{Z}/d\mathbf{Z}$. In particular $\mathcal{A}_\ell(C)$ is an invariant of the pair $(G, \ell : G \to \mathbf{Z}/(\deg d)\mathbf{Z})$ rather than an invariant of the fundamental group alone.

To obtain an invariant of the fundamental group it is convenient to look at the complement to C in an affine plane $\mathbf{C}^2$ where the line at infinity is transversal to C. Then $\pi_1(\mathbf{CP}^2 - C)$ and $\pi_1(\mathbf{C}^2 - C)$ are related via the central extension (cf. [L5]):

$$1 \to \mathbf{Z} \to \pi_1(\mathbf{C}^2 - C) \to \pi_1(\mathbf{CP}^2 - C) \to 1. \quad (4)$$

Moreover, we have the isomorphism

$$\pi_1(\mathbf{C}^2 - C)/\pi_1(\mathbf{C}^2 - C)' = \mathbf{Z}^n \quad (5)$$

where n is the number of irreducible components of C. Identification (5) with $\mathbf{Z}^n$ is obtained by assigning to an element in $\pi_1(\mathbf{C}^2 - C)$ the collection of its linking numbers with the components of C. If $G = \pi_1(\mathbf{C}^2 - C)$ then one defines $\mathcal{A}(C)$ as a group by (2) and endows it with the structure of a module over the ring $R = \mathbf{Q}[G/G'] = \mathbf{Q}[\mathbf{Z}^n]$ resulting from the natural action of G/G' on G'/G''. ($\mathcal{A}(C)$ is closely related to the Alexander module $H_1(\widetilde{\mathbf{C}^2 - C}, pt)$ where $\widetilde{\mathbf{C}^2 - C}$ is the universal cover of $\mathbf{C}^2 - C$, cf. [L3].)

Since the structure of such modules for $n \geq 1$ is rather complicated, even after extending the field from $\mathbf{Q}$ to $\mathbf{C}$, we consider only certain invariants of $\mathcal{A}(C)$. Let $F_i(\mathcal{A}(C)) \subset \mathbf{Q}[\mathbf{Z}^n]$ be the i-th Fitting ideal of $\mathcal{A}(C)$ i.e. the ideal generated by the $(N-i+1) \times (N-i+1)$ minors of the matrix of the map $\Phi : R^M \to R^N$ where $\mathcal{A}(C) = \text{Coker}(\Phi)$. Then each $F_i(\mathcal{A}(C))$ defines a subvariety $V(F_i(\mathcal{A}(C))$ of Spec $(\mathbf{Q}[\mathbf{Z}^n])$ which is the torus $\mathbf{T}^n$. Let $Char_i(C)$ (cf. [L3]) be the support of $V(F_i(\mathcal{A}(C)))$ which is the reduced subvariety of $\mathbf{T}^n$ with the same set of zeros as $V(F_i(\mathcal{A}(C)))$.

According to the Sarnak-Laurent theorem ([AS], [Lau]) points of finite order on any algebraic subvariety X of a torus belong to a finite union of translated subgroups embedded in this subvariety X (a translated subgroup in $\mathbf{T}^n$ is the set of solutions of a system $t_1^{\alpha_1} \cdots t_n^{\alpha_n} = \omega_\beta$ where the $\alpha_i \in \mathbf{Z}$ and ω_β is a root of degree β). These translated subgroups are uniquely determined by X. Hence the collection of translated subgroups of $Char_i(C)$ affording the points of finite order of the latter are invariants of the fundamental group of the complement, and below we give an explicit description of these subgroups.

The translated subgroups of characteristic varieties can be calculated from a presentation using Fox Calculus (cf. [H]). For the module $\mathcal{A}(C)$ of the links of algebraic singularities a calculation of these subgroups is contained implicitly in [Y]. Finally if $N = 1$ the union of these translated subgroups coincides with $Char_i(C)$ as a consequence of the fact that the Alexander polynomial is cyclotomic ([L2]).

In order to describe the translated subgroups of $Char_i(C)$ we shall calculate for any homomorphism $H_1(\mathbf{C}^2 - C) \to \mathbf{Z}/m_1 \oplus \cdots \oplus \mathbf{Z}/m_n$ (which sends a loop into the collection of residues of its linking numbers with the components of C) the first Betti number of a resolution of singularities $\widetilde{V_{m_1,\ldots,m_n}}(C)$ of the corresponding branched abelian cover $V_{m_1,\ldots,m_n}(C)$ of $\mathbf{CP}^2$. This in fact is sufficient to obtain the collection of subgroups we seek due to the following theorem of M.Sakuma:

Theorem 1 ([S]) *Let $C = \bigcup_{i=1}^{i=n} C_i$ be an algebraic curve in $\mathbf{C}^2$ with n components which is transversal to the line at infinity and $V_{m_1,\ldots,m_n}$ (resp. $\widetilde{V_{m_1,\ldots,m_n}}$) be the branched covering of $\mathbf{CP}^2$ (respectively its resolution) defined above. For a torsion point $\omega = (\omega_1, \ldots, \omega_n)$ of $\mathbf{T}^n$ where $\omega_i^{m_i} = 1$ let C_ω be the curve formed by the components of C such that $\omega_i \neq 1$. Then*

$$b_1(\widetilde{V_{m_1,\ldots,m_n}}) = \sum_{\omega \in \mu_{m_1} \times \cdots \times \mu_{m_n}} \max(i \mid \omega \in Char_i(C_\omega)) \tag{6}$$

(here μ_r is the group of roots of unity of degree r).

Note (cf. [L3]) that the first Betti number of the corresponding *unbranched* cover of $\mathbf{C}^2 - C$ is equal to

$$n + \sum_{\omega \in \mu_{m_1} \times \cdots \times \mu_{m_n}} \max(i \mid \omega \in Char_i(C)) \tag{7}$$

where the summation is over all $\omega \neq (1, \ldots, 1)$.

To describe the algebro-geometric calculation of $b_1(\widetilde{V_{m_1,\ldots,m_n}(C)})$ we need to define certain ideals in the local ring of a singular point of C. Let p be such a point and $f_1, \ldots, f_k$ be the local equations of those irreducible components of C in a small neighborhood of p which contains p. Let $I = (i_1, \ldots, i_k \mid l_1, \ldots, l_k)$ $(i_s, l_t \in \mathbf{Z}$ be the array of integers such that $0 \le i_k \le l_k$.

Definition 2 The *ideal of quasiadjunction* (corresponding to the array I) is the ideal $\mathcal{J}_p(I) \subset \mathcal{O}_p$ formed by germs $\phi \in \mathcal{O}_p$ such that $z_1^{i_1} \cdots z_k^{i_k} \phi$ belongs to the adjoint ideal of the surface $V(f_1, \ldots, f_k \mid l_1, \ldots, l_k)$ in $\mathbf{C}^{k+2}$ given by

$$z_1^{l_1} = f_1(x,y), \ldots, z_k^{l_k} = f_k(x,y). \tag{8}$$

We have the following:

Proposition 3 *(1) Let $\mathcal{J}(f_1, \ldots, f_k)$ be the ideal in $\mathcal{O}_p$ generated by*

$$a)\ \frac{\Pi_{i=1}^{i=k} f_i}{f_i f_j} \cdot Jac\left(\frac{f_i, f_j}{x, y}\right), \quad b)\ \frac{f'_{ix}}{f_i} \Pi_{i=1}^{i=k} f_i,$$
$$c)\ \frac{f'_{iy} \cdot \Pi_{i=1}^{i=k} f_i}{f_i}, \quad d)\ \Pi_{i=1}^{i=k} f_i. \tag{9}$$

Then $\mathcal{O}_p/\mathcal{J}(f_1, \ldots, f_k)$ is an Artinian algebra and for any I the ideal quasiadjunction corresponding to I contains $\mathcal{J}(f_1, \ldots, f_k)$.
(2) Let $\mathcal{J}_p(I)$ be the ideal of quasiadjunction corresponding to an array I. Then there exists a polytope $\Delta(\mathcal{J}_p(I))$ in the unit cube $\mathcal{U} = \{0 \le x_i \le 1 \mid i = 1, \ldots, k\}$ (which we shall call the local polytope of quasiadjunction) such that for any array $I' = (i_1, \ldots, i_k \mid l_1, \ldots, l_k)$ the ideal of quasiadjunction corresponding to I' is $\mathcal{J}_p(I)$ if and only if $(\frac{i_1}{l_1}, \ldots, \frac{i_k}{l_k}) \in \Delta$. Polytopes $\Delta(\mathcal{J}(I))$ define the partition of the unit cube into a finite union of non-intersecting polytopes.

In the case $k = 1$ the algebra $\mathcal{O}_p/\mathcal{J}_p(f)$ is just the Milnor algebra of the germ f ([M]), the ideals $\mathcal{J}_p(I)$ are the ideals of quasiadjunction $\mathcal{J}_\kappa$ and $O < \kappa_1 < \cdots < \kappa_{\mu/2} < 1$ (where $\mu = \dim_{\mathbf{C}} \mathcal{O}_p/\mathcal{J}_p(f))$ are the constants of quasiadjunction mentioned earlier. We have $\mathcal{J}_{\kappa_s} = \mathcal{J}((i|l))$ if and only if i, l satisfy $\kappa_{s-1} < \frac{i}{l} \le \kappa_s$ and the polytope of quasiadjunction corresponding to $\mathcal{J}_{\kappa_s}$ is the interval $\kappa_{s-1} < x \le \kappa_s$.

When $k = 2$ the ideal $\mathcal{J}(f, g)$ is generated by $f \cdot g, f g'_x, f g'_y, g f'_x, g f'_y, f'_x g'_y - g'_x f'_y$. For example if the local equation of C is $y(y - x^2)$ (tacnode), then $\mathcal{J}(f, g)$ is the maximal ideal of the local ring. If the array $I = (i_1, i_2 \mid l_1, l_2)$ is such that $\frac{i_1}{l_1} + \frac{i_2}{l_2} \le \frac{1}{2}$ then the corresponding ideal $\mathcal{J}(I)$ of quasiadjunction is the maximal ideal. The polytopes of quasiadjunction are $\{(x, y) \mid x + y \le \frac{1}{2}, 0 \le x \le 1, 0 \le y \le 1$ and $\{(x, y) \mid x + y \ge \frac{1}{2}, 0 \le x \le 1, 0 \le y \le 1\}$.

For the ordinary triple point $xy(x-y)=0$ the local equations of the branches are: $f(x,y)=x, g(x,y)=y, h(x,y)=x-y$. The ideal $\mathcal{J}(f,g,h)$ has as generators $f\cdot g\cdot h, f_x gh, f_y gh, g_x fh, g_y fh, h_x fg, h_y fg, Jac(\frac{f,g}{x,y})\cdot h, Jac(\frac{g,h}{x,y})\cdot f, Jac(\frac{f,h}{x,y})\cdot g$. The polytopes of quasiadjunction are given in the unit cube by $x+y+z>1$ and $x+y+z\leq 1$.

The proof of Proposition 3 is a result of making explicit the conditions for the pull back of a form to be holomorphic on the resolution of (8) which is the normalization of $V(f_1,\ldots,f_k \mid l_1,\ldots,l_k)\times_{\mathbf{C}^2}\widetilde{\mathbf{C}^2(f_1,\ldots,f_k)}$ where $\widetilde{\mathbf{C}^2(f_1,\ldots,f_k)}$ is an embedded resolution of the plane curve singularity $\Pi f_i=0$ (cf. [L4]).

Next let us consider the following partition of the unit cube $\mathcal{U}_{\mathbf{R}^n}$ in the space $\mathbf{R}^n$ (n is the number of irreducible components of C) in which the coordinates of $\mathbf{R}^n$ are labeled by the irreducible component of C. For each local polytope of quasiadjunction $\Delta_{p_i,j}$ corresponding to the singular point p_i near which the components of C labeled $C_{i_1},\ldots,C_{i_k}$ have equations $f_1=\cdots=f_k=0$ consider the polytope $\bar{\Delta}_{p_i,j}\subset\mathcal{U}_{\mathbf{R}^n}$ which is the preimage of $\Delta_{p_i,j}$ for projection of $\mathcal{U}_{\mathbf{R}^n}$ on the subspace with coordinates labeled $i_1,\ldots,i_k$. Let us call two points in $\mathcal{U}_{\mathbf{R}^n}$ equivalent if collections of polytopes $\bar{\Delta}_{p_i,j}$ which contain each of these points coincide. Each equivalence class is a polytope which we call a *global* polytope of quasiadjunction. For example for $n=1$ (i.e. when C is an irreducible curve) such a polytope is an interval between two consecutive elements in the set of rational numbers which is the union of constants of quasiadjunction corresponding to all singularities of the curve.

To each global polytope of quasiadjunction Δ corresponds the sheaf of ideals $\mathcal{J}_\Delta\subset\mathcal{O}_{\mathbf{CP}^2}$ defined by the conditions:
a) $\operatorname{Supp}\mathcal{O}_{\mathbf{CP}^2}/\mathcal{J}_\Delta$ is the set of singular points of C.
b) The stalk $\mathcal{J}_{\Delta,p}$ of $\mathcal{J}_\Delta$ at p is the local ideal of quasiadjunction corresponding to the unique local polytope of quasiadjunction of the singularity p containing Δ.

Definition 4 A global polytope of quasiadjunction Δ is said to be *contributing* if the intersection of the hyperplane $d_1\cdot x_1+\cdots+d_n\cdot x_n=\ell$ for some $\ell\in\mathbf{Z}$ is a face δ (of any dimension) of Δ.

We call δ *the contributing face* of the polytope Δ, ℓ will be called *the level* of Δ, the given hyperplane containing δ will be called *the supporting hyperplane* of the contributing face δ (or of the global polytope of quasiadjunction Δ) and $d(\Delta)=d_1+\cdots+d_{i_n}$ will be called *the total weight* of the supporting hyperplane of the global polytope of quasiadjunction Δ.

For example in the case of an irreducible C a contributing hyperplane is a point coinciding with one of the constants of quasiadjunction κ for one of singularities of C. Its level is $d\cdot\kappa$ where d is the degree of C. In particular a singularity of C affects G'/G'' only if for one of its constants of quasiadjunction κ the product $d\cdot\kappa$ is an integer. For example if the only singularities are cusps, having $\frac{1}{6}$ as the only constant of quasiadjunction, $G'/G'' \otimes \mathbf{Q} = 0$ unless $6 \mid d$, cf. [L2].

Theorem 5 *Let $\widetilde{V_{m_1,\ldots,m_n}}(C)$ be a desingularitation of the abelian cover of $\mathbf{CP}^2$ branched over $C = \bigcup_{i=1}^{i=n} C_i$ (C_i irreducible) and possibly the line at infinity which corresponds to the homomorphism $H_1(\mathbf{C}^2 - C) \to \mathbf{Z}/m_1, \oplus \cdots \oplus \mathbf{Z}/m_n$. Let $C_{i_1,\ldots,i_k} = C_{i_1} \cup \cdots \cup C_{i_k}$ be a curve formed by components $i_1, \ldots, i_k$ of C. Then the first Betti number of $\widetilde{V_{m_1,\ldots,m_n}}(C)$ is equal to*

$$2 \sum_{i_1,\ldots,i_k} \sum_{\Delta(C_{i_1,\ldots,i_k})} N(m_{i_1}, \ldots, m_{i_k}, \delta(\Delta(C_{i_1,\ldots,i_k}))) \cdot$$

$$\dim H^1(\mathbf{CP}^2, \mathcal{J}_{\Delta(C_{i_1,\ldots,i_k})}(d(\Delta(C_{i_1,\ldots,i_k}))) - 3 - \ell(\Delta(C_{i_1,\ldots,i_k}))) \qquad (10)$$

where $N(m_1, \ldots, m_k, \delta)$ is the number of points of the form $(\frac{j_1}{m_1}, \ldots, \frac{j_k}{m_k})$ in the interior of a contributing face $\delta(\Delta)$ of a polytope of quasiadjunction Δ, $\ell(\delta)$ is the level of δ, $d(\delta)$ is its total weight and $\mathcal{J}_\Delta$ is the sheaf of quasiadjunction corresponding to a polytope of quasiadjunction Δ. The summation is over all global polytopes of quasiadjunction Δ (only those admitting a contributing face make a contribution).

We have the following:

Corollary 6 *Translated subgroups yielding the points of finite order of the i-th characteristic variety $Char_i(C)$ are given by:*

$$t_{i_1}^{\alpha_1(\delta)} \cdots t_{i_k}^{\alpha_k(\delta)} = e^{2\pi i \beta(\delta)}, \quad t_j = 1 (j \neq i_1, \ldots, i_k) \qquad (11)$$

where $\alpha_{i_1}(\delta)x_1 + \cdots + \alpha_{i_k}(\delta)x_k = \beta(\delta)$ is the equation of the supporting hyperplane of a contributing face δ for which $\alpha_i(\delta) \in \mathbf{Z}^{>0}$, $\gcd(\alpha_1(\delta), \ldots, \alpha_k(\delta)) = 1$, $\beta(\delta) \in \mathbf{Q}$ and δ is the contributing face of a global polytope of quasiadjunction of a curve formed by the components $C_{i_1}, \ldots, C_{i_k}$ of C for which $dim H^1(\mathbf{CP}^2, \mathcal{J}_\Delta(d(\Delta)) - 3 - \ell(\delta(\Delta))) = i$.

In particular the dimension of the cohomology group $H^1(\mathbf{CP}^2, \mathcal{J}_\Delta(d(\delta) - 3 - \ell(\delta))$ where Δ is a contributing polytope is a topological *invariant of the complement (since it depends only on the fundamental group).*

Example 1 Let us calculate the irregularity of the abelian cover of $\mathbf{CP}^2$ branched over the line arrangement L: $xy(x-y)z = 0$ and corresponding to the homomorphism $H_1(\mathbf{CP}^2 - L) = \mathbf{Z}^3 \to (\mathbf{Z}/n\mathbf{Z})^3$ (the first identification is given by the linking numbers with lines $x = 0, y = 0, x = y$ respectively, $z = 0$ is the line at infinity). The only nontrivial ideal of quasiadjunction is the maximal ideal of the local ring of the point $P : x = y = 0$. The only global ideal of quasiadjunction is the local one corresponding to the polytope cut by $x + y + z \leq 1$. Hence the irregularity of the abelian cover is Card $\{(i,j) \mid 0 < i < n, 0 < j < n, 0 < k < n, \frac{i}{n} + \frac{j}{n} + \frac{k}{n} = 1\} \cdot \dim H^1(\mathcal{J}(3-3-1))$ where $\mathcal{J} = ker(\mathcal{O}_{\mathbf{CP}^2} \to \mathcal{O}_P)$. The sheaf $\mathcal{J}$ has the

Koszul resolution $O \to \mathcal{O}(-2) \to \mathcal{O}(-1) \oplus \mathcal{O}(-1) \to \mathcal{J} \to O$ which yields $H^1(\mathcal{J}(-1)) = H^2(\mathcal{O}_{\mathbf{CP}^2}(-3)) = \mathbf{C}$. Now counting points on $x + y + z = 1$ shows that the irregularity of this abelian cover is equal to $\frac{n^2-3n+2}{2}$.

Example 2 Let us consider the arrangement of lines x_i, $i = 1, \ldots, 6$ formed by the sides of an equilateral triangle (x_1, x_2, x_3) and its medians (x_4, x_5, x_6) arranged so that their vertices are the intersection points of (x_1, x_2, x_4), (x_2, x_3, x_5) and (x_1, x_3, x_6) respectively (cf. [I]) . It has 4 triple and 3 double points. The polytopes of quasiadjunction for (the full) arrangement are the connected components of the partition of $\mathcal{U} = \{(x_1, \ldots, x_6) \mid 0 \leq x_i \leq 1, i = 1, \ldots, 6\}$ by the hyperplanes:

$$x_1 + x_2 + x_4 = 1, x_2 + x_3 + x_5 = 1, x_3 + x_1 + x_6 = 1, x_4 + x_5 + x_6 = 1 \quad (12)$$

(abusing notation we use x_i as the coordinate corresponding to the line x_i). The only hyperplane of the form $H_k : x_1 + x_2 + x_3 + x_4 + x_5 + x_6 = k, k \in \mathbf{Z}$ which contains a face of a global polytope of quasiadjunction (of the full) arrangement is H_2 (which contains the set of solutions of the system of equations formed by all equations in (12)). Moreover the subarrangements formed by the components of this arrangement which have polytope of quasiadjunction admitting contributing faces are triples of lines passing through common point. There are 4 subarrangements of such type for which the supporting hyperplane of a polytope of quasiadjunction is H_1. The contribution into irregularity from the polytope of the first type is equal to $N \cdot \dim H^1(\mathcal{J}(6-3-2))$ where N is the number of solutions to (12) of the form $x_i = \frac{i+1}{n}$ and $\mathcal{J}$ is the ideal of the subvariety of $\mathbf{CP}^2$ formed by triple points of the arrangement. To calculate $\dim H^1(\mathcal{J}(6-3-2))$ notice that 4 triple points form a complete intersection of two quadrics and hence $\mathcal{J}$ admits a resolution:

$$0 \to O(-4) \to O(-2) \oplus O(-2) \to \mathcal{J} \to 0. \quad (13)$$

This yields $H^1(\mathcal{J}(1)) = H^2(\mathcal{O}(-3)) = \mathbf{C}$. The contributions from the subarrangements of the second type were considered in example 1. Note that (as follows from this calculation) the translated planes forming the characteristic varieties are

$$t_1 t_2 t_4 = 1, \quad t_2 t_3 t_5 = 1, \quad t_1 t_3 t_6 = 1 = t_4 t_5 t_6 = 1 \quad (14)$$

(corresponding to the global polytope of quasiadjunction of first type) and

$$t_1 t_2 t_4 = t_3 = t_5 = t_6 = 1, \quad t_2 t_3 t_5 = t_1 = t_4 = t_6 = 1,$$

$$t_1 t_3 t_6 = t_2 = t_4 = t_5 = 1, \quad t_4 t_5 t_6 = t_1 = t_2 = t_3 = 1 \quad (15)$$

corresponding to 4 polytopes of the second type. The number of the points in $\mu_n \times \mu_n \times \mu_n \times \mu_n \times \mu_n \times \mu_n$ which belong to the subgroup (14) and for which $t_1 \neq 1, t_2 \neq 1, t_4 \neq 1$, or, alternatively, the number of solutions to the system (12) of the form

$$\frac{i}{n}, \quad 0 < \frac{i}{n} < 1 \quad (16)$$

is equal to $\frac{(n-1)(n-2)}{2}$. Indeed (12) is equivalent to $x_1 + x_2 + x_4 = 1, x_3 = x_4, x_5 = x_1, x_6 = x_2$ and the number of the solutions of the first of these equations satisfying (16) is $\frac{(n-1)(n-2)}{2}$. Similarly one checks that the number of points on each of the tori (15) with coordinates in μ_n with exactly 3 coordinates non equal to 1 is also $\frac{(n-1)(n-2)}{2}$. Hence for this arrangement the irregularity of $\widetilde{V_{n,n,n}(C)}$ is equal to $5\frac{(n-1)(n-2)}{2}$ (in particular it is 30 for $n = 5$ cf. [I]).

The other two arrangements considered in [I] can be treated similarly. Note that the theorem above also shows that the irregularity of the branched abelian cover $\widetilde{V_{m_1,\dots,m_n}(C)}$ is a polynomial periodic function of $m_1, \dots, m_n$ (cf. [AS],[H]).

References

[AS] S.Adams and P.Sarnak, Betti numbers of congruence groups, *Israel Journal of Mathematics*, vol. 88 (1994), 31-72.

[A] V.I.Arnold, Classification of unimodal critical points, *Funct. anal. and applications*, 7 (1973), 75-76.

[H] E.Hironaka, Multi-polynomial invariants for plane algebraic curves, *Singularities and Complex Geometry, Studies in Advanced Mathematics vol, 5*, Amer. Math. Soc. and International Press, (1996), 67-74.

[I] M-N. Ishida, The irregularity of Hirzebruch's examples of surfaces of general type with $c_1^2 = 3c_2$, *Math. Ann.*, vol.262 (1983), 407-420

[Lau] M.Laurent, Equations diophantine exponentielles, *Inventiones Mathematicae*, vol. 49 (1982), 833-851.

[L1] A.Libgober, Alexander invariants of plane algebraic curves, *Proc. Symp. Pure. Math.*, vol 40, 1983.

[L2] A.Libgober, Alexander polynomial of plane algebraic curves and cyclic multiple planes, *Duke Math.Jour.*, 49 (1982).

[L3] A.Libgober, On the homology of finite abelian covers, *Topology and applications*, vol. 43 (1992), 157-166.

[L4] A.Libgober, Position of singularities of hypersurfaces and the topology of their complements, *Preprint*, 1994, UIC.

[L5] A.Libgober, Groups which cannot be realized as Fundamental groups of the complements to hypersurfaces in $\mathbf{C}^N$, *Algebraic Geometry and Its Applications,* Volume in honor of Prof. S.Abhyankar, Springer Verlag, 1994.

[LV] F.Loeser, M.Vaquie, Le polynôme d'Alexander d'une courbe plane projective, *Topology*, 29 (1990), 163-173.

[M] J.Milnor, Singular points of complex hypersurfaces, Annals of Mathematical Studies 61, *Princeton University Press*, Princeton 1968.

[S] M.Sakuma, Homology of abelian coverings of links and spacial graphs, *Canadian Journal of Math.* vol.17(1) (1995), 201-224.

[St] J.Steenbrink, Mixed Hodge structures on vanishing cohomology, *Real and Complex Singularities, Oslo 76*, Sijthoff & Noordhoff Int. Publ., 525-563.

[Y] M.Yamamoto, Classification of the isolated algebraic singularities by their Alexander polynomials, *Topology*, vol. 23, No. 3, 1989, p.277.

[Z] O.Zariski, On irregularity of cyclic multiple planes, *Ann. of Math.* vol.32, 1931.

Department of Mathematics,
University of Illinois at Chicago,
851 S.Morgan,
Chicago, Ill. 60607,
USA.

e-mail: libgober@math.uic.edu

Elimination of Singularities: Thom Polynomials and Beyond

Osamu Saeki[†] Kazuhiro Sakuma[‡]

Dedicated to Professor C.T.C. Wall on his 60th birthday.

1 Introduction

This is a survey article concerning the authors' papers [47], [33], [50], [51], [52], [53], where the following problem is considered: *Given a smooth map $f : M \to N$ between smooth manifolds, does there exist a smooth map homotopic to f which has no singularities of a prescribed type Σ?* It is known that this problem is almost equivalent to the existence problem of a corresponding jet section $M \to J^r(M,N)$ covering f, which is homotopy theoretic in nature (for example, see [30], [28], [20], [21], [15], [16], [17] etc.). However, this homotopy theoretical problem is usually very difficult to solve, since even the homotopy type of the corresponding fiber in the jet bundle is difficult to determine[1].

The most easily computed part of an obstruction to the existence of a corresponding jet section is the Thom polynomial, which is the homology class represented by the closure of the set $\Sigma(f)$ of the singular points of f of type Σ [58], [31]. In fact, in some cases, it has been shown that the vanishing of the Thom polynomial implies the existence of a map homotopic to f without the prescribed singularities: for example, H. Levine's cusp elimination theorem for maps into the plane [35] and its generalizations to some 0-dimensional singularities by Ando [5], [6]. However, the topological location of $\Sigma(f)$ in the source manifold M can be nontrivial even if the Thom polynomial of Σ vanishes. Thus it is natural to ask whether this (co)homological obstruction to eliminating the singularities of the prescribed type by homotopy is the only one or not. Such a problem has been explicitly posed, for example, in [9, Chapter 4, §5] by Arnol'd, Vasil'ev, Goryunov, and Lyashko. Here we

[†]Partially supported by Grant-in-Aid for Encouragement of Young Scientists (No. 08740057), Ministry of Education, Science and Culture, Japan.

[‡]Partially supported by Grant-in-Aid for Encouragement of Young Scientists (No. 08874004), Ministry of Education, Science and Culture, Japan.

[1]Recently, Ando [8] has completely determined the homotopy type of the fiber corresponding to maps with only fold singularities between equidimensional manifolds.

note that the fold singularity is an exception, since, for example, when M is compact, N is open and $\dim M \geq \dim N$, the fold singularities cannot be eliminated by a trivial reason.

In this paper, we first present an affirmative answer for smooth maps between 4-dimensional manifolds [51]. In this case, the Thom polynomials are the unique obstructions to the elimination of the singularities, including 1- and 2-dimensional ones. This will be shown by using a homotopy principle due to Èliašberg [20] about maps with only fold singularities.

Then we present a remarkable result, which is a negative answer for maps of certain 4-dimensional manifolds into orientable 3-manifolds ([47, Theorem 4], [50, Theorem 4.5]). For such maps, the swallowtail singularities, which are 0-dimensional, can always be eliminated by homotopy by a result of Ando [6]. However, the cusp singularities, which are 1-dimensional, cannot be eliminated, although the corresponding Thom polynomial vanishes. To the authors' knowledge, this is the first negative answer to the question. We will see that this phenomenon is caused essentially by a Rohlin type theorem (see [44], [29]), which is peculiar to 4-dimensional manifolds.

Furthermore, we present a result due to Kikuchi and the first author [33] which suggests that the Thom polynomial may not necessarily be the unique obstruction for a large class of maps. Namely, they have shown that there exists no smooth map $f : M \to N$ with only fold singularities when M is a closed n-dimensional manifold with *odd Euler characteristic* and N is an almost parallelizable p-dimensional manifold with $n \geq p$, provided that $p \neq 1, 3, 7$. This implies that, in the above situation, if a smooth map $f : M \to N$ is *generic* in a certain sense, then f necessarily has a cusp singularity. When $p = 2$, this has already been proved by Thom [58] as a direct consequence of the calculation of the Thom polynomial for cusp singularities. However, for example, it is known that for a map of an n-dimensional closed manifold into $\mathbf{R}^{n-1}$ with only Morin singularities (see §4), the Thom polynomial for cusp singularities always vanishes for all n (see [46], [7]). Thus the above result implies that the Thom polynomial for cusp singularities alone cannot detect the impossibility of eliminating them. Of course, if the Thom polynomial for a higher order singularity does not vanish, then one cannot eliminate the singularities, and hence neither the cusp singularities. However, such an argument is rarely possible, since in general it is very difficult to completely determine the Thom polynomials for higher order singularities. In fact, the above result is closely related to the non-existence problem of elements in the p-stem π_p^S of Hopf invariant one, solved by Adams [1]. We can also generalize the result to maps admitting only Morin singularities of types A_k with $k \leq 3$ when $p \neq 1, 2, 3, 4, 7, 8$ [52].

In the negative cases which have been observed above, the obstruction to the elimination is not exactly the Thom polynomial. Nevertheless, the essential obstruction seems to be homotopy theoretical in nature, like the Rohlin type theorem or the Hopf invariant one problem. In the last part of this paper, we present a result in which smooth structures of manifolds, which are less related to the homotopy theory, play an important role. More specifically, we show that there exist pairs of homeomorphic smooth manifolds such that the elimination of certain singularities is possible for one manifold while it is impossible for the other [53].

Throughout the paper all manifolds and maps are assumed to be C^∞. All homology and cohomology groups are with $\mathbf{Z}_2$ coefficients unless otherwise indicated. The symbol "$\cong$" denotes a diffeomorphism between manifolds or an appropriate isomorphism between algebraic objects.

2 Stable maps between 4-manifolds

Let M be a closed connected *oriented* 4-dimensional manifold and N a stably parallelizable 4-dimensional manifold (not necessarily closed). Let $f : M \to N$ be a stable map; i.e., by definition, f is *stable* if its orbit in the mapping space $C^\infty(M, N)$ (endowed with the C^∞-topology) by the action of $\mathrm{Diff}(M) \times \mathrm{Diff}(N)$ is an open set (see [27]), where $\mathrm{Diff}(M)$ and $\mathrm{Diff}(N)$ denote the diffeomorphism groups of M and N respectively. Note that the set of stable maps is open and dense in $C^\infty(M, N)$ (see [36]). We denote the singular set of f by $S(f)$; i.e., $S(f)$ is the set of those points in M where the rank of the differential of f is strictly less than 4. If p is a point in $S(f)$, there exist local coordinates (x_1, x_2, x_3, x_4) and (y_1, y_2, y_3, y_4) centered at p and $f(p)$ respectively such that f has one of the following normal forms (see [27]):

(1) $y_i = x_i, \quad y_4 = x_4^2$ (fold singularity)
(2) $y_i = x_i, \quad y_4 = x_4^3 + x_1x_4$ (cusp singularity)
(3) $y_i = x_i, \quad y_4 = x_4^4 + x_1x_4^2 + x_2x_4$ (swallowtail singularity)
(4) $y_i = x_i, \quad y_4 = x_4^5 + x_1x_4^3 + x_2x_4^2 + x_3x_4$ (butterfly singularity)
(5) $y_i = x_i, \quad y_3 = x_3^2 + x_1x_4, \quad y_4 = x_4^2 + x_2x_3$ (hyperbolic umbilic)
(6) $y_i = x_i, \quad y_3 = x_3^2 - x_4^2 + x_1x_3 + x_2x_4, \quad y_4 = x_3x_4 + x_2x_3 - x_1x_4$ (elliptic umbilic)

where $y_i = x_i$ for $i = 1, 2, 3$ in cases (1), …, (4) and $i = 1, 2$ in cases (5) and (6). A fold, cusp, swallowtail, or butterfly singularity is also called an A_1-, A_2-, A_3-, or A_4-*type singularity* respectively; and a hyperbolic or an elliptic umbilic a $\Sigma^{2,0}$-*type singularity*. We denote by $A_k(f)$ the set of the A_k-type singular points of f and by $\Sigma^{2,0}(f)$ the set of the hyperbolic and elliptic umbilics of f.

The Thom polynomials for the stable singularities above have been determined by several authors as follows (see [58], [41], [26], [46], [45], [5, Proposition 5.4]):

$$[S(f)]_2^* = \bar{w}_1 = w_1 = 0 \qquad [\overline{A_2(f)}]_2^* = \bar{w}_1^2 + \bar{w}_2 = w_2$$

$$[\overline{A_3(f)}]_2^* = \bar{w}_1^3 + \bar{w}_1 \cdot \bar{w}_2 = 0 \qquad [\overline{A_4(f)}]_2^* = \bar{w}_1^4 + \bar{w}_1 \cdot \bar{w}_3 = 0$$

$$[\overline{\Sigma^{2,0}(f)}]_2^* = \bar{w}_2^2 + \bar{w}_1 \cdot \bar{w}_3 = w_2^2 \qquad ([\overline{\Sigma^{2,0}(f)}]^* = p_1)$$

where $\overline{A_k(f)}$ and $\overline{\Sigma^{2,0}(f)}$ denote the topological closures of $A_k(f)$ and $\Sigma^{2,0}(f)$ in M respectively, the symbol $[X]_2^*$ denotes the Poincaré dual of the $\mathbf{Z}_2$-homology class in M represented by X, and w_i (resp. $\bar{w}_i$) denote the i-th (dual) Stiefel-Whitney class of M. Note that since M is oriented and N is orientable, the set $\Sigma^{2,0}(f)$ has a canonical orientation and $[\overline{\Sigma^{2,0}(f)}]^*$ denotes the Poincaré dual of the $\mathbf{Z}$-homology class represented by $\overline{\Sigma^{2,0}(f)}$. In this case, p_1 denotes the first Pontrjagin class of M.

Thus the Thom polynomials for the 0-dimensional singularities (butterfly singularities and umbilics) are 0 and p_1 respectively. Hence, if p_1 vanishes, then these singularities can be eliminated via a homotopy by results of Ando [5], [6].

Concerning the 1-dimensional singularities, we have the following [51].

Theorem 2.1 *Let M be a closed oriented 4-manifold, N a stably parallelizable 4-manifold and $f : M \to N$ a smooth map. Then there exists a smooth map $g : M \to N$ homotopic to f which has only fold and cusp singularities if and only if the signature of M vanishes (or equivalently $p_1 = 0$).*

The above theorem is proved as follows. The "only if part" follows from the above mentioned results about the Thom polynomials. For the "if part", we first construct a smooth map g_1 homotopic to f which has cusp singularities along an embedded surface such that the Poincaré dual of the $\mathbf{Z}_2$-homology class represented by the surface coincides with w_2. Then we construct a jet section covering g_1 such that on a neighborhood of the embedded surface it coincides with the jet of g_1 and that outside of the neighborhood its image is contained in the part corresponding to the regular points and the fold singularities. Finally we apply the homotopy principle for maps with only fold singularities due to Èliašberg [20].

Recall that the Thom polynomial for the 1-dimensional (swallowtail) singularity set always vanishes. Then, assuming that f already has no 0-dimensional singularities, we can always eliminate the 1-dimensional singularities by Theorem 2.1.

Finally, the Thom polynomial for the 2-dimensional (cusp) singularity set coincides with w_2. If w_2 vanishes, then M is stably parallelizable and the singularity can also be eliminated by a result of Èliašberg [20]. Thus, in our situation, the Thom polynomials are the unique obstructions to eliminating the stable singularities (other than the fold singularities).

Remark 2.2 When $N = S^4$, we can deduce the following attractive result using Theorem 2.1. Let $[M, S^4]$ denote the set of the homotopy classes of continuous maps $f : M \to S^4$. *If $p_1 = 0$ and $w_2 = 0$, then an arbitrary homotopy class of $[M, S^4]$ contains a smooth map $f : M \to S^4$ with only fold singularities, and if $p_1 = 0$ and $w_2 \neq 0$, then an arbitrary homotopy class of $[M, S^4]$ contains a smooth map $f : M \to S^4$ with only fold and cusp singularities.* Recall that $[M, S^4]$ corresponds bijectively to $\mathbf{Z}$ via the mapping degree. The above result is a complete answer to an extension of J. Mather's problem for the dimension pair $(4, 4)$ (see [21, p.1319]).

3 Maps of certain 4-manifolds to 3-manifolds

Let M be a closed connected orientable 4-dimensional manifold and N a connected orientable 3-dimensional manifold. Then it is known that the singularities of stable maps $f : M \to N$ are fold, cusp and swallowtail singularities [27] (for precise definitions, see §4). We denote by $A_k(f)$ $(k = 1, 2, 3)$ the set of the fold, cusp and swallowtail singularities respectively. Then their Thom polynomials have been determined as follows (see [58], [46], [7], [6]):

$$[\overline{A_1(f)}]_2^* = w_2, \qquad [\overline{A_2(f)}]_2^* = 0, \qquad [\overline{A_3(f)}]_2^* = 0.$$

By Ando [6], the 0-dimensional singularity (swallowtail singularities) can always be eliminated by homotopy. As to the 1-dimensional singularity (cusp singularities), we have the following result [47], [50].

Theorem 3.1 *Let M be a closed connected orientable 4-manifold such that $H_*(M; \mathbf{Z}) \cong H_*(\mathbf{C}P^2; \mathbf{Z})$ and N a connected orientable 3-manifold. Then there exists no smooth map $f : M \to N$ with only fold singularities.*

The above result is proved as follows. Suppose that such a map exists. We fix an orientation of M so that the signature of M is equal to 1. The singular set $S(f)$ is an embedded surface and to each component S of $S(f)$ is associated an index, which is defined via f. If the index of S is not zero (i.e., if S consists of indefinite fold singularities), then the structure group of its normal bundle is reduced to the dihedral group of order 8 and it follows that the self-intersection number $S \cdot S$ of S in M vanishes. If the index of S is zero (i.e., if S consists of definite fold singularities), then it is orientable and carries an integral homology class. On the other hand, the second author has shown that the self-intersection number $S(f) \cdot S(f)$ of $S(f)$ in M is always congruent modulo 4 to -1 times the signature of M [54] (see also [50, Theorem 4.5]). He has shown this using a result of Guillou-Marin [29] (see also [44]) which is an extension of Whitney's congruence for surfaces in S^4 [59]. The result is also considered to be a Rohlin type theorem for 4-dimensional manifolds [43]. Thus it follows that there must exist an element $\gamma \in H_2(M; \mathbf{Z})$ whose self-intersection number is congruent modulo 4 to -1. This implies that $k^2 \equiv -1 \pmod 4$ for some integer k. This is a contradiction.

Recall that the Thom polynomial for the cusp singularity always vanishes in our situation. The above theorem shows that, nevertheless, we cannot eliminate the cusp singularity, which is 1-dimensional. In this example, a Rohlin type theorem for 4-dimensional manifolds plays an important role. To the authors' knowledge, this is the first example in which the elimination is impossible in spite of the Thom polynomial being zero.

4 Non-existence of Morin maps and the Hopf invariant one problem

Let $f : M \to N$ be a smooth map of a closed n-dimensional manifold M into a p-dimensional manifold N with $n \geq p$. We say that $q \in M$ is a *Morin singularity of type* A_k ($k = 1, 2, \ldots, p$) if there exist local coordinates $(x_1, x_2, \ldots, x_n)$ centered at q and $(y_1, y_2, \ldots, y_p)$ centered at $f(q)$ such that f has the following normal form (see [38]):

$$y_i \circ f = x_i \ (1 \leq i \leq p-1), \quad y_p \circ f = x_p^{k+1} + \sum_{i=1}^{k-1} x_i x_p^{k-i} \pm x_{p+1}^2 \pm \cdots \pm x_n^2.$$

We set $A_k(f)$ to be the set of the Morin singularities of type A_k of f, $\overline{A_k(f)}$ the topological closure of $A_k(f)$ in M, and $S(f)$ the set of the singularities of f (i.e., $S(f) = \{q \in M : \text{rank } df_q < p\}$). Furthermore, we say that $q \in M$ is a *fold singularity*, a *cusp singularity* or a *swallowtail singularity* if q is a Morin singularity of type A_1, A_2 or A_3 respectively. We call $f : M \to N$ a *Morin map* if f has only Morin singularities (see [13] or [9, p.191]).

Kikuchi and the first author have shown the following result [33], [50].

Theorem 4.1 *Let M be a closed n-dimensional manifold with odd Euler characteristic and N an almost parallelizable p-dimensional manifold ($n \geq p$). If there exists a smooth map $f : M \to N$ with only fold singularities, then $p = 1, 3$ or 7.*

Recall that a smooth manifold N is *almost parallelizable* if the tangent bundle of $N - \{\text{point}\}$ is trivial.

As has been mentioned in the introduction, the above theorem shows that the Thom polynomial for cusp singularities alone cannot detect the impossibility of the elimination.

The proof given in [33] depends heavily on the Brown-Liulevicius theorem [10], which states that if a closed manifold which is not null cobordant immerses into $\mathbf{R}^p$ in codimension one, then it must be cobordant to $\mathbf{R}P^0, \mathbf{R}P^2$ or $\mathbf{R}P^6$. Using different methods, the authors have clarified the geometric background of the above theorem [52]. In fact, it is proved as follows. First note that $S(f)$ is a $(p-1)$-dimensional closed manifold and by the modulo 2 Euler characteristic formula for Morin maps [25], [50], [39], $S(f)$ has odd Euler characteristic. (In particular, p must be odd.) Then consider the map $g = f|S(f) : S(f) \to N$, which is a codimension-1 immersion. Since $N - \{\text{point}\}$ is parallelizable, it can be immersed into $\mathbf{R}^p$ by [40]. Thus we can construct a codimension-1 immersion $h : S(f) \to \mathbf{R}^p$. Modifying h slightly, we may assume that h is an immersion with normal crossings. Then a result of Herbert [32] implies that the number of p-tuple points of h is odd. On the other hand, Eccles [18], [19] has shown that for a given p, a codimension-1 immersion of a closed manifold into $\mathbf{R}^p$ with an odd number of p-tuple points exists if and only if there exists an element of Hopf invariant one in the p-stem π_p^S. Then the result follows from [1].

In [52], Theorem 4.1 has been generalized as follows.

Theorem 4.2 *Let M be a closed n-dimensional manifold with odd Euler characteristic and N an almost parallelizable p-dimensional manifold ($n \geq p$). If there exists a smooth map $f : M \to N$ with at most fold and cusp singularities, then $p = 1, 2, 3, 4, 7$ or 8.*

Theorem 4.3 *Let M be a closed n-dimensional manifold with odd Euler characteristic and N an almost parallelizable p-dimensional manifold ($n \geq p$). If there exists a smooth map $f : M \to N$ with at most fold, cusp and swallowtail singularities and p is even, then $p = 2, 4$ or 8.*

In the situation of Theorem 4.3, when p is odd, we do not know if p must be equal to $1, 3$ or 7. However, the authors have shown that this is true when a certain homological condition about the set of swallowtail singularities is satisfied [52].

Remark 4.4 For each of $p = 1, 3, 7$, we can construct a smooth map $f : M \to \mathbf{R}^p$ with only fold singularities for some closed manifold M with $\dim M \geq p$ and with odd Euler characteristic. For details, see [33]. For $p = 2$, it is known that every closed manifold M with $\dim M \geq 2$ admits a smooth map into $\mathbf{R}^2$ with at most fold and cusp singularities. For $p = 4$, the authors have constructed a smooth map $f : M \to \mathbf{R}^4$ of a closed 4-dimensional manifold M with odd Euler characteristic with only fold and cusp singularities [51]. For $p = 8$, we can imitate this construction using an immersion of $\mathbf{R}P^6$ into $\mathbf{R}^8$ to obtain a smooth map $f : M \to \mathbf{R}^8$ of a closed 8-dimensional manifold M with odd Euler characteristic such that f has only fold and cusp singularities. Summarizing, all the special values of p in Theorems 4.1, 4.2, and 4.3 are in fact realized as target dimensions of the desired Morin maps.

Remark 4.5 In [13], Chess has shown that for a Morin map $f : M \to N$ with $\dim M - \dim N$ odd, the $\mathbf{Z}_2$ homology class $[\overline{A_k(f)}]_2$ represented by $\overline{A_k(f)}$ vanishes in $H_{p-k}(M)$ provided that $k \geq 4$. On the basis of this result, Chess has conjectured that every Morin map $f : M \to N$ with $\dim M - \dim N$ odd is homotopic to a Morin map with at most fold, cusp and swallowtail singularities (see also [9, Chapter 4, §1.5]). Suppose that this conjecture is true and consider a Morin map $f : M \to \mathbf{R}^p$ such that $\dim M - p$ is odd, $p \neq 1, 3, 7$ and that the Euler characteristic of M is odd. Then one can eliminate the Morin singularities of types A_k with $k \geq 4$ homotopically to obtain a smooth map f_1 with at most fold, cusp and swallowtail singularities, but then $[A_3(f_1)]_2$ does not vanish in $H_{p-3}(\overline{A_2(f_1)})$ by [52, Theorem 5.2]. In particular, one cannot eliminate the swallowtail singularities. In [52, Theorem 5.2], if the homological condition is always satisfied, then it is probable that we can construct a counter example of the Chess conjecture. For example, can we define an index modulo 2 of each cusp point so as to assure the homological condition in [52, Theorem 5.2]? For maps into $\mathbf{R}^2$ such an index can be defined by using the intrinsic derivative as is seen in [35].

Remark 4.6 By [46], [7], it is known that for Morin maps $f : M \to N$ with $\dim M - \dim N = 1$, the Thom polynomial for cusp singularities always vanishes. On the other hand, for the same class of maps, the Thom polynomials for Morin singularities of types A_k with $k \geq 4$ also vanish by Chess [13]. As to the Thom polynomial for swallowtail singularities, we see that it is equal to $\bar{w}_1^4 + \bar{w}_1\bar{w}_3$ $(= w_1 w_3)$ by using a formula due to Ando [7]. Thus, if this cohomology class vanishes, we cannot use the Thom polynomials to show that one cannot eliminate a prescribed type of Morin singularities under this dimension assumption.

Remark 4.7 As remarked in [33, Remark 2.3], the second author conjectures that if M is a closed *orientable* n-dimensional manifold with odd Euler characteristic, then M does not admit any smooth map $f : M \to \mathbf{R}^p$ with only fold singularities for $p = 3, 7$. Note that this question makes sense only when n is even. This problem seems to be difficult to solve because of the existence of a Hopf invariant one element in these dimensions. Only a partial answer is known, which has been given by the first author [47, Theorem 4], [50, Theorem 4.5] as has been explained in §3. The reader should also refer to [55, §5], where the 4-dimensional topological background to this problem are explained. For $(n, p) = (4, 3)$, in order to prove the above conjecture affirmatively, assuming that $H_1(M; \mathbf{Z}) = 0$, we see that it suffices to prove the following conjecture. (One uses the classification of symmetric bilinear forms [37], defined over $H^2(M; \mathbf{Z})$ by cup products, and the celebrated theorem of Donaldson [14].

Conjecture. *Let M be a closed 4-dimensional manifold whose integral cohomology ring is isomorphic to that of $k\mathbf{C}P^2 \sharp l\overline{\mathbf{C}P^2}$ for $k + l$ odd, where $\mathbf{C}P^2$ denotes the complex projective plane with the usual complex orientation and $\overline{\mathbf{C}P^2}$ with the opposite orientation. Then there exists no smooth map $f : M \to \mathbf{R}^3$ with only fold singularities.*

The result of the first author quoted above is the case where $(k, l) = (1, 0)$ or $(0, 1)$. Our problem is still open except in this case. To solve it, presumably we need a deeper obstruction than the non-existence of elements of Hopf invariant one.

5 Differentiable structures and special generic maps

Let M be a closed connected 4-dimensional manifold and $f : M \to \mathbf{R}^3$ a smooth map. We say that f is *special generic* if it has only fold singularities of index zero (definite fold singularities) [11], [42], [48], [49], [54], [53]. In this section, we first give a classification theorem of 4-dimensional manifolds with free fundamental groups admitting special generic maps into $\mathbf{R}^3$ as follows [53].

Theorem 5.1 *Let M be a closed connected 4-dimensional manifold with free fundamental group. Then M admits a special generic map into $\mathbf{R}^3$ if and only if M is diffeomorphic to*

$$(\sharp^{r-\varepsilon} S^1 \times S^3)\sharp(\sharp^\varepsilon S^1 \tilde{\times} S^3)\sharp(\sharp^s S^2 \times S^2)\sharp(\sharp^\delta S^2 \tilde{\times} S^2)\sharp\Sigma^4$$

for some $\varepsilon, \delta \in \{0,1\}$ and $s \geq 0$, where r is the rank of the free group $\pi_1(M)$, the connected sum over an empty set is assumed to be the standard 4-sphere, $S^1 \tilde{\times} S^3$ is the nontrivial (and hence nonorientable) S^3-bundle over S^1, $S^2 \tilde{\times} S^2$ is the nontrivial S^2-bundle over S^2, and Σ^4 is a homotopy sphere of the form $\partial(\Delta \times D^2)$ for some compact contractible 3-manifold Δ.

Remark 5.2 In the above theorem, we do not know if Σ^4 is diffeomorphic to the standard 4-sphere or not.

In order to construct the examples mentioned in the introduction, we first recall the construction of the Akbulut manifold. Let Q be Cappell-Shaneson's exotic $\mathbf{R}P^4$ (see [12]); i.e., Q is homeomorphic to $\mathbf{R}P^4$ ([24]), but not diffeomorphic to it. Then Akbulut has found an embedding $\varphi : \mathbf{R}P^2 \hookrightarrow Q\sharp S^2 \times S^2$ such that $\pi_1(Q\sharp S^2 \times S^2 - \varphi(\mathbf{R}P^2)) \cong \mathbf{Z}$ and that the normal bundle of φ is isomorphic to that of a standardly embedded $\mathbf{R}P^2$ in $\mathbf{R}P^4$ [2]. Then we set $M = (Q\sharp S^2 \times S^2 - \mathrm{Int}N) \cup S^1 \tilde{\times} D^3$, where N is a closed tubular neighborhood of $\varphi(\mathbf{R}P^2)$ in $Q\sharp S^2 \times S^2$ and $S^1 \tilde{\times} D^3$ is the nontrivial (and hence nonorientable) D^3-bundle over S^1. This is the so-called *Akbulut manifold* and it is known to be an exotic $S^1 \tilde{\times} S^3 \sharp S^2 \times S^2$ (see [2], [3], [4], [23]). Note that the Akbulut manifold is always nonorientable. Then the following theorem is proved in [53].

Theorem 5.3 *Let M be the Akbulut manifold. Then for every nonnegative integer k, $M\sharp(\sharp^k S^2 \times S^2)$ is homeomorphic to $S^1 \tilde{\times} S^3 \sharp(\sharp^{k+1} S^2 \times S^2)$, while $M\sharp(\sharp^k S^2 \times S^2)$ does not admit any special generic map into $\mathbf{R}^3$.*

Note that $S^1 \tilde{\times} S^3 \sharp(\sharp^{k+1} S^2 \times S^2)$ does admit a special generic map into $\mathbf{R}^3$ by Theorem 5.1. This shows that there exist infinitely many pairs of homeomorphic 4-manifolds such that one of each pair does not admit any special generic map into $\mathbf{R}^3$ while the other does. This shows that the elimination problem of indefinite fold singularities is closely related to the differentiable structure of the source manifolds. Such phenomena can also be found in [48], [49].

Remark 5.4 We note that for the Akbulut manifold M, $M\sharp S^2 \tilde{\times} S^2$ *is diffeomorphic* to $S^1 \tilde{\times} S^3 \sharp S^2 \times S^2 \sharp S^2 \tilde{\times} S^2$ (see [3], [4]) and hence it admits a special generic map into $\mathbf{R}^3$.

Using a result of Kreck [34], we also obtain the following [53].

Theorem 5.5 *Let K denote the underlying smooth 4-dimensional manifold of a K3 surface. Then for every nonnegative integer k, $S^1 \tilde{\times} S^3 \sharp K \sharp (\sharp^k S^2 \times S^2)$ is homeomorphic to $S^1 \tilde{\times} S^3 \sharp (\sharp^{k+11} S^2 \times S^2)$, while $S^1 \tilde{\times} S^3 \sharp K \sharp (\sharp^k S^2 \times S^2)$ does not admit any special generic map into $\mathbf{R}^3$.*

Note that in all the above examples, the source manifolds have infinite cyclic fundamental group and are nonorientable. In the orientable case, we have candidates, which are called *Scharlemann manifolds* [57]. They are constructed as follows. Let $\Sigma = \Sigma(2,3,5)$ be the Brieskorn homology 3-sphere, i.e., the boundary of the Milnor fiber of the link of the isolated singularity defined by $x^2 + y^3 + z^5 = 0$ in $\mathbf{C}^3$, whose fundamental group is isomorphic to the binary icosahedral group. Then we choose $\alpha \in \pi_1(\Sigma)$ such that $\alpha \notin Z(\pi_1(\Sigma))$, the center of $\pi_1(\Sigma)$. We identify α with the corresponding element in $\pi_1(\Sigma \times S^1)$. Remove a tubular neighborhood of an embedded circle representing α in $\Sigma \times S^1$ and attach $S^2 \times D^2$ to its boundary with the trivial framing, i.e., do "spin surgery" along α. The resulting manifold M' is a Scharlemann manifold. Note that we have a number of choices for α. It is known that M' is topologically s-cobordant and hence by Freedman [24], homeomorphic to $S^1 \times S^3 \sharp S^2 \times S^2$. The authors do not know if a Scharlemann manifold M' admits a special generic map into $\mathbf{R}^3$. However, since M' is stably parallelizable, M' does admit a smooth map $g : M' \to \mathbf{R}^3$ with only fold singularities by a result of Èliašberg [21]. This raises the question, from the global singularity theory viewpoint, of whether one can eliminate the indefinite fold singularities of g or not, which is closely related to the smooth structure of the source 4-manifold. If we can prove that M' does not admit any special generic map into $\mathbf{R}^3$, then we obtain an exotic $S^1 \times S^3 \sharp S^2 \times S^2$ and an exotic trivial 2-knot in $S^2 \times S^2$ (see [56]). Note that $M' \sharp S^2 \times S^2$ (resp. $M' \sharp S^2 \tilde{\times} S^2$) does admit a special generic map into $\mathbf{R}^3$, since it is diffeomorphic to $S^1 \times S^3 \sharp (\sharp^2 S^2 \times S^2)$ (resp. $S^1 \times S^3 \sharp S^2 \times S^2 \sharp S^2 \tilde{\times} S^2$), see [22].

References

[1] J. F. Adams, *On the non-existence of elements of Hopf invariant one*, Ann. of Math. **72** (1960), 20–104.

[2] S. Akbulut, *A fake 4-manifold*, Contemporary Math. **35** (1984), 75–141.

[3] S. Akbulut, *A fake $S^3 \tilde{\times} S^1 \sharp S^2 \times S^2$*, Contemporary Math. **44** (1985), 281–286.

[4] S. Akbulut, *Constructing a fake 4-manifold by Gluck construction to a standard 4-manifold*, Topology **27** (1988), 239–243.

[5] Y. Ando, *Elimination of certain Thom-Boardman singularities of order two*, J. Math. Soc. Japan **34** (1982), 241–267.

[6] Y. Ando, *On the elimination of Morin singularities*, J. Math. Soc. Japan **37** (1985), 471–487; Erratum, **39** (1987), 537.

[7] Y. Ando, *On the higher Thom polynomials of Morin singularities*, Publ. RIMS, Kyoto Univ. **23** (1987), 195–207.

[8] Y. Ando, *The homotopy type of the space consisting of regular jets and folding jets in $J^2(n,n)$*, to appear in Japanese J. Math.

[9] V. I. Arnol'd, V. A. Vasil'ev, V. V. Goryunov and O. V. Lyashko, Dynamical systems VI, Singularities: local and global theory, Encyclopaedia Math. Sci., vol.6, Springer-Verlag, Berlin, Heidelberg, New York, 1993.

[10] R. L. W. Brown, *A note on immersions up to cobordism*, Illinois J. Math. **21** (1977), 240–241.

[11] O. Burlet et G. de Rham, *Sur certaines applications génériques d'une variété close à trois dimensions dans le plan*, Enseign. Math. **20** (1974), 275–292.

[12] S. E. Cappell and J. L. Shaneson, *Some new four-manifolds*, Ann. of Math. **104** (1976), 61–72.

[13] D. S. Chess, *A note on the classes* $[S_1^k(f)]$, Proc. Sympos. Pure Math. **40**, Part 1, Amer. Math. Soc., Providence, RI, 1983, pp.221–224.

[14] S. K. Donaldson, *The orientation of Yang-Mills moduli spaces and four-manifold topology*, J. Diff. Geom. **26** (1987), 397–428.

[15] A. A. du Plessis, *Maps without certain singularities*, Comment. Math. Helv. **50** (1975), 363–382.

[16] A. A. du Plessis, *Homotopy classification of regular sections*, Compositio Math. **32** (1976), 301–333.

[17] A. A. du Plessis, *Contact invariant regularity conditions*, Lecture Notes in Math. vol.535, Springer-Verlag, Berlin, Heidelberg, New York, 1976, pp.205–236.

[18] P. J. Eccles, *Multiple points of codimension one immersions*, Lecture Notes in Math. vol.788, Springer-Verlag, Berlin, Heidelberg, New York, 1980, pp. 23–38.

[19] P. J. Eccles, *Codimension one immersions and the Kervaire invariant one problem*, Math. Proc. Camb. Phil. Soc. **90** (1981), 483–493.

[20] J. M. Èliašberg, *On singularities of folding type*, Math. USSR Izv. **5** (1970), 1119–1134.

[21] J. M. Èliašberg, *Surgery of singularities of smooth mappings*, Math. USSR Izv. **6** (1972), 1302–1326.

[22] R. Fintushel and P. S. Pao, *Identification of certain 4-manifolds with group actions*, Proc. Amer. Math. Soc. **67** (1977), 344–350.

[23] R. Fintushel and R. J. Stern, *Another construction of an exotic* $S^1 \tilde{\times} S^3 \sharp S^2 \times S^2$, Contemporary Math. **35** (1984), 269–275.

[24] M. H. Freedman, *The disk theorem for four-dimensional manifolds*, Proc. Int. Congress of Math., August 16–24, 1983, Warszawa, pp.647–663.

[25] T. Fukuda, *Topology of folds, cusps and Morin singularities*, "A Fete of Topology", ed. by Y. Matsumoto, T. Mizutani and S. Morita, Academic Press, New York, 1987, pp. 331–353.

[26] T. Gaffney, *The Thom polynomials of* $\overline{\Sigma^{1111}}$, Proc. Sympos. Pure Math. **40**, Part 1, Amer. Math. Soc., Providence, RI, 1983, pp.399–408.

[27] M. Golubitsky and V. Guillemin, Stable mappings and their singularities, Graduate Texts in Math. no.14, Springer-Verlag, New York, Heidelberg, Berlin, 1973.

[28] M. L. Gromov and J. M. Èliašberg, *Removal of singularities of smooth mappings*, Math. USSR Izv. **5** (1971), 615–639.

[29] L. Guillou et A. Marin, *Une extension d'un théorème de Rohlin sur la signature*, C. R. Acad. Sci. Paris **285** (1977), 95–98.

[30] A. Haefliger, *Lectures on the theorem of Gromov*, Lecture Notes in Math. vol.209, Springer-Verlag, Berlin, Heidelberg, New York, 1971, pp.128–141.

[31] A. Haefliger et A. Kosinski, *Un théorème de Thom sur les singularités des applications différentiables*, Séminaire H. Cartan, E. N. S., 1956/57, Exposé no.8.

[32] R. J. Herbert, Multiple points of immersed manifolds, Mem. Amer. Math. Soc. **34**, no.250, 1981.

[33] S. Kikuchi and O. Saeki, *Remarks on the topology of folds*, Proc. Amer. Math. Soc. **123** (1995), 905–908.

[34] M. Kreck, *Some closed 4-manifolds with exotic differentiable structure*, Lecture Notes in Math. vol.1051, Springer-Verlag, Berlin, Heidelberg, New York, 1984, pp.246–262.

[35] H. Levine, *Elimination of cusps*, Topology **3** (suppl. 2) (1965), 263–296.

[36] J. Mather, *Stability of C^∞-mappings : VI, The nice dimensions*, Proceedings of Liverpool Singularities – Symposium I, Lecture Notes in Math. vol.192, Springer-Verlag, Berlin, Heidelberg, New York, 1971, pp.207–253.

[37] J. Milnor and D. Husemoller, Symmetric bilinear forms, Springer-Verlag, Berlin, Heidelberg, New York, 1973.

[38] B. Morin, *Formes canoniques des singularités d'une application différen tiable*, C. R. Acad. Sci. Paris **260** (1965), 5662–5665, 6503–6506.

[39] I. Nakai, *Characteristic classes and fiber products of smooth mappings*, to appear in Trans. Amer. Math. Soc.

[40] A. Phillips, *Submersions of open manifolds*, Topology **6** (1966), 171–206.

[41] I. R. Porteous, *Simple singularities of maps*, Proceedings of Liverpool Singularities – Symposium I, Lecture Notes in Math. vol.192, Springer-Verlag, Berlin, Heidelberg, New York, 1971, pp.286–307.

[42] P. Porto and Y. K. S. Furuya, *On special generic maps from a closed manifold into the plane*, Topology Appl. **35** (1990), 41–52.

[43] V. A. Rokhlin, *Two-dimensional submanifolds of four-dimensional manifolds*, Functional Anal. Appl. **5** (1971), 39–48.

[44] V. A. Rokhlin, *Proof of Gudkov's hypothesis*, Functional Anal. Appl. **6** (1971), 136–138.

[45] F. Ronga, *Le calcul de la classe de cohomologie entière duale a Σ^k*, Proceedings of Liverpool Singularities – Symposium I, Lecture Notes in Math. vol.192, Springer-Verlag, Berlin, Heidelberg, New York, 1971, pp.313–315.

[46] F. Ronga, *Le calcul des classes duales aux singularités de Boardman d'ordre deux*, Comment. Math. Helv. **47** (1972), 15–35.

[47] O. Saeki, *Notes on the topology of folds*, J. Math. Soc. Japan **44** (1992), 551–566.

[48] O. Saeki, *Topology of special generic maps of manifolds into Euclidean spaces*, Topology Appl. **49** (1993), 265–293.

[49] O. Saeki, *Topology of special generic maps into* $\mathbf{R}^3$, in "Workshop on Real and Complex Singularities", Matemática Contemporânea **5** (1993), 161–186.

[50] O. Saeki, *Studying the topology of Morin singularities from a global viewpoint*, Math. Proc. Camb. Phil. Soc. **117** (1995), 223–235.

[51] O. Saeki and K. Sakuma, *Stable maps between 4-manifolds and elimination of their singularities*, preprint, 1995.

[52] O. Saeki and K. Sakuma, *Maps with only Morin singularities and the Hopf invariant one problem*, to appear in Math. Proc. Camb. Phil. Soc.

[53] O. Saeki and K. Sakuma, *On special generic maps into* $\mathbf{R}^3$, preprint, 1996.

[54] K. Sakuma, *On special generic maps of simply connected* $2n$*-manifolds into* $\mathbf{R}^3$, Topology Appl. **50** (1993), 249–261.

[55] K. Sakuma, *On the topology of simple fold maps*, Tokyo J. Math. **17** (1994), 21–31.

[56] Y. Sato, *Scharlemann's 4-manifolds and smooth 2-knots in* $S^2 \times S^2$, Proc. Amer. Math. Soc. **121** (1994), 1289–1294.

[57] M. Scharlemann, *Constructing strange manifolds with the dodecahedral space*, Duke Math. J. **43** (1976), 33–40.

[58] R. Thom, *Les singularités des applications différentiables*, Ann. Inst. Fourier (Grenoble) **6** (1955-56), 43–87.

[59] H. Whitney, *On the topology of differentiable manifolds*, Lectures in Topology, Michigan Univ. Press, 1940, pp.101–141.

Osamu Saeki
Department of Mathematics
Faculty of Science
Hiroshima University
Higashi-Hiroshima 739
Japan

e-mail: saeki@top2.math.sci.hiroshima-u.ac.jp

Kazuhiro Sakuma
Department of General Education
Kochi National College of Technology
Nankoku-City
Kochi 783
Japan

e-mail: sakuma@ge.kochi-ct.ac.jp

An Introduction to the Image Computing Spectral Sequence

Kevin Houston

1 Introduction

The image computing spectral sequence is a powerful new tool in the analysis of the homology of the image of a finite map. In singularity theory it has been successful in the study of the local topology of images of finite complex analytic maps, see [3], [4], [10] and [21]; a good introduction to the topology of images is given in the survey paper [26]. However, the sequence is surely useful elsewhere, see [15], and it is the intention that this paper introduces the sequence to a wider audience.

As the name suggests the spectral sequence calculates the homology of the image of a finite and proper map. Much work has been done on the study of the topology of fibres of maps and this sequence allows us deep insights into the topology of images. The terms of the sequence are given by the alternating homology of the multiple point spaces of the map, and so Sections 2 and 3 are devoted to a description of these. The sequence is described in Section 4 and the next section gives numerous examples of the sequence in action. Section 6 deals with how to deduce homotopy information about the image using alternating homology. This section is followed by one on the local topology of images of finite complex analytic maps. Most previous results in the study of sufficiently small punctured neighbourhoods of points in complex analytic spaces have relied on finding the number of equations defining the space; the theorems are often strongest for complete intersections as they are defined by the lowest possible number of equations. Images, on the other hand, can require a large number of equations and so sharp results were not forthcoming using traditional theory. The fact that we can now produce many strong results in the study of the local topology of images is a measure of the power of the sequence. The paper concludes with some thoughts on possible future work.

My thanks to Bill Bruce, Victor Goryunov, Neil Kirk, David Mond and Terry Wall for discussions on the material presented here. Thanks are also due to an anonymous referee who made a number of useful suggestions.

2 Alternating Homology of a Complex

Suppose X is a topological space, then the group of permutations on k objects, denoted S_k acts naturally on X^k. Denote by sign the natural sign representation for S_k. The space $Z \subset X^k$ is called S_k-cellular if it is S_k-homotopy equivalent to a cellular complex.

Definition 2.1 *Let*

$$\mathrm{Alt}_{\mathbb{Z}} = \sum_{\sigma \in S_k} \mathrm{sign}(\sigma)\sigma.$$

With this operator we can define alternating homology.

Definition 2.2 *The alternating chain complex of Z, $C_*^{alt}(Z;\mathbb{Z})$ is defined to be the the following subcomplex of the cellular chain complex, $C_n(Z;\mathbb{Z})$, of Z,*

$$C_n^{alt}(Z;\mathbb{Z}) := \mathrm{Alt}_{\mathbb{Z}}\, C_n(Z;\mathbb{Z}).$$

The elements of $C_n^{alt}(Z;\mathbb{Z})$ are called alternating or alternated chains.

Lemma 2.3

$$C_n^{alt}(Z;\mathbb{Z}) \cong \{c \in C_n(Z;\mathbb{Z}) \mid \sigma c = \mathrm{sign}(\sigma)c \textit{ for all } \sigma \in S_k\}.$$

Proof. The generators of this group are those formed by cells in regular orbits, a point further discussed below. From this it is easy to prove the equivalence. □

Definition 2.4 *The alternating homology of Z, $H_*^{alt}(Z;\mathbb{Z})$ is defined to be the homology of $C_*^{alt}(Z;\mathbb{Z})$.*

This notation is different to [3], where $H_i^{alt}(Z;\mathbb{Z})$ denotes the alternating part of integral homology. However, our notation is more in keeping with traditional notation in homology.

If we wish to define alternating homology over general coefficients then we may do so in the usual way by tensoring $C_*^{alt}(Z;\mathbb{Z})$ by the coefficient group.

Example 2.5 Suppose $T = S^1 \times S^1$ denotes the standard torus. Then S_2 acts on T by permutation of the copies of S^1. Using the angular coordinate on S^1 let Z be the points $(z, z+\pi) \in T$, then Z is just a circle with antipodal action. We can give Z a cellular structure by choosing two antipodal points p_1 and p_2 as 0-cells and then the complement of these points will form two 1-cells, e_1 and e_2, upon which S_2 acts by permutation and whose orientation we induce from an orientation of the circle. Then $\sigma(e_1) = e_2$ and $\sigma(e_2) = e_1$, where σ is the non-trivial element of S_2. The group $C_0^{alt}(Z;\mathbb{Z})$ is generated by $c_0 = p_1 - p_2$ and $C_0^{alt}(Z;\mathbb{Z})$ is generated by $c_1 = e_1 - e_2$. The boundary of c_1 is $-p_1 + p_2 - p_1 + p_2 = -2(p_1 - p_2)$. Therefore c_1 is not a cycle and $2(p_1 - p_2)$ is a boundary, hence

$$\begin{aligned} H_0^{alt}(Z;\mathbb{Z}) &= \mathbb{Z}_2, \\ H_1^{alt}(Z;\mathbb{Z}) &= 0. \end{aligned}$$

An alternating homology group is not a subgroup of ordinary homology as the example shows: $H_0^{alt}(Z;\mathbb{Z}) = \mathbb{Z}_2$ is not a subgroup of $H_0(Z;\mathbb{Z}) = \mathbb{Z}$. This is just another way of saying that the alternating part of ordinary homology is not equal to alternating homology.

Definition 2.6 *Suppose X is a topological space then the diagonal of X^k is*

$$Diag(X^k) := \{(x_1,\ldots,x_k) \in X^k | x_i = x_j \text{ for some } i \neq j\}.$$

If the realisation of a non-zero chain lies in the diagonal then the chain is not an alternating chain. Consider any cell in the chain, then some simple transposition will fix the cell, but the sign of such a transposition will be negative, and hence the cell does not alternate. This simple fact has major repercussions when we study the alternating homology of multiple point spaces and is also used to prove the next theorem.

Theorem 2.7 *Suppose $Z \subset W$ are S_k-cellular spaces in X^k such that $W \backslash Z \subset Diag(X^k)$. Then the alternating homologies of the two spaces are isomorphic:*

$$H_i^{alt}(Z;\mathbb{Z}) \xrightarrow{\cong} H_i^{alt}(W;\mathbb{Z}).$$

Proof. The chain complex of Z is a subcomplex of that of W. The cells in W and not in Z are in the diagonal and hence are not alternating. So the natural inclusion $C_i^{alt}(Z;\mathbb{Z}) \to C_i^{alt}(W;\mathbb{Z})$ is an isomorphism of complexes. □

A useful theorem in calculating alternating homology is the following.

Theorem 2.8 *Suppose $Y \subset X^k$ is such that $Y \cap \sigma(Y) = \emptyset$ for all $\sigma \in S_k - \{id\}$ and that Z is the orbit of Y. Then*

$$H_i^{alt}(Z;\mathbb{Z}) \cong H_i(Y;\mathbb{Z})$$

for all i.

Proof. Since $Y \cap \sigma(Y) = \emptyset$ for all $\sigma \in S_k - \{id\}$, each chain in Y corresponds to an alternating chain in Z given by the taking the alternation of the chain in Y. □

As one might expect the only Eilenberg-Steenrod axiom that is not true for alternating homology ($k > 2$), is the point axiom, since the alternating homology of a single point in the diagonal is trivial.

3 Multiple Point Spaces

Multiple point spaces are associated to any finite and proper map. The kth multiple point space is the closure of the set of k-tuples of pairwise distinct points having the same image. These sets are considerably simpler than the sets in the target formed by counting the number of preimages, the former may be non-singular in contrast to the highly singular latter. In effect the multiple point spaces act as a resolution of the image.

The singularity theory viewpoint is that $\mathcal{A}$-equivalence of maps leads to $\mathcal{K}$-equivalence of multiple points spaces. As $\mathcal{K}$-equivalence has been well studied we can transfer results from it to the $\mathcal{A}$-equivalence case. For example, for a finitely $\mathcal{A}$-determined corank 1 map germ, $f : (\mathbb{C}^n, 0) \to (\mathbb{C}^p, 0)$, $n < p$, the Milnor and Tjurina numbers of the multiple point spaces will all be $\mathcal{A}$-invariants of f, see for instance [20] and [16]. In [23] and [8] the $\mathcal{A}$-classification of maps from surfaces to three space is investigated. They could have just as easily worked on the classification of the multiple point spaces, (in fact this can provide a useful check of the classification). The $\mathcal{A}$-classification of corank 1 folding maps is equivalent to the $\mathcal{K}$-classification of double point spaces, (up to diffeomorphisms that preserve fixed point sets), see [23].

Naively, the double point space of a map f should be the set of pairs (x, y) such that $f(x) = f(y)$. This definition includes all pairs (x, x) and in general this leads to the double point space having components of different dimensions. So the following general definition of a multiple point space was introduced to cope with removing these unwanted pieces of the diagonal.

Definition 3.1 *Let $f : X \to Y$ be a finite map of topological spaces then the kth multiple point space of f, denoted $D^k(f)$ is defined by*

$$D^k(f) := closure\{(x_1, \dots, x_k) \in X^k | f(x_1) = \cdots = f(x_k) \text{ for } x_i \neq x_j, i \neq j\}.$$

There exist maps $\epsilon_{i,k}; D^k(f) \to D^{k-1}(f)$ induced from the natural maps $\tilde{\epsilon}_{i,k} : X^k \to X^{k-1}$ given by dropping the ith coordinate from X^k. There also exists maps $\epsilon_k : D^k(f) \to Y$ given by $\epsilon_k(x_1, \dots, x_k) = f(x_1)$.

Definition 3.2 *The kth image multiple point space, $M_k(f)$, is the space $\epsilon_k(D^k(f))$.*

The spaces $M_k(f)$ can be highly singular compared to $D^k(F)$, in the sense that $D^k(f)$ could be non-singular but $M_k(f)$ could have non-isolated singularities.

Example 3.3 Let $f : B^2 \to \mathbb{RP}^2$ be the quotient map that maps the unit disc B^2 to real projective space by antipodally identifying points on the boundary of the disc. Then $D^2(f) \subset B^2 \times B^2$ is just the circle Z in Example 2.5 and so $H_*^{alt}(D^2(f); \mathbb{Z})$ is the same as the alternating homology of Z. The set $D^3(f)$ is empty.

The above definition of multiple point spaces was introduced as a means of avoiding including bits of the diagonal, but consider the following, which we may call the 'idiot's definition of multiple point spaces':

$$ID^k(f) := \text{closure}\{(x_1, \ldots, x_k) \in X^k | f(x_1) = f(x_2) = \cdots = f(x_k)\}.$$

Thus, every point of the diagonal of X^2 is included in $ID^2(f)$ and for all k, $D^k(f) \subset ID^k(f)$, with the excess included in the diagonal. One of the main advantages of the idiot's definition is that it is easy to write down an explicit set of defining equations for each k. This fact has been incredibly useful in studying maps of corank greater than 1, for which defining equations of $D^k(f)$ are not generally known. This fact combined with the following theorem is crucial in much of the recent work on the local topology of images using the image computing spectral sequence.

Theorem 3.4 *The alternating homologies of $D^k(f)$ and $ID^k(f)$ are canonically isomorphic.*

Proof. Since $ID^k(f) \backslash D^k(f) \subset Diag(D^k(f))$ the homologies are the same by Theorem 2.7. □

So it turns out that the idiot's definition is not so idiotic after all. We take the trouble of eliminating extra pieces of the diagonal and then it emerges that the extra bits do not matter when considering alternating homology.

The fact that chains in the diagonal are not alternating is also useful in studying the multiple point spaces of restrictions of maps. Suppose that $F : X \to Y$ is a finite and proper map and that $Z \subset X$. Let $f := F|Z$. It is useful to know the relation between the multiple point spaces of F and f. Since we have $D^k(f) \backslash Diag(X^k) = (D^k(F) \cap Z^k) \backslash Diag(X^k)$ then we have $D^k(f) \subset D^k(F) \cap Z^k$. In general this is not an equality as the next example shows.

Example 3.5 Let $F : \mathbb{R}^2 \to \mathbb{R}^3$ be the parametrization of the Whitney umbrella, $F(x, y) = (x, xy, y^2)$. Then,

$$\begin{aligned} D^2(F) &= \text{closure}\{(x_1, y_1, x_2, y_2) \in \mathbb{R}^2 \times \mathbb{R}^2 \mid (x_1, x_1y_1, y_1^2) = (x_2, x_2y_2, y_2^2); \\ &\qquad (x_1, y_1) \neq (x_2, y_2)\} \\ &= \{(0, y_1, 0, -y_1) \in \mathbb{R}^4\}. \end{aligned}$$

Let $Z = \{y = 0\}$ and $f = F|Z$, i.e. $f(x) = (x, 0, 0)$. So $D^2(f) = \emptyset$, but

$$\begin{aligned} D^2(F) \cap Z^2 &= \{(0, y_1, 0, -y_1)\} \cap \{(x_1, 0, x_2, 0)\} \\ &= \{(0, 0, 0, 0)\} \neq D^2(f). \end{aligned}$$

However, when working with alternating homology this lack of equality is irrelevant.

Theorem 3.6 *Suppose $F : X \to Y$ is finite and proper and that $Z \subset X$. Then if $D^k(f)$ and $D^k(F) \cap Z^k$ are S_k-cellular,*

$$H_i^{alt}(D^k(f);\mathbb{Z}) \cong H_i^{alt}(D^k(F) \cap Z^k;\mathbb{Z}).$$

Proof. The isomorphism follows from Theorem 2.7 since

$$(D^k(F) \cap Z^k)\backslash D^k(f) \subset Diag(X^k).$$

□

4 The Image Computing Spectral Sequence

The image computing spectral sequence was introduced by Goryunov and Mond in [4] as a rational cohomological sequence arising from a finite and proper map, and it was used to study the rational homology of the images of finite complex analytic maps. It was then developed by Goryunov in [3] to work for integer homology when the map was further restricted to being semi-algebraic. The proof involved resolving the image in a similar way to the resolution of the space of knots for Vassiliev invariants, see [28]. Roughly speaking a point in the image with k preimages is replaced by a $(k-1)$-simplex, to form a new space homologically equivalent to the image. The filtration of this space, arising through taking the unions of all simplices of dimension less than k, yields a natural spectral sequence, the terms of which can be written as alternating homology of multiple point spaces. The integral homology spectral sequence works in greater generality than finite maps of semi-algebraic spaces as is obvious from Goryunov's proof in [3]. We find the following theorem in [13].

Theorem 4.1 *Suppose $f : X \to Y$ is a finite and proper continuous map, such that $D^k(f)$ has the S_k-homotopy type of an S_k-cellular complex for all $k > 1$ and $M_k(f)$ has the homotopy type of a cellular complex for all $k > 1$. Then there exists a spectral sequence*

$$E^1_{p,q} = H_q^{alt}(D^{p+1}(f);\mathbb{Z}) \implies H_{p+q+1}(f(X);\mathbb{Z}).$$

The differential is the naturally induced map

$$\epsilon_{1,k_*} : H_i^{alt}(D^k(f);\mathbb{Z}) \to H_i^{alt}(D^{k-1}(f);\mathbb{Z}).$$

The proof of this is given in [13].

5 Examples

The examples here are of a fairly simple nature, such as calculating the homology of real projective spaces. One of the intentions is that the examples illustrate techniques that we can apply in practice.

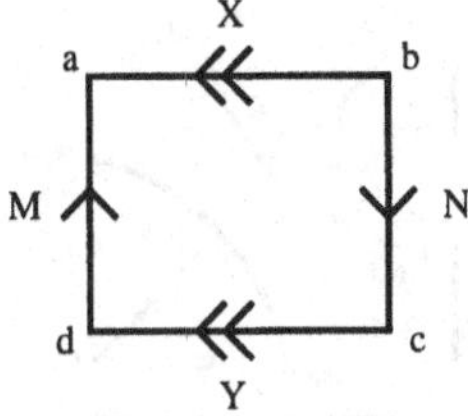

Figure 1: Identifications of a square to form a Klein Bottle.

Example 5.1 We have already calculated the alternating homology of the multiple point spaces for the map $f : B^2 \to \mathbb{RP}^2$, of Example 3.3. The image computing spectral sequence for f is given below.

H_2^{alt}	0	0	0
H_1^{alt}	0	0	0
H_0^{alt}	$\mathbb{Z}$	$\mathbb{Z}_2$	0
	B^2	$D^2(f)$	$D^3(f)$

All the differentials of this sequence must be trivial and so the sequence collapses at E^1 and since there are no extension difficulties we can read off the homology of the image of f, i.e. the real projective plane.

Example 5.2 Suppose PT is the pinched torus given by collapsing a generator of the torus to a point. The corresponding identification map is not finite but the image is homeomorphic to the image of the map $f : S^2 \to Y$ given by identifying the north pole, n, and the south pole, s. Then $D^2(f)$ is just the set $\{(n,s) \sqcup (s,n)\}$ and the sequence is described by the following.

H_2^{alt}	$\mathbb{Z}$	0	0
H_1^{alt}	0	0	0
H_0^{alt}	$\mathbb{Z}$	$\mathbb{Z}$	0
	S^2	$D^2(f)$	$D^3(f)$

The differential $E^1_{1,0} \to E^1_{0,0}$ is trivial and hence the sequences collapses at E^1. We can then read off that

$$H_i(Y;\mathbb{Z}) \cong \begin{cases} \mathbb{Z} & 0 \leq i \leq 2 \\ 0 & i > 2. \end{cases}$$

Example 5.3 The Klein bottle can be given as the image of a quotient map on a square. See Figure 1.

The multiple point spaces are as follows. The double point space is a set of four lines and four distinct points. The four lines arise from the orbits of the lines (X,Y) and (M,N). The points (a,b), (b,a), (c,d) and (d,c) are isolated points of $D^2(f)$. See Figure 2. The lines in $D^2(f)$ contract S_2-

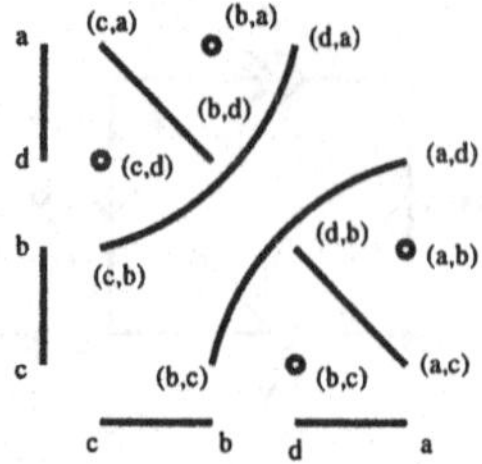

Figure 2: Double point space for Klein Bottle map.

homotopically to points. Thus $H_0^{alt}(D^2(f);\mathbb{Z}) \cong \mathbb{Z}^4$. The set $D^3(f)$ is the orbit under S_3 of the set of points (a,b,c), (a,b,d), (a,c,d) and (b,c,d). The set $D^4(f)$ is just the orbit of (a,b,c,d). Thus the sequence is as follows.

H_2^{alt}	0	0	0	0
H_1^{alt}	0	0	0	0
H_0^{alt}	$\mathbb{Z}$	$\mathbb{Z}^4$	$\mathbb{Z}^4$	$\mathbb{Z}$
	X	$D^2(f)$	$D^3(f)$	$D^4(f)$

We choose as generators of $H_0^{alt}(D^2(f);\mathbb{Z})$:

$$\begin{aligned} e_1 &= \mathrm{Alt}_{\mathbb{Z}}[c,a] = \mathrm{Alt}_{\mathbb{Z}}[b,d], & e_2 &= \mathrm{Alt}_{\mathbb{Z}}[b,a], \\ e_3 &= \mathrm{Alt}_{\mathbb{Z}}[c,d], & e_4 &= \mathrm{Alt}_{\mathbb{Z}}[d,a] = \mathrm{Alt}_{\mathbb{Z}}[c,b]. \end{aligned}$$

We choose as generators of $H_0^{alt}(D^3(f);\mathbb{Z})$:

$$\begin{aligned} f_1 &= \mathrm{Alt}_{\mathbb{Z}}[a,b,c], & f_2 &= \mathrm{Alt}_{\mathbb{Z}}[a,b,d], \\ f_3 &= \mathrm{Alt}_{\mathbb{Z}}[a,c,d], & f_4 &= \mathrm{Alt}_{\mathbb{Z}}[b,c,d]. \end{aligned}$$

We choose $\mathrm{Alt}_{\mathbb{Z}}[a,b,c,d]$ as the generator of $H_0^{alt}(D^4(f);\mathbb{Z})$. It is easy to calculate that

$$\begin{aligned} \epsilon_{1,3}(f_1) &= -e_4 + e_1 - e_2, & \epsilon_{1,3}(f_2) &= e_1 + e_4 - e_2, \\ \epsilon_{1,3}(f_3) &= e_3 + e_4 - e_1, & \epsilon_{1,3}(f_4) &= e_3 - e_1 - e_4, \end{aligned}$$

and hence

$$\begin{aligned} \mathrm{Im}\,\{\epsilon_{1,2} : \mathbb{Z}^4 \to \mathbb{Z}\} &= 0, \\ \mathrm{Ker}\,\{\epsilon_{1,3} : \mathbb{Z}^4 \to \mathbb{Z}^4\} &= \langle f_1 - f_2 + f_3 - f_4 \rangle \mathbb{Z}, \\ \mathrm{Ker}\,\{\epsilon_{1,4} : \mathbb{Z} \to \mathbb{Z}^4\} &= 0. \end{aligned}$$

Noting in particular that $\epsilon_{1,3}(f_2 + f_4) = -e_2 + e_3$ and $\epsilon_{1,3}(f_3 - f_4) = 2e_4$ we calculate that $E^2(f)$ is as follows.

0	0	0	0
$\mathbb{Z}$	$\mathbb{Z} \oplus \mathbb{Z}_2$	0	0

There are no extension problems and we can just read off the homology of the Klein Bottle.

Extension problems can occur even in simple examples of group actions.

Example 5.4 Suppose that we have the circle S^1. Using the angular coordinate $\mathbb{Z}_3 = \{0, 1, 2\}$ acts freely on it by rotation through 120 degrees. We wish to study the image of the quotient map, which we of course know is homeomorphic to S^1. The double point set is defined by

$$\begin{aligned} D^2(f) &= \text{closure}\{(x_1, x_2) \in S^1 \times S^1 \mid f(x_1) = f(x_2), x_1 \neq x_2\} \\ &= \text{closure}\{(x_1, x_2) \in S^1 \times S^1 \mid x_1 = x_2 + \frac{2\pi}{3} \text{ or } x_1 = x_2 + \frac{4\pi}{3}\} \\ &= \{(x_2 + \frac{2\pi}{3}, x_2)\} \cup \{(x_2 + \frac{4\pi}{3}, x_2)\}. \end{aligned}$$

Thus D^2 is a pair of circles in the torus and the action of S_2 is to exchange these circles, so $H_i^{alt}(D^2(f); \mathbb{Z}) = H_i(S^1; \mathbb{Z})$, by Theorem 2.8. Similar reasoning gives that $D^3(f)$ is also a pair of 2 disjoint circles and one can calculate that $H_i^{alt}(D^3(f); \mathbb{Z}) \cong H_i(S^1; \mathbb{Z})$. The set $D^4(f)$ is empty. These alternating homology groups fit into the spectral sequence as follows.

H_2^{alt}	0	0	0	0
H_1^{alt}	$\mathbb{Z}$	$\mathbb{Z}$	$\mathbb{Z}$	0
H_0^{alt}	$\mathbb{Z}$	$\mathbb{Z}$	$\mathbb{Z}$	0
	S^1	$D^2(f)$	$D^3(f)$	$D^4(f)$

The bottom row is easy to deal with. The alternating homology of $D^2(f)$ can be generated by $[\text{Alt}_{\mathbb{Z}}(0, \frac{2\pi}{3})]$ and so $\epsilon_{1,2}([\text{Alt}_{\mathbb{Z}}(0, \frac{2\pi}{3})]) = [0] - [\frac{2\pi}{3}]$, which is trivial in $H_0(S^1)$. So the differential $E^1_{1,0} \to E^1_{0,0}$ is the zero map.

Similarly $H_0^{alt}(D^3(f); \mathbb{Z})$ is generated by $[\text{Alt}_{\mathbb{Z}}(0, \frac{2\pi}{3}, \frac{4\pi}{3})]$. The differential is $\epsilon_{1,2}([\text{Alt}_{\mathbb{Z}}(0, \frac{2\pi}{3}, \frac{4\pi}{3})])$ which is $3[\text{Alt}_{\mathbb{Z}}(0, \frac{2\pi}{3})]$. Thus the differential $E^1_{2,0} \to E^1_{1,0}$ is multiplication by 3.

The map $\epsilon_{1,2}$ projects each disjoint piece of $D^2 = P_1 \cup P_2$ homeomorphically to S^1. So the induced map on homology, $H_1^{alt}(D^2(f); \mathbb{Z}) \to H_1(S^1; \mathbb{Z})$ is

$$\epsilon_{1,2_*}([P_1] - \sigma[P_1]) = \epsilon_{1,2_*}([P_1] - [P_2]) = [S^1] - [S^1] = 0,$$

where σ is the non-trivial element of S_2. The spectral sequence at E^2 is the following.

0	0	0	0
$\mathbb{Z}$	A	B	0
$\mathbb{Z}$	$\mathbb{Z}_3$	0	0

The groups A and B are to be determined. Now, since the differentials are the 'knight's move' on the chessboard and must be trivial it follows that the sequence collapses at E^2 and $E^2 = E^\infty$. This implies that $H_2(f(S^1); \mathbb{Z}) \cong A$

and $H_3(f(S^1);\mathbb{Z}) \cong B$. The domain of f is a one-dimensional space and so the image is too (we will ignore the fact that we know the image is S^1). Thus homology is trivial above degree 1 and so $A \cong B \cong 0$. (It is possible to calculate A and B by studying the differentials, however the method used exemplifies something more general, see Section 4.1 of [15]). This implies that the map $\epsilon_{1,3} : E^1_{2,1} \to E^1_{1,1}$ is an isomorphism.

Thus the sequence at E^2 is

0	0	0	0
$\mathbb{Z}$	0	0	0
$\mathbb{Z}$	$\mathbb{Z}_3$	0	0

From this we can see that the possible choices for $H_1(f(S^1);\mathbb{Z})$ are $\mathbb{Z}$ and $\mathbb{Z}\oplus\mathbb{Z}_3$, (as the image is homeomorphic to S^1 we know it is the former). This example illustrates the point that even in this simple group action setting we have extension problems.

By restricting a map to a subspace we can often compare alternating homology groups of spaces as the next example shows.

Example 5.5 The pseudo projective plane of order d, P_d is the quotient of the unit disk B^2 under the relation (in polar coordinates) of $(1,\theta) \sim (1, \theta + \frac{2\pi}{d})$. Hence P_2 is just the ordinary real projective plane.

For $d = 3$ the quotient map f restricted to the boundary of the disc is the same as the quotient map of the previous example. So apart from the domain the multiple point spaces for the quotient map for P_3 are the same as in that example. Thus there is a good deal of similarity between the spectral sequence there and the sequence $E^1(f)$ below.

H_2^{alt}	0	0	0	0
H_1^{alt}	0	$\mathbb{Z}$	$\mathbb{Z}$	0
H_0^{alt}	$\mathbb{Z}$	$\mathbb{Z}$	$\mathbb{Z}$	0
	B^2	$D^2(f)$	$D^3(f)$	$D^4(f)$

All the differentials are the same as for the above example. By comparison of the two sequences the differentials must coincide for $p > 0$. The only possible difference could have been $E^1_{1,1} \to E^1_{0,1}$ but this is found to be trivial as in the previous example. So the sequence collapses at E^2 to give the following.

0	0	0	0
0	0	0	0
$\mathbb{Z}$	$\mathbb{Z}_3$	0	0

(Note that since the domain and image are two dimensional we cannot deduce directly, as in the last example, that $E^2_{1,1} = 0$.) From the sequence we deduce that $H_1(P_3;\mathbb{Z}) = \mathbb{Z}_3$ and $H_0(P_3;\mathbb{Z}) = \mathbb{Z}$ and zero in other degrees.

Example 5.6 An arrangement in $\mathbb{R}^n$ is simply the inclusion of a number of planes into $\mathbb{R}^n$. (The planes do not have to be hyperplanes). Such a map is ideal for study by the image computing spectral sequence and in fact in [28] p114, Vassiliev used his resolution method (which inspired the proof of existence of the image computing spectral sequence) to study such images.

6 Homotopy results

Rather surprisingly we may use alternating homology and the image computing spectral sequence to deduce homotopy results about the image of a map. For example, the triviality of the fundamental group of the image can sometimes be found using alternating homology.

Let $f : X \to Y$ be a finite and proper map such that $\dim D^2(f) < \dim X$. We generalise a technique from [25], where stable maps from surfaces to 3-space are considered. However the principle is much the same. We attempt to lift loops in the image to loops in the source. The group $H_0^{alt}(D^2(f);\mathbb{Z})$ acts as an obstruction to this. We can arrange the attempted lift so that any discontinuity in the loop occurs above a double point in the image, as outside the double point set in the target the map is a cover, i.e. $f(x_1) = f(x_2)$ for $x_1 \neq x_2$.

The statement $H_0^{alt}(D^2(f);\mathbb{Z}) = 0$ is equivalent to the existence of a path from the point $(x_1, x_2) \in D^2(f)$ to a point (x, x) of the diagonal, this is the content of Lemma 2.6 of [10]. By projecting from $X \times X$ to X on the first component this path projects onto a path from x_1 to x in X and hence to one in Y. By projecting on the second component we can get a path from x to x_2 in X. Combining these two projected paths we get a path from x_1 to x_2 which can be used to repair the original discontinuity. Thus we have repaired the lift above $f(x_1)$, which projects to a loop in Y homotopically equivalent to the original loop.

By performing this process for all discontinuities we get a loop in Y which lifts to a loop in X. The foregoing is stated precisely in the next theorem.

Theorem 6.1 (Corollary 4.19 of [10]) *Suppose $f : X \to Y$ is a proper and finite surjective stratified submersion with X path connected and there exists a point $y \in Y$ with only one preimage. If $H_0^{alt}(D^2(f);\mathbb{Z}) = 0$ then $\pi_1(X)$ surjects onto $\pi_1(Y)$.*

We can also prove triviality of the fundamental group of an image when the group $H_0^{alt}(D^2(f);\mathbb{Z})$ is non trivial. See Section 4.4 of [10].

We shall now use these ideas to prove the well known fact that the dunce cap is contractible.

Example 6.2 Suppose X is a triangle and f is the quotient map given by the identification of the edges, as in Figure 3, to form the dunce cap Y.

We can calculate $D^2(f)$ as the union of the orbits of the lines (A, C), (B, C) and (A, B). This is pictured in Figure 4. We take as an alternating

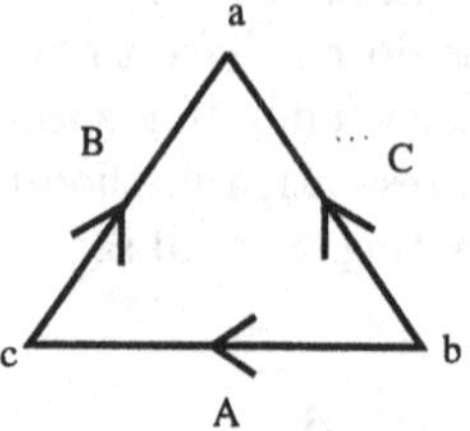

Figure 3: The dunce cap

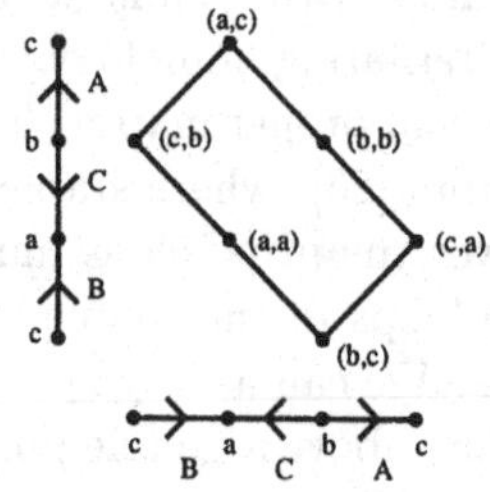

Figure 4: The double point set of the quotient map for the duncecap

1-cycle

$$e_1 = \mathrm{Alt}_{\mathbb{Z}}\{[(b,b) \to (a,c)] + [(a,c) \to (c,b)] + [(c,b) \to (a,a)]\}.$$

So, $H_1^{alt}(D^2(f);\mathbb{Z}) = \mathbb{Z}$. Since any point in $D^2(f)$ is homologous to a point in the diagonal we find $H_0^{alt}(D^2(f);\mathbb{Z}) = 0$.

The triple point space is generated by the identification of the lines A, B and C. Thus $D^3(f)$ is made of triples of points on these lines and we can take as a 1-cell the open interval whose closure starts at (b,c,b) and ends at (c,a,a), denoted $[(b,c,b) \to (c,a,a)]$. The alternating chain $\mathrm{Alt}_{\mathbb{Z}}[(b,c,b) \to (c,a,a)]$ will generate $C_1^{alt}(D^3(f);\mathbb{Z})$ and is a 1-cycle and so can be taken as a generator of $H_1^{alt}(D^3(f);\mathbb{Z})$ as well. Since any point in $D^3(f)$ is homologous to a point in the diagonal we find $H_0^{alt}(D^3(f);\mathbb{Z}) = 0$.

The spectral sequence looks like the following.

H_2^{alt}	0	0	0
H_1^{alt}	0	$\mathbb{Z}$	$\mathbb{Z}$
H_0^{alt}	$\mathbb{Z}$	0	0
	X	$D^2(f)$	$D^3(f)$

We find that

$$\epsilon_{3,1}(\mathrm{Alt}_{\mathbb{Z}}[(b,c,b) \to (c,a,a)]) = e_1$$

and so the differential $E_{2,1}^1 \to E_{2,1}^1$ is an isomorphism. The E^2 terms of the sequence are

0	0	0
0	0	0
$\mathbb{Z}$	0	0

This implies that the reduced homology $\tilde{H}_*(Y;\mathbb{Z})$ is trivial. Now, $H_0^{alt}(D^2(f);\mathbb{Z}) = 0$ implies that $\pi_1(X)$ surjects onto $\pi_1(Y)$ by Theorem 6.1 and so the dunce cap is simply connected. A simply connected, homologically trivial complex is well known to be contractible; this is just a consequence of Hurewicz's theorem.

One of the main methods of turning a homology result into a homotopy one is to use Whitehead's Theorem. We can use it in conjunction with the image computing spectral sequence as follows.

Suppose we have two surjective maps $f_i : X_i \to Y_i$, $i = 1, 2$ for which the image computing spectral sequence exists and there exist maps $g : X_1 \to X_2$ and $h : Y_1 \to Y_2$. Then we can use the spectral sequence to compare the homologies of Y_1 and Y_2. If for some r, $E^r(f_1)$ and $E^r(f_2)$ are isomorphic as sequences then the map $h_* : H_*(Y_1;\mathbb{Z}) \to H_*(Y_2;\mathbb{Z})$ is an isomorphism. If the fundamental groups of Y_1 and Y_2 are trivial then by J.H.C. Whitehead's theorem (see [29], p.181), Y_1 and Y_2 are homotopically equivalent.

7 Singularity Theory

The image computing spectral sequence arose out of the study of singularities of complex analytic maps. In this section we see how to find the multiple point spaces for corank 1 complex analytic maps and outline how the sequence has been used to study the local topology of complex analytic maps.

7.1 Multiple point spaces for corank 1 maps

The way maps behave under deformation is one of the central interests of Singularity Theory. Unfortunately, using the definition given in Section 3, multiple point spaces do not behave well under deformation of maps. By behaving well we mean that for an unfolding $F : X \times S \to Y$ of $f := F|(X \times \{s\})$ for $s \in S$, the following should be a fibre square.

$$\begin{array}{ccc} D^k(f) & \longrightarrow & D^k(F) \\ \downarrow & & \downarrow \\ \{s\} & \longrightarrow & S \end{array}$$

For complex analytic map-germs of corank 1 there exists another definition of multiple point spaces which does behave well under deformation. This was developed by Mond following an idea of Roberts, see [24] and then by Marar and Mond in [20]. We shall now explain the salient points of this theory.

Suppose $f : (\mathbb{C}^n, 0) \to (\mathbb{C}^p, 0)$ is a complex analytic map-germ of corank 1, with $n < p$. Then there exist coordinates on $\mathbb{C}^n$ so that,

$$f(x_1, \ldots, x_{n-1}, y) = (x_1, \ldots, x_{n-1}, h_1(x, y), \ldots, h_{p-n+1}(x, y)).$$

Then we define the kth multiple point space of f, denoted by $\tilde{D}^k(f)$ to distinguish it from our earlier definition, by the equations

$$\begin{vmatrix} 1 & y_1 & \cdots & y_1^{i-1} & h_j(x, y_1) & y_1^{i+1} & \cdots & y_1^{k-1} \\ & & & & \vdots & & & \\ 1 & y_k & \cdots & y_k^{i-1} & h_j(x, y_k) & y_k^{i+1} & \cdots & y_k^{k-1} \end{vmatrix} \Bigg/ \begin{vmatrix} 1 & y_1 & \cdots & y_1^{k-1} \\ & & \vdots & \\ 1 & y_k & \cdots & y_k^{k-1} \end{vmatrix}$$

for $i = 1, \ldots, k-1$ and $j = 1, \ldots, p-n+1$. The space $\tilde{D}^k(f)$ is defined in $\mathbb{C}^{n-1} \times \mathbb{C}^k$ where the coordinates are $(x_1, \ldots, x_{n-1}, y_1, \ldots, y_k)$. These $(k-1)(p-n+1)$ equations are S_k-invariant. Outside of the diagonal the spaces $D^k(f)$ and $\tilde{D}^k(f)$ coincide. (For $\tilde{D}^k(f)$ the diagonal is the set of points in $\mathbb{C}^{n-1} \times \mathbb{C}^k$ that do not lie in regular orbits of S_k). Furthermore for $\tilde{D}^k(f)$, we can, for instance, get isolated points in the diagonal, which is not possible with our original definition of multiple point spaces.

Thus we have a neat description of multiple point spaces of corank 1 maps. We know the number of equations, they are S_k-invariant and the spaces behave well under deformation of f. Even for corank 2 maps a general description of the functions defining the multiple point spaces is not known to me. Such a description would be extremely useful for dealing with the algebra of complex analytic maps. Fortunately in an investigation of the topology we can just use the functions defining the idiot's multiple point space.

The essential parts of Theorem 1.14 of [20] are summed in the next two statements.

Let $f : (\mathbb{C}^n, S) \to (\mathbb{C}^p, 0)$ be a corank 1 finite complex analytic multi-germ with $n < p$. Then f is $\mathcal{A}$-stable if and only if $\tilde{D}^k(f)$ is nonsingular of dimension $p - k(p-n)$ or empty for all $k \geq 2$.

By using the coherent sheaf form of Hilbert's Nullstellensatz we can deduce that f is finitely $\mathcal{A}$-determined if and only if, for all k, $\tilde{D}^k(f)$ is an isolated complete intersection singularity of dimension $p - k(p-n)$, is empty or if $p - k(p-n) < 0$ it consists of at most a single point.

In fact Marar and Mond have a stronger result in [20] involving the restriction of $\tilde{D}^k(f)$ to fixed point sets of the action of S_k on $\mathbb{C}^{n-1} \times \mathbb{C}^k$.

Thus the close connection between the $\mathcal{A}$-equivalence of mappings and the $\mathcal{K}$-equivalence of multiple point spaces can be seen. It would be interesting to find a connection between the $\mathcal{A}_e$-codimension of f and the $\mathcal{K}_e$-codimension of the $D^k(f)$ (under the action of S_k for each k), particularly in the case $p = n+1$.

7.2 Using the Sequence in Local Topology

In this section we outline how the spectral sequence has been useful in the study of local topology of the images of finite complex analytic spaces. Local topology is the study of sufficiently small spheres or local hyperplane sections in a space. This is important as the local knowledge can lead to global knowledge, see [1], [2], [6], [7] and [11]. The local topology of complex hypersurfaces was studied by Milnor in [22] and then by Hamm [5] for isolated complete intersection singularities. The study of local topology of the image of a finite complex analytic map was approached recently by Mond, see [26] for an introduction, where the main local objects are disentanglements of the map. The topology of disentanglements have been studied in [25], [17], [18], [19], [4], [3], [9] and [10]. It became obvious that the image computing spectral sequence could be used to study the topology of sufficiently small spheres in the image of a complex analytic map and this was begun in [10], [11] and [14].

Suppose that Z is a complex analytic space and let $z \in Z$ be a point. Then locally Z embeds into $\mathbb{C}^N$ for some N. The intersection of Z and a sufficiently small sphere of radius ϵ, $L_z = Z \cap S_\epsilon^{2N-1}(z)$, is an analytic invariant of Z and is called the real link of z in Z. (We can define the complex link of the point by taking a local nearby hyperplane section of the point, see [2]). To study the local topology of spaces we need to take the links of strata of a stratification of Z, that is, take a transverse slice to the stratum at z and take the real link of the point in the slice. These stratification technicalities obscure the discussion of how to use the image computing spectral sequence and so we shall ignore them. We outline the basic idea but in practice we take real links of strata. The statements of theorems in [10] are given in terms of rectified homotopical depth which is another technicality I have chosen to ignore.

The best theorem in the study of local topology for calculating specific examples was proved by Hamm in 1969, in [5], following Milnor, [22], and this was that the connectivity of the real link of z was determined by the number of equations needed to define Z at z. This is stated as

$$\pi_i(L_z) = 0 \text{ for } i < N - r - 1,$$

where r is the number of equations needed to define Z in $\mathbb{C}^N$ at z. This result leads to a proof of the Milnor fibre of an isolated complete intersection singularity being homotopically equivalent to a wedge of spheres.

Now suppose that $F : X \to Y$ is a finite and proper complex analytic map where X and Y are manifolds of dimension n and p respectively and $n < p$. (The results of [10] allow more general spaces, but we wish to keep things simple).

Let Z be the image of F and $z \in Z$ be a point with q preimages. The link of z in Z, L_z, is the image of the finite and proper semi-analytic map f given by $f := F|F^{-1}(L_z)$. We can use the image computing spectral sequence to find the homology of L_z. The multiple point spaces of f have the same alternating homology of orbits of links in $D^k(F)$: Firstly, Theorem 3.6 tells us that $D^k(f)$ and $D^k(F) \cap (F^{-1}(L_z))^k$ have the same alternating homology. Let $S_\epsilon(w_1)$ be a sphere centred at a point $w_1 \in \epsilon_k^{-1}(z)$, such that $S_\epsilon(w_1)$ is H-invariant, where H is the isotropy subgroup of S_k at w_1. We can do this for a representative of each orbit of points in $\epsilon_k^{-1}(z)$. Let $S = S_\epsilon(w_1) \cup \cdots \cup S_\epsilon(w_r)$. If we exclude any components in $D^k(F) \cap (F^{-1}(L_z))^k$ that lie wholly in the diagonal then the space $D^k(F) \cap (F^{-1}(L_z))^k$ is an S_k-invariant space which is S_k-homeomorphic to $D^k(F) \cap Orbit(S)$, the orbit of links in $D^k(F)$ mentioned above. The multiple point spaces of F will be defined by functions which are in general not known, so we turn to the idiot's definition of multiple point spaces and consider $ID^k(F)$, and $ID^k(F) \cap Orbit(S)$. The number and form of the defining equations of $ID^k(F)$ are known and we can adapt Hamm's theorem, using Theorem 2.7, to show that

$$H_i^{alt}(D^k(F) \cap Orbit(S); \mathbb{Z}) = 0 \text{ for } 0 < i < nk - p(k-1) - 1.$$

This is the essence of Proposition 3.9 of [10]. The groups $H_0^{alt}(D^k(f) \cap Orbit(S); \mathbb{Z})$ can be calculated to be $\mathbb{Z}^{\frac{q}{(q-k)!q!}}$. For low values of k and i the groups $H_i^{alt}(D^k(F) \cap Orbit(S); \mathbb{Z})$ and $H_i^{alt}(D^k(f) \cap Orbit(S); \mathbb{Z})$ will be isomorphic. We can put this information into our spectral sequence, which will resemble the following example.

H_7^{alt}	$*$	$*$	$*$	$*$	$*$
H_6^{alt}	0	$*$	$*$	$*$	$*$
H_5^{alt}	0	$*$	$*$	$*$	$*$
H_4^{alt}	0	0	$*$	$*$	$*$
H_3^{alt}	0	0	$*$	$*$	$*$
H_2^{alt}	0	0	0	$*$	$*$
H_1^{alt}	0	0	0	$*$	$*$
H_0^{alt}	$\mathbb{Z}^q$	$\mathbb{Z}^{\frac{q!}{(q-2)!2}}$	$\mathbb{Z}^{\frac{q!}{(q-3)!6}}$	$\mathbb{Z}^{\frac{q!}{(q-4)!24}}$	$*$
	X	$D^2(f)$	$D^3(f)$	$D^4(f)$	$D^5(f)$

Here $*$ indicates a possibly non-zero entry. We can easily calculate the differential along the bottom row and and find that the bottom of the sequence clears out to a certain point to leave the following.

$*$	$*$	$*$	$*$	$*$
0	$*$	$*$	$*$	$*$
0	$*$	$*$	$*$	$*$
0	0	$*$	$*$	$*$
0	0	$*$	$*$	$*$
0	0	0	$*$	$*$
0	0	0	$*$	$*$
$\mathbb{Z}$	0	0	$*$	$*$

Thus it can be seen that there is a triangle in the bottom left of this quadrant, which is all zeroes apart from $E^2_{0,0}$. (The $*$ in the bottom row column 4 occurs because there is no guarantee that the image of the differential arising from the map $D^5(f)$ to $D^4(f)$ is isomorphic to the kernel of the differential arising from $D^4(f)$ to $D^3(f)$). This region will of course be the same at E^∞ and so we can describe the homology of the link below a certain level. In the above sequence reduced homology is trivial below degree 4.

Although the group $H_0^{alt}(D^2(f);\mathbb{Z})$ is not trivial, knowledge of the sequence can be used to give homotopy results. See Proposition 4.20 of [10].

8 Future Work

There is much work to be done on the images of finite maps using alternating homology and the image computing spectral sequence. To conclude the paper we will describe some possibly fruitful areas of study.

Finite Group Actions

Even in the simplest examples the multiple point spaces are difficult to describe. However they, and the restrictions to fixed point sets, will give numerous invariants for such actions. See Example 5.4.

Local Topology

The study of local topology of images of finite complex analytic map is far from complete. One important conjecture is the following.

Conjecture 8.1 Conjecture 4.28 of [10]. The complex links of the image of a dimensionally correct complex analytic map are homotopically equivalent to wedges of spheres of varying dimension.

Conjecture 8.2 Conjecture 3.5 of [11]. Suppose $f : X \to \mathbb{C}^p$ is a finite complex analytic map with dim $X < p$ and q the maximal number of preimages of any point $\mathbb{C}^p$, then

$$rhd(f(X)) \geq (rhd(X) - p + 1)q - p + 1$$

where rhd denotes the rectified homotopical depth of a space, see [7].

We can also attempt to generalise the results of [14] to maps not induced from corank 1 maps. However, I have no firm conjectures to make here.

It is conjectured that the complement of bifurcation diagrams of singularities are M-manifolds, see [27]. There Nekrasov uses a resolution like Vassiliev's to prove the conjecture for the A_μ series of singularities. Using the fact that bifurcation spaces are the images of maps is it possible to prove the result in general?

A DeRham Theorem

A DeRham type theorem for alternating homology seems reasonable. Is it possible to define differential forms on the idiot's definition of multiple point spaces and use them to calculate the alternating homology of the true multiple point spaces?

Algebra

In [4] Goryunov and Mond developed the spectral sequence as a cohomological sequence over the rationals using the resolution

$$0 \to \mathbb{Q}_Y \to \mathbb{Q}_X \to \mathrm{Alt}_2 \mathbb{Q}_{D^2(f)} \to \dots .$$

What would be interesting would be to find other sheaves and algebraic structures associated to an image for which we could resolve using structures associated to multiple point spaces and find a similar spectral sequence. For example, through adaption of the argument in [4] we find, for $\mathcal{F}_Y$ a sheaf of vector spaces over $\mathbb{Q}$ on Y, that

$$0 \to \mathcal{F}_Y \to f_* f^*(\mathcal{F}_Y) \to \mathrm{Alt}_2 \epsilon_{2*} \epsilon_2{}^*(\mathcal{F}_Y) \to \dots$$

is exact. The difficulty arises in identifying interesting $\mathcal{F}$ and giving a useful description of $\epsilon_k{}^*\mathcal{F}$.

For f a corank 1 map the precise relation between $\mathcal{A}_e$-normal space of f and $\mathcal{K}_e$-normal spaces of the multiple point spaces needs to be found. For folding maps $\mathcal{A}_e$-codimension is equal to symmetric $\mathcal{K}_e$-codimension of $D^2(f)$, (that is the symmetric part of the $\mathcal{K}_e$-normal space of $D^2(f)$). In general, for map-germs $f : (\mathbb{C}^n, 0) \to (\mathbb{C}^{n+1}, 0)$ is the sum of the symmetric $\mathcal{K}_e$-codimensions of the multiple point spaces equal to the $\mathcal{A}_e$-codimension of f? Such a theorem might follow from the investigation of a resolution of the $\mathcal{A}_e$-normal space, as described above.

In the non-corank 1 case can we use the idiot's definition of multiple point spaces to study the algebra of images?

Abbreviations:
SLNM=Springer Lecture Notes in Mathematics.
AMS=American Mathematical Society.

References

[1] W. Fulton, On the topology of algebraic varieties, in *Proc. Symp. Pure Math. 46*, (AMS, Providence R.I. 1987), 15-46.

[2] M. Goresky and R. MacPherson, *Stratified Morse Theory*, (Springer Verlag, Berlin 1988).

[3] V.V. Goryunov, Semi-simplicial resolutions and homology of images and discriminants of mappings, *Proc. London Math. Soc.*, **70** (1995), 363-385.

[4] V.V. Goryunov and D.M.Q. Mond, Vanishing cohomology of singularities of mappings, *Compositio Math.*, **89** (1993), 45-80.

[5] H.A. Hamm, Lokale topologische Eigenschaften komplexer Räume, *Math. Ann.*, **191** (1971), 235-252.

[6] H.A. Hamm, On the vanishing of local homotopy groups for isolated singularities of complex spaces, *J. Reine Angew. Math.*, **323** (1981), 172-176.

[7] H.A. Hamm and Lê D.T., Rectified homotopical depth and Grothendieck conjectures, in *The Grothendieck Festschrift, Volume II*, Progress in Mathematics 87, (Birkhäuser, Boston 1990), 311-351.

[8] C.A. Hobbs and N.P. Kirk, On the Classification and Bifurcation of Multigerms of Maps from Surfaces to 3-Space, to appear in *Mathematica Scandinavica.*

[9] K. Houston, *Local Topology of Finite Complex Analytic Maps*, PhD Thesis, University of Warwick, 1994.

[10] K. Houston, Local Topology of Images of Finite Complex Analytic Maps, *Topology*, Vol 36, (1997), 1077-1121.

[11] K. Houston, Global Topology of Images of Finite Complex Analytic Maps, *Mathematical Proceedings of the Cambridge Philosophical Society*, (1997), 122, 491-501.

[12] K. Houston, On Singularities of Folding Maps and Augmentations, to appear in *Mathematica Scandinavica.*

[13] K. Houston, A General Image Computing Spectral Sequence, in preparation.

[14] K. Houston, Rectified Homotopical Depth and Image Multiple Point Spaces, to appear in *Proceedings of the American Mathematical Society.*

[15] K. Houston, Images of Finite Maps with One Dimensional Double Point Set, to appear in *Topology and Its Applications.*

[16] K. Houston and N.P. Kirk, On the Classification and Geometry of Corank 1 Map-Germs from Three-Space to Four-Space, these proceedings.

[17] T. De Jong and D. Van Straten, Disentanglements, SLNM 1462, (Springer Verlag, Berlin 1991), 199-211.

[18] W.L. Marar, *Mapping fibrations and multiple point schemes*, PhD Thesis, University of Warwick, (1989).

[19] W.L. Marar, Mapping fibrations, *Manuscripta Math.*, **80** (1993), 273-281.

[20] W.L. Marar and D.M.Q. Mond, Multiple point schemes for corank 1 maps, *J. London Math. Soc.*, (2) **39** (1989), 553-567.

[21] W.L. Marar and D.M.Q. Mond, Real Map-Germs with Good Perturbations, *Topology*, Vol 35, (1996), 157-165.

[22] J. Milnor, *Singular Points of Complex Hypersurfaces*, Ann. of Math. Studies no 61, (Princeton University Press, 1968).

[23] D. Mond, On the classification of germs of maps from $\mathbb{R}^2 \to \mathbb{R}^3$, *Proc. London Math. Soc.* **50** (1985), 333-369.

[24] D. Mond, Some remarks on the geometry and classification of germs of maps from surfaces to 3-space, *Topology*, **26** (1987), 361-383.

[25] D. Mond, Vanishing cycles for analytic maps, SLNM 1462, (Springer Verlag, 1991), 221-234.

[26] D. Mond, Singularities of mappings from surfaces to 3-space, in *Singularity Theory (Proceedings of the College of Singularity Theory, I.C.T.P., Trieste, Italy (1991))*, pp509-526, World Scientific Publ. Co. Singapore 1995.

[27] N.A. Nekrasov, On the cohomology of the complements of the bifurcation diagram of the singularity A_μ, *Functional Analysis and it Applications*, **27**, (1993), 245-250.

[28] V.A. Vassiliev, *Complements of Discriminants of Smooth Maps: Topology and Applications*, Translations of Mathematical Monographs, (AMS, Providence R.I. 1992).

[29] G.W. Whitehead, *Homotopy Theory*, Graduate Texts in Mathematics 61, (Springer Verlag, Berlin 1978).

Mathematics and Statistics Group
Middlesex University
Bounds Green Road
London, N11 2NQ
U.K.

e-mail: k.houston@mdx.ac.uk

On the Classification and Geometry of Corank 1 Map-Germs from Three-Space to Four-Space

Kevin Houston Neil Kirk*

Dedicated to Terry Wall on the occasion of his 60th birthday.

The classification of germs of maps plays an important role in singularity theory. Not only do we obtain specific examples to which existing theory may be applied, but new phenomena often emerge and motivate new ideas. Here we present a classification of the simple corank 1 map-germs $(\mathbb{C}^3,0) \to (\mathbb{C}^4,0)$ up to $\mathcal{A}$-equivalence together with those corank 1 germs of $\mathcal{A}_e$-codimension $\leq$ 4. The classification proceeds inductively at the jet-level using the technique of Complete Transversals developed by Bruce, Kirk and du Plessis [1], and the unipotent determinacy theorems of Bruce, du Plessis and Wall [2]. The calculations are somewhat intensive and are performed using the computer package `Transversal` written by the second author [9, 10]. (This package can deal with numerous calculations central to classification and unfolding theory. Its main aim is to calculate tangent spaces to orbits in a given jet-space and implement the complete transversal classification technique. All the standard equivalence relations are covered but the main concern is for $\mathcal{A}$ classifications.)

We adopt the ideas developed by Mond for map-germs $(\mathbb{C}^2,0) \to (\mathbb{C}^3,0)$. (This classification began as a Ph.D. thesis supervised by Wall [16, 17] and has motivated a lot of work on the $\mathcal{A}$-classification and geometry of map-germs $(\mathbb{C}^n,0) \to (\mathbb{C}^p,0)$ for $n < p$.) In particular, we follow the work of Marar and Mond and, for a map-germ f, consider its disentanglement $\mathrm{Im}(f_t)$, multiple point spaces $D^k(f)$ and image Milnor number (number of vanishing cycles) $\mu_I(f)$, [7, 13, 14, 18, 19]. We describe a general procedure for calculating μ_I which combines the topological techniques discussed by Wall in [22] with the image computing spectral sequence introduced by Goryunov and Mond in [5]. It is a corrected version of a similar argument in [5]. We describe how the procedure applies to the examples in our list, in particular, how one calculates the invariants for general series. For specific cases the same methods may be performed by a computer using a package

*The authors were supported by the EPSRC grants GR/K29227 and GR/J28162; computational work was supported by grant GR/H59855.

such as Singular, [21]. Computer calculations helped us to conjecture the general results for series, and deal with the more 'exotic' normal forms on our list. The 'Mond Conjecture' ($\mathcal{A}_e$-codimension $\leq \mu_I(f)$ with equality if and only if f is weighted homogeneous) is verified for all of the examples in our classification.

We review some of the current techniques relating to classification theory and the topology of map-germs. The topological methods are further exemplified by the detailed examples given at the end of the paper.

Acknowledgements. We would like to thank J.W. Bruce for his help via numerous discussions and suggestions, and V.V. Goryunov and C.T.C. Wall for helpful comments made during the preparation of this article.

1 Classification of Map-Germs

Our main reference for singularity theory will be [23]. We shall use the classification techniques developed in [1] and [2] and adopt the notation given there. The following theorem summarises our main results.

Theorem 1.1. *The list of all $\mathcal{A}$-simple singularities $(\mathbb{R}^3, 0) \to (\mathbb{R}^4, 0)$ is given in Table 1. The list of singularities of $\mathcal{A}_e$-codim ≤ 4 consists of the immersion and the cases A_k, D_k, B_k, C_k, F_4, Q_k, R_k, $(k \leq 4)$, P_1, P_2, P_3^k $(k = 2, 3)$, P_4^1 and $S_{j,k}$ $(k + j \leq 4)$; plus the non-simple germs given in Table 2 (for the latter, by 'codimension ≤ 4' we mean the codimension of the strata). The columns marked 'Det' and 'Cod' give the determinacy degree and $\mathcal{A}_e$-codimension, respectively. (For strata 'Cod' refers to the $\mathcal{A}_e$-codimension of a representative of the strata.)*

The full calculations are very extensive and we omit the details. As a 'proof' we give, in an appendix, a complete summary of the stratification of the jet-space (the 'classification tree') via a series of tables which describe how the classification evolves inductively from the set of $(k-1)$-jets to the corresponding set of k-jets. The size of our classification renders a graphical representation of the stratification (using, say, the 'tree diagrams' akin to those found in the classifications of Mond, [16, 17]) infeasible. However, we feel that our method provides a good substitute. A review of the technical machinery used to perform the classification is discussed below.

We remark that our results also apply to the complex analytic case and provide a corresponding classification of map-germs $(\mathbb{C}^3, 0) \to (\mathbb{C}^4, 0)$. There is, of course, some simplification in this setting due to collapsing of orbits. In particular, many orbits which differ just by a $\pm$ sign in a term collapse to a single orbit. Further simplification of orbits is discussed where appropriate. We present the results initially in the real differentiable case for completeness. When we come to discuss topological invariants we will restrict our attention to the complex analytic case. We will use $\mathbb{K}$ to mean either $\mathbb{R}$ or $\mathbb{C}$ and in this

Label	Singularity	Det	Cod	Conditions
-	$(x, y, z, 0)$	1	0	
A_0	(x, y, z^2, xz)	2	0	
A_k	$(x, y, z^2, z(z^2 \pm x^2 \pm y^{k+1}))$	$k+2$	k	$k \geq 1$
D_k	$(x, y, z^2, z(z^2 + x^2y \pm y^{k-1}))$	k	k	$k \geq 4$
E_6	$(x, y, z^2, z(z^2 + x^3 \pm y^4))$	5	6	
E_7	$(x, y, z^2, z(z^2 + x^3 + xy^3))$	5	7	
E_8	$(x, y, z^2, z(z^2 + x^3 + y^5))$	6	8	
B_k	$(x, y, z^2, z(x^2 \pm y^2 \pm z^{2k}))$	$2k+1$	k	$k \geq 2$
C_k	$(x, y, z^2, z(x^2 + yz^2 \pm y^k))$	$k+1$	k	$k \geq 3$
F_4	$(x, y, z^2, z(x^2 + y^3 \pm z^4))$	5	4	
P_1	$(x, y, yz + z^4, xz + z^3)$	4	1	
P_2	$(x, y, yz + z^5, xz + z^3)$	5	2	
P_3^k	$(x, y, yz + z^6 \pm z^{3k+2}, xz + z^3)$	$3k+2$	$k+1$	$k \geq 2$
P_4^1	$(x, y, yz + z^7 + z^8, xz + z^3)$	8	4	
P_4	$(x, y, yz + z^7, xz + z^3)$	8	5	
Q_k	$(x, y, xz + yz^2, z^3 \pm y^k z)$	$k+1$	k	$k \geq 2$
R_k	$(x, y, xz + z^3, yz^2 + z^4 + z^{2k-1})$	$2k-1$	k	$k \geq 3$
$S_{j,k}$	$(x, y, xz + y^2z^2 \pm z^{3j+2}, z^3 \pm y^k z)$	$m^{(\dagger)}$	$k+j$	$j \geq 1, k \geq 2$

Table 1: Simple Singularities $(\mathbb{R}^3, 0) \to (\mathbb{R}^4, 0)$.

Label	Singularity	Det	Cod	Conditions
I	$(x, y, yz + xz^3 \pm z^5 + az^7, xz + z^4 + bz^6)$	7	5	$a \mp b \neq 0$
II	$(x, y, yz + xz^3 + az^6 + z^7 + bz^8 + cz^9,$ $xz + z^4)$	9	7	(‡)
III	$(x, y, yz + z^5 + z^6 + az^7, xz + z^4)$	7	5	$a \neq 1$
IV	$(x, y, yz + z^5 + az^7, xz + z^4 \pm z^6)$	7	5	$a \neq \pm 1$
V	$(x, y, xz + z^5 + ay^3z^2 + y^4z^2, z^3 \pm y^2z)$	6	5	for all a
VI	$(x, y, xz + z^3, yz^2 + z^5 + z^6 + az^7)$	7	4	$a \neq 1$
VII	$(x, y, xz + z^3, y^2z + xz^2 + az^4 \pm z^5)$	5	5	(§)
VIII	$(x, y, xz + z^4 + az^5 + bz^7, yz^2 + z^4 + z^5)$	7	6	(¶)

Table 2: Non-simple Singularities $(\mathbb{R}^3, 0) \to (\mathbb{R}^4, 0)$ of $\mathcal{A}_e$-codim ≤ 4.

(†) $m = \max(3j + 2, k + 1)$. (§) $a \neq \pm 1, \frac{5}{4}, 0, \frac{1}{2}, \frac{3}{2}$. (¶) $a - b - a^2 \neq 0$.
(‡) finitely-determined for generic a, b, c; see Appendix.

general setting denote the local ring of differentiable/analytic function-germs $(\mathbb{K}^n, 0) \to \mathbb{K}$ by $\mathcal{E}_n$. When we insist on the complex analytic case we will use the notation $\mathcal{O}_n$. The corresponding modules of map-germs will be denoted $\mathcal{E}(n,p)$ and $\mathcal{O}(n,p)$.

1.1 Classification Techniques

We proceed with the classification inductively at the jet-level, classifying in turn all $(k+1)$-jets with a given k-jet. This gives a classification tree whose branches terminate when we obtain finitely determined jets, or when our pre-selected upper bounds on codimension or number of moduli are reached. The classification problem is thus reduced to one of Lie groups (in fact, algebraic groups) acting on affine spaces and techniques such as Mather's Lemma [15, Lemma 3.1] and Complete Transversals may be applied.

The method of Complete Transversals was developed by Bruce, Kirk and du Plessis in a recent paper [1] (following work of Dimca and Gibson in the $\mathcal{K}$ case [3, 4]) and provides an extremely efficient way of listing all representatives of $(k+1)$-jets over a given k-jet. We shall use the version of the results where $\mathcal{G}$ is a unipotent subgroup of $\mathcal{K}$. The special case $\mathcal{G} = \mathcal{A}_1$ was also considered in the work of Ratcliffe, [20]. We refer to [1] for the details but will summarise the essential points below.

Definition 1.2. Let $\mathcal{G}$ be a subgroup of $\mathcal{K}$ such that $L(J^1\mathcal{G})$ acts nilpotently on $\mathbb{K}^{n+p}$. For integers $r \geq 1$ and $s \geq 0$ we define the *nilpotent filtration*

$$M_{r,s}(\mathcal{G}) = \sum_{i \geq s} (L\mathcal{G})^i \cdot (\mathcal{M}_n^r \mathcal{E}(n,p)) + \mathcal{M}_n^{r+1}\mathcal{E}(n,p).$$

(Note that the above sum is finite since $L(J^1\mathcal{G})$ acts nilpotently.) For $r = 0$ we define $M_{0,0}(\mathcal{G})$ to be $\mathcal{M}_n\mathcal{E}(n,p)$ for consistency. The associated *(r,s)-jet-space* $J^{r,s}(n,p)$ is defined to be $\mathcal{M}_n\mathcal{E}(n,p)/M_{r,s}(\mathcal{G})$, and by the *homogeneous* terms of degree (r,s) we mean the image of the space $M_{r,s-1}(\mathcal{G})$ in this quotient and denote this $H^{r,s}$. Given $f \in \mathcal{M}_n\mathcal{E}(n,p)$ we denote its image in $J^{r,s}(n,p)$ by $j^{r,s}f$. We filter the group $\mathcal{G}$ using the normal subgroups $\mathcal{G}_k$ consisting of those germs whose k-jet is equal to that of the identity; the associated *r-jet-group* $J^r\mathcal{G}$ is then defined to be $\mathcal{G}/\mathcal{G}_r$.

For the subgroups $\mathcal{G}$ of $\mathcal{K}$ which are used in practice $J^r\mathcal{G}$ is a Lie group. The idea is to consider the action of $J^r\mathcal{G}$ on $J^{r,s}(n,p)$. For certain $\mathcal{G}$ (for example, when $\mathcal{G}$ is fibrant) this action is well-defined and leads to the following complete transversal result for the nilpotent filtration; see [1].

Theorem 1.3. *Let $\mathcal{G}$ be a fibrant subgroup of $\mathcal{K}$ such that $L(J^1\mathcal{G})$ acts nilpotently on $\mathbb{K}^{n+p}$. Let f be a smooth germ $(\mathbb{K}^n, 0) \to (\mathbb{K}^p, 0)$ and let T be a subspace of $M_{r,s}(\mathcal{G})$ with*

$$M_{r,s}(\mathcal{G}) \subset T + L\mathcal{G} \cdot f + M_{r,s+1}(\mathcal{G}).$$

Then any germ $g : (\mathbb{K}^n, 0) \to (\mathbb{K}^p, 0)$ *with* $g - f \in M_{r,s}(\mathcal{G})$ *is* $\mathcal{G}$*-equivalent to a germ of the form* $f + t + \phi$ *with* $t \in T$ *and* $\phi \in M_{r,s+1}(\mathcal{G})$.

The affine space $f + T$ and the vector subspace T are both referred to as a *complete transversal*.

Since $M_{r,s}(\mathcal{G})$ is a refinement of the standard filtration of $\mathcal{M}_n\mathcal{E}(n,p)$ by degree we obtain an inductive classification process at the jet-level, only in smaller steps than those given by the standard polynomial degree. Indeed, if we take $\mathcal{G} = \mathcal{A}_1$ then the classification proceeds via standard degree (in this case T is a vector subspace of the space H^{k+1} of all homogeneous jets of degree $k+1$). The $\mathcal{A}_1$ classification method is simpler to apply but the use of 'larger' unipotent subgroups $\mathcal{G}$ gives a far more efficient classification procedure, essential during the early stages of the classification. An example should clarify matters.

Example 1.4. Consider the unipotent subgroup of $\mathcal{A}$ having nilpotent Lie algebra

$$L\mathcal{A}_1 \oplus \mathbb{R}\{x\partial/\partial y, x\partial/\partial z, y\partial/\partial z\} \oplus \mathbb{R}\{u_i\partial/\partial u_j \quad \text{for} \quad i > j\}.$$

where (x, y, z) denote coordinates in the source and (u_1, u_2, u_3, u_4) those in the target. Application of the complete transversal theorem, taking $\mathcal{G}$ to be the above group, yields an efficient $\mathcal{A}$ classification method. The spaces $H^{k,s}$ are somewhat tedious to write down: one needs a more explicit identification. It was shown in [1] that this can be achieved by ordering the degree k monomial vectors using a system of source and target weights, $\alpha = (\alpha_1, \ldots, \alpha_n)$ and $\beta = (\beta_1, \ldots, \beta_p)$. The monomial vector $x_1^{k_1} \ldots x_n^{k_n} e_i$ is assigned weight $k_1\alpha_1 + \cdots + k_n\alpha_n - \beta_i$ and then $H^{k,s}$ consists of those homogeneous monomial vectors of (standard) degree k with weight $k+s-1$. This applies in the present example if we assign weights $\alpha = (3, 2, 1)$ and $\beta = (-3, -2, -1, 0)$.

As our example we consider the classification of 3-jets lying over the 2-jet (x, y, yz, xz). Only the first few spaces $H^{3,s}$ feature in the calculations and these are shown in Table 3.

$(3, s)$	Basis for $H^{3,s}$	Weight
$(3,1)$	$\{(0,0,0,z^3)\}$	3
$(3,2)$	$\{(0,0,z^3,0),(0,0,0,yz^2)\}$	4
$(3,3)$	$\{(0,z^3,0,0),(0,0,yz^2,0),(0,0,0,y^2z),(0,0,0,xz^2)\}$	5

Table 3: Generators for the spaces $H^{3,s}$.

Put $f = (x, y, yz, xz)$; a $(3,1)$-transversal is $(x, y, yz, xz+az^3)$, $a \in \mathbb{R}$. Let us apply scaling coordinate changes to simplify matters before proceeding. We obtain the jets $(x, y, yz, xz+z^3)$ and (x, y, yz, xz). In both cases the only higher non-empty $(3, s)$-transversal is the $(3, 2)$-transversal and we obtain the 3-jets $(x, y, yz + bz^3, xz + z^3)$ and $(x, y, yz + bz^3, xz)$, for some $b \in \mathbb{R}$. In the first case it is tempting to scale b to 0 or 1. However, the resulting 3-jets have the same $J^3\mathcal{A}$-codimension, indeed the codimension is found to be constant for all b, suggesting that the family is $\mathcal{A}$-trivial. A simple application of Mather's Lemma shows that this is so. We therefore have the one normal form: $(x, y, yz, xz + z^3)$. Returning to the second jet $(x, y, yz + bz^3, xz)$; here we apply scaling coordinate changes and reduce this to the 3-jets $(x, y, yz + z^3, xz)$ (equivalent to $(x, y, yz, xz + z^3)$ obtained earlier) and (x, y, yz, xz). The complete list of 3-jets over (x, y, yz, xz) is therefore

$$(x, y, yz, xz + z^3), \quad (x, y, yz, xz).$$

If instead we take $\mathcal{G}$ to be $\mathcal{A}_1$, a 3-transversal over (x, y, yz, xz) is

$$(x, y, yz + a_1z^3 + a_2xz^2, xz + a_3z^3 + a_4xz^2 + a_5yz^2) \quad \text{for} \quad a_i \in \mathbb{R}.$$

Of course, this is a vast improvement over trying to classify jets in the affine space $(x, y, yz, xz) + H^3$, but it would still involve a lot of (ad-hoc) work to reduce this $\mathcal{A}_1$-transversal to the two cases above.

The other essential ingredient in the classification process is a good determinacy criterion. A complete solution to the determinacy question was given by Bruce, du Plessis and Wall in [2]. We use the following results.

Theorem 1.5. *Let $\mathcal{G}$ be a strongly closed subgroup of $\mathcal{A}$ such that $L(J^1\mathcal{G})$ acts nilpotently on $\mathbb{K}^{n+p}$. Then a smooth map-germ $f : (\mathbb{K}^n, 0) \to (\mathbb{K}^p, 0)$ is k-$\mathcal{G}$-determined if and only if $\mathcal{M}_n^{k+1}\mathcal{E}(n,p) \subset L\mathcal{G} \cdot f$.*

From [2, Lemma 2.6] we can reduce the above criterion to the following.

Corollary 1.6. *The map-germ f if k-$\mathcal{G}$-determined if and only if*

$$\mathcal{M}_n^{k+1}\mathcal{E}(n,p) \subset L\mathcal{G} \cdot f + \mathcal{M}_n^{k+1} f^*(\mathcal{M}_p)\mathcal{E}(n,p) + \mathcal{M}_n^{2k+2}\mathcal{E}(n,p).$$

The problem is reduced to a finite-dimensional one within the jet-space $J^{2k+1}(n,p)$. In specific examples the jet-space degree may be reduced further by appealing to the terms in $\mathcal{M}_n^{k+1} f^*(\mathcal{M}_p)\mathcal{E}(n,p)$. Thus, establishing the degree of determinacy of a germ is essentially a special case of calculating complete transversals (we have to show that successive complete transversals from degree $k+1$ to (at most) $2k+1$ are empty).

1.2 Details of the Classification

A summary of the classification is given in the appendix. The majority of the calculations were carried out using our computer package `Transversal`, taking $\mathcal{G}$ to be the unipotent subgroup of $\mathcal{A}$ defined in Example 1.4.

2 Geometry and Invariants

For a given map-germ $f : (\mathbb{C}^3, 0) \to (\mathbb{C}^4, 0)$ our main object of study will be the disentanglement of f and its image Milnor number $\mu_I(f)$. The multiple point spaces $D^k(f)$ and their restriction to reflecting hyperplanes $D^k(f)|H$ also provide natural invariants. Indeed, we will show how to calculate $\mu_I(f)$ in terms of the Milnor numbers of the multiple point spaces. The formula is an extension of that derived by Marar, relating to the Euler characteristic of the disentanglement of f, see [12].

The full set of invariants is given in Table 4. The D^4 spaces are empty for all cases except the non-simples I, II, III, IV and VIII and for these cases $\mu(D^4)$, $\mu(D^4|H)$, μ_4^{alt}, $\tau(D^4)$ and $\tau(D^4|H)$ take the values 23, 5, 1, 36 and 6, respectively.

Remarks 2.1. An important conjecture of Mond relates the image Milnor number and the $\mathcal{A}_e$-codimension of a map-germ $f : (\mathbb{C}^n, 0) \to (\mathbb{C}^{n+1}, 0)$: in the nice dimensions we have $\mathcal{A}_e$-codimension$(f) \leq \mu_I(f)$ with equality if and only if f is weighted homogeneous. This has been proved in the case $n = 2$, see [19]. We shall see that it is satisfied for all of the maps on our list. In contrast to the topology there does not seem to be any obvious relation between the Tjurina number of the multiple point spaces and the $\mathcal{A}_e$-codimension of f. However, we do note that in most cases the discrepancy between $\mu_I(f)$ and the $\mathcal{A}_e$-codimension is accounted for by the differences $|\tau(X) - \mu(X)|$ where $X = D^2$, $D^2|H$, D^3 and $D^3|H$. Indeed, this only fails for the cases R_k, VI and VIII; and it clearly holds for the quasihomogeneous case in general if the Mond conjecture is true. We postpone a full investigation of this observation and the Mond conjecture for future work.

2.1 Disentanglements and Multiple Point Spaces

Suppose that $f : (\mathbb{C}^n, 0) \to (\mathbb{C}^p, 0)$ is a finitely $\mathcal{A}$-determined map-germ with $n < p$. In our study, all the $\mathcal{A}$-invariants of f arise from the disentanglement map of f. This map is in essence a stable deformation of f and its image is the analogue of the Milnor fibre of a $\mathcal{K}$-finite map. In contrast to the Milnor fibre the disentanglement of a map is, in general, singular. The map and its image are described in [5], [13] and [7]. Since we are interested in corank 1 maps we shall only formulate the construction for this case. Furthermore, in this case the disentanglement of f always exists and is essentially unique.

Suppose $f : (\mathbb{C}^n, 0) \to (\mathbb{C}^p, 0)$ is a finitely $\mathcal{A}$-determined corank 1 map-germ with $n < p$. Let $F : (\mathbb{C}^n \times \mathbb{C}^r, 0 \times 0) \to (\mathbb{C}^p \times \mathbb{C}^r, 0 \times 0)$ be a non-trivial parametrised unfolding of f with $F(x,t) = (f_t(x), t)$ and $f_0 = f$. We can choose a representative of F, also denoted F, with $F : U \to V \times W$ such that $U \subset \mathbb{C}^n \times \mathbb{C}^r$ and $V \subset \mathbb{C}^p$, $W \subset \mathbb{C}^r$ are open sets. We assume that F is finite and $F^{-1}(0) = 0$. Since F is corank 1 it can be given a canonical Thom stratification, see [13]. If f is not stable then there exists an analytic set $B \subset W$ such that for $t \in B$ the map f_t is not stable (as a multigerm)

Name	$\mu(D^2)$	$\mu(D^2\|H)$	μ_2^{alt}	$\tau(D^2)$	$\tau(D^2\|H)$	$\mu(D^3)$	$\mu(D^3\|H)$	μ_3^{alt}	$\tau(D^3)$	$\tau(D^3\|H)$	μ_I	$\mathcal{A}_e$-codim
A_k	k	k	k	k	k	-	-	-	-	-	k	k
D_k	k	k	k	k	k	-	-	-	-	-	k	k
E_k	k	k	k	k	k	-	-	-	-	-	k	k
B_k	$2k-1$	1	k	$2k-1$	1	-	-	-	-	-	k	k
C_k	$k+1$	$k-1$	k	$k+1$	$k-1$	-	-	-	-	-	k	k
F_4	6	2	4	6	2	-	-	-	-	-	4	4
P_1	0	0	0	0	0	1	1	1	1	1	1	1
P_2	0	0	0	0	0	4	2	2	4	2	2	2
P_3^k	0	0	0	0	0	$6k+1$	3	$k+2$	$6k$	3	$k+2$	$k+1$
P_4^1	0	0	0	0	0	16	4	5	15	4	5	4
P_4	0	0	0	0	0	16	4	5	16	4	5	5
P_k	0	0	0	0	0	(see Example 2.13)					$\frac{1}{6}(k+1)(k+2)$	
Q_k	$k-1$	$k-1$	$k-1$	$k-1$	$k-1$	1	1	1	1	1	k	k
R_k	$2k-3$	1	$k-1$	$2k-3$	1	4	2	2	4	2	$k+1$	k
$S_{j,k}$	$k-1$	$k-1$	$k-1$	$k-1$	$k-1$	$6j+1$	3	$j+2$	$6j+1$	4	$k+j+1$	$k+j$
I	0	0	0	0	0	13	5	5	13	6	6	5
II	0	0	0	0	0	25	7	8	24	8	9	7
III	0	0	0	0	0	13	5	5	13	6	6	5
IV	0	0	0	0	0	13	5	5	13	6	6	5
V	1	1	1	1	1	13	5	5	13	6	6	5
VI	3	1	2	3	1	13	3	4	12	3	6	4
VII	4	2	3	4	2	7	3	3	7	4	6	5
VIII	3	1	2	3	1	13	5	5	13	6	8	6

Table 4: Invariants for map-germs $(\mathbb{C}^3,0) \to (\mathbb{C}^4,0)$.

at some point. There exists ε such that any sphere $S_{\varepsilon'} \subset V \times W$ of radius ε', $0 < \varepsilon' \leq \varepsilon$, centred at the origin in $V \times W$, is transverse to the strata of $F(U)$. Let B_ε denote the open ball of radius ε centred at the origin in $V \times W$. Let $U_t = F^{-1}(B_\varepsilon \cap (V \times \{t\}))$. Then, for small enough $|t|$, with $t \in W - B$ the map $f_t|U_t$ is stable.

Definition 2.2. For such a map, f_t is called the *disentanglement map* (or *stable perturbation*) of f and the image of $f_t|U_t$ is called the *disentanglement* of f.

If f is of corank 1 or (n,p) is in the nice dimensions then the bifurcation set B is a proper analytic subset of W and thus does not separate it. For $|t'|$ small with $t' \in W - B$ the maps f_t and $f_{t'}$ are C^0-$\mathcal{A}$-equivalent and hence their images are homeomorphic. By a versality argument it follows that the topology of these images is independent of the choice of unfolding F and thus only depends on the original map-germ f. If f is not corank 1 and (n,p) not in the nice dimensions then disentanglements do not necessarily exist. (However, it is possible to find a perturbation of f that is topologically stable.)

In [7] it is proved that the disentanglement is homotopically equivalent to a wedge of spheres of possibly varying dimensions. For $p = n+1$, the spheres are all of the same dimension, a fact which was proved in the nice dimensions by Mond in [19]. The number of spheres is an important $\mathcal{A}$-invariant of f.

Definition 2.3. The total number of spheres, denoted $\mu_I(f)$, is called the *image Milnor number* (or *number of vanishing cycles*) of f.

We now consider multiple point spaces and make the following preliminary definition. Given a continuous mapping $f : X \to Y$ of topological spaces we define $D^k(f)$, the kth multiple point space of f, as the closure of the set

$$\{(x_1, \ldots, x_k) \in X^k \mid f(x_1) = \cdots = f(x_k) \quad \text{for} \quad x_i \neq x_j, i \neq j\}.$$

Note that the group of permutations on k objects, S_k, acts on $D^k(f)$. This leads to an action on the rational cohomology of the spaces $D^k(f)$. We will be particularly interested in the subgroup upon which S_k acts via the sign representation. This subgroup is called the alternating cohomology of $D^k(f)$:

$$Alt\ H^i(D^k(f); \mathbb{Q}) = \{c \in H^i(D^k(f); \mathbb{Q}) \mid \sigma c = \operatorname{sign}(\sigma) c \quad \text{for all} \quad \sigma \in S_k\}.$$

However, the above definition of a multiple point space does not behave well under deformation, see [14]. If f is a corank 1 map then it is shown in [14] that the multiple point spaces can be defined in a scheme-theoretic manner by using an adapted form of the Vandermonde determinant. Outside the fixed point set of the action of S_k on $D^k(f)$ the two spaces coincide. We adopt the latter definition in what follows.

Suppose $f : (\mathbb{C}^n, 0) \to (\mathbb{C}^p, 0)$ is of corank 1 and is given in the form

$$(x_1, \ldots, x_{n-1}, z) \mapsto (x_1, \ldots, x_{n-1}, h_1(x, z), \ldots, h_{p-n+1}(x, z)).$$

If $g : \mathbb{C}^n \to \mathbb{C}$ is a function then we define $V_i^k(g) : \mathbb{C}^{n+k-1} \to \mathbb{C}$ to be

$$\left|\begin{matrix} 1 & z_1 & \ldots & z_1^{i-1} & g(x, z_1) & z_1^{i+1} & \ldots & z_1^{k-1} \\ & & & \vdots & & & & \\ 1 & z_k & \ldots & z_k^{i-1} & g(x, z_k) & z_k^{i+1} & \ldots & z_k^{k-1} \end{matrix}\right| \Bigg/ \left|\begin{matrix} 1 & z_1 & \ldots & z_1^{k-1} \\ & & \vdots & \\ 1 & z_k & \ldots & z_k^{k-1} \end{matrix}\right|.$$

Definition 2.4. $D^k(f)$ is defined in $\mathbb{C}^{n+k-1}$ by the ideal generated by the $V_i^k(h_j(x, z))$ for all $i = 1, \ldots, k-1$ and $j = 1, \ldots, p-n+1$. *In what follows we will take coordinates on* $\mathbb{C}^{n+k-1} = \mathbb{C}^{n-1} \times \mathbb{C}^k$ *to be* $(x_1, \ldots, x_{n-1}, z_1, \ldots, z_k)$.

Note that the above generators are S_k-invariant.

Since the defining equations are produced using determinants we can reduce the problem of finding $V_i^k(g)$ to finding the functions $V_i^k(z^m)$ for each m.

Example 2.5. For a corank 1 map-germ $f : (\mathbb{C}^n, 0) \to (\mathbb{C}^p, 0)$ as above, $D^2(f)$ is defined by

$$\frac{h_i(x, z_1) - h_i(x, z_2)}{z_1 - z_2},$$

for $i = 1, \ldots, p - n + 1$.

Example 2.6. For $1 \leq i, j \leq k - 1$ we have $V_i^k(z^j) = \delta_j^i$, where δ_j^i is the Kronecker delta, i.e. $\delta_j^i = 1$ if $i = j$ and zero otherwise.

The last example may be used to prove that a multiple point space is empty. In the case of folding-maps we have $h_1(x, z) = z^2$ so D^k is empty for $k > 2$. Similarly, $V_3^4(x^k z + z^3) = 1$ so none of the simple singularities in our list exhibit quadruple points. Note that some of the singularities in Table 2 do have non-empty D^4 though; this was one of the reasons for extending the classification to such cases.

The main result of [14] is Theorem 2.14 where a description of the multiple point spaces for a finitely $\mathcal{A}$-determined corank 1 map-germ f is obtained, namely that the $D^k(f)$ are isolated complete intersection singularities, (abbr. ICIS). More precisely, f is finitely-determined if and only if for each k, with $p - k(p - n) \geq 0$, $D^k(f)$ is an ICIS of dimension $p - k(p - n)$ or empty, and for those k with $p - k(p - n) < 0$, $D^k(f)$ consists of at most $\{0\}$. Furthermore, f is stable if and only if the spaces are non-singular, or empty.

Suppose that f is finitely $\mathcal{A}$-determined, $p = n + 1$ and H denotes the intersection of m reflecting hyperplanes in $\mathbb{C}^{n+k-1}$. Then $D^k(f)|H$ is an ICIS or empty if $\dim D^k(f) - m \geq 0$, otherwise it consists of at most one point.

The multiple point spaces $D^k(f)$ and $D^k(f')$ for two $\mathcal{A}$-equivalent germs f and f' are $\mathcal{K}$-equivalent by a diffeomorphism that preserves the action of

S_k. In this way $\mathcal{K}$-invariants of the multiple point spaces of f will provide the $\mathcal{A}$-invariants that we study. More generally we can obtain invariants from the restrictions of the $D^k(f)$ to fixed point sets H of the S_k-action, since these restrictions are also ICIS. Two of the most important invariants of an ICIS are the Milnor number μ, and the Tjurina number τ. The Milnor number is a topological invariant and the Tjurina number a deformation theoretic invariant. We shall not discuss these too deeply as good references, such as [11], exist. We shall calculate the Milnor and Tjurina numbers for D^2, D^3, D^4, $D^2|H$, $D^3|H$, where H is a reflecting hyperplane in the requisite space.

For finitely $\mathcal{A}$-determined corank 1 map-germs f the multiple point spaces of the disentanglement map f_t are the Milnor fibres of $D^k(f)$. By a well known result of Hamm such fibres are homotopically equivalent to wedges of spheres of real dimension equal to the complex dimension of the fibre. From this we know that the alternating homology groups for $D^k(f_t)$ must be trivial except for possibly dimensions 0 and $\dim_{\mathbb{C}} D^k(f)$. However, it is easy to show that if the Milnor fibre is connected then the 0th alternating homology group is trivial. Thus only one alternating homology group is of interest. The dimension of this alternating homology group for $D^k(f_t)$ is denoted $\mu_k^{alt}(f)$ and we have $\sum \mu_k^{alt}(f) = \mu_I(f)$. That is, the sum of the $\mu_k^{alt}(f)$ is equal to the image Milnor number. This last fact is the essential content of Theorem 2.6(i) of [5]. As we shall see below we can describe $\mu_k^{alt}(f)$ in terms of the Milnor numbers of $D^k(f)$ and their restriction to fixed point sets of the action of S_k.

2.2 Calculation Techniques for the Invariants

Let X be a topological space with a cellular structure and suppose that the group S_k acts on X in a cellular manner. (That is cells are taken to cells and if a point of a cell is fixed by an element of S_k then the whole cell is fixed). Let X^g denote the fixed point set in X of $g \in S_k$ and let $\chi^{alt}(X) = \sum_i (-1)^i \dim_{\mathbb{Q}} AltH_i(X;\mathbb{Q})$ denote the *alternating Euler characteristic*. First we show how the alternating Euler characteristic is related to the Euler characteristics of the fixed point sets. This is proved in the same manner as Wall's theorem in [22]. Applying this theorem to corank 1 map-germs we obtain the correct statements of [5, Section 3] from which the image Milnor number can then be calculated. The formula (3.2) in [5] breaks down for zero-dimensional complete intersections. The arguments on p.61 need to be adjusted to take into account the cases when X and X^g are zero-dimensional; our arguments avoid this problem.

Theorem 2.7. *Suppose X is a topological space as above. Then,*

$$\chi^{alt}(X) = \frac{1}{k!} \sum_{g \in S_k} \operatorname{sign}(g) \chi(X^g).$$

Proof. Let $[M]$ denote an element of the ring of all $\mathbb{Q}$-linear representations of S_k. One has the equivariant Euler characteristic $\chi_{S_k}(X)$ given by

$$\chi_{S_k}(X) = \sum_q (-1)^q [H_q(X;\mathbb{Q})].$$

For all $g \in S_k$, $\chi_{S_k}(X)(g)$ is equal to the topological Euler characteristic of the fixed point set X^g of g.

Let $\langle\ ,\ \rangle$ denote inner product, then by the general theory of representations,

$$\dim AltH_q(X;\mathbb{Q}) = \langle [H_q(X;\mathbb{Q})], \mathrm{sign}(g)\rangle.$$

Consider the inner product of characters,

$$\begin{aligned}\chi^{alt}(X) &= \langle \chi_{S_k}(X), \mathrm{sign}\rangle \\ &= \frac{1}{|S_k|}\sum_{g\in S_k} \mathrm{sign}(g)\chi_{S_k}(X)(g) \\ &= \frac{1}{k!}\sum_{g\in S_k} \mathrm{sign}(g)\chi(X^g).\end{aligned}$$

□

We now apply the above ideas to multiple point spaces, taking $X = D^k(f_t)$ (a Milnor fibre of $D^k(f)$). Recall the following from [5, Section 3]. Let $\mathcal{P} = (k_1, \ldots, k_1, k_2, \ldots, k_2, \ldots, k_r, \ldots, k_r)$ be a partition of k with $1 \leq k_1 < k_2 < \cdots < k_r$ and k_i appearing α_i times (so $\sum \alpha_i k_i = k$). If $g \in S_k$ has partition $\mathcal{P}$ (that is, in its cycle decomposition there are α_i cycles of length k_i) then X^g is isomorphic to the multiple point space $D^k(f, \mathcal{P})$ defined in [14]. There are $k!/(\Pi_i \alpha_i! k_i^{\alpha_i})$ elements in S_k with partition $\mathcal{P}$ hence, from Theorem 2.7, we have the following.

Corollary 2.8.

$$\chi^{alt}(D^k(f_t)) = \sum_{\mathcal{P}} \frac{\mathrm{sign}(\mathcal{P})}{\Pi_i \alpha_i! k_i^{\alpha_i}} \chi(D^k(f_t, \mathcal{P}))$$

taken over all partitions $\mathcal{P}$ of k in which $1 \leq k_1 < k_2 < \cdots < k_r$.

Note that $\chi^{alt}(D^k(f_t)) = (-1)^{\dim_{\mathbb{C}} D^k(f)} \mu_k^{alt}(f)$ and we therefore have a way of calculating $\mu_k^{alt}(f)$ in terms of the Euler characteristics of the fixed point sets of the multiple point spaces.

Example 2.9. Let h be a map-germ defining a zero-dimensional ICIS X, and suppose S_k acts on the Milnor fibre X_t. Since X_t is a wedge of $\mu(X) = \deg(h) - 1$, 0-spheres (see [11, p.78]), we have

$$\chi^{alt}(X_t) = \frac{\deg(h)}{k!}.$$

Example 2.10. Consider the disentanglement of a finitely $\mathcal{A}$-determined corank 1 map-germ f with $\dim_{\mathbb{C}} D^2(f) \geq 1$. Let H be the hyperplane $z_1 = z_2$ in $\mathbb{C}^{n-1} \times \mathbb{C}^2$ (recall that $(x_1, \ldots, x_{n-1}, z_1, \ldots, z_k)$ denote coordinates on $\mathbb{C}^{n-1} \times \mathbb{C}^k$), let D^2 and $D^2|H$ be the double point spaces of f, and put $d = \dim_{\mathbb{C}} D^2(f)$. Then

$$\begin{aligned}\mu_2^{alt}(f) &= (-1)^d \chi^{alt}(D^2(f_t))\\ &= (-1)^d\Big(\frac{1}{2}\chi(D^2(f_t)) - \frac{1}{2}\chi(D^2(f_t)|H)\Big)\\ &= (-1)^d\frac{1}{2}\Big(1 + (-1)^d\mu(D^2) - (1 + (-1)^{d-1}\mu(D^2|H))\Big)\\ &= \frac{1}{2}\Big(\mu(D^2) + \mu(D^2|H)\Big)\end{aligned}$$

using the fact that a d-dimensional Milnor fibre is a wedge of d-spheres.

Example 2.11. Now consider the case $f : (\mathbb{C}^3, 0) \to (\mathbb{C}^4, 0)$, an $\mathcal{A}$-finite map-germ of corank 1, and denote its multiple point spaces by D^k. In this case D^2 is a surface, D^3 is a curve and D^4 is zero-dimensional. For $D^3 \subset \mathbb{C}^2 \times \mathbb{C}^3$ let H_1 denote a reflecting hyperplane of the form $z_i = z_j$ $(i \neq j)$ and let H_2 denote the intersection of two such reflecting hyperplanes. Then $D^3(f_t)|H_1$ is a zero-dimensional ICIS and $D^3(f_t)|H_2$ is empty. So,

$$\begin{aligned}\mu_3^{alt}(f) &= (-1)\chi^{alt}(D^3(f_t))\\ &= (-1)\Big(\frac{1}{6}\chi(D^3(f_t)) - \frac{1}{2}\chi(D^3(f_t)|H_1)\Big)\\ &= \frac{1}{6}\Big(\mu(D^3) + 3\mu(D^3|H_1) + 2\Big)\end{aligned}$$

and $\mu(D^3|H_1) = \deg(D^3|H_1) - 1$. As D^4 is zero-dimensional we have $\mu_4^{alt}(f) = \frac{1}{24}\deg(D^4)$.

2.2.1 A General Method for Calculating the Invariants

An important method for calculating Milnor numbers of ICIS was devised by Lê; see [11, p.77]. Let X be an ICIS and $X' \supset X$ an ICIS defined by removing one equation then (under favourable circumstances) $\mu(X) + \mu(X')$ is given by the vector space dimension of a quotient ring; the latter being easy to calculate. The idea is therefore to reduce the calculation of the Milnor number of X to that of an ICIS defined by fewer equations and ultimately to that of a hypersurface. In the extreme case of a zero-dimensional ICIS we can appeal to the simpler formula: $\mu(X) = \text{degree (map defining } X) - 1$; see [11, p.78].

We shall not go into the technical details, rather we shall describe how the method applies to the calculations needed in this paper. Let $f : (\mathbb{C}^3, 0) \to (\mathbb{C}^4, 0)$ be an $\mathcal{A}$-finite map-germ of corank 1 and denote its multiple point sets by D^k. Given a map-germ h define $\mathrm{Min}_k(h)$ to be the ideal generated by the $k \times k$ minors of dh. Suppose D^2 and $D^2|H$ are defined in $\mathbb{C}^4$ by the ideals $\langle h_1, h_2\rangle$ and $\langle h_1, h_2, z_1 - z_2\rangle$, respectively. Then

$$\mu(D^2) + \mu(D^2|H) = \dim_{\mathbb{C}} \mathcal{O}_4/(\mathrm{Min}_3(h_1, h_2, z_1 - z_2) + \langle h_1, h_2\rangle)$$

and provided $\mu(h_1)$ is finite we have

$$\mu(h_1) + \mu(D^2) = \dim_{\mathbb{C}} \mathcal{O}_4/(\mathrm{Min}_2(h_1, h_2) + \langle h_1\rangle).$$

In fact, we can always choose the hypersurface h_1 to be non-singular in our examples. A similar formula holds for $\mu(D^3) + \mu(D^3|H)$, only now we may exploit the fact that $D^3|H$ is zero-dimensional and $\mu(D^3|H) = \dim_{\mathbb{C}} \mathcal{O}_5/I - 1$ where I is the ideal defining $D^3|H$.

The above gives a general method for calculating the invariants of any given singularity on our list. The technique reduces the calculation to one of calculating the vector space dimension of a quotient ring and is easily carried out on a computer using a 'standard basis algorithm' such as that provided with `Singular`. The calculation of the Tjurina number can similarly be dealt with using computer packages such as `Singular` or `Transversal`. For series we can only use the computer to deal with specific members of the series and conjecture the general result. Through simple changes of coordinates we see that these series are of the following types: $(x, y, z^2, h(x, y, z^2))$ and $(x, y, x^k z + z^3, yz^m + g(x, z))$ where h and g are functions, k is a natural number and $m \in \{1, 2\}$. The former is an example of a folding-map and is easily dealt with. In this case $V_1^2(z^2) = z_1 + z_2$ and $V_2^3(z^2) = 1$. So D^3 is empty, D^2 is defined in $\mathbb{C}^3$ by $V_1^2(h)(x, y, z, -z) = 0$ and $D^2|H$ is equivalent to the space defined by $V_1^2(h)(x, y, 0, 0) = 0$. (See [8] for general results on folding-maps $(\mathbb{C}^n, 0) \to (\mathbb{C}^p, 0)$.) The remainder of this paper deals with general results for calculating the invariants, starting with a discussion on the quasihomogeneous case.

Before moving on we mention the following application of the above ideas which, we believe, has not appeared in previous work on such maps.

Example 2.12. Corank 1 map-germs $f : (\mathbb{C}^2, 0) \to (\mathbb{C}^3, 0)$ have been studied extensively by Mond. In addition to his many formulae, we can use the fact that $D^2|H$ is an ICIS defined in D^2 by an extra linear equation to deduce that

$$\mu_2^{alt}(f) = \frac{1}{2} \dim_{\mathbb{C}} \mathcal{O}_3 \Big/ \left\langle \frac{\partial V_1}{\partial x}\left(\frac{\partial V_2}{\partial z_1} + \frac{\partial V_2}{\partial z_2}\right) - \frac{\partial V_2}{\partial x}\left(\frac{\partial V_1}{\partial z_1} + \frac{\partial V_1}{\partial z_2}\right), V_1, V_2 \right\rangle$$

where $f(x, z) = (x, g_1(x, z), g_2(x, z))$, D^2 is defined in $\mathbb{C}^3$, taking (x, z_1, z_2) as coordinates, and $V_i = V_1^2(g_i)$.

2.2.2 Weights and Degrees for Corank 1 Map-germs

If f is quasihomogeneous then so are its multiple point spaces and we can calculate their Milnor numbers using weights and degrees, see [6]. For a

$$\mu(D^2) = -1 + \frac{(d_1-w_0)(d_2-w_0)}{w_0^2 w_1 w_2}\Big(w_0(w_0+2w_1+2w_2)+w_1w_2 - (d_1+d_2-2w_0)(2w_0+w_1+w_2)+(d_1-w_0)^2 + (d_1-w_0)(d_2-w_0)+(d_2-w_0)^2\Big),$$

$$\mu(D^2|H) = 1 + \frac{(d_1-w_0)(d_2-w_0)}{w_0w_1w_2}\Big(d_1+d_2-3w_0-w_1-w_2\Big),$$

$$\mu(D^3) = 1 + \frac{\Pi_i(d_i-w_0)(d_i-2w_0)}{w_0^2w_1w_2}\Big(2(d_1+d_2)-9w_0-w_1-w_2\Big),$$

$$\deg(D^3|H) = \frac{\Pi_i(d_i-w_0)(d_i-2w_0)}{w_0^2w_1w_2},$$

$$\deg(D^4) = \frac{\Pi_i(d_i-w_0)(d_i-2w_0)(d_i-3w_0)}{w_0^4w_1w_2}.$$

Table 5: Weights and Degrees Formulae for Quasihomogeneous Map-germs.

quasihomogeneous ICIS of positive dimension it is well known that τ and μ are equal. Thus for quasihomogeneous f we obtain the Tjurina numbers of D^2, $D^2|H$ and D^3 also. We miss only $\tau(D^3|H)$ and $\tau(D^4)$, as these spaces are zero dimensional ICIS.

Let $f:(x,y,z)\mapsto(x,y,g_1(x,y,z),g_2(x,y,z))$ be a finitely determined quasihomogeneous map. It is useful to list the weights and degrees formulae for the maps under study; see Table 5. Here z, x, y have weights w_0, w_1, w_2 and g_1, g_2 have degrees d_1, d_2, respectively. Compare with the formulae in [5]. We can now calculate the alternating Milnor numbers using Examples 2.10 and 2.11.

Example 2.13. For the singularity P_k (see the remarks in the appendix) we have $(w_0,w_1,w_2)=(1,2,k+2)$ and $(d_1,d_2)=(k+3,3)$. The space D^2 is non-singular, $\mu(D^3)=1+(k+1)(k-1)$ and $\deg(D^3|H)=k+1$ so by Example 2.11, $\mu_3^{alt}=\frac{1}{6}(k+1)(k+2)$. Note that if k is divisible by 3 then this number is a fraction; for such values of k the map f is not finitely determined. Since D^2 is non-singular and D^4 is empty we have $\mu_I=\mu_3^{alt}$.

Example 2.14. For the singularity Q_k we have $(w_0,w_1,w_2)=(k,k+2,2)$ and $(d_1,d_2)=(2k+2,3k)$. From this we obtain $\mu(D^2)=\mu(D^2|H)=k-1$ and therefore $\mu_2^{alt}=k-1$. We find that $\mu(D^3)=1$ and $\deg(D^3|H)=2$, so $\mu_3^{alt}=1$. (We shall deduce these facts using a different method in the following sections.) Again D^4 is empty and we conclude that $\mu_I=\mu_2^{alt}+\mu_3^{alt}=k$.

2.2.3 Double Point Space Invariants

From now on we will concentrate on map-germs of the form $(x,y,z)\mapsto(x,y,x^kz+z^3,yz^m+g(x,z))$, as introduced in Section 2.2.1. Strictly speaking, we will concentrate on enough subclasses to cover all of the series on our list. We begin with the case $m = 1$ as we can deal with this in its full generality.

Theorem 2.15. *Suppose f is finitely $\mathcal{A}$-determined and of the form*

$$(x, y, z) \mapsto (x, y, x^k z + z^3, yz + g(x, z)).$$

Then $\mu(D^2) = \mu(D^2|H) = \tau(D^2) = \tau(D^2|H) = k - 1$. Hence, $\mu_2^{alt} = k - 1$.

Proof. The double point space is defined in $\mathbb{C}^4$ by

$$x^k + z_1^2 + z_1 z_2 + z_2^2, \qquad y + \frac{g(x, z_1) - g(x, z_2)}{z_1 - z_2}.$$

By removing the y coordinate we see that we are dealing with an A_{k-1} hypersurface singularity, the Milnor and Tjurina numbers are therefore both equal to $k - 1$. Restricting to $z_1 - z_2$ we find $D^2|H$ is an A_{k-1} singularity also. Finally, μ_2^{alt} is given as in Example 2.10. □

The above covers all of the series except R_k. The case $m = 2$, $k = 1$ with g a function of z only will complete our treatment. In the following let $V(g) = V_1^2(g)$.

In examples we often find that the element $z_1 + z_2$ is a member of the ideal we are studying. This means we can remove z_2 and replace $V(z^k)(z_1, z_2)$ in the ideal by $V(z^k)(z, -z)$. Therefore, the following lemma proves useful.

Lemma 2.16. $V(z^k)(z, -z) = \begin{cases} z^{k-1} & \textit{if } k \textit{ is odd,} \\ 0 & \textit{if } k \textit{ is even.} \end{cases}$

We also have,

$$\frac{\partial V(z^k)}{\partial z_1}(z, -z) = [\tfrac{k}{2}] z^{k-2} \quad \textit{and} \quad \frac{\partial V(z^k)}{\partial z_2}(z, -z) = (-1)^k [\tfrac{k}{2}] z^{k-2}.$$

The square brackets indicate that we take the integer part of the number.

Proof. It is a simple matter to show that $V(z^k) = \sum_{i=0}^{k-1} z_1^i z_2^{k-i-1}$, therefore

$$V(z^k)(z, -z) = (-1)^{k-1} z^{k-1} \sum_{i=0}^{k-1} (-1)^i,$$

giving the required result. In a similar manner we deduce

$$\frac{\partial V(z^k)}{\partial z_1}(z, -z) = (-1)^{k-1} z^{k-2} \sum_{i=0}^{k-1} i(-1)^i.$$

and the conclusion follows as $\sum_{i=0}^{k-1} i(-1)^i = (-1)^{k-1}[\frac{k}{2}]$. The other case is dealt with similarly. □

Theorem 2.17. *Suppose $f : (\mathbb{C}^3, 0) \to (\mathbb{C}^4, 0)$ is finitely $\mathcal{A}$-determined and of the form*

$$(x, y, z) \mapsto (x, y, xz + z^3, yz^2 + g(z)),$$

where g has no term of degree less than 3. Then

$$\tau(D^2) = \mu(D^2) = \{\textit{lowest odd degree in } g\} - 2$$

and $\tau(D^2|H) = \mu(D^2|H) = 1$. Hence, $\mu_2^{alt} = \frac{1}{2}(\{\textit{lowest odd degree in } g\} - 1)$.

Proof. The double point space D^2 is defined in $\mathbb{C}^4$ by

$$x + z_1^2 + z_1 z_2 + z_2^2, \qquad y(z_1 + z_2) + V(g).$$

So, eliminating x, we see that D^2 is equivalent to the hypersurface germ defined by $y(z_1 + z_2) + V(g)$. Let $g(z) = \sum_{i=3} a_i z^i$, that is a_3 is the first (possibly) non-zero coefficient. The Milnor number of D^2 is given by the dimension of the following space:

$$\begin{aligned}
&\mathcal{O}_3 \Big/ \left\langle z_1 + z_2, y + \frac{\partial V(g)}{\partial z_1}, y + \frac{\partial V(g)}{\partial z_2} \right\rangle \\
&\cong \mathcal{O}_2 \Big/ \left\langle y + \frac{\partial V(g)}{\partial z_1}(z, -z), y + \frac{\partial V(g)}{\partial z_2}(z, -z) \right\rangle \\
&\cong \mathcal{O}_2 \Big/ \left\langle y + \sum_{i=3} a_i [\tfrac{i}{2}] z^{i-2}, y + \sum_{i=3} a_i (-1)^i [\tfrac{i}{2}] z^{i-2} \right\rangle \\
&\cong \mathcal{O}_2 \Big/ \left\langle 2y + 2 \sum_{i \text{ even}} a_i [\tfrac{i}{2}] z^{i-2}, 2 \sum_{i \text{ odd}} a_i [\tfrac{i}{2}] z^{i-2} \right\rangle \\
&\cong \mathcal{O}_1 \Big/ \left\langle \sum_{i \text{ odd}} a_i [\tfrac{i}{2}] z^{i-2} \right\rangle.
\end{aligned}$$

Thus $\mu(D^2) = \{\text{lowest odd degree in } g\} - 2$. The calculation for $\tau(D^2)$ is similar.

The space $D^2|H$ is given by $z_1 - z_2$ and $y(z_1 + z_2) + V(g)$ and hence is isomorphic to the hypersurface singularity in $\mathbb{C}^2$ given by $V(g)(z, z) + 2yz$. This is a quadratic singularity and hence $\mu(D^2|H) = \tau(D^2|H) = 1$. The calculation of μ_2^{alt} then follows from Example 2.10. □

Example 2.18. Consider the singularity R_k, where $g(z) = z^4 + z^{2k-1}$. Then $\mu(D^2) = 2k - 3$ and $\mu(D^2|H) = 1$, hence $\mu_2^{alt} = k - 1$.

2.2.4 Triple Point Space Invariants

To begin with let f be finitely $\mathcal{A}$-determined of the form

$$(x, y, z) \mapsto (x, y, x^k z + z^3, yz^m + g(x, z)),$$

as introduced in Section 2.2.1. In the following let $V_i(g) = V_i^3(g)$ for $i = 1, 2$, and for $m \in \{1, 2\}$ define $\bar{m}$ to be its 'opposite', that is $\bar{m} = 3 - m$.

Lemma 2.19. *For a corank 1 map-germ, D^3 lies in $\mathbb{C}^2 \times \mathbb{C}^3 = \mathbb{C}^5$ and its defining equations can be calculated using the following inductive form:*

$$V_i(z^{k+3}) = z_1z_2z_3V_i(z^k) - (z_1z_2 + z_1z_3 + z_2z_3)V_i(z^{k+1}) + (z_1 + z_2 + z_3)V_i(z^{k+2}).$$

where $k \geq 0$. The initial terms are

$$V_1(1) = V_2(1) = V_2(z) = V_1(z^2) = 0 \quad and \quad V_1(z) = V_2(z^2) = 1.$$

Proof. The proof is a straightforward tedious calculation. □

Corollary 2.20. *The defining equations for $D^3(f)$ are given by*

$$x^k - (z_1z_2 + z_1z_3 + z_2z_3), \qquad z_1 + z_2 + z_3,$$
$$\delta_1^m y + V_1(g), \qquad \delta_2^m y + V_2(g).$$

Thus for $D^3|H(f)$ the equations become

$$x^k + 3z^2, \quad \delta_1^m y + V_1(g)(z, z, -2z), \quad \delta_2^m y + V_2(g)(z, z, -2z).$$

The presence of $z_1 + z_2 + z_3$ allows us to reduce the number of defining equations and the number of variables by one each. Furthermore, the S_3-invariance is replaced by S_2-invariance. To proceed further we cannot deal with such general f and will therefore restrict to more specific map-germs.

Example 2.21. Consider the singularity Q_k; here $m = 1$ and $g(x, z) = xz^2$. The singularities of the triple point space and its restriction to a reflecting hyperplane are both equivalent to a quadratic singularity.

With a little more work we can also deal with the other series containing the term $x^k z + z^3$, namely $S_{j,k}$ where $m = 1$ and $g(x, z) = x^2z^2 + z^{3j+2}$. The following lemma will be useful in the sequel for dealing with $D^3|H$.

Lemma 2.22. $V_1(z^k)(z, z, -2z) = b_k z^{k-1}$ *and* $V_2(z^k)(z, z, -2z) = c_k z^{k-2}$*, where b_k and c_k are constants defined by $b_{k+3} = 3b_{k+1} - 2b_k$ and $c_{k+3} = 3c_{k+1} - 2c_k$. The initial terms are $b_0 = c_0 = c_1 = b_2 = 0$ and $b_1 = c_2 = 1$.*

Proof. It is easy to deduce from Lemma 2.19 that

$$V_i(z^{k+3})(z, z, -2z) = -2z^3V_i(z^k)(z, z, -2z) + 3z^2V_i(z^{k+1})(z, z, -2z).$$

Then we apply induction. □

Theorem 2.23. *Consider the singularity $S_{j,k}$. Then $\tau(D^3) = \mu(D^3) = 6j + 1$ and $\tau(D^3|H) = 4$, $\mu(D^3|H) = 3$.*

Proof. From Corollary 2.20 and Lemma 2.22 we see that D^3 is defined in $\mathbb{C}^3$ by

$$x^k + z_1^2 + z_1z_2 + z_2^2, \qquad x^2 + V_2(z^{3j+2})(z_1, z_2, -z_1 - z_2)$$

and $D^3|H$ in $\mathbb{C}^2$ by $x^k + 3z^2$ and $x^2 + c_{3j+2}z^{3j}$. With a little more work and appealing to Lemma 2.19 we find that a basis for $N\mathcal{K}_e(D^3)$ is given by e_1, xe_1 and the vector e_2 multiplied by the monomials

$$1, z_1, \dots, z_1^{3j-1}, z_2, z_1z_2, \dots, z_1^{3j-2}z_2.$$

Similarly, a basis for $N\mathcal{K}_e(D^3|H)$ is given by $(1,0)$, $(0,1)$, $(x,0)$, $(0,z)$. This gives us the Tjurina numbers. For $\mu(D^3|H)$ we calculate the degree of the map defining $D^3|H$.

$$\deg(D^3|H) = \dim_{\mathbb{C}} \mathcal{O}_2/\langle x^k + 3z^2, x^2 + c_{3j+2}z^{3j}\rangle = 4.$$

This implies that $\mu(D^3|H) = 3$. To calculate $\mu(D^3)$ we use Lê's method (see Section 2.2.1) which gives $\mu(D^3) + \mu(D^3|H) = 6j + 4$. □

For the remaining series it is enough to restrict to the case $k = 1$ with g some function of z only. In the following let f be given by

$$(x, y, z) \mapsto (x, y, xz + z^3, yz^m + g(z)),$$

where $m = 1, 2$. By Corollary 2.20, D^3 and $D^3|H$ are isomorphic to hypersurfaces defined by $V_{\bar{m}}(g)(z_1, z_2, -z_1 - z_2)$ and $V_{\bar{m}}(g)(z, z, -2z)$ in $\mathbb{C}^2$ and $\mathbb{C}$ respectively.

Theorem 2.24. *If f is finitely $\mathcal{A}$-determined and g has no terms of degree less than $2 + \bar{m}$, then*

$$\tau(D^3|H) = \mu(D^3|H) = \{\textit{lowest degree in } g\} + m - 4.$$

Proof. The space $D^3|H$ is a hypersurface in $\mathbb{C}$ defined by $V_{\bar{m}}(g)(z, z, -2z)$. We therefore have

$$\begin{aligned} \mu(D^3|H) &= \deg(D^3|H) - 1 \\ &= \dim_{\mathbb{C}} \mathcal{O}_1/\langle V_{\bar{m}}(g)\rangle - 1 \\ &= \{\text{lowest degree in } g\} - \bar{m} - 1 \\ &= \{\text{lowest degree in } g\} + m - 4, \end{aligned}$$

using Lemma 2.22 and noting that $b_k \neq 0$ for $k \geq 3$, $c_k \neq 0$ for $k \geq 4$ (hence the clause that g has no terms of degree less than $2 + \bar{m}$). In a similar way we have

$$\tau(D^3|H) = \dim_{\mathbb{C}} \mathcal{O}_1 \Big/ \Big\langle \frac{\partial V_{\bar{m}}(g)}{\partial z}, V_{\bar{m}}(g) \Big\rangle = \dim_{\mathbb{C}} \mathcal{O}_1/\langle V_{\bar{m}}(g)\rangle - 1. \quad \square$$

For the D^3 invariants we cannot be so general. We will deal with the cases R_k and P_3^k explicitly. The only series not dealt with in this section is P_k, but this is quasihomogeneous. We are therefore in a position to complete Table 4.

Theorem 2.25. *(i) Consider the singularity R_k; here $m = 2$ and $g(z) = z^4 + z^{2k-1}$. Then $\tau(D^3) = \mu(D^3) = 4$.*

(ii) Consider the singularity P_3^k; here $m = 1$ and $g(z) = z^6 + z^{3k+2}$. Then $\tau(D^3) = 6k$ and $\mu(D^3) = 6k + 1$.

Proof. For (i) the space D^3 is defined by $V_1(g)(z_1, z_2, -z_1 - z_2)$ and

$$V_1(z^4 + z^{2k-1})(z_1, z_2, -z_1 - z_2) = -z_1^2 z_2 - z_1 z_2^2 + \{\text{Terms in degree } 2k - 2\}.$$

This is a D_4 hypersurface singularity and $\tau(D^3) = \mu(D^3) = 4$. The calculation for (ii) is similar though somewhat more complicated. We will just note that a basis of $N\mathcal{K}_e(D^3)$ is given by

$$1, z_2^2, z_1, \ldots, z_1^{3k-1}, z_2, z_1 z_2, \ldots, z_1^{3k-2} z_2$$

and similarly for $N\mathcal{R}_e(D^3)$ but with the extra term z_1^{3k}. □

Appendix: Classification Summary

The following tables give an $\mathcal{A}$-invariant stratification of the space of all jets, subject to the strata having codimension less than or equal to 7. Thus, by [23, p.510], we obtain a list of finitely-determined germs up to $\mathcal{A}_e$-codimension 4. For each k, a list of $\mathcal{A}$-invariant strata for the jet-space $J^k(3, 4)$ is given, along with the $J^k\mathcal{A}$-codimension; when the corresponding jet is k-determined this is the $\mathcal{A}$-codimension of the singularity. These strata are the $J^k\mathcal{A}$-orbits, or unions of such orbits when moduli occur, which project down onto the $J^{k-1}\mathcal{A}$-strata obtained at the previous level. (The coefficients a, b, c appearing in some jets are moduli. In such cases the $J^k\mathcal{A}$-codimension of a representative of the stratum is given; the $J^k\mathcal{A}$-codimension of the stratum is just this stated codimension minus the number of moduli.) For ease of reference between the $J^{k-1}\mathcal{A}$-orbits and the $J^k\mathcal{A}$-orbits we employ the following system. Each $J^{k-1}\mathcal{A}$-orbit is labelled with a capital letter, for example 'A'. This label will appear in the first column of the table of $J^k\mathcal{A}$-orbits, thus highlighting the $J^k\mathcal{A}$-orbits lying over 'A'.

There are several occasions when we shall not list the $J^k\mathcal{A}$-orbits over a given $(k-1)$-jet. These include cases where the jet is determined or is a stem and the orbits lying over it form a natural 'series'. Determined jets are indicated by the appearance of the determinacy degree, and series by the appearance of the symbol •, in the column marked 'det'; in addition, the name of the singularity (from Tables 1 and 2 appears in the 'label' column. The symbol '–' will appear in the 'label' column to indicate that the classification is taken no further for a particular jet (for example jets which exceed the codimension or corank bounds).

1-Jets				
	det	stratum	codim	label
	1	$(x, y, z, 0)$	0	-
	-	$(x, y, 0, 0)$	2	A
	-	$(x, 0, 0, 0)$	6	-
	-	$(0, 0, 0, 0)$	12	-

2-Jets				
	det	stratum	codim	label
A	2	(x, y, z^2, xz)	2	A_0
	•	$(x, y, z^2, 0)$	4	A*
	-	(x, y, yz, xz)	4	B
	-	$(x, y, xz, 0)$	5	C
	-	$(x, y, 0, 0)$	8	-

A: classification of the jets over $(x, y, z^2, 0)$ gives the folding-map type series. See the following remarks.

3-Jets				
	det	stratum	codim	label
B	-	$(x, y, yz, xz + z^3)$	4	A
	-	(x, y, yz, xz)	6	B
C	3	$(x, y, xz + yz^2, z^3 \pm y^2 z)$	5	Q_2
	•	$(x, y, xz + yz^2, z^3)$	6	Q_k
	-	$(x, y, xz, z^3 \pm y^2 z)$	6	C
	-	$(x, y, xz + z^3, yz^2)$	6	D
	-	(x, y, xz, z^3)	7	E
	-	$(x, y, xz + z^3, y^2 z + xz^2)$	7	F
	-	(x, y, xz, yz^2)	7	G
	-	$(x, y, xz + z^3, y^2 z)$	8	-
	-	$(x, y, xz + z^3, xz^2)$	8	-
	-	$(x, y, xz + yz^2, y^2 z + xz^2)$	8	-
	-	$(x, y, xz + z^3, 0)$	9	-
	-	$(x, y, xz + yz^2, y^2 z)$	9	-
	-	$(x, y, xz + yz^2, xz^2)$	9	-
	-	$(x, y, xz, y^2 z + xz^2)$	9	-
	-	$(x, y, xz + yz^2, 0)$	10	-
	-	$(x, y, xz, y^2 z)$	10	-
	-	(x, y, xz, xz^2)	10	-
	-	$(x, y, xz, 0)$	11	-

4-Jets				
	det	stratum	codim	label
A	4	$(x, y, yz + z^4, xz + z^3)$	4	P_1
	-	$(x, y, yz, xz + z^3)$	5	A
B	-	$(x, y, yz + xz^3, xz + z^4)$	6	B
	-	$(x, y, yz, xz + z^4)$	7	C
	-	$(x, y, yz, xz \pm xz^3)$	8	D*
	-	$(x, y, yz - xz^3, xz + yz^3 \pm xz^3)$	8	D*
	-	$(x, y, yz \pm xz^3, xz + yz^3)$	8	-
	-	$(x, y, yz - \frac{1}{4}xz^3, xz + yz^3 \pm xz^3)$	9	-
	-	$(x, y, yz, xz + yz^3)$	9	-
	-	(x, y, yz, xz)	11	-
C	•	$(x, y, xz + y^2z^2, z^3 \pm y^2z)$	6	$S_{j,2}$
	-	$(x, y, xz, z^3 \pm y^2z)$	7	E
D	•	$(x, y, xz + z^3, yz^2 + z^4)$	6	R_k
	-	$(x, y, xz + z^3, yz^2)$	7	F
E	•	$(x, y, xz + y^2z^2, z^3 + y^3z)$	7	$S_{j,3}$
	•	$(x, y, xz + y^2z^2, z^3)$	8	$S_{j,k}$
	-	$(x, y, xz, z^3 + y^3z)$	8	-
	-	(x, y, xz, z^3)	9	-
F	-	$(x, y, xz + z^3, y^2z + xz^2 + az^4)$ $a \neq -1, \frac{5}{4}$	8	G
	-	$(x, y, xz + z^3, y^2z + xz^2 - z^4 + yz^3 \pm xz^3)$	8	-
	-	$(x, y, xz + z^3, y^2z + xz^2 - z^4 + yz^3 - \frac{1}{4}xz^3)$	9	-
	-	$(x, y, xz + z^3, y^2z + xz^2 + \frac{5}{4}z^4 + yz^3 \pm xz^3)$	8	-
	-	$(x, y, xz + z^3, y^2z + xz^2 + \frac{5}{4}z^4 + yz^3 - \frac{1}{4}xz^3)$	9	-
	-	$(x, y, xz + z^3, y^2z + xz^2 - z^4 \pm xz^3)$	8	-
	-	$(x, y, xz + z^3, y^2z + xz^2 - z^4)$	9	-
	-	$(x, y, xz + z^3, y^2z + xz^2 + \frac{5}{4}z^4 \pm xz^3)$	8	-
	-	$(x, y, xz + z^3, y^2z + xz^2 + \frac{5}{4}z^4)$	9	-
G	-	$(x, y, xz + z^4, yz^2 + z^4)$	7	H
	-	$(x, y, xz, yz^2 + z^4)$	8	-
	-	$(x, y, xz + z^4, yz^2)$	8	-
	-	(x, y, xz, yz^2)	9	-

D: These are $\mathcal{A}$-equivalent over $\mathbb{C}$. (Applying Mather's Lemma we see that the family $(x, y, yz + axz^3, xz + yz^3 - xz^3)$ splits into 3 orbits corresponding to $a > -\frac{1}{4}$, $a = -\frac{1}{4}$ and $a < -\frac{1}{4}$. Over $\mathbb{C}$ we can reduce the first and last cases to $a = 0$ and then another application of Mather's Lemma allows us to remove the yz^3 term.) We do not pursue any of these cases further on codimension grounds.

5-Jets				
	det	stratum	codim	label
A	5	$(x, y, yz + z^5, xz + z^3)$	5	P_2
	-	$(x, y, yz, xz + z^3)$	6	A
B	-	$(x, y, yz + xz^3 \pm z^5, xz + z^4)$	6	B
	-	$(x, y, yz + xz^3, xz + z^4)$	7	C
C	-	$(x, y, yz + y^5, xz + z^4)$	7	D
	-	$(x, y, yz, xz + z^4)$	8	-
E	-	$(x, y, xz + z^5 + ay^3z^2, z^3 \pm y^2z)$ for all a	8	E
	-	$(x, y, xz + y^3z^2, z^3 \pm y^2z)$	8	-
	-	$(x, y, xz, z^3 \pm y^2z)$	9	-
F	-	$(x, y, xz + z^3, yz^2 + z^5)$	7	F
	-	$(x, y, xz + z^3, yz^2)$	8	-
G	5	$(x, y, xz + z^3, y^2z + xz^2 + az^4 \pm z^5)$ $a \neq \pm 1, \frac{5}{4}, 0, \frac{1}{2}, \frac{3}{2}$	8	VII
	-	$(x, y, xz + z^3, y^2z + xz^2 + az^4)$ $a \neq \pm 1, \frac{5}{4}$	9	-
H	-	$(x, y, xz + z^4 + az^5, yz^2 + z^4 + z^5)$ for all a	8	G
	-	$(x, y, xz + z^4 + z^5, yz^2 + z^4)$	8	-
	-	$(x, y, xz + z^4, yz^2 + z^4)$	9	-

6-Jets				
	det	stratum	codim	label
A	•	$(x, y, yz + z^6, xz + z^3)$	6	P_3^k
	-	$(x, y, yz, xz + x^3)$	7	A
B	-	$(x, y, yz + xz^3 \pm z^5, xz + z^4 + az^6)$ for all a	7	B
C	-	$(x, y, yz + xz^3 + az^6, xz + z^4)$ $a \neq -4$	8	C
	-	$(x, y, yz + xz^3 - 4z^6, xz + z^4 \pm z^6)$	8	-
	-	$(x, y, yz + xz^3 - 4z^6, xz + z^4)$	9	-
D	-	$(x, y, yz + z^5 + z^6, xz + z^4 \pm z^6)$	7	D*
	-	$(x, y, yz + z^5, xz + z^4 \pm z^6)$	7	E*
	-	$(x, y, yz + z^5 + z^6, xz + z^4 - \frac{3}{8}z^6)$	8	-
	-	$(x, y, yz + z^5, xz + z^4)$	8	-
E	6	$(x, y, xz + z^5 + ay^3z^2 + y^4z^2, z^3 \pm y^2z)$ for all a	8	V
	6	$(x, y, xz + z^5 + ay^3z^2, z^3 \pm y^2z)$ $27a^2 \pm 4 \neq 0$	9	-
F	-	$(x, y, xz + z^3, yz^2 + z^5 + z^6)$	7	F
	-	$(x, y, xz + z^3, yz^2 + z^5)$	8	-
G	-	$(x, y, xz + z^4 + az^5, yz^2 + z^4 + z^5)$ for all a	8	G

D,E: these are $\mathcal{A}$-equivalent over $\mathbb{C}$. (Applying Mather's Lemma we see that the family $(x, y, yz + z^5 + az^6, xz + z^4 - z^6)$ is trivial for $3a^2 - 8 \neq 0$.)

7-Jets				
	det	stratum	codim	label
A	-	$(x,y,yz+z^7,xz+z^3)$	7	A
	-	$(x,y,yz+xz^6,xz+z^3)$	8	B
	-	$(x,y,yz,xz+z^3)$	9	-
B	7	$(x,y,yz+xz^3 \pm z^5+bz^7,xz+z^4+az^6)$ $a \mp b \neq 0$	8	I
C	-	$(x,y,yz+xz^3+az^6+z^7,xz+z^4)$ $a \neq -4,2$	8	C
	-	$(x,y,yz+xz^3+az^6,xz+z^4)$ $a \neq -4,2$	9	-
D	7	$(x,y,yz+z^5+z^6+az^7,xz+z^4)$ $a \neq 1$	8	III
E	7	$(x,y,yz+z^5+az^7,xz+z^4 \pm z^6)$ $a \neq \pm 1$	8	IV
F	7	$(x,y,xz+z^3,yz^2+z^5+z^6+az^7)$ $a \neq 1$	8	VI
G	7	$(x,y,xz+z^4+az^5+bz^7,yz^2+z^4+z^5)$ $a-b-a^2 \neq 0$	9	VIII

8-Jets				
	det	stratum	codim	label
A	8	$(x,y,yz+z^7+z^8,xz+z^3)$	7	P_4^1
	8	$(x,y,yz+z^7,xz+z^3)$	8	P_4
B	-	$(x,y,yz+xz^6+az^8,xz+z^3)$ for all a	9	-
C	-	$(x,y,yz+xz^3+az^6+z^7+bz^8,xz+z^4)$ $a \neq -4,2$	9	A

9-Jets				
	det	stratum	codim	label
A	9	$(x,y,yz+xz^3+az^6+z^7+bz^8+cz^9,$ $xz+z^4)$ (see below)	10	II^*

II: This jet is finitely-determined for generic a, b, c. More precisely, determinacy holds provided (a,b,c) does not lie on a finite set of proper algebraic varieties in $\mathbb{R}^3$ (or $\mathbb{C}^3$). The defining equations for these varieties were, to some extent, determined by computer but were extremely complicated and we do not describe them here. Instead, since finite-determinacy is an open condition, we can prove 'generic determinacy' for the above family by exhibiting determinacy for a given member of the family. The corresponding calculation is computationally *far* less intensive and was verified using `Transversal`: taking $(a,b,c) = (12,3,4)$ we find that the corresponding germ is 9-determined.

Remarks. (i) Map-germs f having 2-jet $(x,y,z^2,0)$ are the so-called 'folding-maps' and can be written in the form $(x,y,z^2,zh(x,y,z^2))$ where h is a function on a manifold with boundary. The $\mathcal{A}$-classification of map-germs f is reduced to one of $\mathcal{K}_\delta$-classification of function-germs h; that is, $\mathcal{K}$-classification where the source coordinate changes preserve the hyperplane $z = 0$. The latter classifications are known and we can, essentially, just read off the normal forms from existing lists. Details of the above were given for the case $(\mathbb{R}^2, 0) \to (\mathbb{R}^3, 0)$ by Mond in [17] and generalised by Wilkinson in [24].

(ii) Classification of jets over the 5-jet $(x, y, yz, xz + z^3)$ leads to further branching with the appearance of the series P_3^k and more complicated branching at the 7-jet-level. The jet is therefore not a stem. However, further classification reveals the series P_k (see below) as part of the more complicated structure lying over $(x, y, yz, xz + z^3)$. This provides another example of a series to study in Section 2 so we list it here. (The additional terms in this series fall into neither the class of simple map-germs, nor those of $\mathcal{A}_e$-codimension ≤ 4.)

$$P_k: \quad (x, y, yz + z^{k+3}, xz + z^3) \quad (k+3)\text{-determined},$$

for $k \geq 1$, k not a multiple of 3.

(iii) As the classification progresses, germs are ruled out as non-simple when adjacencies with modular families occur. The calculations are straightforward but tedious. To prove the remaining germs are simple we calculate the adjacencies which arise in an $\mathcal{A}$-versal unfolding, or at least calculate all *possible* adjacencies. For the folding-maps these adjacencies follow immediately from existing results. For the remaining cases the calculations are straightforward but will be reported in a follow-up article.

References

[1] Bruce, J.W., Kirk, N.P., du Plessis, A.A., Complete transversals and the classification of singularities, *Nonlinearity*, **10** (1997), 253–275.

[2] Bruce, J.W., du Plessis, A.A., Wall, C.T.C., Determinacy and unipotency, *Invent. Math.*, **88** (1987), 521–554.

[3] Dimca, A., Gibson, C.G., Contact unimodular germs from the plane to the plane, *Quart. J. Math. Oxford (2)*, **34** (1983), 281–295.

[4] Dimca, A., Gibson, C.G., Classification of equidimensional contact unimodular map-germs, *Math. Scand.*, **56** (1985), 15–28.

[5] Goryunov, V.V., Mond, D.M.Q., Vanishing cohomology of singularities of mappings, *Compositio Math.*, **89** (1993), 45–80.

[6] Greuel, G.-M., Hamm, H., Invarianten quasihomogener vollständiger Durchschnitte. *Invent. Math.* **49**, 1 (1978), 67–86.

[7] Houston, K., Local topology of images of finite complex analytic maps, *Topology*, **36**, 5 (1997), 1077–1121.

[8] Houston, K., On singularities of folding maps and augmentations, to appear in *Math. Scand.*.

[9] Kirk, N.P., Transversal: A Maple Package for Singularity Theory, User Manual, 3rd edition, Preprint, *University of Liverpool*, (1998).

[10] Kirk, N.P., Computational aspects of classifying singularities, Preprint, *University of Liverpool*, (1997).

[11] Looijenga, E.J.N., *Isolated Singular Points on Complete Intersections*, London Math. Soc. Lecture Note Series 77, (Cambridge University Press, 1984).

[12] Marar, W.L., The Euler characteristic of the disentanglement of the image of a corank 1 map germ, in *Singularity Theory and its Applications, Warwick 1989*, D. Mond and J. Montaldi (eds.), Lecture Notes in Math. 1462, Springer Verlag, Berlin, Heidelberg, New York, (1991), 212–220.

[13] Marar, W.L., Mapping fibrations, *Manuscripta Math.*, **80** (1993), 273–281.

[14] Marar, W.L., Mond, D.M.Q., Multiple point schemes for corank 1 maps, *J. London Math. Soc.* (2), **39** (1989), 553–567.

[15] Mather, J.N., Stability of C^∞-mappings IV: Classification of stable germs by **R**-algebras, *Publ. Math.*, IHES, **37** (1970), 223–248.

[16] Mond, D.M.Q., The classification of germs of maps from surfaces to 3-space, with applications to the differential geometry of immersions, Ph.D. thesis, *University of Liverpool*, (1982).

[17] Mond, D.M.Q., On the classification of germs of maps from $\mathbf{R}^2$ to $\mathbf{R}^3$, *Proc. London Math. Soc.* (3), **50** (1985), 333–369.

[18] Mond, D.M.Q., Some remarks on the geometry and classification of germs of maps from surfaces to 3-space, *Topology*, **26**, 3 (1987), 361–383.

[19] Mond, D.M.Q., Vanishing cycles for complex analytic maps, in *Singularity Theory and its Applications, Warwick 1989*, D. Mond and J. Montaldi (eds.), Lecture Notes in Math. 1462, Springer Verlag, Berlin, Heidelberg, New York, (1991), 221–234.

[20] Ratcliffe, D., Stems and series in $\mathcal{A}$-classification, *Proc. London Math. Soc.* (1), **70** (1995), 181–213.

[21] The SINGULAR project; directed and coordinated by G.-M. Greuel, G. Pfister and H. Schoenemann, Department of Mathematics, University of Kaiserslautern. Latest information available from `http://www.mathematik.uni-kl.de/~zca/Singular`

[22] Wall, C.T.C., A note on symmetry of singularities, *Bull. London Math. Soc.*, **12** (1980), 169–175.

[23] Wall, C. T. C., Finite determinacy of smooth map-germs, *Bull. London Math. Soc.*, **13** (1981), 481–539.

[24] Wilkinson, T.C., The geometry of folding maps, Ph.D. Thesis, *University of Newcastle Upon Tyne*, (1991).

K. Houston
Mathematics and Statistics Group
Middlesex University
Bounds Green Road
London
N11 2NQ
UK

e-mail: k.houston@mdx.ac.uk

N.P. Kirk
Department of Mathematical Sciences
The University of Liverpool
P.O. Box 147
Liverpool
L69 3BX
UK

e-mail: neilpk@liverpool.ac.uk

Multiplicities of Zero-Schemes in Quasihomogeneous Corank-1 Singularities $\mathbf{C}^n \to \mathbf{C}^n$

W.L. Marar J.A. Montaldi M.A.S. Ruas

Abstract

How many cusps does a swallowtail have,
After it becomes a stable map,
And how many swallowtails does a butterfly have,
After it . . . (with apologies to B. Dylan)

Introduction

Consider the map

$$\begin{array}{rccc} F: & \mathbf{C}^2 & \to & \mathbf{C}^2 \\ & (x,y) & \mapsto & (x, y^4 + xy), \end{array}$$

(which is a section of the swallowtail singularity) and its perturbation

$$F_\varepsilon(x,y) = (x, y^4 + xy + \varepsilon y^2).$$

The singular set of F is given by $4y^3 + x = 0$, and the discriminant $\Delta(F)$ of F (the image of its singular set) is a curve with a singular point at the origin. The singular set of F_ε is also a smooth curve, but its image $\Delta(F_\varepsilon)$ is a curve with 2 cusps (A_2-points) and a double point (an $A_{(1,1)}$-point) — see Figure 1.

It turns out (and is well-known) that the number of cusps and double points is independent of the perturbation, provided the perturbation is a

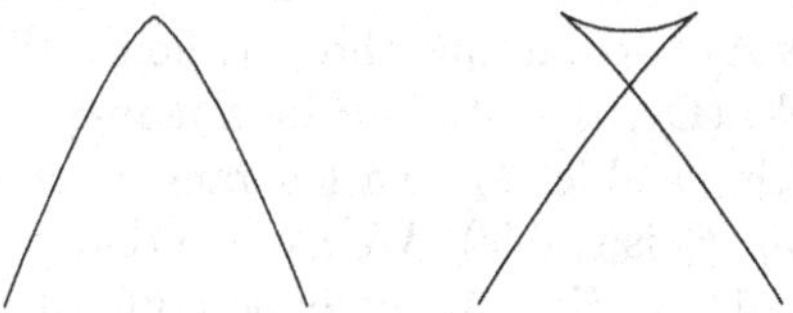

Figure 1: Discriminants of F and F_ε — the swallowtail.

stable map. T. Fukuda and G. Ishikawa [3] show that the number of cusps is given by the dimension of a local algebra associated to F, and independently J. Rieger [15] gives formulae for both the number of cusps and the number of double points in the case that F is of corank 1 — see also [16]. T. Gaffney and D. Mond [6] give formulae for both the number of cusps and the number of double points for a general $\mathcal{A}$-finitely-determined map-germ $\mathbf{C}^2 \to \mathbf{C}^2$.

In this paper, we consider the analogous problem for map-germs $F : \mathbf{C}^n \to \mathbf{C}^n$; that is, given such a map-germ, consider a perturbation which is stable, and ask how many occurrences of each isolated feature in $\Delta(F_\varepsilon)$ there are. The features are the *zero-schemes* of the title, and the numbers are the *multiplicities*. We are able to give answers in the case that *F is of corank 1*. In particular, if F is weighted homogeneous, then we give a closed formula (Theorem 1) for these numbers in terms of the weights and degrees of F. However, unlike Fukuda, Ishikawa and Rieger, we do not consider the case of real map-germs.

The final section 3 of the paper uses this result to give a formula for the multiplicities of the strata in the generalized swallowtail discriminant (Theorem 9).

A 3-dimensional example analogous to the swallowtail one above can be obtained by taking a section of the butterfly:

$$\begin{array}{rcl} F: \mathbf{C}^3 & \to & \mathbf{C}^3 \\ (x_1, x_2, y) & \mapsto & (x_1, x_2, y^5 + x_1 y^2 + x_2 y). \end{array}$$

Here the singular set is a smooth surface in $\mathbf{C}^3$, whose image $\Delta(F)$ is a surface with a cuspidal edge and a more degenerate point at the origin. A stable perturbation (or stabilization) F_ε can be given by

$$F_\varepsilon(x_1, x_2, y) = (x_1, x_2, y^5 + x_1 y^2 + x_2 y + \varepsilon y^3).$$

A schematic illustration of $\Delta(F_\varepsilon)$ is given in Figure 2. The interesting isolated features (zero-schemes) of $\Delta(F_\varepsilon)$ are the 2 swallowtail points (A_3-points), and the 2 points where a cuspidal edge passes through a smooth sheet ($A_{(2,1)}$-points). There could in principle be a further isolated feature, namely a triple point of $\Delta(F_\varepsilon)$ where three smooth sheets intersect ($A_{(1,1,1)}$-points), but such a singularity does not occur in this example. The purpose of this paper is to be able to predict these numbers from the form of F, without studying F_ε explicitly. For example, if y^5 were replaced by y^6 in the butterfly example above, then according to Theorem 1, any stabilization would have one $A_{(1,1,1)}$-point, six $A_{(2,1)}$-points and three A_3-points. See Example 2 below.

In general, let $F : (\mathbf{C}^n, 0) \to (\mathbf{C}^n, 0)$ be a map-germ with a degenerate (non-stable) singularity, and let F_ε be a 1-parameter *stabilization* of F (that is, for $\varepsilon \neq 0$, the map F_ε is stable). We assume that F is of corank 1 (that is, dF_0 has rank $n-1$). If F is $\mathcal{A}$-finitely-determined, then the singularity of F at 0 splits up into a number of non-degenerate zero-dimensional stable singularities of F_ε, which we now describe.

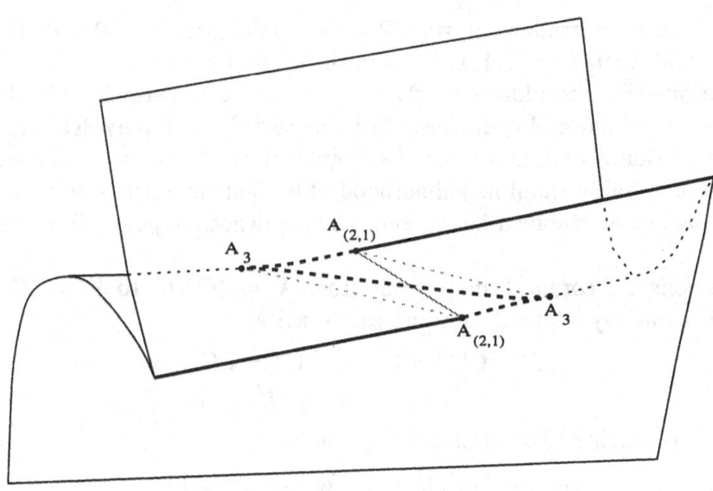

Figure 2: Discriminant of F_ε ($\varepsilon < 0$) — the butterfly. (thick lines are cuspidal edges, grey lines are self-intersections; broken lines are hidden)

A stable map-germ $G : (\mathbf{C}^n, 0) \to (\mathbf{C}^n, 0)$ has an A_k singularity ($k \leq n$) if it is left-right equivalent to the germ,

$$(x_1, \dots, x_{n-1}, y) \mapsto (x_1, \dots, x_{n-1}, y^{k+1} + x_1 y^{k-1} + \dots + x_{k-1} y).$$

Moreover, any stable corank 1 map-germ is an A_k for some natural number k. As is easily seen from this normal form, the set of points in $\mathbf{C}^n$ where a stable map has an A_k singularity is a submanifold of codimension k (given by $x_1 = \dots = x_{k-1} = y = 0$). The image of this set is then an immersed submanifold of codimension k. It turns out that a map with only corank 1 singularities is stable if and only if these submanifolds in the discriminant are in general position [11, (1.6)].

Definition Suppose the map $G : \mathbf{C}^n \to \mathbf{C}^n$ is stable (and defined on some open subset of $\mathbf{C}^n$). Let z be in the image of G, and put $S = G^{-1}(z) = \{s_1, \dots, s_d\}$. Suppose G has an A_{r_j} singularity ($r_j \geq 0$) at s_j (for $j = 1, \dots, d$). In the image, the corresponding submanifolds consisting of A_{r_j} singularities intersect at z, for $j = 1, \dots, d$. Then z represents a *zero-scheme* if and only if this intersection is zero-dimensional. Since G is stable, these submanifolds are in general position so this occurs if and only if $r_1 + \dots + r_d = n$. That is, after suppressing those r_j equal to zero, $\mathcal{P} = (r_1, \dots, r_\ell)$ is a partition of n. We call such a multi-singularity an $A_\mathcal{P}$-singularity.

For example, in the case $n = 2$, the two possibilities of zero-schemes are a cusp, with $\mathcal{P} = (2)$, and a double-fold, with $\mathcal{P} = (1, 1)$; for $n = 3$ the three possibilities are a swallowtail, with $\mathcal{P} = (3)$, a fold-cusp, with $\mathcal{P} = (2,1)$ and a triple fold, with $\mathcal{P} = (1,1,1)$ — as in the examples above.

The question we address is, given an $\mathcal{A}$-finite map-germ $F : (\mathbf{C}^n, 0) \to (\mathbf{C}^n, 0)$ (i.e. of finite $\mathcal{A}$-codimension or equivalently $\mathcal{A}$-finitely determined), and a partition $\mathcal{P}$ of n, how many $A_\mathcal{P}$ singularities are there in a stabilization of F, in a suitably small neighbourhood of 0? This number is independent of the particular stabilization chosen, and we denote it $\#A_\mathcal{P}(F)$ or simply $\#A_\mathcal{P}$.

We consider corank-1 map-germs from $X = (\mathbf{C}^n, 0)$ to $Y = (\mathbf{C}^n, 0)$. Choosing linearly adapted coordinates, we write

$$\begin{array}{rccc} F: & \mathbf{C}^{n-1} \times \mathbf{C} & \to & \mathbf{C}^{n-1} \times \mathbf{C} \\ & (x,y) & \mapsto & (x, f(x,y)). \end{array} \tag{1}$$

When F is weighted homogeneous, we put,

$$\begin{array}{rclrcl} w_0 & = & \mathrm{wt}(y), & w_i & = & \mathrm{wt}(x_i), \\ d & = & \mathrm{degree}(f), & w & = & \prod_{i=1}^{n-1} w_i. \end{array} \tag{2}$$

Let $\mathcal{P} = (r_1, \dots, r_\ell)$ be a partition of n, with $r_1 \geq r_2 \geq \cdots \geq r_\ell \geq 1$, and call ℓ the *length* of $\mathcal{P}$. Define $N(\mathcal{P})$ to be the order of the subgroup of the permutation group S_ℓ which fixes $\mathcal{P}$. Here S_ℓ acts on $\mathbf{R}^\ell$ by permuting the coordinates. For example, for $\mathcal{P} = (4,4,2,2,2,1,1,1)$ we have $N(\mathcal{P}) = (2!)(3!)^2 = 72$.

Theorem 1 *Let $F : (\mathbf{C}^n, 0) \to (\mathbf{C}^n, 0)$ be a corank-1 weighted-homogeneous $\mathcal{A}$-finite map-germ, with weights and degrees as above. For any stabilization of F, and any partition $\mathcal{P}$ of n,*

$$\#A_\mathcal{P}(F) = \frac{w_0^{n-1}}{N(\mathcal{P})w} \prod_{j=1}^{n+\ell-1} \left(\frac{d}{w_0} - j \right),$$

where ℓ is the length of $\mathcal{P}$, and $N(\mathcal{P})$ is defined above.

Example 2 Let $F : \mathbf{C}^3 \to \mathbf{C}^3$ be defined by

$$F(x_1, x_2, y) = (x_1, x_2, y^6 + x_1 y^2 + x_2 y).$$

This is weighted homogeneous, with weights and degrees given by $(w_1, w_2, w_0) = (4,5,1)$ and $d = 6$, so that $d/w_0 = 6$, and $w = w_1 w_2 = 20$.

As already described above, the three types of zero-schemes that occur stably in dimension 3 are given by the partitions $\mathcal{P} = (3)$ (a swallowtail point), $\mathcal{P} = (2,1)$ (a cusp-fold point) and $\mathcal{P} = (1,1,1)$ (a triple fold point). The number of each of these occurring in a stabilization of F can be found from the formula of Theorem 1:

$$\begin{array}{rcl} \#A_{(3)} & = & \dfrac{1}{1 \times 20}(6-1)(6-2)(6-3) = 3 \\ \#A_{(2,1)} & = & \dfrac{1}{1 \times 20}(6-1)\cdots(6-4) = 6 \\ \#A_{(1,1,1)} & = & \dfrac{1}{6 \times 20}(6-1)\cdots(6-5) = 1, \end{array}$$

as claimed earlier.

If the map-germ F is not weighted homogeneous, but is still $\mathcal{A}$-finite, then the multiplicities $\#A_{\mathcal{P}}$ can be computed as the dimensions of certain local algebras, see Corollary 5 and Example 8 below.

1 The $A_{\mathcal{P}}$ schemes

Associated to $X = \mathbf{C}^{n-1} \times \mathbf{C}$ and a partition $\mathcal{P}$ of n we will be considering various spaces. In particular,

$$\begin{aligned} X_\ell &= \mathbf{C}^{n-1} \times \mathbf{C}^\ell, \\ X^\ell &= \mathbf{C}^{n-1} \times \mathbf{C}^{\ell+n}, \end{aligned}$$

where $\ell = \text{length}(\mathcal{P})$. The first of these spaces is used in this section, while the second is used in §2. We will also be considering a versal deformation $\widetilde{F}$ of F, with base $\mathbf{C}^d$, and then we denote $\widetilde{X_\ell} = \mathbf{C}^d \times X_\ell$, and similarly $\widetilde{X^\ell} = \mathbf{C}^d \times X^\ell$.

Let $\widetilde{F} : \widetilde{X} \to \widetilde{Y}$ be an $\mathcal{A}_e$-versal unfolding of F (with base $\mathbf{C}^d$), so that

$$\widetilde{F}(u,x,y) = (u,x,\widetilde{f}(x,y,u)) = (u,\widetilde{F}_u(x,y)).$$

Any stabilization F_ε of F can be induced from the versal deformation $\widetilde{F}$, so from now on we consider only this versal deformation.

For each partition $\mathcal{P} = (r_1,\ldots,r_\ell)$ of n we consider (following ideas of Gaffney [5]) the subscheme $\widetilde{V}(\mathcal{P})$ of $\widetilde{X_\ell} := \mathbf{C}^d \times \mathbf{C}^{n-1} \times \mathbf{C}^\ell$, defined by

$$\widetilde{V}(\mathcal{P}) := \text{clos}\left\{ (u,x,y_1,\ldots,y_\ell) \in \widetilde{X_\ell} \;\middle|\; \begin{array}{l} \bullet\ y_i \neq y_j \\ \bullet\ \widetilde{F}(u,x,y_i) = \widetilde{F}(u,x,y_j) \\ \bullet\ \widetilde{F}_u \text{ has a singularity of type} \\ \qquad A_{r_j} \text{ at } (u,x,y_j) \end{array} \right\},$$

where 'clos' means the analytic closure in $\widetilde{X_\ell}$.

Let $\pi = \pi_{\mathcal{P}} : \widetilde{V}(\mathcal{P}) \to \mathbf{C}^d$ be the restriction to $\widetilde{V}(\mathcal{P})$ of the Cartesian projection $\widetilde{X_\ell} \to \mathbf{C}^d$. For generic $u \in \mathbf{C}^d$, the fibre $\pi^{-1}(u)$ consists of those 'multi-points' (also known as 'sets') where $\widetilde{F}_u$ has an $A_{\mathcal{P}}$ multi-germ. We are thus interested in the degree of $\pi_{\mathcal{P}}$.

Proposition 3 *If $\mathcal{P} = (r_1,\ldots,r_\ell)$ is a partition of n, then*

$$\#A_{\mathcal{P}} = \frac{1}{N(\mathcal{P})}\text{degree}(\pi(\mathcal{P})).$$

PROOF Let $\mathbf{y} = (y_1, \ldots, y_\ell) \in \widetilde{V}(\mathcal{P})$ and $\sigma \in S_\ell$. We have

$$\mathbf{y}_\sigma := (y_{\sigma(1)}, \ldots, y_{\sigma(\ell)}) \in \widetilde{V}(\mathcal{P})$$

if and only if $r_{\sigma(j)} = r_j$ for each $j = 1, \ldots, \ell$. There are $N(\mathcal{P})$ such σ. The points $\mathbf{y}$ and $\mathbf{y}_\sigma$ are distinct, but the corresponding sets $\{y_1, \ldots, y_\ell\}$ are the same, and it is the sets that are counted in $\#A_\mathcal{P}$. □

Let $\widetilde{\mathcal{I}}(\mathcal{P})$ be the ideal in $\mathcal{O}_{\widetilde{X}_\ell}$ defining $\widetilde{V}(\mathcal{P})$, and put

$$\mathcal{I}(\mathcal{P}) = (\widetilde{\mathcal{I}}(\mathcal{P}) + \langle u_1, \ldots, u_d \rangle) / \langle u_1, \ldots, u_d \rangle \subset \mathcal{O}_{X_\ell},$$

corresponding to the intersection of $\widetilde{V}(\mathcal{P})$ with $\{0\} \times X_\ell = X_\ell$. The main theorem follows from the remaining two propositions of this section.

It follows from the definition of $\widetilde{\mathcal{I}}(\mathcal{P})$, that *at generic points of* $\widetilde{V}(\mathcal{P})$ (i.e. where $y_i \neq y_j$),

$$\widetilde{\mathcal{I}}(\mathcal{P}) = \left\langle (\partial_y \tilde{f})_1, \ldots, (\partial_y^{r_1} \tilde{f})_1, \ldots, (\partial_y \tilde{f})_\ell, \ldots, (\partial_y^{r_\ell} \tilde{f})_\ell \right\rangle + \left\langle \tilde{f}_1 - \tilde{f}_2, \ldots, \tilde{f}_1 - \tilde{f}_\ell \right\rangle, \tag{3}$$

where $\tilde{f}_k$ denotes $\tilde{f}$ evaluated at (u, x, y_k), for $1 \leq k \leq \ell$, and $(\partial_y^i \tilde{f})_k$ denotes the ith partial derivative of $\tilde{f}$ with respect to y at the point (u, x, y_k), for $1 \leq k \leq \ell$ and $1 \leq i \leq r_k$.

Proposition 4 *Suppose* $\widetilde{V}(\mathcal{P})$ *is non-empty.*
(a) $\widetilde{V}(\mathcal{P})$ *is smooth of dimension* d*;*
(b) $\pi(\mathcal{P}) : \widetilde{V}(\mathcal{P}) \to \mathbf{C}^d$ *is finite and* $\pi^{-1}(\pi(0)) = \{0\}$*;*
(c) the degree of $\pi(\mathcal{P})$ *coincides with* $\dim_\mathbf{C} \mathcal{O}_{X_\ell}/\mathcal{I}(\mathcal{P})$.

It follows from this proposition that the ideal $\mathcal{I}(\mathcal{P})$ is a complete intersection.

PROOF (a) Since $\widetilde{F}$ is versal, it follows a fortiori that it is a stable map, and then part (a) follows immediately from [9, Proposition 2.13].

(b) The projection $\pi_\mathcal{P} : \widetilde{V}(\mathcal{P}) \to \mathbf{C}^d$ is a finite mapping. In fact, for a generic $u \in \mathbf{C}^d$, the fibre $\pi^{-1}(u)$ is finite and consists of those 'multi-points' where $\widetilde{F}_u$ has an $A_\mathcal{P}$ multi-germ. The germ $\widetilde{F}_0 = F$ is $\mathcal{A}$-finite. So, by the Mather-Gaffney geometric criterion ([4] or [17, Theorem 2.1]), it is stable away from zero. Thus, $\pi^{-1}(\pi(0)) = \{0\}$.

(c) Since $\widetilde{V}(\mathcal{P})$ is smooth and hence is Cohen-Macaulay at zero, the degree of $\pi_\mathcal{P}$ coincides with $\dim_\mathbf{C} \mathcal{O}_{X_\ell}/\mathcal{I}(\mathcal{P})$ [8, Prop. 5.12]. □

Note that combining Propositions 3 and 4(c) gives a method for computing the multiplicities even in the case that F is not weighted homogeneous, provided we can compute $\mathcal{I}(\mathcal{P})$:

Corollary 5

$$\#A_\mathcal{P} = \frac{1}{N(\mathcal{P})} \dim_\mathbf{C} \left(\frac{\mathcal{O}_{X_\ell}}{\mathcal{I}(\mathcal{P})} \right).$$

In Section 2 we show how to compute $\mathcal{I}(\mathcal{P})$ and we give an example of how this applies. We also prove the following, which combined with the corollary above, proves Theorem 1.

Proposition 6 *If F is weighted homogeneous, with weights and degree as in equation (2), then*

$$\dim_{\mathbf{C}}\left(\frac{\mathcal{O}_{X_\ell}}{\mathcal{I}(\mathcal{P})}\right) = \frac{1}{w_0^\ell w}\prod_{j=1}^{n+\ell-1}(d - jw_0).$$

2 Multiple point schemes

Nearby the $(A_{r_1} + \cdots + A_{r_\ell}) = A_{(r_1,\ldots,r_\ell)}$ multi-germs, there are points in the target with $(r_1+1)+(r_2+1)+\cdots+(r_\ell+1) = (n+\ell)$ preimages. We shall follow D. Mond [14] and define an $(n+\ell)$-tuple scheme in $X^\ell = \mathbf{C}^{n-1} \times \mathbf{C}^{n+\ell}$, which on the appropriate diagonal specializes to the ideal defining $A_{(r_1,\ldots,r_\ell)}$ multi-germs (Proposition 7 below).

As usual, given a corank-1 map-germ $F : \mathbf{C}^n \to \mathbf{C}^n$ we choose linearly adapted coordinates on $\mathbf{C}^n$ so that $F(x,y) = (x, f(x,y))$ as in equation (1). Having chosen such coordinates on $\mathbf{C}^n$, we denote the coordinates of X^ℓ by

$$(x, \mathbf{y}) = (x, y_1^0, \ldots, y_1^{r_1}, y_2^0, \ldots, y_2^{r_2}, \ldots, y_\ell^0, \ldots, y_\ell^{r_\ell}).$$

We define an ideal $\mathcal{J}(f,\mathcal{P}) \subset \mathcal{O}_{X^\ell}$ by

$$\mathcal{J}(f,\mathcal{P}) = \langle h_i \mid i = 1, \ldots, n+\ell-1\rangle,$$

with

$$h_i = V^{-1}.\begin{vmatrix} 1 & y_1^0 & \cdots & (y_1^0)^{i-1} & f_1^0 & (y_1^0)^{i+1} & \cdots & (y_1^0)^{n+l-1} \\ \vdots & \vdots & & \vdots & \vdots & \vdots & & \vdots \\ 1 & y_1^{r_1} & \cdots & (y_1^{r_1})^{i-1} & f_1^{r_1} & (y_1^{r_1})^{i+1} & \cdots & (y_1^{r_1})^{n+l-1} \\ \vdots & \vdots & & \vdots & \vdots & \vdots & & \vdots \\ 1 & y_\ell^0 & \cdots & (y_\ell^0)^{i-1} & f_\ell^0 & (y_\ell^0)^{i+1} & \cdots & (y_\ell^0)^{n+l-1} \\ \vdots & \vdots & & \vdots & \vdots & \vdots & & \vdots \\ 1 & y_\ell^{r_\ell} & \cdots & (y_\ell^{r_\ell})^{i-1} & f_\ell^{r_\ell} & (y_\ell^{r_\ell})^{i+1} & \cdots & (y_\ell^{r_\ell})^{n+l-1} \end{vmatrix},$$

where $V = V(y_1^0, \ldots, y_1^{r_1}, \ldots, y_\ell^0, \ldots, y_\ell^{r_\ell})$ is the Vandermonde determinant and

$$f_k^i = f(x, y_k^i).$$

It follows from Cramer's rule that the ideal $\mathcal{J}(f,\mathcal{P})$ defines the set of points in X^ℓ where all the f_k^i coincide [14]. (Note that in the h_i some superscripts are indices, while others represent powers!)

For the versal deformation $\widetilde{F}$, one defines the ideal $\mathcal{J}(\tilde{f},\mathcal{P})$ in $\mathcal{O}_{\tilde{X}^\ell}$ in exactly the same way, with $\tilde{f}_k^i = \tilde{f}(u, x, y_k^i)$.

In X^ℓ there is a diagonal of particular interest, namely,

$$\Delta(\mathcal{P}) = \{(x,\mathbf{y}) \in X^\ell \mid y_k^i = y_k^j,\ \forall i,j = 1,\dots,r_k,\ \forall k = 1,\dots,\ell\},$$

which can be parametrized in the obvious way by $(x, y_1, \dots, y_\ell)$:

$$(x,\mathbf{y}) = (x, y_1, \dots, y_1, y_2, \dots, y_2, \dots, y_\ell, \dots, y_\ell), \tag{4}$$

with y_i occurring $r_i + 1$ times. This corresponds to an embedding j_ℓ of X_ℓ into X^ℓ. Of course, there is a similar embedding of $\widetilde{X}_\ell$ in $\widetilde{X}^\ell$. A generic point of $\Delta(\mathcal{P})$ is one of the form (4) with $y_i \neq y_j$, for $i \neq j$. We often simply write Δ in place of $\Delta(\mathcal{P})$.

Let $\mathcal{I}_{\Delta(\mathcal{P})}$ be the ideal defining $\Delta(\mathcal{P})$, that is

$$\mathcal{I}_{\Delta(\mathcal{P})} = \left\langle y_k^i - y_k^0, \mid i = 1,\dots,r_k,\ k = 1,\dots,\ell \right\rangle,$$

and let $\mathcal{J}_\Delta(f,\mathcal{P})$ be the $\mathcal{O}_{X^\ell}$ ideal defined by

$$\mathcal{J}_\Delta(f,\mathcal{P}) = \mathcal{J}(f,\mathcal{P}) + \mathcal{I}_{\Delta(\mathcal{P})}.$$

It was shown in [9] that at a generic point of $V(\mathcal{J}_\Delta(f,\mathcal{P}))$, f has a singularity of type A_{r_j} at (x, y_j), and $f(x,y_1) = \cdots = f(x,y_l)$ (see proof of Proposition 7(c) below).

Proposition 7 *(a) The ideal $\mathcal{J}(\tilde{f},\mathcal{P})$ is reduced, and the multiple point variety $V(\mathcal{J}(\tilde{f},\mathcal{P})) \subset \widetilde{X}^\ell$ is smooth of dimension $d+n$ (or is empty);*
(b) $\mathcal{J}_\Delta(f,\mathcal{P})$ is a complete intersection singularity;
(c) Let $j_\ell : X_\ell \hookrightarrow X^\ell$ be the embedding with image $\Delta(\mathcal{P})$ given in (4). Then the surjection $j_\ell^* : \mathcal{O}_{X^\ell} \to \mathcal{O}_{X_\ell}$ satisfies $j_\ell^*(\mathcal{J}_\Delta(f,\mathcal{P})) = \mathcal{I}(\mathcal{P})$ and consequently induces an isomorphism

$$j_\ell^* : \frac{\mathcal{O}_{X^\ell}}{\mathcal{J}_\Delta(f,\mathcal{P})} \xrightarrow{\simeq} \frac{\mathcal{O}_{X_\ell}}{\mathcal{I}(\mathcal{P})}.$$

PROOF (a) The dimension is clear: for each value of (u, x, Y) in the target there are finitely many points (u,x,y) which map to this under $\widetilde{F}$. The smoothness is less obvious, but follows from [9].
(b) The ideals $\langle u_1,\dots,u_d\rangle$ and $\mathcal{I}_\Delta$ have d and n generators respectively, and the intersection of $V(\mathcal{J}(f,\mathcal{P}))$ with the diagonal $\Delta(\mathcal{P})$ is reduced to a single point (the origin) so that for dimensional reasons the ideal is a complete intersection.
(c) It is proved in [9, Lemma 2.7] that at generic points of $\Delta(\mathcal{P})$ one has,

$$\begin{aligned}\mathcal{J}_\Delta(f,\mathcal{P}) = \big\langle (\partial_y f)_1, \dots, (\partial_y^{r_1} f)_1, \dots, (\partial_y f)_\ell, \dots, (\partial_y^{r_\ell} f)_\ell \big\rangle \\ + \langle f(x,y_i) - f(x,y_1); 2 \le i \le \ell \rangle + \mathcal{I}_{\Delta(\mathcal{P})},\end{aligned}$$

where the $(\partial_y^i f)_k$ are as defined in equation (3). It follows that generically $j_\ell^* \mathcal{J}_\Delta(f,\mathcal{P}) = \mathcal{I}(\mathcal{P})$. Part (c) then follows from the fact that two reduced complete intersection ideals that coincide generically are in fact the same. □

PROOF OF PROPOSITION 6 According to Proposition 7(c) it is enough to compute $\dim(\mathcal{O}_{X^\ell}/\mathcal{J}_\Delta(f,\mathcal{P}))$, and if f is weighted homogeneous this last can be computed by Bezout's theorem [12] since $\mathcal{J}_\Delta(f,\mathcal{P})$ is a complete intersection.

The generators of $\mathcal{J}_\Delta(f,\mathcal{P})$ are the h_j and the $y_k^i - y_k^0$. For each $j = 1,\ldots,n+\ell-1$ one has

$$\text{degree}(h_j) = d - jw_0,$$

while the other generators have degree w_0. The product of all the degrees of the generators is therefore

$$\left(\prod_{j=1}^{n+\ell-1}(d-jw_0)\right) w_0^n.$$

Since $\mathcal{J}_\Delta(f,\mathcal{P})$ is a weighted homogeneous complete intersection (Proposition 7(b)), we can apply Bezout's theorem [12], whence its colength is

$$\frac{1}{w_0^{\ell+n}w}\left(\prod_{j=1}^{n+\ell-1}(d-jw_0)\right) w_0^n = \frac{1}{w_0^\ell w}\prod_{j=1}^{n+\ell-1}(d-jw_0),$$

as required. □

Example 8 Let $f : \mathbf{C}^3 \to \mathbf{C}^3$ be the non-weighted-homogeneous map-germ given by

$$f(x_1,x_2,y) = (x_1, x_2, y^5 + x_1 y + {x_2}^2 y^2 + x_2 y^3).$$

(this is denoted 5_2 in the classification in [10]: note that this is not equivalent to a weighted-homogeneous map since the discriminant Milnor number and the $\mathcal{A}_e$-codimension do not coincide [2]).

Using MAPLE (see the Appendix below for the programme) we computed the three ideals $\mathcal{I}(\mathcal{P})$ for the three possible partitions. First we computed $\mathcal{J}(f,\mathcal{P})$, then substituted $\mathcal{I}_\Delta$. By Proposition 7 this gives $\mathcal{I}(\mathcal{P})$, and one then deduces the multiplicity from Corollary 5. The results are

$$\begin{aligned}
\mathcal{I}((2,1)) &= \big\langle -3\,y_1^2y_2^2 - 2\,y_2^3y_1 + x_1,\ 3\,y_1^2y_2 + 6\,y_2^2y_1 + y_2^3 + x_2^2, \\
&\qquad -y_1^2 - 6\,y_1y_2 - 3\,y_2^2 + x_2,\ 2\,y_1 + 3\,y_2\big\rangle \\
\mathcal{I}((3)) &= \big\langle 15\,y_1^4 + x_1,\ -20\,y_1^3 + x_2^2,\ 10\,y_1^2 + x_2\big\rangle \\
\mathcal{I}((1,1,1)) &= \langle 1\rangle .
\end{aligned}$$

It follows that

$$\begin{aligned} \#A_{(2,1)} &= 3 \\ \#A_{(3)} &= 3 \\ \#A_{(1,1,1)} &= 0. \end{aligned}$$

Note that $\#A_{(3)}$ is given in [10], but the values of the other two invariants are new.

Applying Theorem 1 or the method above to the corank-1 simple germs classified by Marar and Tari [10] enables us to 'complete' their Table 1 by giving the new invariants $\#A_{(1,2)}$ and $\#A_{(1,1,1)}$. It turns out that these are all zero, except for $\#A_{(1,2)}(5_k)$ for $k = 1, 2, 3$. The results are:

$$\#A_{(1,2)}(5_1) = 2, \qquad \#A_{(1,2)}(5_2) = \#A_{(1,2)}(5_3) = 3.$$

In particular, all the simple germs $f : (\mathbf{C}^3, 0) \to (\mathbf{C}^3, 0)$ satisfy

$$\#A_{(1,1,1)}(f) = 0.$$

3 Multiplicities of strata in generalized swallowtails

In this final section, we use Theorem 1 to give a simple formula for the local multiplicity of the closure of each stratum in the discriminant of an isolated A_k singularity.

Consider the stable A_k map $F : \mathbf{C}^k \to \mathbf{C}^k$,

$$\begin{aligned} F(x_1, \ldots, x_{k-1}, y) &= (X_1, \ldots, X_{k-1}, Y) \\ &= (x_1, \ldots, x_{k-1}, y^{k+1} + x_1 y^{k-1} + \cdots + x_{k-1} y). \end{aligned}$$

This map is clearly weighted homogeneous, with weights $\mathrm{wt}(x_i) = \mathrm{wt}(X_i) = i + 1$, $\mathrm{wt}(y) = 1$ and $\mathrm{wt}(Y) = k + 1$. The discriminant $\Delta(F)$ is stratified by the various $A_{\mathcal{P}}$ multi-germs, where $\mathcal{P} = (r_1, \ldots, r_\ell)$ is a partition of any $n \leq k + 1 - \ell$. Denote this stratum by $\Delta_{\mathcal{P}}$ and its closure by $Z_{\mathcal{P}}$. $Z_{\mathcal{P}}$ is an algebraic subvariety of $\mathbf{C}^k$ of dimension $D = k - n$.

Note that if $n > k+1-\ell$ then $\Delta_{\mathcal{P}}$ is empty, as observed by Goryunov [7, §4.3]. Indeed, close to $\Delta_{\mathcal{P}}$ there are points with at least $\sum_i (r_i + 1) = (n + \ell)$ preimages; however F has multiplicity $k+1$ so that $n+\ell \leq k+1$ (Goryunov's $D(\mu_1, \ldots, \mu_k)$ corresponds to our $\Delta_{\mathcal{P}}$ for $\mathcal{P} = (\mu_1 + 1, \ldots, \mu_k + 1)$.)

Theorem 9 *The multiplicity of $Z_{\mathcal{P}}$ at the origin is given by,*

$$\frac{1}{N(\mathcal{P})}(D+1)D(D-1)\cdots(D-\ell+2),$$

where $D = \dim(Z_{\mathcal{P}})$ and $N(\mathcal{P})$ is defined in the introduction.

To prove this, we first need a lemma on the geometric structure of A_k discriminants.

Lemma 10 *Let $Z_{\mathcal{P}}$ be as above, and let (z_i) be any sequence of points in $Z_{\mathcal{P}}$ converging to 0. Then*

$$T_0 Z_{\mathcal{P}} := \lim_{i\to\infty} T_{z_i} Z_{\mathcal{P}} = \{(\mathbf{X}, Y) \mid X_{k-n+1} = X_{k-n+2} = \cdots = X_{k-1} = Y = 0\}.$$

PROOF As is well-known and easy to see, the discriminant of F coincides with the discriminant of the orbit map $\sigma_0 : \mathbf{C}^k_s \to \mathbf{C}^k_t$ for the action of the permutation group S_{k+1}, where $\mathbf{C}^k_s$ is identified with the subspace of $\mathbf{C}^{k+1}$ the sum of whose coordinates vanishes, and S_{k+1} acts on $\mathbf{C}^{k+1}$ by permuting the coordinates. Consider the extension σ of σ_0 to $\mathbf{C}^{k+1}$ defined as usual by,

$$\begin{aligned} \sigma : \mathbf{C}^{k+1} &\to \mathbf{C}^{k+1} \\ (y_1, \ldots, y_{k+1}) &\mapsto \Big(\sum_i y_i, \sum_{i<j} y_i y_j, \ldots, y_1 \cdots y_{k+1}\Big). \end{aligned}$$

Clearly, $\mathbf{C}^k_t$ is to be identified with the subspace of $\mathbf{C}^{k+1}$ with vanishing first coordinate. It will be more convenient for computations to change coordinates in the target of σ so that σ takes the form

$$\tilde{\sigma}(y_1, \ldots, y_{k+1}) = \Big(\sum_i y_i, \sum_i y_i^2, \sum_i y_i^3, \ldots, \sum_i y_i^{k+1}\Big).$$

Note that the linear subspaces of the form $T_0 Z_{\mathcal{P}}$ are preserved by the differential at the origin of this change of coordinates; indeed this differential is a diagonal matrix.

Denote by $\tilde{\Delta}$ the discriminant of $\tilde{\sigma}$.

Given the partition $\mathcal{P} = (r_1, \ldots, r_\ell)$ of n, the stratum $\tilde{\Delta}_{\mathcal{P}}$ is the image under $\tilde{\sigma}$ of $\Sigma_{\mathcal{P}} \subset \mathbf{C}^{k+1}$. Let $D + 1 = \dim(\tilde{\Delta}_{\mathcal{P}})$ (so $D = \dim(Z_{\mathcal{P}})$ as in the theorem). It is convenient to extend $\mathcal{P}$ by $D + 1 - \ell$ zeros, so that $r_j = 0$ for $j = \ell + 1, \ldots, D + 1$. The stratum $\Sigma_{\mathcal{P}} \subset \mathbf{C}^{k+1}$ is parametrized by

$$(y_1, \ldots, y_{D+1}) \mapsto (y_1, \ldots, y_1, y_2, \ldots, y_2, \ldots, y_\ell, \ldots, y_\ell, y_{\ell+1}, \ldots, y_{D+1}),$$

where y_j occurs with multiplicity $r_j + 1$, and the y_j are distinct.

Write $\tilde{\sigma}_{\mathcal{P}}$ for the restriction of $\tilde{\sigma}$ to $\Sigma_{\mathcal{P}}$. Using this parametrization of $\Sigma_{\mathcal{P}}$, $\tilde{\sigma}_{\mathcal{P}}$ has the form,

$$\tilde{\sigma}_{\mathcal{P}}(y_1, \ldots, y_{D+1}) = \Big(\sum_i (r_i + 1) y_i, \sum_i (r_i + 1) y_i^2, \ldots, \sum_i (r_i + 1) y_i^{k+1}\Big).$$

Thus, at a point $y \in \Sigma_{\mathcal{P}}$, the differential of $\tilde{\sigma}_{\mathcal{P}}$ is

$$d\tilde{\sigma}_{\mathcal{P}}(y) = \begin{bmatrix} r_1 + 1 & \cdots & r_{D+1} + 1 \\ 2(r_1 + 1) y_1 & \cdots & 2(r_{D+1} + 1) y_{D+1} \\ \vdots & & \vdots \\ (k+1)(r_1 + 1) y_1^k & \cdots & (k+1)(r_{D+1} + 1) y_{D+1}^k \end{bmatrix}.$$

Notice that the top $(D+1)\times(D+1)$ minor is equal to

$$(D+1)!\left(\prod(r_i+1)\right)V(y_1,\ldots,y_{D+1}),$$

where V is the Vandermonde determinant, which is non-vanishing on $\widetilde{\Delta}_{\mathcal{P}}$. Consequently, at points of $\widetilde{\Delta}_{\mathcal{P}}$, the tangent space to $\widetilde{\Delta}_{\mathcal{P}}$ projects isomorphically onto $\mathbf{C}^{D+1}$ (defined by the vanishing of the last $k-D$ coordinates).

Finally, since $\tilde{\sigma}$ is weighted-homogeneous, and the last $k-D$ components are of strictly higher degree than the first $D+1$, it follows that in the limit as

$$(y_1,\ldots,y_{D+1})\to(0,\ldots,0),$$

the tangent space to $\widetilde{\Delta}_{\mathcal{P}}$ tends to $\mathbf{C}^{D+1}$. Intersecting source and target with $\mathbf{C}^k_s$ and $\mathbf{C}^k_t$ respectively shows that the same is true of the tangent space to $\Delta_{\mathcal{P}}$, as required. □

PROOF OF THEOREM 9 It follows from this lemma that the multiplicity at 0 of $Z_{\mathcal{P}}$ is given by the intersection multiplicity of $Z_{\mathcal{P}}$ with the n-dimensional subspace

$$\{(\mathbf{X},Y)\mid X_1=\cdots=X_{k-n}=0\},$$

which is complementary to the unique limiting tangent space $T_0Z_{\mathcal{P}}$, and it remains for us to compute this multiplicity.

To this end, consider the map $g:\mathbf{C}^n\to\mathbf{C}^n$ defined by

$$g(u_1,\ldots,u_{n-1},y)=(u_1,\ldots,u_{n-1},y^{k+1}+u_1y^{n-1}+\cdots+u_{n-1}y),$$

which is induced from F by the immersion $\gamma:\mathbf{C}^n\to\mathbf{C}^k$,

$$\gamma(u_1,\ldots,u_{n-1},y)=(0,\ldots,0,u_1,\ldots,u_{n-1},y),$$

in the sense that $F\circ\gamma=\gamma\circ g$.

By the lemma, this inclusion is transverse to $\Delta(F)$ away from the origin, so that it is $\mathcal{K}_{\Delta(F)}$-finite, and consequently, g is $\mathcal{A}$-finite (Damon [1]). Moreover, a stabilization g_ε of g is obtained by perturbing the embedding γ to an embedding γ_ε transverse to $\Delta(F)$, and *a fortiori* transverse to $Z_{\mathcal{P}}$. If γ_ε is transverse to $Z_{\mathcal{P}}$, then $\mathrm{image}(\gamma_\varepsilon)\cap Z_{\mathcal{P}}=\mathrm{image}(\gamma_\varepsilon)\cap\Delta_{\mathcal{P}}$ is a finite set (for dimensional reasons).

The points of this intersection are precisely the image under γ_ε of the points in $\mathbf{C}^n$ (the image of g_ε) over which g_ε has an $A_{\mathcal{P}}$ singularity. Since g is weighted homogeneous, the number of such points is given by Theorem 1. A simple computation then proves Theorem 9. □

Appendix: A MAPLE Programme

The MAPLE programme used for computing $\mathcal{I}(\mathcal{P})$ is short and simple, so can be included here. It runs (at least) on MapleV Release 4.

```
> restart;
> with(linalg);
```

Define function f, and partition $\mathcal{P}$:

```
> f := y^5 + x[1]*y + x[2]^2*y^2 + x[2]*y^3 ;
> P := [1,2];
```

Find dimension of space and length of partition and check that $\mathcal{P}$ is indeed a partition of n:

```
> n := nops(indets(f));
> ell := nops(P);
> if convert(P,'+') <> n
>   then print('ERROR, P should be a partition of n')
> fi;
```

A trick to get indices for the multiple point scheme:

```
> Y := array(1..ell,0..max(op(P)));
> YY := [seq(seq(Y[i,j],j=0..P[i]),i=1..ell)];
> V:=factor(det(vandermonde(YY)));
```

Define the generators h_i of the multiple point scheme:

```
> h := proc(i::integer)
>   local W, j;
>   W := vandermonde(YY);
>   for j to nops(YY) do
>     W[j,i+1] := subs(y=YY[j], f)
>   od;
>   simplify(factor(det(W))/V)
> end;
```

The ideal $\mathcal{J}(f, \mathcal{P})$:

```
> J := [seq(h(i), i=1..n+ell-1)]:
```

Equations for the diagonal $\Delta(P)$:

```
> Delta := {seq( seq(Y[i,j]=y[i], j=0..P[i]), i=1..ell)};
```

Now compute $\mathcal{J}_\Delta$, restricted to Δ — in other words $\mathcal{I}(\mathcal{P})$:

```
> IP := subs(Delta, J);
```

Acknowledgements

This research was partially supported by grants from FAPESP and the CNPq. We would like to thank the referee for making several useful suggestions which we have incorporated. We have also benefited from discussions with T. Fukui, D. Mond and J. Rieger.

References

[1] J. Damon, $\mathcal{A}$-equivalence and the equivalence of sections of images and discriminants. In *Singularity Theory and its Applications, Part I*, Springer L.N.M. **1462** (1991), 93–121.

[2] J. Damon & D. Mond, $\mathcal{A}$-codimension and the vanishing topology of discriminants. *Invent. Math.* **106** (1991), 217–242.

[3] T. Fukuda & G. Ishikawa, On the number of cusps of stable perturbations of a plane-to-plane singularity, *Tokyo J. Math.*, **10** (1987), 375–384.

[4] T. Gaffney, Properties of finitely determined germs, Ph.D thesis, Brandeis University, 1976.

[5] T. Gaffney, Multiple points and associated ramification loci. Proceedings of A.M.S. Symposia in Pure Mathematics **40:I** (1983), 429–437.

[6] T. Gaffney & D. Mond, Weighted homogeneous maps from the plane to the plane. *Math. Proc. Camb. Phil. Soc.* (1991) **109**, 451–470.

[7] V. Goryunov, Semi-simplicial resolutions and homology of images and discriminants of mappings. *Proc. London Math. Soc.* **70** (1995), 363–385.

[8] E.J.N. Looijenga, *Isolated singular points on complete intersections*, London Math. Soc. Lecture Note Series **77**, Cambridge University Press, (1984).

[9] W.L. Marar & D. Mond, Multiple point schemes for corank 1 maps. *J. London Math. Soc.* **39** (1989), 553–567.

[10] W.L. Marar & F. Tari, On the geometry of simple germs of corank 1 maps from $\mathbf{R}^3$ to $\mathbf{R}^3$ *Math. Proc. Camb. Phil. Soc.* (1996) **119**, 469–481.

[11] J.N. Mather, Stability of C^∞ mappings, IV: Classification of stable germs by $\mathbf{R}$-algebras. *Publ. Math. IHES* **37** (1969), 223–248.

[12] J. Milnor & P. Orlik, Isolated singularities defined by weighted homogeneous polynomials. *Topology* **9** (1970), 385–393.

[13] D. Mond & R. Pellikaan, Fitting ideals and multiple points of analytic mappings. Springer L.N.M **1414** (1989), 553–567.

[14] D. Mond, Some remarks on the geometry and classification of germs of maps from surface to 3-space, *Topology* **26** (1987), 361–383.

[15] J. Rieger, Families of maps from the plane to the plane, *J. London Math. Soc.* **36** (1987), 351–369.

[16] J. Rieger, Versal topological stratification and the bifurcation geometry of map-germs of the plane, *Math. Proc. Camb. Phil. Soc.*, **107** (1990), 127–147.

[17] C.T.C. Wall, Finite determinacy of smooth map-germs. *Bull. London Math. Soc.* **13** (1981), 481–539.

Marar & Ruas
ICMSC-USP
Caixa Postal 668
13560-970 São Carlos
S.P. Brasil

Montaldi
Institut Nonlinéaire de Nice
Université de Nice - Sophia Antipolis
06560 Valbonne
France

e-mail:
walmarar@icmsc.sc.usp.br
montaldi@inln.cnrs.fr
maasruas@icmsc.sc.usp.br

Butterflies and Umbilics of Stable Perturbations of Analytic Map-Germs $(\mathbf{C}^5, 0) \to (\mathbf{C}^4, 0)$

Toshizumi Fukui*

1 Introduction

Let $f : (\mathbb{C}^5, 0) \to (\mathbb{C}^4, 0)$ be an analytic map-germ. If f is a stable-germ, it is equivalent one of the following forms (with their labels, names and Thom-Boardman symbols):

$$
\begin{array}{ll}
f(x,y,a,b,c) = (x^2+y^2, a, b, c) & A_1, (\text{fold}, \Sigma^2) \\
f(x,y,a,b,c) = (x^3+ax+y^2, a, b, c) & A_2, (\text{cusp}, \Sigma^{2,1}) \\
f(x,y,a,b,c) = (x^4+ax^2+bx+y^2, a, b, c) & A_3, (\text{swallowtail}, \Sigma^{2,1,1}) \\
f(x,y,a,b,c) = (x^5+ax^3+bx^2+cx+y^2, a, b, c) & A_4, (\text{butterfly}, \Sigma^{2,1,1,1}) \\
f(x,y,a,b,c) = (x^2y+y^3+ay^2+bx+cy, a, b, c) & D_4, (\text{umbilic}, \Sigma^{2,2})
\end{array}
$$

By J.Mather's work [9-12], a representative of a map-germ $f : (\mathbb{C}^5, 0) \to (\mathbb{C}^4, 0)$ can be approximated by locally infinitesimally stable maps. We call such an approximation a stable perturbation of f. In this paper, we consider the following problem: How many butterflies and umbilics will appear in a stable perturbation of a given map-germ $f : (\mathbb{C}^5, 0) \to (\mathbb{C}^4, 0)$? Let $b(f_u)$, $u(f_u)$ denote the number of A_4, D_4 points of a stable perturbation f_u of f. If f is $\mathcal{K}$-finite, then there is a $\mathcal{K}$-versal unfolding of f, and so $b(f_u)$ and $u(f_u)$ do not depend on a generic choice of u. We denote these numbers by $b(f)$, $u(f)$, and it is these we now study.

One main consequence of this paper is the following: Assume that $f : (\mathbb{C}^5, 0) \to (\mathbb{C}^4, 0)$ is a $\mathcal{K}$-finite germ. If $c_{2,1,1,1}(f) < \infty$ and f has kernel rank 2, then $b(f) + 4u(f) = c_{2,1,1,1}(f)$. The number $c_{2,1,1,1}(f)$ is defined in the next section.

2 Thom-Boardman Singularities

In this section, we review the Thom-Boardman singularity set Σ^I following the work of B.Morin [14]. Let $\mathbb{K}$ denote the field of real or complex numbers

*This work is partly supported by Grant-in-Aid for encouragement of young scientists.

$\mathbb{C}$. Let U be an open subset of $\mathbb{K}^n$ and V of $\mathbb{K}^p$, $J = J^r(U,V)$ the jet space of order r and $\pi_n : J \to U$, $\pi_p : J \to V$ the natural projection. Let $\mathcal{F}$ denote the foliation in J whose leaves are fibers of π_p, and $\mathcal{G}$ the foliation whose leaves are images of jet sections $j^r f : U \to J$ where f are polynomial maps of degree less than or equal to r. Let I be a Boardman symbol of length $k(=: |I|)$, i.e. $I = (i_1, i_2, ..., i_k)$ is a k-tuple of integers with $n \geq i_1 \geq i_2 \geq ... \geq i_k \geq 0$. We set $tI = (i_1, ..., i_{k-1})$ if $|I| = k \geq 2$ and $tI = \emptyset$ if $|I| = 1$. Then we define inductively the *Thom-Boardman submanifold* Σ^I in J by

$$\Sigma^I = \{z \in \Sigma^{tI} : \dim(T_z\mathcal{F} \cap T_z\mathcal{G} \cap T_z\Sigma^{tI}) = i_k, k = |I|\},$$

where $T_z\mathcal{F}, T_z\mathcal{G}$ denote the tangent spaces of the leaves of $\mathcal{F}, \mathcal{G}$ including $z \in J$. Here we understand that $\Sigma^\emptyset = J$. This is well-defined because Σ^I is nonsingular. See [2,(6.1)], [14, p.15, p.97]. We denote the codimension of Σ^I in J by $\nu_I(n,p)$. From [1, p46], [2, (6.5)], [14, p15], it follows that $\nu_I(n,p)$ is

$$(p-n+i_1)\mu(I)-(i_1-i_2)\mu(i_2,...,i_k)-(i_2-i_3)\mu(i_3,...,i_k)-\cdots-(i_{k-1}-i_k)\mu(i_k)$$

where $\mu(i_1,...,i_k)$ denotes the number of k-tuples $(j_1,...,j_k)$ of integers so that $i_r \geq j_r \geq 0$ $(0 \leq r \leq k)$, $j_1 \geq j_2 \geq ... \geq j_k \geq 0$, and $j_1 > 0$.

Let B be the ring of differentiable functions which are constant on each leaf of $\mathcal{F}$, and $\mathcal{D}$ the submodule of the vector fields of J, whose elements are tangent to all leaves of $\mathcal{G}$. For a Boardman symbol $I = (i_1, ..., i_k)$, we denote by Δ^I the ideal generated by Δ^{tI} and the determinants $\det(d_i\varphi_j)_{1\leq i,j\leq n-i_k+1}$ where $d_i \in \mathcal{D}, \varphi_j \in B + \Delta^{tI}$. Here $\Delta^\emptyset$ is defined to be the zero ideal.

It is sometimes convenient to write down an explicit coordinate system of J. Let $x = (x_1, ..., x_n)$ denote a coordinate system of $U \subset \mathbb{K}^n$, and $y = (y_1, ..., y_p)$ that of $V \subset \mathbb{K}^p$. Then we can write down the canonical coordinate functions on J, namely the functions X_i, Y_j, Z^j_σ for $1 \leq i \leq n$, $1 \leq j \leq p$, and $\sigma = (\sigma_1, ..., \sigma_m)$ an m-tuple of non-decreasing positive integers with $1 \leq m \leq r$. Here we denote m by $|\sigma|$. These are defined by

$$X_i = x_i \circ \pi_n,\ Y_j = y_j \circ \pi_p,\ Z^j_\sigma(j^r f(p)) = \frac{\partial^{|\sigma|}(y_j \circ f)}{\partial x_{\sigma_1} \cdots \partial x_{\sigma_m}}(p),$$

where f is the germ at p of any map from U to V. We understand $Y_j = Z^j = Z^j_\emptyset$. We define an $(m+1)$-tuple $\sigma(i)$ and vector fields D_i $(1 \leq i \leq n)$ by

$$Z^j_{\sigma(i)}(j^r f(p)) = \frac{\partial^{|\sigma|+1}(y_j \circ f)}{\partial x_i \partial x_{\sigma_1} \cdots \partial x_{\sigma_m}}(p), \quad D_i = \frac{\partial}{\partial X_i} + \sum_{\sigma:|\sigma|<r} \left(\sum_{j=1}^{p} Z^j_{\sigma(i)} \frac{\partial}{\partial Z^j_\sigma} \right).$$

These D_i $(1 \leq i \leq n)$ generate $\mathcal{D}$. For a Boardman symbol $I = (i_1, ..., i_k)$, Δ^I is the ideal generated by Δ^{tI} and subdeterminants of order $n - i_k + 1$ of the matrix ${}^t(D_iY_j, D_ig_s)$ where $g = (g_1, ..., g_t)$ is a system of generators of Δ^{tI}. For the proof of this fact, see Lemma 2 in I.(b) and Lemma 3 in I.(c) in [14]. Let I' denote the smallest (in the lexicographic order) Boardman symbol J which is larger (in the lexicographic order) than I with $|J| \leq |I|$. Note that I' is not defined if $I = (n, ..., n)$.

Theorem 2.1 *Let Z_I be the zero locus of Δ^I in the jet space J. Then Z_I is nonsingular along Σ^I, and, as underlying topological spaces,*

$$Z_I = \Sigma^I \cup Z'_I \text{ (disjoint union)}$$

where $Z'_I = Z_{I'}$ if I' is defined, and $Z'_I = \emptyset$ if I' is not defined.

Reference for Proof. See [14], Théorème on p.15, Corollaire on p.97. □

Let $f : (\mathbb{K}^n, 0) \to (\mathbb{K}^p, 0)$ be a $\mathbb{K}$-analytic map-germ and $\mathcal{O} = \mathcal{O}_n$ the ring of $\mathbb{K}$-analytic function-germs of $(\mathbb{K}^n, 0)$. For a Boardman symbol $I = (i_1, ..., i_k)$, we define a set-germ $\Sigma^I(f)$ in $(\mathbb{K}^n, 0)$ and an ideal $\Delta^I(f)$ by $\Sigma^I(f) = (j^k f)^{-1}(\Sigma^I)$, $\Delta^I(f) = (j^k f)^*(\Delta^I)$. We set $c_I(f) = \dim_{\mathbb{K}} \mathcal{O}_n/\Delta^I(f)$. This number $c_I(f)$ was first considered by J.Nuño Ballesteros and M.Saia in [15]. As a consequence of (7.1) and (7.1.3) in [6], we have the following:

Lemma 2.2 *Let f be a $\mathcal{K}$-finite map germ with $c_I(f) < \infty$. Then, the local intersection number of $j^r f(U)$ and Z_I at $j^r f(0)$ does not exceed $c_I(f)$, and equality holds iff Z_I is Cohen-Macaulay at $j^r f(0)$.*

See [17], for the definition of a $\mathcal{K}$-finite map germ. Using (1.2), it is possible to analyze map-germs. This idea first appeared in [3] for map germs from the plane to the plane. After [3], T.Gaffney and D.Mond analyzed several classes of map germs using this idea in [7, 13],

A map-germ $f : (\mathbb{K}^n, 0) \to (\mathbb{K}^p, 0)$ is said to be a *singularity of class* Σ^I if $0 \in \Sigma^I(f)$. Then $0 \in \Sigma^i(f)$ iff f has kernel rank i at 0. A map-germ is said to be *generic* if its jet extension is transverse to each Boardman submanifold. If f is generic, then the $\Sigma^I(f)$'s are nonsingular and $\Sigma^{i_1,\dots,i_k}(f) = \Sigma^{i_k}(f|\Sigma^{i_1,\dots,i_{k-1}})$. Any map-germ may be approximated as accurately as one wishes by a generic map. See [1, p.45], [2], [12], and [14], for the proof.

We return to our situation. We set $(n, p) = (5, 4)$. In this case, the list of Boardman symbol I with $\mathrm{cod}\Sigma^{\mathrm{I}} \leq 5$ is the following:

I	(2)	(2,1)	(2,1,1)	(2,1,1,1)	(2,2)
$\mathrm{cod}\Sigma^{\mathrm{I}}$	2	3	4	5	5

If I is one of the Boardman symbols above, then it is easy to see $\mathrm{ht}(\Delta^{\mathrm{I}}) = \mathrm{cod}\Sigma^{\mathrm{I}}$. A general element in such a Boardman submanifold Σ^I is equivalent to the jet of the map defined in the first paragraph of the paper. Thus, we are interested in the numbers $c_{2,1,1,1}(f)$ and $c_{2,2}(f)$, if they are finite. In [4,5], we prove the following:

Proposition 2.3 *Let $f : (\mathbb{C}^5, 0) \to (\mathbb{C}^4, 0)$ be a $\mathcal{K}$-finite map-germ with $c_{2,2}(f) < \infty$. Then $u(f) \leq c_{2,2}(f)$, and equality holds if f has kernel rank ≤ 2 at 0.*

By (1.1), we have $Z_{2,1,1,1} = \Sigma^{2,1,1,1} \cup Z_{2,2}$, as sets. Thus the number $c_{2,1,1,1}$ is a reflection of butterflies and umbilics.

Proposition 2.4 *Let $f : (\mathbb{C}^5, 0) \to (\mathbb{C}^4, 0)$ be a $\mathcal{K}$-finite map-germ with $c_{2,1,1,1}(f) < \infty$. Then $b(f) + 4u(f) \leq c_{2,1,1,1}(f)$, and equality holds iff $Z_{2,1,1,1}$ is Cohen-Macaulay at $j^r f(0)$.*

Proof. We first see that $c_{2,1,1,1}(f) = 4$, if $f : (\mathbb{C}^5, 0) \to (\mathbb{C}^4, 0)$ is a stable map-germ with $0 \in \Sigma^{2,2}(f)$. This is D_4-singularity, and is given by $f(x_1, ..., x_5) = (\frac{1}{2}x_1^2 x_2 + \frac{1}{6}x_2^3 + \frac{1}{2}x_2^2 x_3 + x_2 x_4 + x_1 x_5, x_3, x_4, x_5)$. By elementary computation, we have $\Delta^2(f) = (x_1 x_2 + x_5, \frac{1}{2}(x_1^2 + x_2^2) + x_2 x_3 + x_4)$, $\Delta^{2,1}(f) = \Delta^2(f) + (x_2^2 + x_2 x_3 - x_1^2)$, $\Delta^{2,1,1}(f) = \Delta^{2,1}(f) + (x_1(4x_2 + 3x_3), x_2(4x_2 + 3x_3))$, and $\Delta^{2,1,1,1}(f) = (x_1, x_2, x_3)^2 + (x_4, x_5)$. Thus, $c_{2,1,1,1}(f) = 4$. The remaining assertion is a consequence of (2.2). □

3 Cohen-Macaulay Property of $Z_{2,1,1,1}$ Along Σ^2

The purpose of this section is to show the following

Theorem 3.1 *$Z_{2,1,1,1}$ is Cohen-Macaulay along Σ^2.*

Corollary 3.2 *Suppose that $f : (\mathbb{C}^5, 0) \to (\mathbb{C}^4, 0)$ is a $\mathcal{K}$-finite map-germ with kernel rank ≤ 2. If $c_{2,1,1,1}(f)$ is finite, then $b(f) + 4u(f) = c_{2,1,1,1}(f)$.*

Proof. This is immediate from (2.2) and (3.1). □

We assume that f has kernel rank 2 at 0. The map-germ $f : (\mathbb{C}^5, 0) \to (\mathbb{C}^4, 0)$ is expressed by $f(x_1, x_2, x_3, x_4, x_5) = (g(x_1, x_2, x_3, x_4, x_5), x_3, x_4, x_5)$ for suitable chosen systems of coordinates. We take the coordinate system (X_i, Y_j, Z_σ^j) of the jet space $J^4(\mathbb{C}^5, \mathbb{C}^4)$ determined by these coordinates. Then $\Delta^2 = (Z_1^1, Z_2^1)$, and $\Delta^{2,1} = \Delta^2 + (H)$, where $H = \det(Z_{ij}^1)_{i,j=1,2}$. Setting $H_i = D_i H$, we have

$$H_1 = Z_{111}^1 Z_{22}^1 + Z_{11}^1 Z_{122}^1 - 2Z_{12}^1 Z_{112}^1,$$

and

$$H_2 = Z_{112}^1 Z_{22}^1 + Z_{11}^1 Z_{222}^1 - 2Z_{12}^1 Z_{122}^1.$$

Then $\Delta^{2,1,1} = (Z_1^1, Z_2^1) + I_2(Z_{ij}^1, H_i)$. Set

$$A^1 = \begin{vmatrix} Z_{12}^1 & H_1 \\ Z_{22}^1 & H_2 \end{vmatrix}, \quad A^2 = \begin{vmatrix} Z_{11}^1 & H_1 \\ Z_{12}^1 & H_2 \end{vmatrix}.$$

Then, it is easy to see that

$$\Delta^{2,1,1,1} = (Z_1^1, Z_2^1) + I_2\begin{pmatrix} Z_{11}^1 & Z_{12}^1 & H_1 & D_1A^1 & D_1A^2 \\ Z_{12}^1 & Z_{22}^1 & H_2 & D_2A^1 & D_2A^2 \end{pmatrix}.$$

Here, $I_2(M)$ denote the ideal generated by subdeterminants of order 2 of the matrix M. Note that the last term does not depend on Z_i^1, $i = 1, 2$. Since $\mathrm{ht}(\Delta^{2,1,1,1}) = 5$, this last term has height 3. Because of the results in [8], it is enough to establish the following:

Lemma 3.3 *We have the following identity of ideals.*

$$I_2\begin{pmatrix} Z_{11}^1 & Z_{12}^1 & H_1 & D_1A^1 & D_1A^2 \\ Z_{12}^1 & Z_{22}^1 & H_2 & D_2A^1 & D_2A^2 \end{pmatrix} = I_2\begin{pmatrix} G & H_1 & H_2 \\ H_1 & Z_{11}^1 & Z_{12}^1 \\ H_2 & Z_{12}^1 & Z_{22}^1 \end{pmatrix},$$

where

$$G = \tfrac{1}{3}((Z_{22}^1)^2 Z_{1111}^1 - 4Z_{12}^1 Z_{22}^1 Z_{1112}^1 + 2Z_{11}^1 Z_{22}^1 Z_{1122}^1 \\ + 4(Z_{12}^1)^2 Z_{1122}^1 - 4Z_{11}^1 Z_{12}^1 Z_{1222}^1 + (Z_{11}^1)^2 Z_{2222}^1).$$

Proof. The ideal on the right hand side is generated by H, A^i, K_{ij}, where

$$K_{ij} = \begin{vmatrix} G & H_i \\ H_j & Z_{ij}^1 \end{vmatrix}, \quad (i, j = 1, 2).$$

By routine calculation, we have the following identities:

$$\begin{aligned}
\begin{vmatrix} Z_{11}^1 & D_1A^2 \\ Z_{12}^1 & D_2A^2 \end{vmatrix} &= 3K_{11} - 3Z_{111}^1 A^1 + 3Z_{112}^1 A^2 \\
&\quad -H(Z_{11}^1 Z_{1122}^1 - 2Z_{12}^1 Z_{1112}^1 + Z_{22}^1 Z_{1111}^1 - Z_{111}^1 Z_{122}^1 + (Z_{112}^1)^2), \\
\begin{vmatrix} Z_{11}^1 & D_1A^1 \\ Z_{12}^1 & D_2A^1 \end{vmatrix} &= \begin{vmatrix} Z_{12}^1 & D_1A^2 \\ Z_{22}^1 & D_2A^2 \end{vmatrix} = 3K_{12} - 3Z_{112}^1 A^1 + 3Z_{122}^1 A^2 \\
&\quad -H(Z_{11}^1 Z_{1222}^1 - 2Z_{12}^1 Z_{1122}^1 + Z_{22}^1 Z_{1112}^1 - Z_{111}^1 Z_{222}^1 + Z_{112}^1 Z_{122}^1), \\
\begin{vmatrix} Z_{12}^1 & D_1A^1 \\ Z_{22}^1 & D_2A^1 \end{vmatrix} &= 3K_{22} + 3Z_{122}^1 A^1 - 3Z_{222}^1 A^2 \\
&\quad -H(Z_{22}^1 Z_{1122}^1 - 2Z_{12}^1 Z_{1222}^1 + Z_{11}^1 Z_{2222}^1 - Z_{112}^1 Z_{222}^1 + (Z_{122}^1)^2), \\
\begin{vmatrix} H_1 & D_1A^1 \\ H_2 & D_2A^1 \end{vmatrix} &= 3Z_{222}^1 K_{11} - 6Z_{122}^1 K_{12} + 3Z_{112}^1 K_{22} \\
&\quad -A^1(Z_{11}^1 Z_{1222}^1 + 3Z_{12}^1 Z_{1122}^1 - Z_{22}^1 Z_{1112}^1 + Z_{111}^1 Z_{222}^1 - Z_{112}^1 Z_{122}^1) \\
&\quad -A^2(Z_{11}^1 Z_{2222}^1 - 2Z_{12}^1 Z_{1222}^1 - 2Z_{112}^1 Z_{222}^1 + 2(Z_{122}^1)^2) \\
&\quad +HZ_{1122}^1(Z_{11}^1 Z_{222}^1 - 2Z_{12}^1 Z_{122}^1 + Z_{22}^1 Z_{112}^1), \\
\begin{vmatrix} H_1 & D_1A^2 \\ H_2 & D_2A^2 \end{vmatrix} &= 3Z_{122}^1 K_{11} - 6Z_{112}^1 K_{12} + 3Z_{111}^1 K_{22} \\
&\quad -A^1(Z_{22}^1 Z_{1111}^1 - 2Z_{12}^1 Z_{1112}^1 - 2Z_{111}^1 Z_{122}^1 + 2(Z_{112}^1)^2) \\
&\quad -A^2(Z_{22}^1 Z_{1112}^1 + 3Z_{12}^1 Z_{1122}^1 - Z_{11}^1 Z_{1222}^1 + Z_{111}^1 Z_{222}^1 - Z_{112}^1 Z_{122}^1) \\
&\quad +HZ_{1122}^1(Z_{22}^1 Z_{111}^1 - 2Z_{12}^1 Z_{112}^1 + Z_{11}^1 Z_{122}^1),
\end{aligned}$$

and

$$\begin{vmatrix} D_1A^1 & D_1A^2 \\ D_2A^1 & D_2A^2 \end{vmatrix} = \begin{vmatrix} 3K_{11} + 4H(Z^1_{11}Z^1_{1122} - 2Z^1_{12}Z^1_{1112} + Z^1_{22}Z^1_{1111}) & Z^1_{111} & Z^1_{112} \\ 3K_{12} + 4H(Z^1_{11}Z^1_{1222} - 2Z^1_{12}Z^1_{1122} + Z^1_{22}Z^1_{1112}) & Z^1_{112} & Z^1_{122} \\ 3K_{22} + 4H(Z^1_{11}Z^1_{2222} - 2Z^1_{12}Z^1_{1222} + Z^1_{22}Z^1_{1122}) & Z^1_{122} & Z^1_{222} \end{vmatrix}$$
$$- H\begin{vmatrix} 0 & Z^1_{11} & Z^1_{12} & Z^1_{22} \\ Z^1_{11} & Z^1_{1111} & Z^1_{1112} & Z^1_{1122} \\ Z^1_{12} & Z^1_{1112} & Z^1_{1122} & Z^1_{1222} \\ Z^1_{22} & Z^1_{1122} & Z^1_{1222} & Z^1_{2222} \end{vmatrix} + 4H(Z^1_{12})^2\begin{vmatrix} Z^1_{1112} & Z^1_{1122} \\ Z^1_{1122} & Z^1_{1222} \end{vmatrix} + H^2\begin{vmatrix} Z^1_{1111} & Z^1_{1122} \\ Z^1_{1122} & Z^1_{2222} \end{vmatrix}$$
$$- H((Z^1_{111}Z^1_{222})^2 - 3(Z^1_{112}Z^1_{122})^2 + 4Z^1_{111}(Z^1_{122})^3 + 4(Z^1_{112})^3Z^1_{222}$$
$$- 6Z^1_{111}Z^1_{112}Z^1_{122}Z^1_{222}).$$

Considering the identities above modulo H, A^1, A^2, we see these imply (3.3).

Remark 3.4 To check the identities above it is enough to see (the left hand side) – (the right hand side) = 0. This is quite elementary, but tedious to establish by hand. The author used the computer algebra system Mathematica to check these facts. It is not necessary to check all the identities above. Switching 1 and 2, we see the third identity comes from the first, the fifth from the fourth, and the first equality in the second identities.

4 A Consequence for a Map-Germ $(\mathbf{C}^{n+5}, 0) \to (\mathbf{C}^4, 0)$

Let n be a non-negative integer. It is not difficult to generalize the results above to a map-germ $f : (\mathbb{C}^{n+5}, 0) \to (\mathbb{C}^4, 0)$ of class $\Sigma^{n+2,2}$. We first list the Boardman symbols with $\mathrm{cod}\Sigma^{\mathrm{I}} \leq \mathrm{n} + 5$ for maps $(\mathbb{C}^{n+5}, 0) \to (\mathbb{C}^4, 0)$.

I	$(n+2)$	$(n+2,1)$	$(n+2,1,1)$	$(n+2,1,1,1)$	$(n+2,2)$
$\mathrm{cod}\Sigma^{\mathrm{I}}$	$n+2$	$n+3$	$n+4$	$n+5$	$n+5$
name	fold	cusp	swallowtail	butterfly	umbilic

The names are for a generic element in Σ^I. We similarly define the numbers $b(f)$, $u(f)$ for a map germ $f : (\mathbb{C}^{n+5}, 0) \to (\mathbb{C}^4, 0)$. Now we assume that $f : (\mathbb{C}^{n+5}, 0) \to (\mathbb{C}^4, 0)$ is a map-germ of class $\Sigma^{n+2,2}$, i.e. defined by the following normal form in some coordinate systems:

$$f(x_1, x_2, ..., x_{n+5}) = \left(g(x_1, x_2, x_{n+3}, x_{n+4}, x_{n+5}) + \sum_{i=3}^{n+2} x_i^2, x_{n+3}, x_{n+4}, x_{n+5} \right),$$

where g is a function in the variables $(x_1, x_2, x_{n+3}, x_{n+4}, x_{n+5})$.

Then we have the following:

$$\Delta^{n+2}(f) = (g_1, g_2, x_3, ..., x_{n+2}), \quad \text{where } g_i = \frac{\partial g}{\partial x_i},$$

$$\Delta^{n+2,1}(f) = \Delta^{n+2}(f) + (H), \quad \text{where } H = \begin{vmatrix} g_{11} & g_{12} \\ g_{21} & g_{22} \end{vmatrix}, \; g_{ij} = \frac{\partial^2 g}{\partial x_i \partial x_j},$$

$$\Delta^{n+2,2}(f) = \Delta^{n+2} + (g_{11}, g_{12}, g_{22}),$$

$$\Delta^{n+2,1,1,1}(f) = \Delta^{n+2}(f) + I_2 \begin{pmatrix} G & H_1 & H_2 \\ H_1 & g_{11} & g_{12} \\ H_2 & g_{21} & g_{22} \end{pmatrix},$$

where $H_i = \frac{\partial H}{\partial x_i}$, $g_{ijk} = \frac{\partial^3 g}{\partial x_i \partial x_j \partial x_k}$, $g_{ijk\ell} = \frac{\partial^4 g}{\partial x_i \partial x_j \partial x_k \partial x_\ell}$, and $G = \frac{1}{3}((g_{22})^2 g_{1111} - 4g_{12}g_{22}g_{1112} + 2g_{11}g_{22}g_{1122} + 4(g_{12})^2 g_{1122} - 4g_{11}g_{12}g_{1222} + (g_{11})^2 g_{2222})$. If f is $\mathcal{K}$-finite and both $c_{n+2,1,1,1}(f)$, $c_{n+2,2}(f)$ are finite, then

$$b(f) + 4u(f) = c_{n+2,1,1,1}(f), \quad \text{and} \quad u(f) = c_{n+2,2}(f).$$

Thus, we get an algebraic formula for $b(f)$ and $u(f)$.

To end this section, we ask the following question hoping for an affirmative answer.

Question 4.1 *Is $Z_{n+2,1,1,1}$ Cohen-Macaulay along Σ^{n+2}?*

5 An Example

Here, we present a simple example to see what we get using our method. Let $f : (\mathbb{C}^5, 0) \to (\mathbb{C}^4, 0)$ be the map germ defined by

$$f(x, y, a, b, c) = (x^2 y + \tfrac{1}{4} y^4 + 2ax + by + cy^2, a, b, c).$$

By elementary calculation, we have $\Delta^2(f) = (xy + a, x^2 + y^3 + b + 2cy)$, $\Delta^{2,2}(f) = \Delta^2(f) + (x, y, c)$, $H = 2(3y^3 + 2cy - 2x^2)$, $G = 8y^2$, $\Delta^{2,1,1,1}(f) = \Delta^2(f) + I_2$, where

$$I_2 = I_2 \begin{pmatrix} 8y^2 & -8x & 18y^2 + 4c \\ -8x & 2y & 2x \\ 18y^2 + 4c & 2x & 3y^2 + 2c \end{pmatrix} = I_2 \begin{pmatrix} 4y^2 & -4x & 9y^2 + 2c \\ -4x & y & x \\ 9y^2 + 2c & x & \frac{3}{2}y^2 + c \end{pmatrix}$$

$$= I_2 \begin{pmatrix} 4y^2 & -4x & 9y^2 + 2c \\ -4x & y & x \\ 2c & 10x & -\frac{75}{2}y^2 - \frac{7}{2}c \end{pmatrix} = I_2 \begin{pmatrix} 4y^2 & -4x & 2c \\ -4x & y & 10x \\ 2c & 10x & -\frac{75}{2}y^2 - 8c \end{pmatrix}$$

$$= I_2 \begin{pmatrix} y^2 & -2x & c \\ -2x & y & 10x \\ c & 10x & -\frac{75}{2}y^2 - 8c \end{pmatrix}$$

$$= (y^3 - 4x^2, 5xy^2 + cx, 20x^2 + cy, \tfrac{75}{2}y^4 + 8cy^2 + c^2, 5xy^2 + 2cx, \tfrac{75}{2}y^3 + 8cy + 100x^2)$$

$$= (x^2, xy^2, y^3, cx, cy, c^2).$$

Obviously, $c_{2,2}(f) = 1$, and $c_{2,1,1,1}(f) = 6$. Thus, we obtain $b(f) = 2$, and $u(f) = 1$.

6 Observations using Macaulay

It is now common to carry out many computations using some computer algebra systems. One such system for commutative algebra is Macaulay (see [16]). Using this we found two observations, which are interesting from our point of view. Unfortunately, Macaulay works only with positive characteristic (=31991, is the default), and I do not know any proof of the observations below. It is a good question to find a mathematical interpretation of them. We consider a map germ $f : (\mathbb{C}^5, 0) \to (\mathbb{C}^4, 0)$.

Observation 6.1 *The quotient ideal* $\Delta^{2,1,1,1} : \Delta^{2,2}$ *defines a Cohen-Macaulay space along* Σ^2.

Observation 6.2 $Z_{2,2}$ *is a Cohen-Macaulay space along* Σ^3.

To see this, use Proposition 7 in Appendix A.5 of [16]. If this holds, we obtain $u(f) = c_{2,2}(f)$ for a $\mathcal{K}$-finite germ f with kernel rank 3 at 0 and $c_{2,2}(f) < \infty$.

7 Real Germs

Let $f : (\mathbb{C}^n, 0) \to (\mathbb{C}^p, 0)$ be a holomorphic map germ defined over the reals. Then it induces a real map germ $(\mathbb{R}^n, 0) \to (\mathbb{R}^p, 0)$, denoted by f_R.

Lemma 7.1 *Let* f *be a* $\mathcal{K}$*-finite map-germ with* $c_I(f) < \infty$, f_u *a generic perturbation of* $f_0 = f$, *which means transversality of its jet section to* Σ^I. *Then the number of real points of the intersection between the image of the jet section of* f_u *and* Z_I *near* $j^4 f(0)$ *does not exceed* $c_I(f)$. *Moreover, if* Z_I *is Cohen-Macaulay there, then these two numbers are congruent modulo* 2.

Proof. Obvious. □

Using this, we obtain some consequences for real map germs, and now state one. Let $f_R : (\mathbb{R}^5, 0) \to (\mathbb{R}^4, 0)$ be a map germ with kernel rank 2 at 0. Let $f_{u,R}$ denote a stable perturbation of $f_{0,R} = f_R$, and $b(f_{u,R})$, $u(f_{u,R})$ denote the numbers of A_4, D_4 points of $f_{u,R}$. If f_R is $\mathcal{K}$-finite and $c_{2,1,1,1}(f_R) < \infty$, then $b(f_{u,R}) + 4u(f_{u,R}) \leq c_{2,1,1,1}(f_R)$, and $b(f_{u,R}) \equiv c_{2,1,1,1}(f_R) \pmod 2$. It is left to the reader to write down other consequences.

References

[1] V.I.Arnold, S.M.Gusein-Zade and A.N.Varchenko' *Singularities of differentiable maps I*, Monographs in Mathematics **82**, Birkhäuser, 1985.

[2] J.Boardman, Singularities of Differentiable maps, *Inst. Hautes Études Sci. Publ. Math.*, vol. 33, 1967, 21–57.

[3] T.Fukuda and G.Ishikawa, On the number of cusps of stable perturbations of plane-to-plane singularities, *Tokyo J. of Math.*, vol. 10, 1987, 375–384.

[4] T.Fukui, J.Nuño Ballesteros and M.Saia, On the number of singularities in generic deformations of map germs (preprint).

[5] T.Fukui, J.Nuño Ballesteros and M.Saia,Counting singularities in stable perturbations of map germs, *RIMS kokyuroku* 926, 1995.

[6] W.Fulton, *Intersection theory*, Ergebnisse der Mathematik und ihrer Grenzgebiete, 3. Folge Band 2, Springer-Verlag, 1984.

[7] T.Gaffney and D.Mond, Cusps and double folds of germs of analytic maps $\mathbb{C}^2 \to \mathbb{C}^2$, *J. London Math. Soc.*, 43, 1991, 185–192

[8] R.Kutz, Cohen-Macaulay rings and ideal theory in rings of invariants of algebraic groups, *Trans. Amer. Math. Soc.*,194, 1974, 115–129.

[9] J.N.Mather, Stability of C^∞-mappings IV: Classification of stable germs by $\mathbb{R}$-algebras, *Publ. Math. Inst. Hautes Etudes Sci.*,37, 1970, 228–248.

[10] J.N.Mather, Stability of C^∞-mappings VI: The nice dimensions, in *Proceeding of Liverpool singularities symposium, I*, Lect. Notes in Math. 192, Springer-Verlag, 1971, 207–253.

[11] J.N.Mather, Stable map-germs and algebraic geometry, in *Manifolds–Amsterdam*, Lect. Notes in Math. 197, Springer-Verlag, 1971, 176–193.

[12] J.N.Mather, Generic projection, Ann. of Math., 98, 1973, 226–245

[13] D.Mond, Vanishing cycles for analytic maps,in *Singularity theory and its application, Warwick 1989, Part I*, Lect. Note in Math. 1462, Springer-Verlag, 1991, 221–234.

[14] B.Morin, Calcul jacobian, *Ann. scient. Éc. Norm. Sup.*, 8, 1975,1–98.

[15] J.Nuño Ballesteros and M.Saia, An invariant for map germs (preprint).

[16] M.Stillman, M.Stillman, and D.Bayer, *Macaulay User Manual*, (ftp math.harvard.edu)

[17] C.T.C.Wall, Finite determinacy of smooth map-germs, *Bull. London Math. Soc.*,13, 1981, 481–539

Department of Mathematics,
Faculty of Science,
Saitama University,
255 Shimo-Okubo,
Urawa, 338
Japan

email: tfukui@rimath.saitama-u.ac.jp

Singular Phenomena in Kinematics

P.S. Donelan C.G. Gibson

Les principes généraux de la theorie des systèmes articulés n'ont pas encore été éclairais et on est reduit à un ensemble de recherches isolées et de résultats curieux sans liens apparents entre eux. C'est une raison de plus pour engager les géomètres à éclairer cette question encore obscure: les progrès de la science se font parfois par les côtés les plus inattendus.

Gabriel Koenigs, *Leçons de Cinématique.*

1 Introduction

Although the kinematics of rigid bodies has been an essential ingredient of engineering for millenia, its serious mathematization required both the impetus of the Industrial Revolution, and significant developments in geometry. Mechanisms for converting circular into reciprocating rectilinear motion were widely adopted in industrial machinery. For many purposes approximate linear motion is an acceptable substitute for exact linear motion. A well known example is the Watt 4–bar of Fig. 1 discovered in 1784. The device comprises three smoothly jointed bars moving with one degree of freedom (dof), the mid–point of the middle (or coupler) bar describing the *Watt curve.* The curve has a self–crossing with two branches, one having 5–point contact with its tangent, so representing an excellent approximation to a straight line.

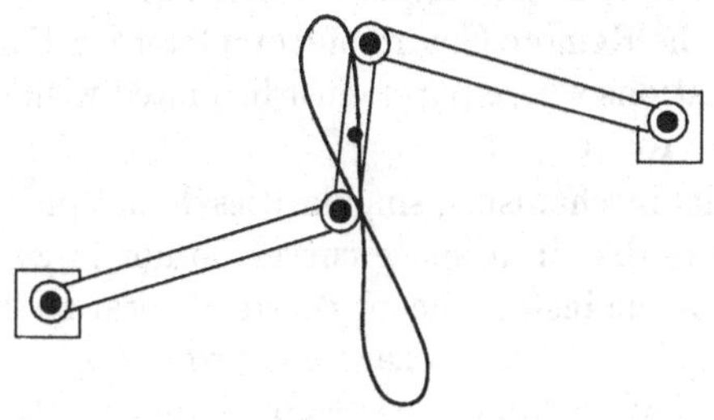

Figure 1: The Watt 4–bar mechanism.

A central feature of curves traced by planar mechanisms is that they exhibit singularities. Generally one expects *nodes*, as in the Watt curve. The engineering significance of a node is that the mechanism can transport an object to a "working position", leave it there, and then retrieve it after a prescribed rotation of a crank. Nodes can degenerate to *cusps* which likewise have engineering interest, namely that they provide "dwell", the property that the tracing point comes instantaneously to rest. In kinematic terms, the tracing point coincides with the instantaneous centre of rotation of the moving plane. As the mechanism moves so the instantaneous centre traces out a curve, the *moving centrode*. This is the first example of a *bifurcation curve*, representing choices of tracing points which give rise to trajectories exhibiting exceptional singular behaviour.

The Watt mechanism belongs to the family of planar 4–bars whose geometry is discussed in Section 2. The joints are simply revolute hinges. Spatial linkages use a greater variety of joints [35, page 8] such as the *spherical* or S–joint (realized in engineering by a ball joint), the *revolute* or R–joint (hinge), the *prismatic* or P–joint (slider) and the *helical* or H–joint (screw). R–, P– and H–joints have one dof while S–joints have three. A simple example is the spatial 4–bar or RRRR device, analogous to the planar 4–bar. In general such devices are rigid, for generically the dof of the joints of a closed loop must be at least 7 for mobility. However there are exceptions, such as the planar 4–bar (when the hinges are concurrent at a point at infinity) and the spherical 4–bar (when the hinges are concurrent at a finite point). And there exist fascinating mobile RRRR devices (the Bennett mechanisms [4, 5]) discovered at the turn of the century, in which the hinges fail to be concurrent, but belong to the reguli of a hyperboloid. One broad class of motions with several dof arises on composing a number of simpler motions, yielding *serial* devices such as robot arms, in which the joints are either revolute or prismatic. A second approach is to apply several simultaneous constraints, to obtain *parallel* devices. For example, by constraining k fixed points $P_1, \ldots, P_k$ on the body to move on corresponding surfaces $S_1, \ldots, S_k$ in the ambient space we obtain a motion with $6-k$–dof. Motions with three dof are of particular interest. The simplest example is when S_1, S_2, S_3 are planes, producing the classical *Mannheim* motions. And when S_1, S_2, S_3 are spheres we obtain the Remote Centre of Compliance (RCC) device [18], used in assembly applications where parts must be mated with clearance less than the accuracy of the robot.

For 1–dof spatial mechanisms, singularities do not play a major role. The underlying reason is that in 3–space curves do not in general self–intersect and the analogue of the instantaneous centre of rotation is an instantaneous screw axis. However the RSSR linkage has two dof, so traces a surface (the *queeroid* of [51]) which exhibits cross caps. Until relatively recently, rigid body motions with several dof were largely of theoretical interest. However the rapid development of robot technology and computer control systems has created a surge of interest. Of special interest, both for mechanical reasons and for the development of effective control algorithms, are singularities of such motions. Thus a central problem is to understand the trajectory singularities, a problem which can be attacked at several levels. Which singularities would one expect generically, or within a given broad class, or for a specific family? The systematic study of singular phenomena in kinematics has its genesis in a detailed study [30] of the singularities exhibited by the planar 4–bar, and the paper [14] on generic singularities of planar Euclidean motions. The resulting "singular viewpoint" lends coherence to a number of interesting kinematical topics, and enables a significant extension of the classical core of the subject. This survey represents a condensed version of a series of lectures on 'Singular Phenomena in Kinematics' given in the Instituto de Ciências Matemáticas de São Carlos, University of São Paulo, Brazil.

2 Geometry of The Planar Four–Bar

Studying the motion of a specific family of mechanical or robotic devices proceeds generally in two steps. The first is to identify a *configuration space* for the motion, typically a real variety whose points correspond one–to–one with the configurations, and to determine the natural map from this configuration space into the Lie group of rigid displacements of the ambient space. This is the *kinematic mapping* of theoretical robotics, with image the *kinematic image*. The second step is to compose the natural action of the group on the ambient space with the motion to view the trajectories as images of the kinematic image under a family of submersions. The singular behaviour will vary with the *design parameters* for the family, representing the various distances and angles specifying a given device. In the space of design parameters one expects to find a hypersurface (for historical reasons we dub it the *Grashof hypersurface*) giving rise to degenerate singular behaviour. Here is a brief account [30] of a classic example, the family of planar 4–bars. Each 4–bar comprises a fixed base to which are attached two links (or bars), free to rotate about a fixed point in the base, and a *coupler link* joining them to form a closed quadrilateral.

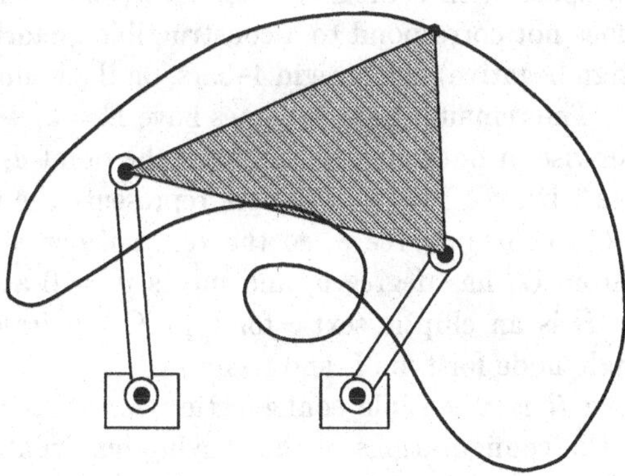

Figure 2: A 4–bar coupler curve.

The design parameters for the 4–bar are the link lengths d_1, d_2, d_3, d_4. The link directions are represented by unit complex numbers z_1, z_2, z_3, z_4. Write $z_k = x_k + iy_k$ with x_k, y_k real for $1 \le k \le 4$; assuming the base to be fixed, $x_4 = -1$, $y_4 = 0$. Using the closure of the quadrilateral the configuration space is represented by five equations in six real unknowns, defining a curve in $\mathbb{R}^6$:

$$\begin{aligned} d_1x_1 + d_2x_2 + d_3x_3 &= d_4 \\ d_1y_1 + d_2y_2 + d_3y_3 &= 0 \\ x_k^2 + y_k^2 &= 1 \quad (1 \le k \le 3). \end{aligned}$$

This variety has a real point (i.e. the quadrilateral constructs) if and only if $e_4 \le e_1 + e_2 + e_3$, where e_1, e_2, e_3, e_4 are the link lengths, in increasing order of magnitude. For geometry, we complexify and projectivize to obtain the equations:

$$\begin{aligned} d_1x_1 + d_2x_2 + d_3x_3 &= d_4w \\ d_1y_1 + d_2y_2 + d_3y_3 &= 0 \\ x_k^2 + y_k^2 &= w^2 \quad (1 \le k \le 3), \end{aligned} \tag{1}$$

defining a *linkage curve* C in complex projective 6–space $P\mathbb{C}^6$. C meets the hyperplane $w = 0$ in complex conjugate lines L, $\overline{L}$. Computations show that the finite singularities (on $w = 0$) occur at three pairs of complex conjugate points on L, $\overline{L}$. Infinite singularities (off $w = 0$) occur if and only if $y_1 = y_2 = y_3 = 0$, representing the flattening of the quadrilateral: in that case substitution in equation (1) yields the *Grashof relations*

$$d_1 \pm d_2 \pm d_3 \pm d_4 = 0.$$

Four–bars which fail to satisfy these relations are *generic*. Scaling the parameters so $d_4 = 1$, the relations define eight planes (whose union is the Grashof surface) in the 3–space with coordinates d_1, d_2, d_3 of which one plane (all signs positive) does not correspond to a constructible quadrilateral. Four of the others (one sign negative) define rigid 4–bars, on the boundary of the constructible region. The remaining three planes have $E = e_1 + e_4 - e_2 - e_3 = 0$ and intersect pairwise in lines, and mutually at the point $d_1 = d_2 = d_3 = 1$. That yields Table 1 below. The type symbol represents the number of finite singularities on C. C has degree 8, so the *residual curve* R, obtained by removing L, $\overline{L}$ from C, has degree 6, and meets $w = 0$ at the six infinite singular points. R is an elliptic sextic for type O, an irreducible rational sextic with a single node for type I, and so on.

The real curve R is a sensible configuration space, as its points correspond 1–1 with the configurations of the moving quadrilateral. Indeed we have a family of configuration spaces parametrized by the design parameters which can be thought of as a single *big configuration space*, and a key question is how the geometry changes within the family. Basic degenerations

type	relation	name
O	$e_1 + e_4 \neq e_2 + e_3$	generic
I	$e_1 + e_4 = e_2 + e_3$	circumscriptible
I′	$e_1 + e_2 + e_3 = e_4$	rigid
II	$e_1 = e_2 \neq e_3 = e_4$	parallelogram
III	$e_1 = e_2 = e_3 = e_4$	rhombus

Table 1: Five basic types of 4–bar.

take place on the Grashof surface. For generic 4–bars the R is a compact smooth 1–manifold: for $E > 0$ it is topologically a circle and for $E < 0$ a union of two circles. In fact the connected components of the complement of the Grashof surface provide further refinement, enumerated originally by Hain [31]. They classify 4–bars in terms of the 'rocking' or 'cranking' capabilities of the input/output links, in particular whether they rock 'inwards' or 'outwards'. A similar analysis shows that the Grashof hypersurface associated to the planar n–bar loop reduces (as in the case $n = 4$) to a union of hyperplanes. The topology of the configuration space has been studied by Niemann, in his unpublished thesis [47], via a natural Morse function on the big configuration space, yielding an explicit formula for the Betti numbers: for $n = 5$ the configuration space is a compact surface of genus ≤ 4, and for $n = 6$ it is a compact 3–manifold with a finitely presented fundamental group.

The second step in the process is to study the trajectories of *tracing points* P in the *moving plane* determined by the coupler bar. We think of the trajectory as a projection of R. Write $P = d_1 z_1 + k z_2$ for some complex number $k = k_1 + ik_2$ with k_1, k_2 real, and think of P having homogeneous coordinates $p = (p_1, p_2, p_3)$ where

$$p_1 = d_1 x_1 - k_2 y_2 + k_1 x_2, \quad p_2 = d_1 y_1 - k_2 x_2 + k_1 y_2, \quad p_3 = w.$$

These formulas define a family of projections $\pi_k : \mathbb{PC}^6 - V \to \mathbb{PC}^2$ with V the centre of projection, i.e. the 3–space given by $p_1 = p_2 = p_3 = 0$. Under this projection L, $\overline{L}$ map to the circular points at infinity $I = (1 : i : 0)$, $\overline{I} = (1 : -i : 0)$ and the image of the restriction $\pi_k | R$ is the *complex coupler curve* C_k in $\mathbb{PC}^2$. To understand this better, write W for the 4–space in $\mathbb{PC}^6$ defined by the linear equations in (1). W intersects V in a line L_1 which (bar exceptional cases) does not meet R. As $R \subseteq W$ it suffices to consider the projections $\pi_k : W - L_1 \to \mathbb{PC}^2$ given by the same forms p_1, p_2, p_3. The key observation is that L_1 meets both L and $\overline{L}$. This has the following consequence. Write q_1, q_2, q_3 for the quadrics in W obtained by intersecting W with $x_1^2 + y_1^2 = w^2$, $x_2^2 + y_2^2 = w^2$, $x_3^2 + y_3^2 = w^2$ and consider the resulting net $\lambda_1 q_1 + \lambda_2 q_2 + \lambda_3 q_3$. Given a point in W, the condition for a quadric in the net to pass through that point is a linear one in λ_1, λ_2, λ_3 defining a pencil of quadrics in W: taking the given point to be any point of L_1 not on L, $\overline{L}$

we see that a quadric in the pencil meets L_1 in three points, hence contains L_1. *There is therefore a pencil of quadrics in the net containing* L_1. Its intersection is a Segre quartic surface S. Provided the pencil is generic (the tracing points for which the pencil fails to be generic form a "bad" curve in the moving plane) S contains sixteen lines, each meeting five other mutually skew lines. L, $\overline{L}$, projecting to I, $\overline{I}$, are two of the lines in S meeting L_1, and there will be three others L_2, L_3, L_4, projecting to points I_2, I_3, I_4. In this way we obtain the classical [36] birational isomorphism between S, with the six lines deleted, and the plane, with the conic F through I, $\overline{I}$, I_2, I_3, I_4 deleted. F is the *circle of singular foci* well-known in the engineering literature [35]. In particular, the isomorphism restricts to one between R and C_k, ensuring that C_k has the same geometric genus, and that off F it has the same number and type of singularities as R (none in the generic case). The only other singularities of C_k are on F: I, $\overline{I}$ are ordinary triple points, and each of I_2, I_3, I_4 is either a node or a cusp. In the exceptional cases one also learns a great deal about the possible singularities on a trajectory from the geometry of the associated degenerate Segre surface [11]. We conclude that for a general tracing point, C_k has five singular points, namely two ordinary triple points at I, $\overline{I}$, and three nodes on the circle F. Any transition from one type of mechanical behaviour to another will have to take place on a "bifurcation" curve of tracing points in the moving plane. The key aim of the singular approach is to understand such behaviour better.

3 The Singular Viewpoint

The planar 4–bar example shows that the singular viewpoint can be applied at two (intimately connected) levels. At one level we consider the motion as a whole, a map from a configuration space (with possible singularities) into the group of rigid body displacements. For robot arms the motion is typically [48, 52] from a 6–dimensional configuration space to the 6–dimensional group of displacements of 3–space. For spatial motions with three or fewer parameters there are generically no singularities; but for planar motions the group is 3–dimensional so much more is known about stable singularities. At a second level we are interested in the trajectory maps whose images are the trajectories in the ambient space of points in the moving body. We consider trajectories as maps in their own right and then establish connections with the underlying motion.

3.1 Equivalence of Motions

Formally, we proceed as follows. Let $SE(p)$ be the special Euclidean group, i.e. the Lie group of proper rigid motions of $\mathbb{R}^p$ with its standard Euclidean structure. By an *n–dof motion* of $\mathbb{R}^p$ we mean a smooth mapping $\lambda : N \to SE(p)$, where N is a smooth manifold of dimension n. The $\mathcal{A}$-type of a motion germ has a significant effect on which singularities can be exhibited by trajectories. However $\mathcal{A}$–equivalence fails to take into account the group structure on the target $SE(p)$. Given that the position of a rigid body is represented by an element of $SE(p)$ only up to choices of body and space coordinates, one is led naturally to the following definition.

Definition 3.1 Let $\lambda_1 : N_1, x_1 \to SE(p), g_1$ and $\lambda_2 : N_2, x_2 \to SE(p), g_2$ be (germs of) n–dof motions of $\mathbb{R}^p$. Then λ_1, λ_2 are *$\mathcal{I}$-equivalent* (isometry equivalent) if there exist an invertible germ $\phi : N_1, x_1 \to N_2, x_2$ and elements h, k in $SE(p)$ such that $\lambda_2(x) = h.\lambda_1(\phi(x)).k$.

$\mathcal{I}$–equivalent motion germs are $\mathcal{A}$–equivalent, but the former is a much finer relation, too fine to yield a tractable basis for a classification. $\mathcal{I}$–equivalence is implicit in the classical kinematics literature where much appears about the *instantaneous* invariants [6] in the very special cases of planar and spatial 1–dof motions. One objective (Section 4) is to find first order $\mathcal{I}$–invariants of general motion germs, particularly for spatial motions.

3.2 A Transversality Lemma for Trajectories

The action of $SE(p)$ on $\mathbb{R}^p$ yields a smooth map $M_\lambda : N \times \mathbb{R}^p \to \mathbb{R}^p$ given by $(t, w) \mapsto \lambda(t)(w)$. We think of M_λ as a composite of the action with the motion as follows. Let $\alpha : SE(p) \times \mathbb{R}^p \to \mathbb{R}^p$ denote the action: then $M_\lambda = \alpha \circ (\lambda \times 1)$. For a fixed *tracing point* $w \in \mathbb{R}^p$ there is a smooth map $M_{\lambda,w} : N \to \mathbb{R}^p$ defined by $t \mapsto \lambda(t)(w)$, called the *trajectory* of w under λ. Thus M_λ may be thought of as a p–parameter family of trajectories. Given positive integers r and k this induces a multijet extension ${}_rj^kM_{\lambda,w} : N^{(r)} \to {}_rJ^k(N, \mathbb{R}^p)$, and since $M_{\lambda,w}$ depends smoothly on w we obtain the following mapping, where the subscript '1′ indicates we are taking jets with respect to the first component only,

$$ {}_rj_1^kM_\lambda : N^{(r)} \times \mathbb{R}^p \longrightarrow {}_rJ^k(N, \mathbb{R}^p). $$

Lemma 3.1 *Let $\mathcal{S}$ be a finite stratification of ${}_rJ^k(N, \mathbb{R}^p)$. The set of n–dof motions $\lambda : N \to SE(p)$ with ${}_rj_1^kM_\lambda$ transverse to $\mathcal{S}$ is residual in $C^\infty(N, SE(p))$, endowed with the Whitney C^∞ topology.* (Basic Transversality Lemma [23].)

This falls within the ambit of Montaldi's result on composite maps [43] since the action is submersive, yielding a simpler proof than the original. It is important to bear in mind that this is a result of general *kinematics*, i.e. it is concerned with generic motions of a rigid body. However one of the ultimate objectives is to apply the general framework to *robotics*, i.e. mechanically generated motions of a rigid body. Mathematically, the step from kinematics to robotics is a major one; in general kinematics a motion can be rendered generic by arbitrarily small deformations within the *infinite dimensional* space of all motions, but in robotics the motion can only be deformed in the *finite dimensional* space of design parameters. Thus a serious underlying problem of the subject is to study the genericity of examples. At present there are no general results in this direction. Current research suggests that the genericity question is a very subtle one, and that even quite simple examples can fail to be generic.

3.3 Classification

In the above there is a codimensional restriction on the relevant strata of $\mathcal{S}$. Let X be an $\mathcal{A}$-invariant smooth submanifold of ${}_rJ^k(n,p)$, giving rise in a natural way to an $\mathcal{A}$-invariant smooth submanifold Y of ${}_rJ^k(N,\mathbb{R}^p)$. Suppose that ${}_rj_1^kM_\lambda$ is transverse to $\mathcal{S}$, and that Y is one of the strata. Then a necessary condition for the pull–back of Y under ${}_rj_1^kM_\lambda$ to be non–void is that Y has codimension $\leq rn+p$, the dimension of the domain. However by local triviality of the multijet bundle the codimensions of X, Y in their respective multijet spaces differ by $(r-1)p$. Thus the condition on X is that its codimension should be $\leq 2p+r(n-p)$. Given n, p the guiding principle for classification is to list strata in the multijet space up to this value. In particular, for a generic motion, non–stable $\mathcal{A}$–simple multigerms of trajectories will be of $\mathcal{A}_e$–codimension $\leq p$. Within robotics the important cases are $p=2$ and $p=3$ where one hopes for complete solutions. For $p=2$ there is a complete listing (Section 5) for all dimensions n, with a natural trichotomy provided by the cases $n=1$, $n=2$ and $n\geq 3$: for $p=3$ the listings are complete for $n\leq 3$, and then only (Section 6) via computer algebra.

3.4 Bifurcation

For any family of smooth mappings a key role is provided by the *bifurcation set* of parameters for which the corresponding mapping exhibits a non–stable multigerm. Thus for the p–parameter family of trajectories $M_\lambda : L\times\mathbb{R}^p \to \mathbb{R}^p$ it comprises the set of tracing points $w\in\mathbb{R}^p$ for which the trajectory $M_{\lambda,w}$ exhibits a non–stable multigerm. The bifurcation set attached to the motion of a robotic device is fundamental, representing the boundary between qualitatively different types of mechanical behaviour. The kinematics literature does contain attempts to determine the (local and global) structure of bifurcation sets associated to mechanically simple examples, such as Müller's extensive analysis of the 4–bars in [45, 46]. The basic approach has been to seek polynomial equations: however that runs into the general difficulty that the zero locus may be larger than the bifurcation set. One of the strengths of the singular approach is that the unfolding theory determines the local structure of the bifurcation set up to diffeomorphism (or in a few cases just homeomorphism). More precisely we have the next result, using the Basic Transversality Lemma and the standard result in singularity theory [40, page 190], that any versal unfolding of an $\mathcal{A}$–finite germ is equivalent to a suspension of a miniversal unfolding. Thus the local study of the bifurcation set of M_λ reduces to the analysis of miniversal unfoldings for the relevant normal forms.

Theorem 3.1 *Let $\lambda : N \to SE(p)$ be a generic Euclidean motion. Then the germ (at a finite set of points) of any trajectory is versally unfolded by M_λ, and the unfolding is equivalent to a suspension of a miniversal unfolding of the associated normal form. In particular, the local structure of the bifurcation set for M_λ is determined by that for the miniversal unfolding.*

A major goal of the subject is to establish good algorithms leading to computer graphic renderings of bifurcation sets, a programme which has hardly begun. The idea of the *instantaneous singular set* discussed in Section 4 leads in principle to an algorithmic approach for *some* spatial bifurcation sets, but we are not aware of any resulting computer implementations.

4 First Order Invariants

Before considering the singularities on trajectories it is worth detailing what information can be ascertained concerning the locus of tracing points possessing singular trajectories. One of the oldest intuitions of kinematics is that when a plane lamina moves with one dof there is at any instant a point, the instantaneous centre of rotation, whose trajectory is singular at that instant. Likewise, when a spatial body moves with one dof there is at any instant a line, the instantaneous axis of rotation. However, in the spatial case points on the axis do not have singular trajectories unless a certain $\mathcal{I}$–invariant (the pitch) is at that instant zero. The step from planar to spatial motions is not simply an increase of one in the dimension of the ambient space: there is a more fundamental difference, namely that in the classical list of simple Lie groups, the rotation subgroups of $SE(2)$ and $SE(3)$ lie in different series.

4.1 Instantaneous Singular Sets

For a 1–manifold N, there are $\mathcal{I}$–invariant stratifications of the jet bundles $J^k(N, SE(p))$ with $k = 1, 2$ to which transversality of the jet extension of a motion implies the following [13]. For p even, there is a unique instantaneous centre, except at a discrete set of points in N where the instantaneous rotational component vanishes; moreover the locus of instantaneous centres will be an immersion in $\mathbb{R}^p$. For p odd, there is a unique instantaneous axis of rotation for each point in N and the locus of those lines form a non–cylindrical ruled surface. Moreover there is a discrete set in N at which the axis consists of singular points of trajectories. By the Basic Transversality Lemma these properties are generic. The generalization to motions with several dof is hardly touched upon in the classical literature. We introduce the following:

Definition 4.1 Given a motion germ $\lambda : N, x \to SE(p), e$ the *instantaneous singular set* (ISS) of λ at x is defined to be $I_x = \{q \in \mathbb{R}^p \mid M_{\lambda,q} \text{ singular at } x\}$.

For 1–dof planar (spatial) motions this yields the instantaneous centre of rotation (axis at pitch zero) alluded to above. A more explicit determination of the ISS can be arrived at as follows. There is no loss in generality in assuming that $\lambda(x) = e$, the identity element of $SE(p)$. Now if α is the action of $SE(p)$ on $\mathbb{R}^p$ then for $q \in \mathbb{R}^p$, let $\alpha_q : SE(p) \to \mathbb{R}^p$ be the map $\alpha_q(g) = \alpha(g, q) = g.q$. We may write the trajectory as a composite: $M_{\lambda,q} = \alpha_q \circ \lambda$. Assuming λ itself to be non–singular and applying the Chain Rule and the Rank Theorem, the required condition for $q \in I_x$ is

$$\dim(\operatorname{im} T_x\lambda \cap \ker T_e\alpha_q) > \max\{0, n-p\}. \tag{2}$$

The kernel of $T_e\alpha_q$ is the tangent space at e to the subgroup of $SE(p)$ fixing q. In the planar case this condition is easy to interpret. Generically, for $n = 2$ the ISS is an affine line, the locus of points q for which the (2–dimensional) image of the tangent map of λ has non–trivial intersection with the tangent space to the one–parameter subgroup of $SE(2)$ fixing q. For $n \geq 3$, singular trajectories are only possible when the motion itself has a singularity. The spatial case requires a study of first-order $\mathcal{I}$–invariants.

4.2 Screw Systems

In the spatial case, the generalisation of the 1–dof instantaneous screw about an invariant axis to motions with several dof is known as a 'screw system'. The most significant early reference, virtually never quoted in the kinematics literature, is Klein's 1872 paper [38]. This contains the germ of a key idea, namely the way in which screw systems sit relative to the "Klein complex". Ball's 1900 treatise [3], concentrated almost exclusively on the least degenerate cases. The screw system concept was revived in the engineering community by Hunt [35], who pointed out that the mechanically interesting cases were virtually always special in some sense, and proposed a classification. However, the equivalence relation underlying the listing was not explicit, raising doubts about its completeness. This situation was rectified in [27], and a classification was described intrinsically by the present authors in [15] following the original philosophy of Klein. In [16] we showed that the listing gave rise to natural Whitney stratifications, and gave complete descriptions of the specialisations. These results are outlined below.

In the definition of $\mathcal{I}$–equivalence it is no restriction to suppose that $g_1 = g_2 = e$, in which case $\mathcal{I}$–equivalence requires $k = h^{-1}$. Let $\lambda : N, x \to SE(3), e$ be a motion germ. Write $se(3)$ for the Lie algebra of $SE(3)$. Changes of coordinates at the source $x \in N$ leave the image of the differential $T_x\lambda : T_xN \to se(3)$ invariant, while the derivative at the target $e \in SE(3)$ of conjugation gives rise to the *adjoint action* of the group on its Lie algebra. This action is linear so induces an action on subspaces of $se(3)$

of given dimension k: thus if two motion germs are $\mathcal{I}$–equivalent the images of their derivatives lie in the same orbit under this action. The objects of interest are the vector subspaces of $se(3)$, representing images of derivatives of motions. A k–dimensional subspace ($1 \leq k \leq 6$) of $se(3)$ is called a *motor system* or *k–system*, and can be viewed as an element of the Grassmannian $G(k, se(3))$ of k–dimensional subspaces. Alternatively, we may think of a k–system as a $(k-1)$–dimensional projective subspace of the projective Lie algebra $Pse(3)$, the *screw space*: in that case, projective subspaces are *screw systems*.

We need to express the action more explicitly. $SE(3)$ is isomorphic to a semi–direct product $SO(3) \times_s T(3)$, with $T(3)$ the translation subgroup, so we can identify an element of $SE(3)$ with a pair (A, a) where $A \in SO(3)$ and $a \in T(3)$; as a result the Lie algebra $se(3)$ is isomorphic to a semi–direct product $so(3) \times_s t(3)$ of Lie algebras, where we can identify $t(3)$ with $\mathbb{R}^3$. Thus elements of $se(3)$ can be represented by a pair (B, b) where B is a 3×3 skew–symmetric matrix and $b \in \mathbb{R}^3$, hence by a pair of 3–vectors (u, v), where u is the unique vector for which $Bx = u \wedge x$ ($u = 0$ when $B = 0$), and $v = b$. The latter are *motor coordinates* for $se(3)$, or *screw coordinates* for $Pse(3)$. In these coordinates the adjoint action is given by

$$(A, a).(u, v) = (Au, Av - Au \wedge a).$$

Two basic invariants are classical, namely the *Killing form* $\langle u, u \rangle$, and the *Klein form* $\langle u, v \rangle$, where $\langle\ ,\ \rangle$ represents the standard scalar product on $\mathbb{R}^3$. The group $SE(p)$ behaves like a geometrically reductive group, in that the ring of invariant polynomials for the adjoint action is finitely generated. In [15] we gave explicit generators, the form depending on the parity of p. When $p = 3$ the ring is generated by the Klein and Killing forms, a result assumed in the subject but for which we are not aware of any prior reference.

Since the Killing and Klein forms have degree 2, the ratio $h = \langle u, v \rangle / \langle u, u \rangle$ is an invariant of the action induced by the adjoint action on the screw space, called the *pitch*. When $u = 0$ we define $h = \infty$. The concept of "pitch" is directly related to its usual physical meaning as follows. Each element of the Lie algebra defines a Killing vector field on $\mathbb{R}^3$ whose orbits are setwise invariant under projectivisation in $se(3)$. In general ($h \neq 0, \infty$), the orbits are helices about an invariant axis (the instantaneous screw axis) with common pitch h. When $h = 0$ the helices degenerate to circles about a fixed axis, corresponding to infinitesimal rotation, and when $h = \infty$ the helices degenerate to lines parallel to the axis, corresponding to instantaneous translation. There is an obvious connection with the R–, H– and P–joints of the introduction.

It is natural and profitable to foliate the screw space by pitch. Formally, define the *pitch quadric*

$$Q_h(u; v) = \langle u, v \rangle - h\langle u, u \rangle$$

whose zero set is the union of the screws of pitch h, and those of pitch ∞. Note that Q_∞ is degenerate: it is a projective 2–plane (the vertex of the *complex* quadric Q_∞) contained in all the Q_h. A special role is played by the quadric Q_0 containing all screws of pitch zero. Mathematically, Q_0 is the Klein quadric in real projective 5–space, whose points represent lines in projective 3–space, where now $(u; v)$ correspond to Plücker line coordinates. One simply identifies a screw of pitch zero with its fixed axis. Under this identification, subsets of Q_0 correspond to ruled subsets of $\mathrm{P}\mathbb{R}^3$ as in Table 2. In particular the Klein quadric has a double ruling by families of 2–planes, namely the classical α– and β–planes. The importance of Q_0 in engineering terms is that R–joints are more widely used than others.

Q_0	$\mathrm{P}\mathbb{R}^3$
point	line
α–plane	bundle of lines through a point
β–plane	bundle of lines in a plane
line	planar pencil of lines
conic	ruled hyperboloid

Table 2: Correspondence between Q_0 and $\mathrm{P}\mathbb{R}^3$.

The quadrics Q_h form a highly degenerate pencil, the *Klein complex*, central to an understanding of screw systems. The key intuition is that any screw system will intersect the Klein complex in a pencil of quadrics of lower dimension, whose projective type provides an invariant. Thus a 2–system will intersect the complex in a pencil of points, a 3–system in a pencil of conics, and so on. The details of the classification appear in [15, 16]. The most interesting case is provided by 3–systems. The general type is labelled IA_1, for which a projective basis can be put in the normal form

$$(1, 0, 0; h_\alpha, 0, 0) \quad (0, 1, 0; 0, h_\beta, 0) \quad (0, 0, 1; 0, 0, h_\gamma).$$

The moduli h_α, h_β, h_γ are the *principal* pitches: namely those for which the 3–system and the pitch quadric intersect in a reducible conic; we may assume them to be ordered, say $h_\alpha > h_\beta > h_\gamma$. A refinement of the classification emphasises the role of screws of pitch zero. Technically this is achieved by weakening the adjoint action, allowing Euclidean similarities in place of rigid displacements. This enables us to reduce the number of pitch moduli by one. The subtypes of the IA_1 type (in an obvious notation) are as follows:

$$\mathrm{IA}_1^{+++}, \mathrm{IA}_1^{++0}, \mathrm{IA}_1^{++-}, \mathrm{IA}_1^{+0-}, \mathrm{IA}_1^{+--}, \mathrm{IA}_1^{0--}, \mathrm{IA}_1^{---}.$$

Let us now return to the determination of instantaneous singular sets and concentrate on this case $p = 3$. They can be derived from the condition (2) and the refined classification of 3–systems. The image of the derivative of the motion is, projectively, just the associated screw system while the kernel of $T_e\alpha_q$ is projectively the α–plane A_q in Q_0 consisting of the bundle of lines through q. Therefore the ISS is the set of q for which the screw system intersects A_q in a projective subspace of dimension $\geq \max\{0, n-3\}$. Moreover, this number (plus one) gives the corank of the singularity on the trajectory of q. The ISS associated to a given screw system type can now be determined via a straightforward computation based on the normal form. The details will appear in [12]. For example, in the most general case of a 3–dof spatial motion $\lambda : N, x \to SE(3), e$ the screw system is of type IA_1 with no principal pitch zero; its intersection with Q_0 is an irreducible conic (or empty) and hence (Table 2) the ISS is a ruled hyperboloid in $\mathbb{R}^3$.

4.3 Regularity and Genericity

Standard techniques of singularity theory show [16] that the partitions of the Grassmannians $G(k, se(3))$ given by the listing process for screw systems are smooth, mostly involving several moduli: indeed they are Whitney regular stratifications. (There is even a complete description of the hierarchy of specializations.) As a result there is a corresponding transversality lemma for spatial motions. An n–dof motion $\lambda : N \to SE(3)$ gives rise to a jet extension $j^1\lambda : N \to J^1(N, SE(3))$. There is no reason to suppose *a priori* that λ is regular, however in the case $n \leq 3$, on an open and dense set in $C^\infty(N, SE(3))$ motions are immersive. In any case, we may partition $J^1(N, SE(3))$ into the sets Σ^k consisting of jets of rank k with $0 \leq k \leq n$. A stratification of $G(k, se(3))$ induces a stratification $\mathcal{S}_k$ of Σ^k via (left) translation in the group, and hence a stratification $\mathcal{S}$ of the jet bundle. λ is *1-generic* when it is transverse to $\mathcal{S}$. The Thom Transversality Theorem gives:

Theorem 4.1 *The 1-generic motions $\lambda : N \to SE(3)$ form a residual set in $C^\infty(N, SE(3))$, endowed with the Whitney C^∞ topology.*

Indeed, the regularity of the stratifications ensures that for $n \leq 3$ the 1-generic motions form an open and dense set. A 1-generic motion will exhibit only strata of codimension $\leq n$, which restricts considerably the types of screw systems one expects to encounter. Explicit lists of these screw systems can be found in [15].

5 Trajectories of Planar Motions

Descriptions of generic n–dof planar motions, including complete listings of the mono– and multigerms which can appear on their trajectories, fall into three distinct cases, $n = 1$, $n = 2$ and $n \geq 3$. We give an account of the trajectory singularities and their connections with the $\mathcal{A}$–type and ISS of the motion.

5.1 One dof Planar Motions

The case $n = 1$ is significant since it formalises intuitions familiar to any working kinematician. An account for an engineering audience appeared in [21]. One simplifying feature is that, since the motion is a map from a one–dimensional space into the 3–dimensional group $SE(2)$, general motions do not themselves exhibit singularities.

Theorem 5.1 (**[23]**) *For a residual set of 1-dof motions* $\lambda : N \to SE(2)$ *of the plane, any multigerm of a trajectory is* $\mathcal{A}$*-equivalent to one of the normal forms in Table 3 below.*

type	description	normal form	cod	versal unfolding
A_0	smooth point	$(s, 0)$	0	
A_2	ordinary cusp	(s^2, s^3)	1	$(s^2, s^3 + as)$
A_4	ramphoid cusp	(s^2, s^5)	2	$(s^2, s^5 + as^3 + bs)$
A_1	node	$(s, 0; 0, t)$	0	
A_3	tacnode	$(s, 0; t, t^2)$	1	$(s, a; t, t^2)$
A_5	flecnode	$(s, 0; t, t^3)$	2	$(s, a; t, t^3 + bt)$
D_5	cusp plus line	$(s, 0; t^3, t^2)$	2	$(s, a; t^3 + bt, t^2)$
D_4	triple point	$(s, 0; 0, t; u, u)$	1	$(s, 0; 0, t; u, u + a)$
D_6	tacnode plus line	$(s, 0; 0, t; u, u^2)$	2	$(s, as + b; 0, t; u, u^2)$
$\tilde{E}_7$	quadruple point	$(t, 0; 0, s; u, u; v, \lambda v)$	3	$(t, 0; 0, s; u + a, u;$ $v + b, \lambda v)$

Table 3: Local models for 1–dof planar motions.

The listing of monogerms first appeared as part of the listing of $\mathcal{A}$–simple monogerms of parametrized plane curves in [8]. It was an observation of Bruce that for (irreducible) monogerms there is a 1–1 correspondence between $\mathcal{A}$–equivalence classes and $\mathcal{K}$–equivalence classes of defining equations for the image. Thus the type symbols in Table 3 are the $\mathcal{K}$–types of germs at the origin of smooth functions on the plane whose zero–set is the image of the corresponding multigerm. All the local models are *smoothly* equivalent to the unfoldings of the trajectory germs obtained by varying the position of the tracing point except the $\mathcal{A}$–unimodular family $\tilde{E}_7$ (with modulus $\lambda \neq 0, 1$) for which we only have *topological* equivalence. For multigerms, the case when both branches are *immersive* appeared in [56]; pictures of the versal unfoldings appear in [23]. The ramphoid cusp (A_4) illustrates the danger (mentioned in Section 3.4) of seeking local pictures of bifurcation sets via defining equations. In the unfolding (a, b)–plane, the line $b = 0$ (with $a \neq 0$) represents ordinary cusps, whilst the *half* parabola $4b = a^2$, $a < 0$ represents tacnodes. However an algebraic curve containing these in its zero locus has to have equation $b(4b - a^2) = 0$, including a half parabola of no physical relevance. Beyond the list in Table 3 lie germs of higher codimension. The first such families are those corresponding to the $\mathcal{K}$–type of the germ $t, 0 \mapsto t^3, 0$; the $\mathcal{A}$–types are labelled E_{6k}, E_{6k+2} in [8]. It was shown in [13] that these types can only occur when the centrode curve (the locus of ISS) itself has a singularity. This link between singularity type and the ISS presages similar results for 2–dof motions.

There are three bifurcation curves associated to a 1–dof motion, corresponding to the multigerms of codimension 1: the cusp, tacnode and triple point curves. For the planar 4–bar, the cusp curve, not explicitly available from the analysis in Section 2, is nevertheless easy to render because it is traced by the instantaneous centre of rotation, the point of intersection of the first and third links. A robust algorithm for the tacnode curve results from an understanding [28] of the geometry in Section 2. The triple point curve is empty, except when the Grashof relations are satisfied, when it can be determined explicitly. At isolated points on these bifurcation curves one expects to encounter codimension 2 singularities. Further analysis [11] of the 4–bar geometry permits only the ramphoid cusp and flecnode.

5.2 Two dof Planar Motions

The study of 2–dof planar motions lies beyond the confines of classical kinematics. The trajectories are 2–dimensional subsets of the plane. In particular, they exhibit singularities at their boundaries: from the roboticist's point of view these are the degeneracies which can appear on "workspace boundaries". An account of the results for an engineering audience appears in [22].

Theorem 5.2 ([24]) *For a residual set of 2–dof motions $\lambda : N \to SE(2)$ of the plane, any monogerm of a trajectory is $\mathcal{A}$–equivalent to one of the normal forms in Table 4.*

The stable monogerms (immersion, fold and cusp) were listed in the seminal paper of Whitney [55], corank 1 germs were listed independently by Arnold [2], du Plessis [49] and Gaffney [20], while the corank 2 germs which appear in [49, page 134], require some work to tie them into this listing. The sobriquets for the corank 2 germs reflect the forms of their versal unfoldings: for the sharksfin these are essentially sections of the hyperbolic umbilic, and for the deltoid sections of the elliptic umbilic. A complete listing of generic multigerm types is given in [24]: it comprises 11 distinct bigerms, 5 trigerms and a unimodal family of quadrigerms.

Restrictions on the type of trajectory germs possible are imposed by the $\mathcal{A}$-type of the motion germ itself. A generic 2–dof planar motion is an immersed surface in $SE(2)$, except for possibly a discrete set of cross cap points. At an immersive point of the motion *only* corank 1 germs can appear on the trajectories: conversely, putting the motion germ into a normal form, we see that *any* corank 1 germ can appear. At a cross cap point we have the following answer, which appeared independently in [10].

name	normal form	cod	versal unfolding
immersion	(x, y)	0	
fold	(x, y^2)	1	
cusp	$(x, xy + y^3)$	1	
lips	$(x, y^3 + x^2y)$	1	$(x, y^3 + x^2y + ay)$
beaks	$(x, y^3 - x^2y)$	1	$(x, y^3 - x^2y + ay)$
swallowtail	$(x, xy + y^4)$	1	$(x, xy + y^4 + ay^2)$
goose	$(x, y^3 + x^3y)$	2	$(x, y^3 + x^3y + ay + bxy)$
butterflies	$(x, xy + y^5 \pm y^7)$	2	$(x, xy + y^5 \pm y^7 + ay^2 + by^3)$
gulls	$(x, xy^2 + y^4 + y^5)$	2	$(x, xy^2 + y^4 + y^5 + ay + by^3)$
sharksfin	$(x^2 + y^3, y^2 + x^3)$	2	$(x^2 + y^3 + ay, y^2 + x^3 + bx)$
deltoid	$(x^2 - y^2 + x^3, xy)$	2	$(x^2 - y^2 + x^3 + ax, xy + bx)$

Table 4: Monogerms for 2–dof planar motions.

Theorem 5.3 **([29])** *If a motion has a cross cap, the only generic singularity types which can appear on trajectories are fold, cusp, swallowtail, butterfly, sharksfin and deltoid.*

That all these types can appear is established by putting the cross cap into its standard normal form (s, st, t^2) and then, for each generic type, constructing an explicit submersive germ $\mathbb{R}^3, 0 \to \mathbb{R}^2, 0$ whose composite with the standard cross cap is of the given type. That the remaining types (the lips, beaks, goose and gulls) cannot appear is established similarly: take the composite of an arbitrary submersive germ with the standard cross-cap, and follow the classification of planar germs.

For 2–dof planar motions there are three bifurcation curves associated to monogerm types (lips, beaks and swallowtail) and five for multigerms. For the lips and beaks types (those having non–smooth critical sets) the bifurcation curves can be described as envelopes of 2–parameter families of lines.

Theorem 5.4 **([29])** *Let $\lambda : N \to SE(2)$ be an immersive, generic motion. The envelope of the ISS lines of points in N is the union of the closures of the lips and beaks bifurcation curves.*

This result opens up the possibility of rendering the lips and beaks bifurcation curves using computer algebra programs to determine an equation for the envelope. An interesting is the *double 4–bar*, in which one 4–bar is mounted on the coupler link of a second, providing a 2–dof motion exhibiting most of the types in Table 4. Computer renderings of typical critical images appear in [57, 29]. It is an example of a *composite* motion, i.e. two 1–dof motions (one for each 4–bar) composed in $SE(2)$, and typical of motions arising from serial devices such as robot arms. The behaviour of composite motions is simpler than that of general motions; for instance the deltoid cannot appear as a trajectory singularity for composite 2–dof planar motions. Corank 2 singularities admit a neat description for composite planar motions, namely that the moving instantaneous centre for the first motion should coincide with the fixed instantaneous centre for the second. This has a simple mechanical interpretation for the double 4–bar [29].

5.3 Planar Motions with ≥ 3 dof

The results of Section 5.2 can be extended to higher dof motions via a Splitting Lemma [50] analogous to Thom's result for functions [53, pages 59–60]. Like Thom's result it is established by an inductive argument on the k–jet using the Complete Transversal Lemma [9] for the induction step.

Theorem 5.5 *Let $F : \mathbb{R}^n, 0 \to \mathbb{R}^2, 0$ be an $\mathcal{A}$-finite germ of corank 1. Then F is $\mathcal{A}$-equivalent to a germ $(x, f(x, y_1, \ldots, y_r) \pm t_1^2 \pm \cdots \pm t_s^2)$ where $f(x, y_1, \ldots, y_r)$ has zero 1–jet and $f(0, y_1, \ldots, y_r)$ has zero 2–jet.* (Splitting Lemma.)

We say F is *stably* $\mathcal{A}$–equivalent to $(x, f(x, y_1, \ldots, y_r))$. The kernel of the differential of F has dimension $(n-1)$, and the cokernel has dimension 1, so the second intrinsic derivative [41, page 213] is represented by a single quadratic form, of rank s. Thus r and s are $\mathcal{A}$–invariants of F. For $n \geq 3$ *only* corank 1 singularities appear generically so essentially one obtains the same list of corank 1 germs as in the 2–dof case, plus two $\mathcal{A}$–unimodular families. More precisely:

Theorem 5.6 *For a residual set of n-dof planar motions $\lambda : N \to SE(2)$ with $n \geq 3$, any trajectory monogerm is stably $\mathcal{A}$-equivalent either to one of the corank 1 normal forms in Table 4, or to one of the $\mathcal{A}$-unimodular families $(x, xy + y^3 + by^2z + z^3 \pm y^5)$ where the modulus b is distinct from two exceptional values $b = 0$, $b = b_0$ with $4b_0^3 + 27 = 0$.*

The $\mathcal{A}$–unimodular families (distinguished by $\pm$) are dubbed the *ephemera*. (In computer renderings of versal unfoldings for the ephemera [17], one picture bears a striking resemblance to *Ephemera danica*, the common may–fly.) For a fixed value of b the germs can be distinguished by the determinant of their discriminant matrices [19]. The ephemera are best understood as 1–parameter unfoldings (with x as the unfolding parameter) of the umbilic $y^3 + by^2z + z^3 \pm y^5$. For $b > b_0$ this is a D_4^-, and for $b < b_0$ a D_4^+, in the Arnold notation. The full bifurcation sets for the umbilics are well–understood [54] and enable the specializations to the ephemera to be determined. The fold, cusp, lips, beaks and swallowtail all specialize to the ephemera; the double and triple folds are the only multigerm specializations. Can ephemera appear as germs of trajectories associated to generic n–dof planar motions? Unlike the case $n < 3$, a singular point on a trajectory can *only* arise from a singular point of the motion. For a general n–dof planar motion λ with $n \geq 3$ the motion germ is stable, and by results of Morin [44], $\mathcal{A}$–equivalent to 2–parameter unfoldings of the fold, cusp or swallowtail types. Here is the answer.

Theorem 5.7 ([17]) *At a singular motion germ of (a) fold type, (b) cusp type, (c) swallowtail type only the following trajectory germs can appear: (a) $\mathcal{A}$–simple types, i.e. those stably $\mathcal{A}$–equivalent to the corank 1 germs in Table 4; (b) fold, cusp and ephemera types; (c) fold, swallowtail and ephemera types. Each such trajectory type can be exhibited for some motion germ of the given type.*

6 The Spatial Case

The study of n–dof spatial motions is a core concern of engineering robotics. From the mathematical point of view it is more challenging than the planar case, in that the study of the motion itself begins to play a substantial role. There is as yet no complete listing of trajectory singularities, but we will describe briefly what is known.

6.1 One dof Spatial Motions

The case $n = 1$ gives rise [23] to the short list in Table 5: the two monogerms represent the beginning of the list of $\mathcal{A}$–simple monogerms of space curves obtained in [25].

type	description	normal form	cod	versal unfolding
A_0	smooth point	$(s,0,0)$	0	
A_2	simple cusp	$(s^2,s^3,0)$	2	(s^2,s^3+as,bs)
A_1	node	$(s,0,0;t,t,0)$	1	$(s,0,0;t,t,a)$
A_3	tacnode	$(s,0,0;t,t^2,0)$	3	$(s,a,b+cs;t,t^2,0)$
L_3^3	triple point	$(s,0,0;0,t,0;0,0,u)$	3	$(s,a,0;0,t,b;c,0,u)$

Table 5: Multigerms for 1–dof spatial motions.

Bifurcation takes place on a *nodal surface* on which lies a *cuspidal curve*, and exceptional points off the curve corresponding

to tacnodes and triple points: however as we have seen in Section 4 the cuspidal curve is forced by the geometry of the motion to be a union of lines. The versal unfoldings have immediate geometric content. Perhaps the most interesting case is that of the tacnode, where the bifurcation set is the Whitney umbrella in the unfolding (a,b,c)–space defined by $ac^2 = b^2$, with "handle" the a–axis. All the unfoldings represent a parabola and a line. Off the umbrella the line and the parabola fail to meet. On the umbrella (but off the handle) they meet in a single point. On the handle the line and parabola are coplanar and for $a > 0$ they meet in two distinct points, while for $a < 0$ they fail to meet.

6.2 Two dof Spatial Motions

Mond [42] classified the singularities of smooth mappings from a surface into a 3–manifold. This gives rise to the listing of trajectory monogerms for 2–dof spatial motions in Table 6. All the germs are $\mathcal{A}$–simple except the $\mathcal{A}$–unimodular family $P_3(c)$ (there being four exceptional values of the modulus: $c = 0, \frac{1}{2}, 1, \frac{3}{2}$). The only generic singularity type is the *cross cap* type S_0. These have been observed, for example, as stable phenomena on coupler surfaces of the RSSR linkage [51].

type	normal form	cod	type	normal form	cod
S_0	(x, y^2, xy)	0	B_3^+	$(x, y^2, x^2y + y^7)$	3
S_1^+	$(x, y^2, y^3 + x^2y)$	1	B_3^-	$(x, y^2, x^2y - y^7)$	3
S_1^-	$(x, y^2, y^3 - x^2y)$	1	C_3^+	$(x, y^2, xy^3 + x^3y)$	3
S_2	$(x, y^2, y^3 + x^3y)$	2	C_3^-	$(x, y^2, xy^3 - x^3y)$	3
S_3^+	$(x, y^2, y^3 + x^4y)$	3	H_2	$(x, y^3, xy + y^5)$	2
S_3^-	$(x, y^2, y^3 - x^4y)$	3	H_3	$(x, y^3, xy + y^8)$	3
B_2^+	$(x, y^2, x^2y + y^5)$	2	$P_3(c)$	$(x, xy + y^3, xy^2 + cy^4)$	3
B_2^-	$(x, y^2, x^2y - y^5)$	2			

Table 6: Monogerms for 2–dof spatial motions.

The only *generic* multigerms are double points, along curves in the plane, and triple points, at discrete points on those curves. The numbers C of cross caps, and T of triple points, in a generic versal unfolding provide basic invariants. The geometry is largely determined by the *double point* curve in the source. For germs $(x, y^2, yp(x, y^2))$ this curve has equation $p(x, y^2) = 0$ so is easy to study. Bifurcations were studied in [42, 26] giving rise to computer generated renderings. For instance, the unimodular germs $P_3(c)$ have versal unfoldings $(x, xy + y^3 + ay, xy^2 + cy^4 + by + dy^3)$. For a fixed value of the modulus c we obtain in (a, b, d)–space a cubic surface of cross caps, and a quartic surface of triple points. As $P_3(c)$ is weighted homogeneous, the intersection of these surfaces can be studied via the plane section $d = 1$, where they give rise to cuspidal cubic curves. Their intersections depend not just on the four exceptional values of c listed above, but also two further values $c = \frac{4}{3}$ and $c = 2$ (see [26, Figure 16]). The listing of higher codimension multigerms initiated in [42] is continued in [33, 34]. However for multigerms with several branches the number of moduli increases rather quickly.

6.3 Spatial Motions with ≥ 3 dof

Results here are limited; the best understood case is $n = 3$. The stable monogerms are classical (regular, fold, cusp and swallowtail). The first listing beyond this appears in Arnold's work on the evolution of galaxies, where the germs are gradients of smooth functions in three variables. Bruce [7] considered arbitrary germs, and obtained a list up to codimension 1: all these are $\mathcal{A}$-simple of corank 1, so appear in the recent listing [39]. Beyond this the listing is a lengthy technical exercise. A list up to codimension 3 was given by Hawes [32] using the computer program TRANSVERSAL developed in Liverpool by Kirk [37]. There are over 50 types, including germs of higher modality whose unfoldings are unstudied. It is probably impractical (and unprofitable) to continue this process much further. The evidence is that the motions which arise in engineering robotics have special features (such as being composite) limiting the range of possible trajectory germs and suggesting new lines of development.

References

[1] Arnold, V. I. (1976) Wavefront Evolution and Equivariant Morse Lemma. *Comm. Pure Appl. Math.* **29** 557–582.

[2] Arnold, V. I. (1979) Indices of Singular points of 1–Forms on a Manifold with Boundary, Convolution of Invariants of Reflection Groups, and Singular Projections of Smooth Surfaces. *Russian Math. Surveys*, **34**(2), 3-38.

[3] Ball, R. S. (1900) The Theory of Screws, Cambridge University Press.

[4] Bennett, G. T. (1903) A New Mechanism. *Engineering*, **76**, 777–778.

[5] Bennett, G. T. (1913) The Skew Isogram Mechanism. *Proc. London Math. Soc. (2nd series)*, **13**, 151–173.

[6] Bottema, O. and Roth, B. (1990) Theoretical Kinematics. Dover Publications.

[7] Bruce, J. W. (1986) A Classification of 1–Parameter Families of Map–Germs $(\mathbb{R}^3, 0) \to (\mathbb{R}^3, 0)$ with Applications to Condensation Problems. *Jour. Lond. Math. Soc.*, **33**(2), 375–384.

[8] Bruce, J. W. and Gaffney, T. J. (1982) Simple Singularities of Mappings $(\mathbb{C}, 0) \to (\mathbb{C}^2, 0)$. *J. London Math. Soc.*, **26**, 465–474.

[9] Bruce, J. W., Kirk, N. P. and du Plessis, A. A. (1997) Complete Transversals and the Classification of Singularities, *Nonlinearity.* **10**, 253–275.

[10] Bruce, J. W. and West, J. M. (1998) Functions on a Cross Cap. *Math. Proc. Camb. Phil. Soc.*, **123**, 19–39.

[11] Cocke, M. W., Donelan, P. S. and Gibson, C. G. On the Genericity of Planar 4–Bar Motions. In preparation.

[12] Cocke, M. W., Donelan, P. S. and Gibson, C. G. Instantaneous Singular Sets for Spatial Motions. In preparation.

[13] Donelan, P. S. (1988) Generic Properties in Euclidean Kinematics, *Acta Applicandae Mathematicae*, **12**, 265–286.

[14] Donelan, P. S. (1993) On the Geometry of Planar Motions. *Quarterly J. of Math. Oxford*, **44**(2), 165–181.

[15] Donelan, P. S. and Gibson, C. G. (1991) First–Order Invariants of Euclidean Motions. *Acta Applicandae Mathematicae*, **24**, 233–251.

[16] Donelan, P. S. and Gibson, C. G. (1993) On the Hierarchy of Screw Systems. *Acta Applicandae Mathematicae*, **32**, 267–296.

[17] Donelan, P. S., Gibson, C. G. and Hawes, W. (1997) Trajectory Singularities of General Planar Motions. Preprint, University of Liverpool.

[18] Dorf, R. C. (Ed.) (1988) International Encyclopedia of Robotics, John W. Wiley and Sons.

[19] Gaffney, T., du Plessis, A. A. and Wilson, L. C. (1990) Map–Germs Determined by their Discriminant. To appear in the *Proceedings of the Hawaii–Provence Singularities Conferences*, Travaux en Cours, Hermann, Paris.

[20] Gaffney, T. (1983) The Structure of $T\mathcal{A}(f)$, Classification and an Application to Differential Geometry. *AMS Proceedings of Symposia in Pure Mathematics, Singularities, Part 1*, **40**, 409–427.

[21] Gibson, C. G. (1992) Kinematic Singularities – A New Mathematical Tool. *Third International Workshop on Advances in Robot Kinematics*, Ferrara, Italy, 209–215.

[22] Gibson, C. G., Hawes, W. H. and Hobbs, C. A. (1994) Local Pictures for General Two–Parameter Motions of the Plane. In *Advances in Robot Kinematics and Computational Geometry*, Kluwer Academic Publishers, 49–58.

[23] Gibson, C. G. and Hobbs, C. A. (1995) Local Models for General One–Parameter Motions of the Plane and Space. *Proc. Royal Society of Edinburgh*, **125A**, 639–656.

[24] Gibson, C. G. and Hobbs, C. A. (1996) Singularity and Bifurcation for General Two–Dimensional Planar Motions. *New Zealand Journal of Mathematics.* **25** 141–163.

[25] Gibson, C. G. and Hobbs, C. A. (1992) Simple Singularities of Space Curves. *Math. Proc. Camb. Phil. Soc.*, **113**, 297–310.

[26] Gibson, C. G., Hobbs, C. A. and Marar, W. L. (1997) On Versal Unfoldings of Singularities for General Two–Dimensional Spatial Motions. *Acta Applicandae Mathematicae*, **47**(2) (221–242.

[27] Gibson, C. G. and Hunt, K. H. (1992) Geometry of Screw Systems. *Mech. Machine Theory*, **12**, 1–27.

[28] Gibson, C. G., Marsh, D. and Xiang, Y. (1996) An Algorithm to Generate the Transition Curve of the Planar Four–Bar Mechanism. *Mech. Machine Theory*, **31**, 381–395.

[29] Gibson, C. G., Marsh, D. and Xiang, Y. (1997) Singular Aspects of Generic Planar Motions with Two Degrees of Freedom. To appear in *Mechanism and Machine Theory.*

[30] Gibson, C. G. and Newstead, P. E. (1986) On the Geometry of the Planar 4–Bar Mechanism. *Acta Applicandae Mathematicae*, **7**, 113–135.

[31] Hain, K. (1964) Das Spektrum des Gelenkvierecks bei veränderlicher Gestell–Länge. *VDI Forsch Ing.-Wes.*, **30**, 33–42.

[32] Hawes, W. (1995) Multi–Dimensional Motions of the Plane and Space. PhD Thesis, University of Liverpool.

[33] Hobbs, C. A. (1993) Kinematic Singularities of Low Dimension. PhD Thesis, University of Liverpool.

[34] Hobbs, C. A. and Kirk, N. P. (1997) On the Classification and Bifurcation of Multigerms of Maps from Surfaces to 3–Space. To appear in *Mathematica Scandinavica.*

[35] Hunt, K. H. (1978) Kinematic Geometry of Mechanisms, Clarendon Press, Oxford.

[36] Jessop, C. M.(1916) Quartic Surfaces with Singular Points, Cambridge University Press.

[37] Kirk, N. P. (1993) Computational Aspects of Singularity Theory. PhD Thesis, University of Liverpool.

[38] Klein, F. (1871) Notiz Betreffend dem Zusammenhang der Liniengeometrie mit der Mechanik starrer Körper. *Math. Ann.*, **4**, 403–415.

[39] Marar, W. L. and Tari, F. (1995) On the Geometry of Simple Germs of Corank 1 Maps from $\mathbb{R}^3$ to $\mathbb{R}^3$. Preprint, University of Liverpool.

[40] Martinet, J. (1982) Singularities of Smooth Functions and Maps. London Mathematical Society Lecture Note Series, **58**, Cambridge University Press.

[41] Mather, J. N. (1970) Stability of C^∞ Mappings: VI The Nice Dimensions. *Springer Lecture Notes in Mathematics, Proceedings of Liverpool Singularities Symposium I*, **192**, 207–253.

[42] Mond, D. M. Q. (1985) On the Classification of Germs of Maps from $\mathbb{R}^2$ to $\mathbb{R}^3$. *Proc. London Math. Soc.*, **50**(3), 333–369.

[43] Montaldi, J. A. (1991) On Generic Composites of Maps. *Bull. London Math. Soc.*, **23**, 81–85.

[44] Morin, B. (1965) Formes Canoniques des Singularités d'une Application Différentiable. *Compt. Rendus Acad. Sci. Paris*, **260**, 5662–5665, 6503–6506.

[45] Müller R. (1889) Über die Doppelpunkte der Koppelkurve *Z. Math. Phys.*, **34**, 303–305 & 372–375, [trans. Tesar D. in *(Translations of) Papers on Geometrical Theory of Motion applied to Approximate Straight Line Motion,* Kansas State University Bulletin, **46** (1962), Special Report No. 21.]

[46] Müller R. (1903) Über einige Kurven, die mit der Theorie des ebenen Gelekvierecks in Zusammenhang stehen *Z. Math. Phys.*, **48**, 224–248, [trans. Tesar D. in *(Translations of) Papers on Geometrical Theory of Motion applied to Approximate Straight Line Motion,* Kansas State University Bulletin, **46** (1962), Special Report No. 21.]

[47] Niemann, S. H. (1978) Geometry and Mechanics. PhD Thesis, Oxford University.

[48] Pai D. K. and Leu M. C. (1992) Genericity and Singularities of Robot Manipulators. *IEEE Trans. Robotics and Automation*, **8**, 545–559.

[49] du Plessis, A. A. (1980) On the Determinacy of Smooth Map–Germs. *Invent. Math.*, **58**, 107–160.

[50] Rieger, J. H. and Ruas, M. A. S. (1987) Classification of $\mathcal{A}$–simple germs from k^n to k^2. *Comp. Math.*, **79**, 99–108.

[51] Robertson, G. D. and Torfason, L. E. The Queeroid – A New Kinematic Surface. *Proc. Fourth World Congress on the Theory of Machines and Mechanisms*, 717–719.

[52] Tchoń, K. and Muszyński, R. (1995) Singularities of Non–Redundant Robot Kinematics. To appear in *Int. J. Robotics Research.*

[53] Thom, R. (1975) Structural Stability and Morphogenesis, W. A. Benjamin, Inc. Reading, Massachusetts.

[54] Wall, C. T. C. (1980) Affine Cubic Functions III. The Real Plane. *Math. Proc. Camb. Phil. Soc.*, **87**(1), 1–14.

[55] Whitney, H. (1955) On Singularities of Mappings of Euclidean Space I: Mappings of the Plane to the Plane. *Ann. Math.*, **62**, 374–410.

[56] Wilkinson, T. (1991) The Geometry of Folding Maps, PhD Thesis, University of Newcastle.

[57] Xiang, Y. (1995) Bifurcation of Singularities of Planar Mechanisms with One– or Two–Degrees of Freedom. PhD Thesis, Napier University.

P.S. Donelan
School of Mathematical and Computing Sciences
Victoria University of Wellington
PO Box 600, Wellington
New Zealand.

e-mail: peter.donelan@vuw.ac.nz

C.G. Gibson
Mathematical Sciences
University of Liverpool
PO Box 147, Liverpool L69 3BX
United Kingdom.

e-mail: c.g.gibson@liverpool.ac.uk

Singularities of Developable Surfaces

Go-o Ishikawa

Dedicated to the 60th birthday of Professor C.T.C. Wall.

Abstract

In this survey article we explain recent results concerning singularities appearing on the tangent developables of space curves via the notion of projective duality. First we examine, from our viewpoint, the classical local classification and the Hartman-Nirenberg theorem on developable surfaces. Then we introduce results on the local differentiable and topological classification of tangent developables of space curves. Lastly we present some related problems and questions, for instance concerning finite determinacy.

1 Introduction

A smooth surface in Euclidean three space is called **a developable surface** if its Gaussian curvature vanishes everywhere. By Gauss' fundamental theorem, a developable surface is locally isometric to a planar domain. This motivates the name "developable" (see [58]). Let $z = f(x, y)$ be a smooth surface. If we set $r = z_{xx}, s = z_{xy}, t = z_{yy}$, then the surface is a developable surface if and only if the Monge-Ampère equation $rt - s^2 = 0$ is satisfied on the surface [47], [34].

Developable surfaces are well-known objects from the last century. For instance, analytic developable surfaces are locally classified into three types: *cones, cylinders* and *tangent developables.* In the smooth case, we can rather freely change the three types through planar trapezia to make various (not necessarily analytic) non-singular portions of smooth developable surfaces.

A tangent developable is part of the union of the one-parameter family of tangent lines to a fixed space curve. In classical algebraic geometry, tangent developables play a significant role in the duality theory of space curves (cf.[10], [20]). Tangent developables of space curves form a special class of ruled surfaces, namely, developable ruled surfaces, and of necessity have singularities at least along the original curve, while ruled surfaces generically have no non-isolated singularities.

Recently, developable surfaces have attracted attention through their relation with computer science. They are widely used in industry, and are fundamental objects in computer-aided design [24]. Though singularities should be avoided in practical situations, the appearance of singularities in developable surfaces is essential to their nature.

It has become clear that it is useful, even in the smooth category, to use the viewpoint of projective duality, or Legendre transformation [3], and modern singularity theory, for the full understanding of the structure of developable surfaces, both locally and globally [54].

The tangent developable of a space curve γ has constant tangent space along each generating line. These are called osculating planes to γ, and form a space curve γ^* in the dual space; the dual curve [54]. The tangent developable of γ is the projection to the original space of the projective conormal bundle over γ^* in the dual projective space. In other words, the tangent developable of γ is **the dual surface** of the dual curve γ^*.

In fact the tangent developable of a curve γ is regarded as the envelope of the one-parameter family of osculating planes if the dual curve γ^* is non-singular. If γ^* is singular, then the envelope consists of the tangent developable and the osculating plane to the base point [29].

Conversely, if a one-parameter family of planes is given, then its envelope contains a tangent developable as a component, provided the corresponding curve in the dual projective space is of *finite type* as defined in §2. For instance the envelope of the family of normal planes to a space curve is the tangent developable of the focal curve [52].

Thus we can regard a tangent developable as a degenerate Legendre singularity or a wave front set of a possibly singular Legendre variety in the sense of Legendre singularity theory [2], [3].

We are then led to natural questions such as: *What kinds of singularities appear in tangent developables of curves? Can we classify the singularities at least locally?*

In a neighborhood of a non-singular point, the tangent developable has the parametrisation $(t, s) \mapsto \boldsymbol{x}(t) + s\dot{\boldsymbol{x}}(t)$. We shall give the parametrisation of the tangent developable near a possibly singular point (of finite type) of the curve, which also covers the non-singular case. See §4.

When we try to answer the questions above, several equivalence relations come to mind for space curve-germs and space surface-germs: isometric equivalence, affine equivalence, differentiable equivalence, and topological equivalence.

The classification under affine equivalence seems to be the most natural one for tangent developables since the tangent lines to a space curve are invariant under affine equivalence. However isometric and affine equivalences are too strong to get a finite list after classification, even for the simplest singularities, namely, cuspidal edges.

Therefore, in the following sections, we consider the local differentiable or topological classifications. We pursue the classification of these geometric objects up to diffeomorphisms or homeomorphisms by means of geometric invariants. In fact we examine the condition that the leading terms of the parametrisation determine the equivalence class of the tangent developable.

Determinacy is, here as in other classifications, the key to our results [43], [60].

After recalling the basic notions and the history of the study in §2, we explain some classical results concerning developable surfaces using projective duality in §3. The parametrisations of tangent developables are given in §4. Then we give the differentiable classification in §5 and the main idea of the proof in §6. We also present the topological classification in §7. A method to illustrate the topological equivalence classes of tangent developables is given in §8. The key ideas of the topological classification theorem are exhibited in §9. We finish by presenting several open questions in §10.

In [37], [38] there are some general results on the bifurcation problem of curves and their tangent developables. We also find unexpected applications of tangent developables, for example, in [59], [51], [23], [25]. See also the excellent expositions [9], [52] on singularities of tangent developables and on the other related subjects.

In this article smoothness means infinite differentiability. For instance a smooth manifold is a C^∞ manifold, and a smooth mapping is a C^∞ mapping.

The author is grateful to Professors P. Giblin, S. Izumiya, T. Morimoto for helpful comments and suggestion, and to the organisers of the symposium for giving him the opportunity to survey the present subject. The author also sincerely thanks Professor C.T.C. Wall who drew my attention to the topological classification problem.

2 Basic notions and history

To review the history of the classification, first we recall some fundamental notions for space curves.

Let M be a one-dimensional smooth manifold (parameter space), $\gamma : M \to \mathbf{R}^3$ a germ of smoothly parametrised space curve and $t_0 \in M$. Represent γ as $\boldsymbol{x}(t) = (x_1(t), x_2(t), x_3(t))$, using local smooth coordinates of M centred at t_0. Then we say that γ *of finite type* at $t_0 \in M$ if the infinite sequence of vectors $\dot{\boldsymbol{x}}(t_0), \ddot{\boldsymbol{x}}(t_0), \ldots, \boldsymbol{x}^{(k)}(t_0), \ldots,$ generate three space. Then, after some smooth coordinate changes on (M, t_0), and some affine coordinate change of $(\mathbf{R}^3, \gamma(t_0))$, if necessary, we write the curve γ in the form (standard form)

$$x_1(t) = t^m + o(t^m),\ x_2(t) = t^{m+s} + o(t^{m+s}),\ x_3(t) = t^{m+s+r} + o(t^{m+s+r}),$$

for some positive integers m, s, r. The triplet $(m, m+s, m+s+r)$ is independent of the choice of affine local coordinates, and it is called the *type* of the curve-germ γ: type$(\gamma) = (a_1, a_2, a_3) = (m, m+s, m+s+r)$. In fact, if γ is of finite type at t_0, then there exists a unique line (resp. a unique plane) having the highest contact with γ at t_0. We call it **the tangent line** (resp. **the osculating plane**) to γ at t_0. Then a_1 (resp. a_2, a_3) is the order of contact of γ with the origin $\gamma(t_0)$ (resp. with the tangent line, with the osculating plane) at t_0. For coordinates in the standard form, the tangent line (resp. the osculating plane) is given by x_1-axis (resp. (x_1, x_2)-plane). A curve-germ is of finite type if and only if the curve does not have infinite tangency with any affine plane.

We treat curves which are of finite type at each point. Note that only curve-germs of finite type appear in a generic multi-dimensional family of curves ([54] Theorem 2.2).

We call a point t_0 **an ordinary point** if type(γ) at t_0 is $(1,2,3)$, in other words, if the Wronskian $|\dot{x}(t), \ddot{x}(t), \dddot{x}(t)|$ does not vanish at t_0. Otherwise we call it **a special point**. Special points are isolated, since the Wronskian has a zero of finite order at each special point. Moreover we remark that the parametrised curve is nonsingular at t_0 if and only if $m = 1$.

To date, several results are known on the local differentiable classification of developables. At an ordinary point, it is classically known that the tangent developable of a space curve has a cuspidal edge along the curve itself. In fact the tangent developable has a parametrisation which is smoothly right-left equivalent to the map-germ $(x,t) \mapsto (x, t^2, t^3)$ at the origin. (See [9], [52].)

For the simplest case of special points, namely in the case of type $(1,2,4)$, Cleave [11], Gaffney, du Plessis [17] and Scherbak [54] showed that the tangent developable has a parametrisation which is smoothly right-left equivalent to the map-germ $(x,t) \mapsto (x, t^2, xt^3)$, which is called **the folded umbrella** [1]. In this case the tangent developable has, besides the cuspidal edge, a self-intersection locus through the special point.

Subsequently Mond gave a differentiable classification of tangent developable of curves of type $(1,2,2+r)$, $r \leq 5$, and of type $(1,3,4)$ in [44], [45]. For instance, for the type $(1,3,4)$, the tangent developable has singularities along the original curve and along the tangent line to the origin, and has two branches of double point loci: It is called the **Mond surface**.

On the other hand, Arnold [1] observed that the tangent developable of a singular curve of type $(2,3,4)$ is differentiably equivalent to the swallowtail from Legendre singularity theory [3]. See also [54].

In [56], Scherbak showed that the tangent developable of a curve of type $(1,3,5)$ is, in the complex analytic category, holomorphically equivalent to the variety of irregular orbits of the finite reflection group H_3 in $\mathbf{C}^3$.

For the topological classification, Mond [45] proved that the tangent developable of a curve of type $(1,2,2+r)$, $r = 1,2,\ldots$, is homeomorphic to the plane if r is odd and to the cone on the figure eight on the sphere if r is even. However no generalisation of this result of Mond concerning the topological classification was known.

3 Projective duality

Denote by $\mathbf{RP}^3$ the real projective space of dimension 3, and by $\mathbf{RP}^{3*}$ the dual projective space, which is identified with the space of projective planes in $\mathbf{RP}^3$. Consider the 5-dimensional incidence manifold:

$$Q = \{(p,q) \in \mathbf{RP}^3 \times \mathbf{RP}^{3*} \mid p \in q^*\},$$

where $q^* \subset \mathbf{RP}^3$ is the plane defined by $q \in \mathbf{RP}^{3*}$. The pair of fibrations $\pi_1 : Q \to \mathbf{RP}^3$ and $\pi_2 : Q \to \mathbf{RP}^{3*}$ define a contact structure, that is, a completely non-integrable distribution of codimension one, $D = \mathrm{Ker}\pi_{1*} \oplus \mathrm{Ker}\pi_{2*}$ on Q. In fact the natural identifications of Q to the manifolds of contact elements, that is, the projective cotangent bundles $PT^*\mathbf{RP}^3$ and to $PT^*\mathbf{RP}^{3*}$ are contact diffeomorphisms respectively [2], [34].

Now we recall the fundamental duality theorem from [54]. Let $\gamma : (M, t_0) \to \mathbf{RP}^3$ be a smooth curve-germ. The osculating planes to γ, considered as projective planes, form a curve in the dual projective space: $\gamma^* : (M, t_0) \to \mathbf{RP}^{3*}$. We call γ^* the dual of γ. Then Arnold-Scherbak [54] gave an elegant proof of the following:

Lemma 3.1 *If γ is of type $(m, m+s, m+s+r)$ at t_0, then the dual γ^* is also of finite type and of type $(r, r+s, r+s+m)$ at t_0.*

To see this directly, first take homogeneous coordinates X_0, X_1, X_2, X_3 such that $x_i = X_i/X_0, i = 1,2,3$, give a standard form of γ. Next take homogeneous coordinates Y_0, Y_1, Y_2, Y_3 of $\mathbf{RP}^{3*}$ so that the equation

$$Y_0X_3 + Y_1X_2 + Y_2X_1 + Y_3X_0 = 0,$$

gives the incident relation. Then the dual curve

$$\gamma^*(t) = (y_1(t), y_2(t), y_3(t)), \quad y_i = Y_i/Y_0,$$

is obtained by solving the system of equations

$$\begin{aligned} x_3(t) + x_2(t)y_1 + x_1(t)y_2 + y_3 &= 0, \\ \dot{x}_3(t) + \dot{x}_2(t)y_1 + \dot{x}_1(t)y_2 &= 0, \\ \ddot{x}_3(t) + \ddot{x}_2(t)y_1 + \ddot{x}_1(t)y_2 &= 0, \end{aligned}$$

first for $t \neq 0$, and then by extending the solutions to $t = 0$ smoothly.

Now we review some classical results. Any smooth surface S in $\mathbf{R}^3 \subset \mathbf{RP}^3$ lifts, with respect to π_1, to a Legendre surface, namely, an integral manifold of D; $\tilde{S} = \{(p, T_pS) \in Q \mid p \in S\}$. Here the tangent space T_pS is regarded as a projective plane in $\mathbf{RP}^3$. Then we project $\tilde{S}$ in the other direction; $\pi_2 : \tilde{S} \to \mathbf{RP}^{3*}$. The essential point is that S is a developable surface if and only if the rank of the projection π_2 is at most one everywhere on $\tilde{S}$. Then we have in [21], [34]:

Lemma 3.2 *If S is a smooth developable surface in $\mathbf{R}^3$, then there exists a smooth one-dimensional foliation of S by straight lines such that S has a constant tangent space along each leaf.*

Using this fact we see that the classical local classification of developables is valid for the class of analytic surfaces. In fact, at each point in S, we take a germ of an analytic transversal to the foliation passing through a point in S. Then, by lifting to $\tilde{S}$ and composing with π_2, we have an analytic curve β in the dual projective space, the projective conormal bundle of which coincides with the Legendre lift $\tilde{S}$ at least locally. If β collapses to a point q, then S is contained in the plane q^*, the projective dual to q. If β is contained in a plane Π, then S is a cone made from lines passing through the point $\Pi^* \in \mathbf{RP}^3$. When Π^* lies in the plane at infinity, S becomes a cylinder. If β is not contained in any plane, then S is the tangent developable of the dual curve $\gamma = \beta^*$ in $\mathbf{RP}^3$.

We can consequently regard cones and cylinders as degenerations of tangent developables. A tangent developable of a space curve, which is not contained in any plane, is the dual surface of the dual curve to the original space curve. The dual curve, in this case, is not a plane curve. Then *a cone is the dual surface of a plane curve and a cylinder is the dual surface of a curve lying on the plane at infinity.*

If we start from a smooth space curve β in $\mathbf{RP}^{3*}$, which is not a plane curve, but has infinite contact with a plane, then we find many examples of smooth developable surfaces which are not governed by the classical local classification.

The same argument as above can be used to prove the classical Hartman-Nirenberg theorem [21], [58], [55]: *A properly embedded connected smooth developable surface S in $\mathbf{R}^3$ is necessarily a cylinder.* In fact, near each point on S, the π_2-projection β is contained in a plane Π such that Π^* lies on the plane at infinity $\mathbf{RP}^3 - \mathbf{R}^3$. Moreover, near a point where the rank of the projection is equal to one, β is not contained in any line; if it were, then S would reduce to a line. Since S is connected, we see the global projection $\pi_2(\tilde{S})$ is contained in a plane. See [34] for details and for the higher dimensional case.

4 Parametrisations of tangent developables

Let γ be a space curve of type $(m, m+s, m+s+r)$ at t_0. Then the dual γ^* is of type $(r, r+s, r+s+m)$ at t_0 by Lemma 3.1. We fix non-zero constants a, b, c, and represent γ^* as

$$\gamma^* : y_1 = at^r, \; y_2 = bt^{r+s} + o(t^{r+s}), \; y_3 = ct^{r+s+m} + o(t^{r+s+m}),$$

for some smooth coordinate t of (M, t_0), and affine coordinates y_1, y_2, y_3 on $(\mathbf{R}P^{3*}, \gamma^*(0))$.

We set

$$F(t, x) = y_3(t) + y_2(t)x_1 + y_1(t)x_2 + x_3.$$

Then we have a parametrisation of the tangent developable of γ (resp. the curve γ itself), solving the equation $F = \partial F/\partial t = 0$ (resp. $F = \partial F/\partial t = \partial^2 F/\partial t^2 = 0$), first for $t \neq 0$, and then extending the solution to $t = 0$.

Setting $x_1 = x$, we have $x_2 = -\dot{y}_3/\dot{y}_1 - x\dot{y}_2/\dot{y}_1$, from $\partial F/\partial t = 0$. Further from $F = 0$, we see that $x_3 = x_3(t,x)$ satisfies $\partial x_3/\partial t = -y_1 \partial x_2/\partial t$.

If we set $a = -(r+s)(r+s+m), b = -r(r+s+m), c = -r(r+s)$, to simplify the expression, we obtain the parametrisation $f = \mathrm{dev}(\gamma) : (\mathbf{R}^2, 0) \to (\mathbf{R}^3, 0)$ defined by

$$\begin{aligned}
x_1 &= x, \\
x_2 &= t^{s+m} + t^{s+m+1}\varphi(t) + x(t^s + t^{s+1}\psi(t)), \\
x_3 &= (r+s)(r+s+m)\int_0^t u^r \frac{\partial x_2(u,x)}{\partial u} du \\
&= (r+s)(s+m)t^{r+s+m} + \cdots + x\{s(r+s+m)t^{r+s} + \cdots\},
\end{aligned}$$

for some smooth functions-germs $\varphi(t)$ and $\psi(t)$. See [33].

From this explicit parametrisation, we see that the tangent developable of γ is singular along the original curve γ, and further along the tangent line at a special point if $s \geq 2$, as well as at points of the self-intersection loci. Note that the parametrisation $\mathrm{dev}(\gamma)$ has very degenerate singularities even in the sense of Legendre singularity theory, because the singularities along $\{t = 0\}$ are more degenerate than the ordinary cuspidal edge if $s \geq 3$ or $r \geq 2$.

Consequently to classify the singularities arising from tangent developables, we are naturally led to the problem of the determinacy of $\mathrm{dev}(\gamma)$ by its leading terms.

5 Differentiable classification

Two smooth map-germs $f, f' : (\mathbf{R}^2, 0) \to (\mathbf{R}^3, 0)$ are called **differentiably (right-left) equivalent** if there exist diffeomorphism-germs $\sigma : (\mathbf{R}^2, 0) \to (\mathbf{R}^2, 0)$ and $\tau : (\mathbf{R}^3, 0) \to (\mathbf{R}^3, 0)$ such that $\tau \circ f = f' \circ \sigma$.

Generalising all known results on the local differentiable classification of tangent developables, we have the following determinacy result in [29], [31]:

Theorem 5.1 *The local differentiable equivalence class of the tangent developable is determined by the type* $\mathbf{A}$ *of a space curve, namely,*

$$\#\left(\{\mathrm{dev}(\gamma) \mid \mathrm{type}(\gamma) = \mathbf{A}\}/\underset{\mathrm{diff}}{\sim}\right) = 1$$

if and only if $\mathbf{A}$ *is one of the following:*

$$(1,2,2+r); (r : \text{positive integer}), (2,3,4), (1,3,4), (3,4,5), (1,3,5).$$

For example, if $\mathbf{A} = (1,3,5)$, then the normal form of the tangent developable under differentiable equivalence is given by

$$(t,x) \mapsto (x,\ t^3 + xt^2,\ 12t^5 + 10xt^4),$$

which we call the **Scherbak surface**.

A higher dimensional generalisation of Theorem 5.1 is given in [29], [31]. The complex analytic analogue to Theorem 5.1 holds in the same way [30].

6 Technical digression

Two approaches are useful when considering differentiable normal forms of singularities of tangent developables. One is that of Legendre singularity theory [3] as in [1], [54], [56], and the other is by analysing directly the parametrisation as in [9], [52], [44], [45]. In fact we need both viewpoints to complete the differentiable classification in [29], [31].

Recall the parametrisation $\mathrm{dev}(\gamma) = (x_1, x_2, x_3)$ in §4 satisfies

$$x_1 = x, \quad \partial x_3/\partial t = (r+s)(r+s+m)t^r \partial x_2/\partial t.$$

Then we see the exterior derivative of the third component satisfies

$$dx_3 = \{\partial x_3/\partial x + at^r \partial x_2/\partial x\}dx_1 - at^r dx_2,$$

where $a = -(r+s)(r+s+m)$. In this sense the tangent developables are governed by some algebraic structure.

When we trivialise a family of tangent developables of curves of constant type using the infinitesimal method [43], we try to solve the corresponding variational equations in terms of vector fields. Then we encounter a certain module describing the ramification of a finite mapping [26], [46]. This object also appears in the classification of isotropic mappings [27], and in the characterisation of certain stabilities [32] in symplectic geometry. We can also apply it to classify differential equations [22].

Let $g = (g_1, \dots, g_p) : (\mathbf{R}^n, 0) \to (\mathbf{R}^p, 0)$ be a finite smooth map-germ $(n \le p)$. We define **the ramification module** R_g of g to be the set of smooth function-germs $e : (\mathbf{R}^n, 0) \to \mathbf{R}$ such that the exterior differential de of e is a functional linear combination of $dg_1, \dots, dg_p$: $de = a_1 dg_1 + \cdots + a_n dg_n$, for some smooth function-germs $a_i : (\mathbf{R}^n, 0) \to \mathbf{R}, 1 \le i \le n$. This is a variant of the ring of functions constant along components of fibres of a map-germ $(\mathbf{R}^n, 0) \to (\mathbf{R}^p, 0)$ when $n > p$, [42]. We see R_g is a module over the algebra of all smooth function-germs $(\mathbf{R}^p, 0) \to \mathbf{R}$ via the pull-back by g.

For the parametrisation of a tangent developable, setting $g = (x_1, x_2)$, we see that x_3 belongs to the ramification module R_g.

We study the structure of the ramification module in [26], [28], [27], [30], [31]. The key point is that R_g is a differentiable algebra in the sense of Malgrange [41] so that we can apply Malgrange's preparation theorem for differentiable algebras to solve the variational equation. We expect that this object will have other applications in the analysis of various geometrical objects.

7 Topological classification

The result on the differentiable classification turns our concern to the topological classification of tangent developables.

Two smooth map-germs $f, f' : (\mathbf{R}^2, 0) \to (\mathbf{R}^3, 0)$ are called **topologically (right-left) equivalent** if there exist homeomorphism-germs $\sigma : (\mathbf{R}^2, 0) \to (\mathbf{R}^2, 0)$ and $\tau : (\mathbf{R}^3, 0) \to (\mathbf{R}^3, 0)$ such that $\tau \circ f = f' \circ \sigma$.

Proceeding from the pioneering work of Mond [45], we have in [33]:

Theorem 7.1 *Let #top denote the number of topological equivalence classes:*

$$\#\mathrm{top} = \#\left(\{\mathrm{dev}(\gamma) \mid \mathrm{type}(\gamma) = (\mathrm{m}, \mathrm{m}+\mathrm{s}, \mathrm{m}+\mathrm{s}+\mathrm{r})\}/\underset{\mathrm{top}}{\sim}\right).$$

If s, r are not both even, then #top = 1.
If m is odd, s, r are both even, then #top = 1, when $m = 1$, and #top = 2, when $m \geq 3$.
If m, s, r are all even, then #top is uncountably infinite.

Moreover the topological types are determined simply by the parities of types except for the case m, s, r are all even, or s, r are both even, m is odd and ≥ 3. Furthermore the homeomorphisms can be taken to preserve the original curves and the tangent lines to the base points.

In particular we see that two non-singular space curves ($m = 1$) of the same type have homeomorphic tangent developables.

Note there are two kinds of determinacies. Topological determinacy with respect to higher order perturbations of the original curve, and topological determinacy by the parity of the type, namely, the orders of leading terms of the curve. Note also the similarity between Theorem 7.1 and the following well-known fact: Non-flat smooth function-germs $(\mathbf{R}, 0) \to (\mathbf{R}, 0)$ are topologically equivalent to the function-germ $y = x$ or $y = x^2$ depending on the parity of the order. Furthermore the number of topological types of flat functions is uncountably infinite. Here $f : (\mathbf{R}, 0) \to (\mathbf{R}, 0)$ is *flat* if its infinite jet $j^\infty f(0) = 0$.

For simplicity we divide all types into eight classes by the parities of (m, s, r) and exhibit them with the names of their topological equivalence classes below. For pictures of the tangent developable in each case, see [33].

	$s \setminus r$	odd	even
m: odd	odd	(i) cuspidal edge	(ii) folded umbrella
m: odd	even	(iii) Mond surface	(iv) (pseudo) Scherbak surface
m: even	odd	(v) swallowtail	(vi) folded pleat
m: even	even	(vii) butterfly	(viii) not determined

8 Section and Projection

To illustrate the topological types of tangent developables, it is useful the use the method of "section and projection", which originated with Cayley [10], and later clarified by Piene [49] and Scherbak [54]. This is also used to establish the indeterminacy result in Theorem 7.1 [33].

Let $f = \text{dev}(\gamma) : (\mathbf{R}^2, 0) \to (\mathbf{R}^3, 0)$ be the parametrisation of the tangent developable of γ in its normal form as given in §4. Since f is a finite map-germ, the germ at 0 of the image $f(\mathbf{R}^2)$ is well-defined, which is denoted by $\text{DEV}(\gamma)$.

Set $\Pi_c = \{x_1 = c\}$, for sufficiently small $c \neq 0$. Note that the plane Π_c intersects transversely to γ. Take a section $\text{DEV}_c(\gamma) = \Pi_c \cap \text{DEV}(\gamma)$. Consider the line ℓ_t obtained from the osculating planes O_t to γ at t by cutting by the plane Π_c. Then the section $\text{DEV}_c(\gamma)$ is the envelope of the family of lines ℓ_t. We see that the dual ℓ_t^* of ℓ_t is the line connecting two points O_t^* and Π_c^* in the dual projective space and, the dual of $\text{DEV}_c(\gamma)$ is the set $\{\ell_t^* \mid t \in (\mathbf{R}, 0)\}$ in the projective plane consists of lines passing through the point Π_c^*. Thus we see the section is the planar projective dual [8], [61] to the projection of γ^* from the corresponding point in $\mathbf{RP}^{3*}$.

If (y_1, y_2, y_3) represents γ^*, then the projection is given by the plane curve $(y_1, cy_2 + y_3)$, where $\text{ord}(y_1, y_2, y_3) = (r, r + s, r + s + m)$.

Example 8.1 First let $\text{type}(\gamma) = (1,2,3)$. Then $\text{type}(\gamma^*) = (1,2,3)$. Thus we have a projection $(t + \cdots, c(t^2 + \cdots) + t^3 + \cdots)$. This plane curve has exactly one inflection point. Therefore the dual has one cuspidal point, which is a section of $\text{DEV}(\gamma)$. So the tangent developable has a cuspidal edge.

Next let $\text{type}(\gamma) = (1,2,4)$, $\text{type}(\gamma^*) = (2,3,4)$. Then, depending on the sign of c, the projection is a beak-like cusp with no double points and no double tangents, or, with one double point and one double tangent, which is self-dual in each case. Then we have a folded umbrella as the tangent developable.

Consider the more complicated case where $\text{type}(\gamma) = (2,4,5)$, $\text{type}(\gamma^*) = (1,3,5)$. Then for a sign of c, the projection has three inflection points and three double tangents, therefore its dual has three cuspidal edges and three self-intersections. For another sign of c, the projection has only one inflection point and no double tangents. Thus we have a butterfly [36] as the tangent developable.

Similarly, we get topological pictures for several other cases. However, to obtain a rigorous classification, we need to directly analyse the parametrisations as in the next section.

9 Topological triviality

Consider the typical parametrisation $f_0 : (\mathbf{R}^2, 0) \to (\mathbf{R}^3, 0)$ defined by

$$f_0(t,x) = (x,\ t^{s+m} + xt^s,\ (r+s)(s+m)t^{r+s+m} + s(r+s+m)xt^{r+s}),$$

setting $\varphi(t) = 0, \psi(t) = 0$ for f.

To describe the double point loci, set $f_0(t_1, x) = f_0(t_2, x)$. Eliminating x we have

$$d_0(t_1,t_2) = \begin{vmatrix} t_2^s - t_1^s, & t_2^{s+m} - t_1^{s+m} \\ s(r+s+m)(t_2^{r+s} - t_1^{r+s}), & (r+s)(s+m)(t_2^{r+s+m} - t_1^{r+s+m}) \end{vmatrix}$$

which must vanish for such t_1 and t_2. Then we set

$$\begin{aligned} g(z) &= (1/rm)d_0(1,z) \\ &= z^{r+2s+m} - \frac{(r+s)(s+m)}{rm} z^{r+s+m} + \frac{s(r+s+m)}{rm} z^{r+s} \\ &\quad + \frac{s(r+s+m)}{rm} z^{s+m} - \frac{(r+s)(s+m)}{rm} z^s + 1. \end{aligned}$$

We call $g(z)$ **the characteristic polynomial** of $\mathbf{A} = (m, m+s, m+s+r)$. For this very explicit polynomial, we easily see the following:

(0) $z = \xi$ is a real root of the equation $g(z) = 0$ if and only if $t_2 = \xi t_1$ is a real branch of $\{d_0(t_1,t_2) = 0\}$.

(1) $z = 1$ is a multiple root of $g(z) = 0$ of multiplicity 2.

(2) If ξ is a root, then also $1/\xi$ is a root of $g(z) = 0$.

(3) If there exist a triple point of f_0 then there exist non-trivial ($\neq 1$) roots ξ, η of $g(z) = 0$ such that the product $\xi\eta$ is also a non-trivial root of $g(z) = 0$.

Recall the theorem of Descartes: *The number of real positive roots of a real algebraic equation $g = 0$, counted with multiplicities, is at most the number of sign-changes of terms of g.*

Our $g(z)$ has coefficients whose signs vary as $+-++-+$. Therefore we see $g(z) = 0$ has at most 2 solutions except for the multiple root $z = 1$. Since $g''(1) > 0$, $\lim_{z\to\infty} g(z) = \infty$, $g(0) = 1$, we see $g(z) = 0$ has no real positive root other than $z = 1$.

For negative roots of $g(z) = 0$, we apply the theorem of Descartes to $g(-z)$, and check all seven classes. For instance, for the case (vii), the number of sign-changes is equal to 3 if $r > m$, and 5 if $r < m$. Further $g(-1) = 0$, $(dg/dz)(-1) = -2(s+m)s/r < 0$ and $\lim_{z\to-\infty} g(z) = -\infty$. Therefore we see there exist just 3 negative roots. The arguments for the remaining cases are similar.

So we obtain the following: The real roots other than 1 of the equation $g(z) = 0$ are all negative and simple in all seven cases from (i) to (vii). In each case, the number of them is equal to (i): 0, (ii): 1, (iii): 2, (iv): 3, (v): 1, (vi): 0, (vii): 3.

In particular, we see that -1 is a root of $g(z) = 0$ just in the cases (ii), (iv), (v) and (vii). However if s and r are both even and m is odd, then there do not exist points (x, t) and $(x, -t)$ near the origin having the same image under f_0. Nevertheless, it is possible that one extra component of the double point loci emerges after a perturbation of f_0 from the root -1 of the characteristic polynomial in the case (iv). If $m = 1$, then it is impossible, but if $m \geq 3$, then it is in fact the case. Thus determinacy fails for (iv), $m \geq 3$. However the position of the extra component of the double point loci is restricted, and we conclude that exactly two topological classes appear: **Scherbak surface** and **pseudo Scherbak surface**.

In the case (viii) : m, s and r are all even, the polynomial $g(z)$ has -1 as a multiple root. This implies essentially the indeterminacy of the topological type of the parametrisation $\text{dev}(\gamma)$.

Now we apply the Kuiper-Kuo theorem [39], [40]to $d(t_1, t_2)$ divided by $(t_1 - t_2)^2$, provided it has isolated singularities, that is, when the type **A** does not belong to the class (iv), $m \geq 3$ or (viii). Then we see no bifurcation occurs by a type-preserving deformation of γ. So we construct a stratification of the corresponding unfolding of $\text{dev}(\gamma)$. In this sense, the study of isolated singularities is applied to that of non-isolated singularities.

Then we further show, using standard techniques that the stratification is Thom regular and we apply Thom's isotopy lemma [18] to obtain the topological triviality of the family of tangent developables of space curves of constant type. In the process of the proof, we discover some general local properties for the tangent developables. Except in the case (viii), the tangent developable has, near the base point, no double points on the curve itself and on the tangent line at the base point, no points of self-tangency, no triple points, and a topological cone structure.

10 Open problems and questions

We summarise the situation we are considering using the correspondence

$$\{\text{Space Curves}\} \longrightarrow \{\text{Singular Space Surfaces}\},$$

obtained by taking tangent developables. We are then concerned with, not just singular surfaces themselves, but the interrelation between these two objects, for instance, the determinacy of the "shape" of tangent developables by leading terms of curves.

We conclude the present survey by giving several promising open problems and questions.

(1) Give the local differentiable classification, or holomorphic classification over the complex numbers, of tangent developables of curves of type other than those in Theorem 5.1, for instance, of type $(1, 3, 6)$.

Extending the usual notion of determinacy [60], we recognise that the notion of determinacy remains significant in our classification problem. A space curve-germ $\gamma : (\mathbf{R}, 0) \to (\mathbf{R}^3, 0)$ is called *finitely differentiably* (resp. *topologically*) *determined relatively to the developable* if there exists a positive integer k such that any curve-germ $\gamma' : (\mathbf{R}, 0) \to (\mathbf{R}^3, 0)$ with $j^k\gamma'(0) = j^k\gamma(0)$ has a tangent developable which is differentiably (resp. topologically) equivalent to that of γ.

We see γ is finitely differentiably determined relatively to the developable if the type of γ at 0 is one of types in Theorem 5.1. Also we see, from Theorem 7.1, that γ is finitely topologically determined relatively to the developable if, for the type $(m, m+s, m+s+r)$ of γ at 0, s and r are not both even, or, s and r are both even and $m = 1$. Then we ask the following:

(2) For any γ of type $(m, m+s, m+s+r)$, and for any positive integer k, does there exist a finitely differentiably (or at least topologically) determined space curve-germ γ' relatively to the developable with $j^k\gamma'(0) = j^k\gamma(0)$?

The general methods used to show the topological triviality of a family of complex varieties are (a) by finding an algebraic torsion to construct controlled vector fields, and (b) by showing the constancy of some invariants, which implies regularity of stratifications. See for instance, [12], [13], [15], [16]. Then the next question is:

(3) Can the general framework above be applied to the local topological classification of tangent developables in the complex analytic case?

Recently there have appeared several global results on the topology of tangent developables of generic space curves. For a generic space curve, the tangent developable has, as topological singular points, (folded) umbrellas and generic triple points, as well as transversal double points, and therefore it is homeomorphic to the image of a stable mapping locally at each point [4]. See also [5], [53], [6], [34], [57]. Note that Piene [50] has explored the global enumerative geometry of tangent developables of algebraic space curves over the complex (or indeed a general algebraically closed) field.

There are also several results on the topology of dual surfaces of generic space curves, that is, the tangent developables of the dual curves of generic space curves [14], [48], [7].

From the local classification given in the previous sections, it is natural to consider the global topology of tangent developables of more degenerate curves. In other words,

(4) In the global context, does the tangent developable of a generic (in the sense of Tougeron) curve have a (locally topological) cone structure?

References

[1] V.I. Arnold, *Lagrangian manifolds with singularities, asymptotic rays, and the open swallowtail*, Funct. Anal. Appl., **15-4** (1981), 235–246.

[2] V.I. Arnold, *Singularities of Caustics and Wave Fronts*, Kluwer Academic Publishers, Dordrecht, 1990.

[3] V.I. Arnol'd, S.M. Gusein-Zade, A.N. Varchenko *Singularities of Differentiable Maps I,* Birkhäuser, 1985

[4] J.J. nuño Ballesteros, *On the number of the triple points of the tangent developable,* Geom. Dedicata **47-3** (1993), 241–254.

[5] J.J. nuño Ballesteros, M.C.R. Fuster, *Global bitangency properties of generic closed space curves,* Math. Proc. Camb. Phil. Soc., **112** (1992), 519–526.

[6] J.J. nuño Ballesteros, O. Saeki, *Singular surfaces in 3-manifolds, the tangent developable of a space curve and the dual of an immersed surface in 3-space,* in Real and Complex Singularities, ed. W.L. Marar, Pitman Research Notes in Math., **333**, Pitman, 1995, pp. 49–46.

[7] T. Banchoff, T. Gaffney, C. McCrory, *Counting tritangent planes of space curves,* Topology, **24–1** (1985), 15–23.

[8] J.W. Bruce, *On contact of hypersurfaces,* Bull. London Math. Soc. **13** (1981), 51–54.

[9] J.W. Bruce, P.J. Giblin, *Curves and Singularities,* 2nd. ed., Cambridge Univ. Press, Cambridge, 1992.

[10] A. Cayley, *Mémoire sur les coubes à double courbure et les surfaces développables,* Journal de Mathématique Pure et Appliquées (Liouville), **10** (1845), 245–250. = The Collected Mathematical Papers vol.I, pp. 207–211.

[11] J.P. Cleave, *The form of the tangent developable at points of zero torsion on space curves,* Math. Proc. Camb. Phil. Soc. **88** (1980), 403–407.

[12] J. Damon, *Finite determinacy and topological triviality I,* Invent. math., **62** (1980), 299–324.

[13] J. Damon, *Topological triviality and versality for subgroup of $\mathcal{A}$ and $\mathcal{K}$,* Memoir of Amer. Math. Soc., **75–389** (1988).

[14] M. Freedman, *Planes triply tangent to curves with nonvanishing torsion,* Topology, **19** (1980), 1–8.

[15] T. Gaffney, *Integral closure of modules and Whitney equisingularity,* Invent. math., **107** (1992), 301–322.

[16] T. Gaffney, *Polar multiplicities and equisingularity of map germs,* Topology, **32–1** (1993), 185–223.

[17] T. Gaffney, A. du Plessis, *More on the determinacy of smooth map-germs,* Invent. Math. **66** (1982), 137–163.

[18] C.G. Gibson, K. Wirthmüller, A.A. du Plessis, E.J.N. Looijenga, *Topological Stability of Smooth Mappings,* Lecture Notes in Math., **552**, Springer-Verlag, 1976.

[19] A.B. Givental, *Lagrangian imbeddings of surfaces and unfolded Whitney umbrella,* Funkt. Anal. Prilozhen, **20-3** (1986), 35–41.

[20] P. Griffiths, J. Harris, *Algebraic geometry and local differential geometry,* Ann. Sci. Éc. Norm. Sup., **12** (1979), 355–432.

[21] P. Hartman, L. Nirenberg, *On spherical image whose Jacobians do not change sign,* Amer. J. Math., **81** (1959), 901–920.

[22] A. Hayakawa, G. Ishikawa, S. Izumiya, K. Yamaguchi, *Classification of generic integral diagrams and first order ordinary differential equations,* International J. Math., **5–4** (1994), 447–489.

[23] F. Hinterleitner, *Examples of separating coordinates for the Kline-Gordon equation in $2+1$-dimensional flat space-time,* J. Math. Phys, **37–6** (1996), 3032–3040.

[24] J. Hoschek, H. Pottmann, *Interpolation and approximation with developable B-spline surfaces,* in Mathematical Mathods for curves and surfaces, ed. by M. Dæhlen, T. Lyche, L.L. Schumacker, Vanderbilt Univ. Press, 1995.

[25] T. Inoue, L.L. Wegge, *On the geometry of the production possibility frontier,* Internat. Economic Review **27–3** (1986), 727–737.

[26] G. Ishikawa, *Families of functions dominated by distributions of C-classes of map-germs,* Ann. Inst. Fourier **33–2** (1983), 199–217.

[27] G. Ishikawa *The local model of an isotropic map-germ arising from one dimensional symplectic reduction,* Math. Proc. Camb. Phil. Soc., **111–1** (1992), 103–112.

[28] G. Ishikawa, *Parametrization of a singular Lagrangian variety,* Trans. Amer. Math. Soc., **331–2** (1992), 787–798.

[29] G. Ishikawa, *Determinacy of envelope of the osculating hyperplanes to a curve,* Bull. London Math. Soc., **25** (1993), 603–610.

[30] G. Ishikawa, *Parametrized Legendre and Lagrange varieties,* Kodai Math. J., **17–3** (1994), 442–451.

[31] G. Ishikawa, *Developable of a curve and its determinacy relatively to the osculation-type,* Quarterly J. Math., **46** (1995), 437–451.

[32] G. Ishikawa, *Symplectic and Lagrange stabilities of open Whitney umbrellas,* Invent. math., **126-2** (1996), 215–234.

[33] G. Ishikawa, *Topological classification of the tangent developables of space curves,* Hokkaido Univ. Preprint Series, **341** (1996).

[34] G. Ishikawa, T. Morimoto, *Solution surfaces of Monge-Ampère equations,* Hokkaido Univ. Preprint Series, **376** (1997).

[35] S. Izumiya, W.L. Marar, *On topologically stable singular surfaces in a 3-manifold,* J. of Geom., **52** (1995), 108–119.

[36] S. Janeczko, M. Roberts, *Classifications of symmetric caustics I: symplectic equivalence,* Singularity theory and its application II, ed. by M. Roberts, I. Stewart, Lecture Notes in Math. **1463**, Springer-Verlag (1991), pp. 193–219.

[37] M.É. Kazaryan, *Singularities of the boundary of fundamental systems, flat points of projective curves, and Schubert cells,* Itogi Nauki Tekh., Ser. Sorrem. Probl. Mat. (Contemporary Problems of Mathematics) **33**, VITINI, 1988, pp. 215–232.

[38] M.É. Kazaryan, *Flattening of projective curves, singularities of Schubert stratifications of Grassmannians and flag varieties, and bifurcations of Weierstrass points of algebraic curves,* Uspekhi Mat. Nauk **46–5** (1991), 79–119 = Russian Math. Surveys **46–5** (1991), 91–136.

[39] N.H. Kuiper, *C^1-equivalence of functions near isolated critical points,* Symposium in Infinite Dimensional Topology, (Baton Rouge, 1967).

[40] T.C. Kuo, *On C^0 sufficiency of functions near isolated critical points,* Topology, **8** (1969), 167–171.

[41] B. Malgrange, *Ideals of Differentiable Functions,* Oxford Univ. Press, 1966.

[42] B. Malgrange, *Frobenius avec singularités, 2. Le cas général,* Invent. math., **39** (1977), 67–89.

[43] J.N. Mather, *Stability of C^∞ mappings III: Finitely determined map-germs,* Publ. Math. I.H.E.S., **35** (1968), 127–156.

[44] D. Mond, *On the tangent developable of a space curve,* Math. Proc. Camb. Phil. Soc. **91** (1982), 351–355.

[45] D. Mond, *Singularities of the tangent developable surface of a space curve,* Quart. J. Math. Oxford, **40** (1989), 79–91.

[46] D. Mond, *Deformations which preserve the non-immersive locus of a map-germ,* Math. Scand., **66** (1990), 21–32.

[47] T. Morimoto, *La géométrie des équations de Monge-Ampère,* C. R. Acad. Sc. Paris, **289** (1979), 25–28.

[48] T. Ozawa, *The number of triple tangencies of smooth space curves,* Topology, **24–1** (1985), 1–13.

[49] R. Piene, *Numerical characters of a curve in projective n-space,* Real and Complex Singularities, ed. by P. Holm, Sijthoff & Noordhoff International Publishers, (1977), pp. 475–495.

[50] R. Piene, *Cuspidal projections of space curves,* Math. Ann., **256** (1981), 95–119.

[51] W.F. Pohl, *The self-linking number of a closed space curve,* J. Math. Mech., **17–10** (1968), 975–987.

[52] I.R. Porteous, *Geometric Differentiation, for the Intelligence of Curves and Surfaces,* Cambridge Univ. Press, Cambridge, 1994.

[53] O. Saeki, *Separation by a codimension-*1 *map with a normal crossing point,* Geom. Dedicata, **57–3** (1995), 235–247.

[54] O.P. Scherbak, *Projectively dual space curves and Legendre singularities,* Trudy Tbiliss. Univ. **232–233** (1982), 280–336. = Sel. Math. Sov. **5-4** (1986), 391–421.

[55] S. Sternberg, *Lectures on Differential Geometry,* 2nd ed., Printice-Hall, N.J., 1983.

[56] O.P. Scherbak, *Wavefront and reflection groups,* Russian Math. Surveys **43–3** (1988), 149–194.

[57] B. Shapiro, *Discriminants of convex curves are homeomorphic,* Preprint, 1996.

[58] J.J. Stoker, *Differential Geometry,* Pure and Applied Math. **20**, Wiley-Interscience, New York 1969.

[59] G. Thorbergsson, *Tight analytic surfaces,* Topology, **30–3** (1991), 423–428.

[60] C.T.C. Wall, *Finite determinacy of smooth map-germs,* Bull. London Math. Soc., **13** (1981), 481–539.

[61] C.T.C. Wall, *Duality of singular plane curves,* J. London Math. Soc., **50** (1994), 265–275.

Go-o ISHIKAWA,
Department of Mathematics,
Hokkaido University,
Sapporo 060, JAPAN

e-mail: ishikawa@math.sci.hokudai.ac.jp

Singularities of Solutions for First Order Partial Differential Equations

Shyuichi Izumiya

1 Introduction

This is a survey article on recent results about the singularities for the solution of the Cauchy problem of first order partial differential equations in the following forms:

$$\begin{cases} \dfrac{\partial y}{\partial t} + H(t, x_1, \ldots, x_n, \dfrac{\partial y}{\partial x_1}, \ldots, \dfrac{\partial y}{\partial x_n}) = 0 \quad (t > 0) \\ y(0, x_1, \ldots, x_n) = \phi(x_1, \ldots, x_n), \end{cases} \tag{H}$$

and

$$\begin{cases} \dfrac{\partial y}{\partial t} + \displaystyle\sum_{i=1}^{n} \dfrac{\partial f_i(y)}{\partial x_i} = 0 \quad (t > 0) \\ y(0, x_1, \ldots, x_n) = \phi(x_1, \ldots, x_n), \end{cases} \tag{C}$$

where H, f_i's and ϕ are C^∞-functions.

The former equation is called *a Hamilton-Jacobi equation* and the latter *a single conservation law.* These equations play an important role in various fields (e.g., calculus of variations [41], optimal control theory [17], differential games [16] for (H), and gas dynamics [46] and oil reservoir problems [20] for (C)).

For sufficiently small t the solution y is classically determined using the characteristic method. Although y is initially smooth there is, in general, a critical time beyond which characteristics cross. After the characteristics cross, the solution y is multi-valued, that is singularities appear. In §2 (respectively, §4) we give a survey of the classification of those singularities for (H) (respectively, (C)).

The notion of *viscosity solutions* [11] (respectively, *entropy solutions* [36]) has provided the right weak setting for the study of (H) (respectively, (C)). Existence and uniqueness of the solution of (H) in the viscosity sense (respectively, (C) in the entropy sense) have been established in [11] (respectively, [36]). The single-valued viscosity (respectively, entropy) solution coincides with the smooth geometric solution until the first critical time. After the characteristics cross, the viscosity (respectively, entropy) solution develops *shock waves* i.e., curves across which the gradient of the viscosity solution (respectively, the value of the entropy solution) is discontinuous.

The method of constructing the weak solution by selecting the proper single-valued branch was introduced by Tsuji [43,44] for (H). For a convex Hamiltonian function, Nakane [39] has constructed the semi-concave solution (i.e, the viscosity solution) past the first critical time, and Bogaevskii [6] has classified the generic perestroikas of shocks of the viscosity solutions for $n = 1, 2, 3$. The viscosity solution of (H) for non-convex Hamiltonian in a neighborhood of the first critical time has been constructed by Kossioris [34,35] by selecting a continuous single-valued branch of the graph of the geometric solution, in which the shock curves of the viscosity solution correspond to the intersection of the branches of the graph of the multi-valued geometric solution. In order to study the evolution of the shock curves we follow the evolution of the intersections of the branches defining the shock. After that we solve the local Riemann problem for each stage. In the case when $n = 1$, we have developed the above arguments in [31]; the results are described in §3. However for $n \geq 2$, we have no idea how to solve the Riemann problem.

On the other hand, the study of shocks of entropy solutions of (C) has a long history. For example, the case $n = 1$ has been studied by Ballow [3], Guckenheimer [21], Chen [10], Jennings [32] and Dafermos [14] etc. However, there are few results in the case when $n \geq 2$. Guckenheimer [22], Wagner [46] and Nakane [38] have studied entropy solutions for (C) in the higher dimensional case.

Acknowledgment. The author wishes to thank Terry and Sandra Wall for their hospitality and encouragement during the author's stay at their home in the summer of 1996. The author has been inspired very much by Terry's survey articles on singularity theory, he will be very happy if this survey will assist the next generation in the same way as Terry's surveys have helped his own.

All maps considered here are class C^∞ unless stated otherwise.

2 Geometric Solutions for Hamilton-Jacobi Equations

In this section we give a survey of the geometric framework for the study of (H) which was described in [25,26,27,28,29]. We consider the 1-jet bundle $J^1(\mathbb{R} \times \mathbb{R}^n, \mathbb{R})$ and the canonical 1-form Θ on that space. Let $(t, x_1, \ldots, x_n)$ be the canonical coordinate system on $\mathbb{R} \times \mathbb{R}^n$ and $(t, x_1, \ldots, x_n, y, s, p_1, \ldots, p_n)$ the corresponding coordinate system on $J^1(\mathbb{R} \times \mathbb{R}^n, \mathbb{R})$. Then, the canonical 1-form is given by $\Theta = dy - \sum_{i=1}^n p_i \cdot dx_i - s \cdot dt = \theta - s \cdot dt$. We define the natural projection $\Pi : J^1(\mathbb{R} \times \mathbb{R}^n, \mathbb{R}) \to (\mathbb{R} \times \mathbb{R}^n) \times \mathbb{R}$ by

$\Pi(t,x,y,s,p) = (t,x,y)$. *A Hamilton-Jacobi equation* is a hypersurface

$$E(H) = \{(t,x,y,s,p) \in J^1(\mathbb{R} \times \mathbb{R}^n, \mathbb{R}) \mid s + H(t,x,p) = 0\}$$

in $J^1(\mathbb{R} \times \mathbb{R}^n, \mathbb{R})$. *A geometric (multi-valued) solution* of $E(H)$ is a Legendrian submanifold L in $J^1(\mathbb{R} \times \mathbb{R}^n, \mathbb{R})$ lying in $E(H)$. In this case the wave front set $W(i) = \Pi(L)$ is "the graph" of the geometric solution which is generally a hypersurface with singularities.

To study (H) we need the following framework: For any $c \in \mathbb{R}$ near 0, we define

$$E(H)_c = \{(c,x,y,-H(c,x,p),p) \mid (x,y,p) \in J^1(\mathbb{R}^n, \mathbb{R})\}.$$

Then, $E(H)_c$ is a $(2n+1)$-dimensional submanifold of $J^1(\mathbb{R}\times\mathbb{R}^n, \mathbb{R})$ and $\Theta|E(H)_c = dy - \sum_{i=1}^n p_i dx_i$ gives a contact structure on $E(H)_c$. We define a mapping $\iota_c : J^1(\mathbb{R}^n, \mathbb{R}) \to E(H)_c$ by $\iota_c(x,y,p) = (c,x,y,-H(c,x,p),p)$. The mapping ι_c is a contact diffeomorphism.

We say that *a geometric Cauchy problem (with initial condition L') associated with the time parameter(GCPT) is given for an equation* $E(H)$ if there is given an n-dimensional submanifold $i : L' \subset E(H)$ with $i^*\Theta = 0$ and $i(L') \subset E(H)_c$ for some $c \in (\mathbb{R}, 0)$. Since $X_H \notin TE(H)_c$, we have $X_H \notin TL'$, where X_H is the characteristic vector field given by

$$X_H = \frac{\partial}{\partial t} + \sum_{i=1}^n \frac{\partial H}{\partial p_i}\frac{\partial}{\partial x_i} + (\sum_{i=1}^n p_i \frac{\partial H}{\partial p_i} - H)\frac{\partial}{\partial y} - \frac{\partial H}{\partial t}\frac{\partial}{\partial s} - \sum_{i=1}^n \frac{\partial H}{\partial x_i}\frac{\partial}{\partial p_i}.$$

By using the classical characteristic method, we can show that there exists a unique geometric solution around L'. It is clear that the classical Cauchy problem (H) is a GCPT. The geometric part of our investigation is to classify the generic perestroikas of wave fronts (graphs) of geometric solutions of (H) along the time parameter.

In order to study the singularities of the geometric solution we introduce the notion of one-parameter Legendrian unfoldings [25,26]. Let R be an $(n+1)$-dimensional smooth manifold, $\mu : (R, u_0) \to (\mathbb{R}, t_0)$ be a submersion germ and $\ell : (R, u_0) \to J^1(\mathbb{R}^n, \mathbb{R})$ be a smooth map germ. We say that the pair (μ, ℓ) is *a Legendrian family* if $\ell_t = \ell|\mu^{-1}(t)$ is a Legendrian immersion germ for any $t \in (\mathbb{R}, t_0)$. By definition, for any (μ, ℓ) Legendrian family, there exists a unique element $h \in C^\infty_{u_0}(R)$ such that $\ell^*\theta = h \cdot d\mu$, where $C^\infty_{u_0}(R)$ is the ring of smooth function germs at u_0. So we define a map germ $\mathcal{L} : (R, u_0) \to J^1(\mathbb{R}\times\mathbb{R}^n, \mathbb{R})$ by $\mathcal{L}(u) = (\mu(u), x\circ\ell(u), y\circ\ell(u), h(u), p\circ\ell(u))$. We can easily show that $\mathcal{L}$ is a Legendrian immersion germ. If we fix 1-forms Θ and θ, the Legendrian immersion germ $\mathcal{L}$ is uniquely determined by the Legendrian family (μ, ℓ). We call $\mathcal{L}$ *a Legendrian unfolding associated with the Legendrian family* (μ, ℓ). It is clear that the solution germ at each point of the GCPT for (H) is a Legendrian unfolding.

In order to study the generic classification of perestroikas of singularities (i.e., how a singularity is generated, how one type can change into another and how different types of singularities interact), we now study how various branches of the multi-valued graph $W_t = (\{t\} \times \mathbb{R}^n \times \mathbb{R}) \cap W(i)$ intersecting at a point bifurcate in time for an arbitrary Hamiltonian $H(t,x,p)$. For this purpose, we classify the perestroikas of singularities of multi-Legendrian unfoldings which are expressed in terms of multi-germs.

Let $\mathcal{L}_i : (R, u_0) \to (J^1(\mathbb{R} \times \mathbb{R}^n, \mathbb{R}), z_i)$ $(i = 1, \dots, r)$ be Legendrian unfoldings with $\Pi(z_i) = 0$ where $z_1, \dots, z_r$ are distinct. We call $(\mathcal{L}_1, \dots, \mathcal{L}_r)$ *a multi-Legendrian unfolding.* Let $(\mathcal{L}_1, \dots, \mathcal{L}_r)$ and $(\mathcal{L}'_1, \dots, \mathcal{L}'_r)$ be multi-Legendrian unfoldings. We say that these are $P_{(r)}$*-Legendrian equivalent* if there exist contact diffeomorphism germs

$$K_i : (J^1(\mathbb{R} \times \mathbb{R}^n, \mathbb{R}), z_i) \to (J^1(\mathbb{R} \times \mathbb{R}^n, \mathbb{R}), z'_i) \quad (i = 1, \dots, r)$$

of the form

$$K_i(t,x,y,s,p) = (\phi_1(t), \phi_2(t,x,y), \phi_3(t,x,y), \phi_4^i(t,x,y,s,p), \phi_5^i(t,x,y,s,p))$$

and a diffeomorphism germ $\Psi : (R, u_0) \to (R, u'_0)$ such that $K_i \circ \mathcal{L}_i = \mathcal{L}'_i \circ \Psi$ for any $i = 1, \dots, r$. It is clear that if two multi-Legendrian unfoldings are $P_{(r)}$-Legendrian equivalent, then there exists a diffeomorphism germ

$$\Phi : (\mathbb{R} \times (\mathbb{R}^n \times \mathbb{R}), 0) \to (\mathbb{R} \times (\mathbb{R}^n \times \mathbb{R}), 0)$$

of the form $\Phi(t,x,y) = (\phi_1(t), \phi_2(t,x,y), \phi_3(t,x,y))$ such that $\Phi(\cup_{i=1}^r W(\mathcal{L}_i)) = \cup_{i=1}^r W(\mathcal{L}_i)$. Thus the above equivalence describes how bifurcations of wavefronts (i.e. graphs of solutions) interact. We can define the notion of stability with respect to the $P_{(r)}$-Legendrian equivalence in the same way as for the ordinary Legendrian stability (see [2,47]). Thanks to Arnol'd-Zakalyukin theory ([2,47]), we can construct multi-generating families of multi-Legendrian unfoldings and give a classification of $P_{(r)}$-Legendrian stable Legendrian unfoldings by using the classification of multi-families of function germs in Zakalyukin [47]. However, we only present the classification list for $n = 1$. For the case $n = 2, 3$, see [2,27,28,47].

Theorem 2.1 [28,47] *Suppose that $n = 1$. Then a generic multi-Legendrian unfolding is $P_{(r)}$-Legendrian equivalent to one of the multi-Legendrian unfoldings in the following list :*

$$\begin{aligned}
{}^0A_1 &: (t, u, 0, 0, 0) \; ; \\
{}^0A_2 &: (t, 3u^2, 2u^3, 0, u) \; ; \\
{}^1A_3 &: (t, 4u^3 + 2ut, 3u^4 + u^2 t, -u^2, u). \\
{}^0({}^0A_1 {}^0A_1) &: ((t, u, -u, 0, -1), (t, u, u, 0, 1)) \; ; \\
{}^1({}^0A_1 {}^0A_1) &: ((t, u, t \pm u^2, 1, \pm 2u), (t, u, 0, 0, 0)) \; ; \\
{}^1A_2 {}^0A_1 &: ((t, 3u^2 - t, 2u^3, u, u), (t, u, -u, 0, -1)). \\
{}^0A_1 {}^0A_1 {}^0A_1 &: ((t, u, t - u, 1, -1), (t, u, 0, 0, 0), (t, u, u, 0, 1)).
\end{aligned}$$

When we consider the geometric solution, we can drop the germ ${}^1({}^0A_1\,{}^0A_1)$ from the above list because the geometric solution is a one-to-one immersion into the unfolded 1-jet space. These multi-germs have been called *standard perestroikas of fronts* by Bogaevskii [7]. The standard perestroikas of fronts in the case $n = 1$ are depicted in Figure 1.

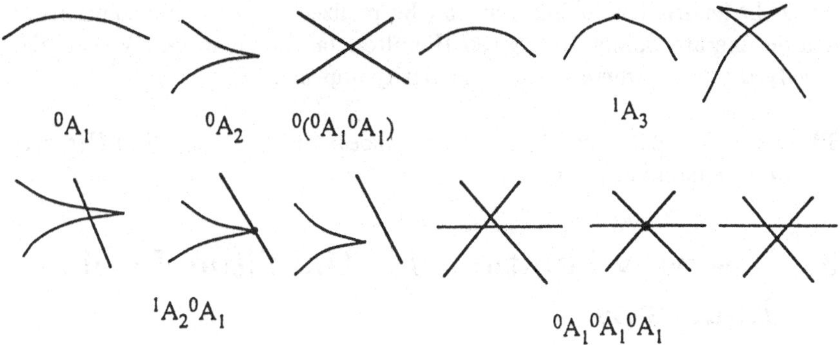

Figure 1.

We now consider the types of singularities that geometric solutions to a given (fixed) Hamilton-Jacobi equation might exhibit. For this purpose we refer to the following theorem of Bogaevskii [7].

Theorem 2.2 [7] *Suppose that $n \le 3$. There exists a dense subset $\mathcal{O} \subset C^\infty(\mathbb{R}^n, \mathbb{R})$ with the following properties: For any $\phi \in \mathcal{O}$, each multi-Legendrian unfolding given by the geometric solution for GCPT of (H) is $P_{(r)}$-Legendrian equivalent to one of the standard perestroikas of fronts. Here we use the Whitney fine topology on the space $C^\infty(\mathbb{R}^n, \mathbb{R})$.*

This theorem shows that the standard perestroikas of fronts is the full list for the generic perestroikas of the graphs of geometric solutions for a fixed Hamilton-Jacobi equations. However, we have to consider the realization problem for the standard perestroikas of fronts as a geometric solution for a given Hamilton-Jacobi equation. For this purpose, we need a kind of non-degeneracy condition on the Hamiltonian function. We say that a Hamiltonian function $H(t, x, p)$ is *non-degenerate at* (t_0, x_0, p_0) if $\frac{\partial^2 H}{\partial p_i \partial p_j}(t_0, x_0, p_0) \neq 0$ for some $1 \le i, j \le n$. This condition is weaker than the condition that $H(t, x, p)$ is convex (or concave) with respect to the $(p_1, \ldots, p_n)$-variables at (t_0, x_0, p_0) for $n \ge 2$. The following theorem is a realization theorem for generic singularities for a given Hamilton-Jacobi equation.

Theorem 2.3 [28,29] *Let $H(t, x, p)$ be a non-degenerate Hamiltonian function germ at (t_0, x_0, p_0) and $\mathcal{L} : (R, u_0) \to (J^1(\mathbb{R} \times \mathbb{R}^n, \mathbb{R}), (t_0, x_0, y_0, s_0, p_0))$ be a $P_{(1)}$-Legendrian stable Legendrian unfolding associated with (μ, ℓ). Then* *there exists a Legendrian unfolding $\mathcal{L}'$ which is a geometric solution of the Hamilton-Jacobi equation $s + H(t, x, p) = 0$ such that $\mathcal{L}$ and $\mathcal{L}'$ are $P_{(1)}$-Legendrian equivalent.*

We remark that the ${}^1A_3$ singularity (even for general n) describes how a singularity appears from a smooth solution. These are $P_{(1)}$-Legendrian stable Legendrian unfoldings, so can be realized as geometric solutions at non-degenerate points for a given Hamilton-Jacobi equation. We can also specify the point where the ${}^1A_3$-singularity appears.

Theorem 2.4 [29] *If an ${}^1A_3$-singularity appears at (t_0, x_0, p_0), then $H(t, x, p)$ is non-degenerate at (t_0, x_0, p_0).*

3 Viscosity Solutions for Hamilton-Jacobi Equations

For convenience, we stick to the case when the Hamiltonian $H(p_1, \ldots, p_n)$ depends only on the momentum variables in this section. Some of the results can be extended to more general Hamiltonians. Viscosity solutions for non-linear equations of first order have been introduced by Crandall and Lions, see [11].

A continuous function $y_v \in C([0, \infty) \times I\!R^n)$ is *a viscosity solution* of (H) provided

$$\frac{\partial \psi}{\partial t}(t, x) + H(\frac{\partial \psi}{\partial x_1}(t, x), \ldots, \frac{\partial \psi}{\partial x_n}(t, x)) \leq 0, \quad \text{(resp., } \geq 0)$$

for any $\psi \in C^1([0, \infty) \times I\!R^n)$ for which $y_v - \psi$ attains a local maximum (resp., local minimum) at the point $(t, x) \in \mathcal{O}$ and the function y_v satisfies the initial condition $\lim_{t \to 0+} y_v(t, x) = \phi(x)$.

Firstly we consider the case when the Hamiltonian $H(p_1, \ldots, p_n)$ is uniformly convex (or concave). In this case we refer to the following result of Bardi-Evans [4].

Theorem 3.1 [4] *Assume that the Hamiltonian $H(p_1, \ldots, p_n)$ is uniformly convex, then*

$$y(t, x) \equiv \inf_q \sup_p \{\phi(q) + \langle p, x - q \rangle - H(p)t\}$$

is the unique viscosity solution of (H), *where $\langle \ , \ \rangle$ is the canonical inner product on $I\!R^n$.*

We remark that the above exact formula of the viscosity solution is called *the Hopf-Lax type formula*; there are some results related to this formula in

[18,23,37]. We now consider the family of functions $F(t,x,p,q) = \phi(q) + \langle p, x-q\rangle - H(p)t$. Since $H(p)$ is uniformly convex, we have

$$\sup_p \{\phi(q) + \langle p, x-q\rangle - H(p)t\} = F|\Sigma_p(F),$$

where

$$\Sigma_p(F) = \left\{(t,x,p,q) \mid \frac{\partial F}{\partial p_i} = x_i - q_i - \frac{\partial H}{\partial p_i}(p)\,t = 0, \quad i = 1,\dots,n\right\}.$$

If $\phi(q)$ has a minimum, then we have

$$\inf_q \sup_p \{\phi(q) + \langle p, x-q\rangle - H(p)t\} = \min_q \left\{\phi(q) + \left\langle p, \frac{\partial H}{\partial p}(p)\right\rangle - H(p)t\right\}$$

so the viscosity solution of (H) in this case is the minimum function of a certain family of smooth functions. In [6], Bogaevskii has classified the perestroikas of shocks of viscosity solutions for $n = 1, 2, 3$ by using this fact. The picture for $n = 2$ is given in Figure 2.

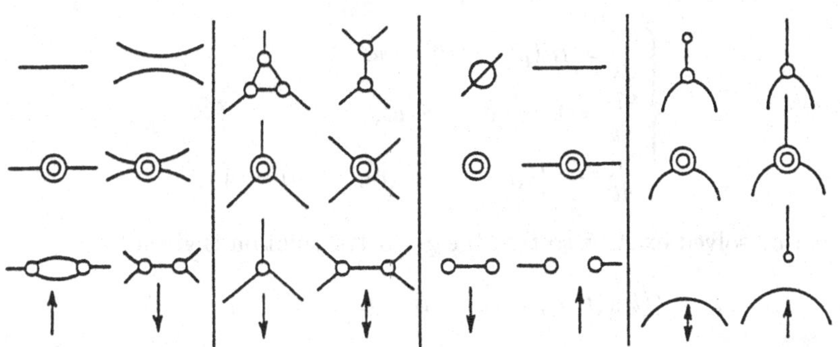

Figure 2.

In [9] Chaperon defined the notion of minimax solutions for (H) using the notion of global generating families. Joukovskaia [33] has classified generic singularities of the minimax solutions for $n = 1, 2$. If the Hamiltonian is convex, it is known that the minimax solution coincides with the viscosity solution.

For a general (non-convex) Hamiltonian, the situation is quite different. For convenience, we only consider the case when $n = 1$. Since $H(p)$ is not assumed to be uniformly convex (or concave), we cannot use Theorem 3.1, so the situation will be quite complicated even in this case. We next state the criterion for a solution in a form which is useful for the construction of the solution. To this end, assume that $\mathcal{O} \subset (0,\infty) \times \mathbb{R}$ is open and that there is a smooth curve $t \to x = \chi(t)$, $t \in (t_1, t_2) \subset \mathbb{R}^+$, with negative slope, which divides $\mathcal{O}$ into two open sets $\mathcal{O}^+$ and $\mathcal{O}^-$, $\mathcal{O} = \Gamma \cup \mathcal{O}^+ \cup \mathcal{O}^-$, $\Gamma = \{(t,x) : x = \chi(t)\}$, where $\mathcal{O}^+$ lies on the right of Γ. Then we have the following theorem.

Theorem 3.2 [12,34] *Let $y_v \in C(\mathcal{O})$ and $y_v = y_v^+$ in $\mathcal{O}^+ \cup \Gamma, y_v = y_v^-$ in $\mathcal{O}^- \cup \Gamma$ where $y_v^\pm \in C^1(\mathcal{O}^\pm \cup \Gamma)$. Then y_v is a viscosity solution of* (H) *in $\mathcal{O}$ if and only if the following conditions hold:*
a) y_v^+ and y_v^- are classical solutions of (H) *in $\mathcal{O}^+$ and $\mathcal{O}^-$ respectively,*
b) if $y_{v,x}^- > y_{v,x}^+$ (resp., $y_{v,x}^- < y_{v,x}^+$) across Γ, then

$$H((1-\lambda)y_{v,x}^+ + \lambda y_{v,x}^-) - (1-\lambda)H(y_{v,x}^+) - \lambda H(y_{v,x}^-) \le 0 \quad (resp., \ge 0),$$

where $\lambda \in [0,1]$ and $y_{v,x}^\pm = \frac{\partial y_v^\pm}{\partial x}$. That is, the graph of H lies respectively below or above the line segment joining the points $(y_{v,x}^+, H(y_{v,x}^+))$ and $(y_{v,x}^-, H(y_{v,x}^-))$.

The condition b) will be referred to in the sequel as the *viscosity criterion.* The hypersurface Γ in the neighbourhood of which y_v has the properties specified in the above theorem is the *shock curve.* In this case the characteristic equation (i.e., the equation corresponding to the characteristic vector field) is given as follows:

$$\begin{cases} \dfrac{dx}{dt} = H'(p), \quad x(0) = u, \\ \dfrac{dp}{dt} = 0, \quad p(0) = \phi'(u), \\ \dfrac{dy}{dt} = -H(p) + p \cdot H'(p), \quad y(0) = \phi(u). \end{cases}$$

It can be solved exactly, so that the geometric solution is given by

$$L_{\phi,t} = \{(t, x(t,u), y(t,u), -H(p(t,u)), p(t,u)) \mid u \in \mathbb{R}\},$$

where

$$\begin{cases} x(t,u) = u + tH'(\phi'(u)), \\ p(t,u) = \phi'(u), \\ y(t,u) = t\{-H(\phi'(u)) + \phi'(u)H'(\phi'(u))\} + \phi(u). \end{cases}$$

Before the first critical time, when characteristics cross in the (t,x)–plane, $W_t = \Pi(L_{\phi,t})$ is the graph of the viscosity solution y_v. After the characteristics cross, W_t becomes singular. Theorem 2.1 describes the generic singularities of W_t. The first singularity appears in the form of ${}^1A_3$. See Figure 3a, where we give the shape of the singularity appearing. By Theorem 2.4, these appear at the convex or the concave points of the Hamiltonian function. Away from the singularity, the viscosity solution is given by W_t. In [31,34] we have constructed the unique viscosity solution beyond the first critical time by selecting a single-valued branch of W_t. Assume that the singularity of type ${}^1A_3$ appears at the point (t_0, x_0, p_0). After the critical time t_0, the wave front W_t is three-valued on an interval $(x_1(t), x_2(t))$; see Figure 3b. We can choose the viscosity solution past t_0 by selecting a continuous single-valued branch of W_t as in Figure 3c. In view of Theorem 2.5 the viscosity criterion is satisfied across $\chi(t)$ while y_v is a classical solution away from $\chi(t)$. Hence, by uniqueness of the viscosity solution, this gives the viscosity solution of (H) beyond t_0.

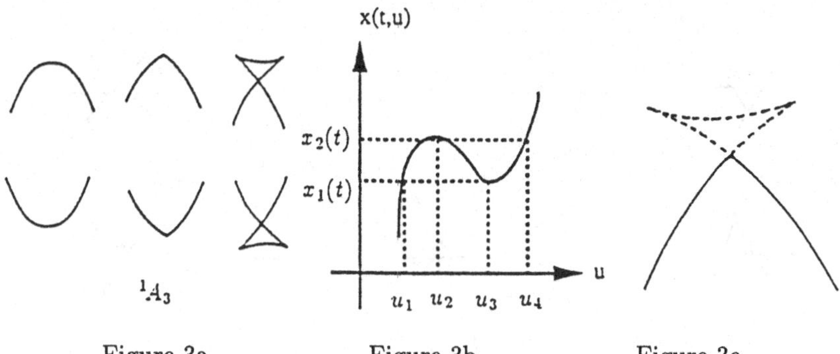

Figure 3a. Figure 3b. Figure 3c.

By this construction, we have extended the viscosity solution beyond the first critical time t_0. According to Theorem 2.5 the shock is generated in a convex or concave domains of $H(p)$, so the viscosity criterion is automatically satisfied. The graph of the viscosity solution past the first critical time is depicted by a full line in Figure 1c, where we assume that H is convex in a neighborhood of the ${}^1A_3$ singularity appearing. The shock corresponds to the intersection of the two branches and it is called a *genuine shock*. The genuine shock is defined as the intersection of two incoming characteristics (or waves) and its speed is given by the Rankine-Hugoniot condition

$$\chi'(t) = \frac{H(y^+_{v,x}(t, \chi(t))) - H(y^-_{v,x}(t, \chi(t)))}{y^+_{v,x}(t, \chi(t)) - y^-_{v,x}(t, \chi(t))},$$

where $\chi'(t) = \frac{d\chi}{dt}(t)$.

Therefore in order to follow the evolution of the shock we have to study the following questions:

a) How do different branches of the multi-valued graph of W_t intersecting at one point bifurcate in time?

b) If the two branches initially defining the shock continue to cross, is the viscosity criterion satisfied across the intersection?

The normal forms of the generic perestroikas of different branches of W_t are given in Theorem 2.1. If the viscosity criterion is satisfied at the time $t_\alpha = t_0 + \varepsilon$, we can choose the correct branch of the graphs of the geometric solutions as viscosity solutions. See Figure 4 (all pictures can be turned upside down).

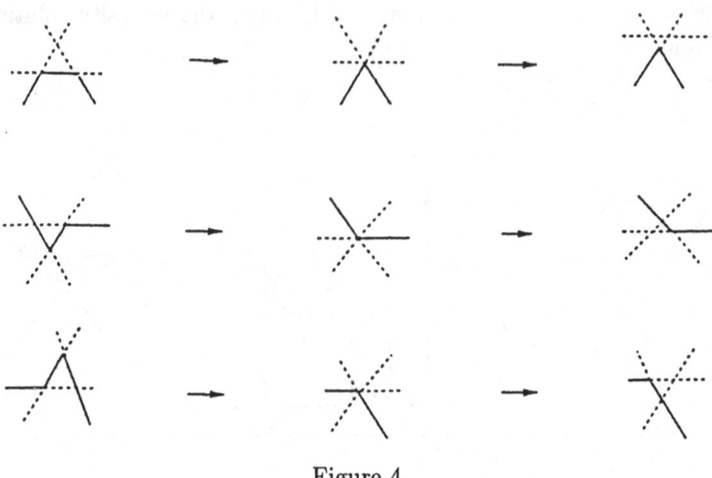

Figure 4.

We will now investigate how the viscosity criterion can be violated across the intersection of two branches. Assume that a generated shock is defined by two intersecting branches y^- and y^+. We denote by y^- (resp. y^+) the branch representing the viscosity solution for $x < \chi(t)$ (resp. $x > \chi(t)$). If the two branches continue to intersect they evolve according to ${}^0({}^0A_1\,{}^0A_1)$. We denote by $\chi(t)$ the intersection of the two branches. It is clear that for generic Hamiltonian functions $H(p)$, H has only Morse type critical points and no tritangent lines. So we assume that the Hamiltonian has the above properties. By Theorems 2.1 and 2.2, we have the following theorem.

Theorem 3.3 [31] *For a generic initial function ϕ, if the viscosity criterion is violated at t_α, then only the following 3 cases may occur:*
(1) The normal form is ${}^0({}^0A_1\,{}^0A_1)$ and $\overline{P^+P^-}$ is tangent to the graph of $H(p)$ at only one of the points P^+, P^- and the line is not tangent to the graph at other points between these points.
(2) The normal form is ${}^0({}^0A_1\,{}^0A_1)$ and $\overline{P^+P^-}$ is not tangent to the graph of $H(p)$ at each point P^+, P^- and there exists only one another point between these points at where the above line is tangent to the graph.
(3) The normal form is ${}^1A_2\,{}^0A_1$ and $\overline{P^+P^-}$ is tangent to the graph of $H(p)$ at only one of the points P^+, P^- and the line is not tangent to the graph at other points between these points.
Here,

$$P^+ = (y_x^+(t_\alpha, \chi(t_\alpha)), H(y_x^+(t_\alpha, \chi(t_\alpha))),$$
$$P^- = (y_x^-(t_\alpha, \chi(t_\alpha)), H(y_x^-(t_\alpha, \chi(t_\alpha))),$$

and $\overline{P^+P^-}$ denotes the line through P^+, P^- in the $(p, H(p))$-plane.

In [31] we have solved the local Riemann problems and constructed viscosity solutions for each case in the above theorem. We considered there another five cases. However, Bogaevskii pointed out that these are not generic. Furthermore, he has shown that the case 3) cannot occur if the viscosity criterion is satisfied before the perestroika time t_α. So we may consider cases 1) and 2). Here we only mention the case 1) in order to introduce the technique which is used in [31,34]. We assume that the graph of the viscosity solution at the time $t \le t_\alpha$ is as depicted in Figure 5a.

Figure 5a.

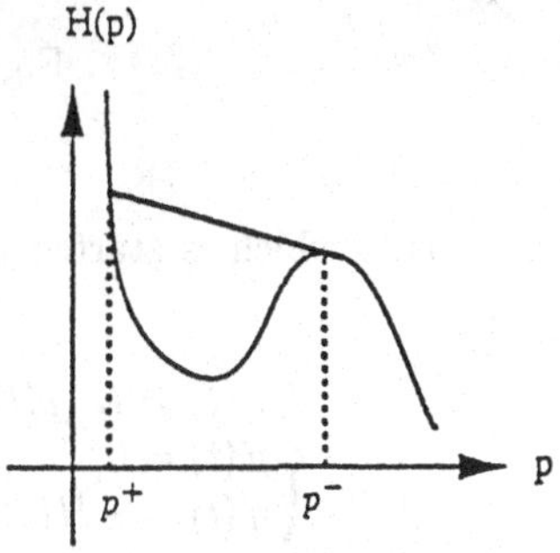

Figure 5b.

For convenience, we assume that $\overline{P^+P^-}$ is tangent to the graph of $H(p)$ at the point $(y_x^-(t_\alpha, \chi(t_\alpha)), H(y_x^-(t_\alpha, \chi(t_\alpha)))$ and $H''((y_x^-(t_\alpha, \chi(t_\alpha))) < 0$ (see Figure 5b). As we have already mentioned the genuine shocks satisfy the Rankin-Hugoniot condition. So we should construct new characteristics which satisfy both the Rankin-Hugoniot condition and the viscosity criterion. In this case we have

$$H'(y_x^-(t_\alpha, \chi(t_\alpha))) = \frac{H(y_x^+(t_\alpha, \chi(t_\alpha))) - H(y_x^-(t_\alpha, \chi(t_\alpha)))}{y_x^+(t_\alpha, \chi(t_\alpha)) - y_x^-(t_\alpha, \chi(t_\alpha))} = \chi'(t_\alpha).$$

If

$$H'(y_x^-(t, \chi(t))) < \frac{H(y_x^+(t, \chi(t))) - H(y_x^-(t, \chi(t)))}{y_x^+(t, \chi(t)) - y_x^-(t, \chi(t))}$$

for $t_\alpha \le t < t_\alpha + \varepsilon$ for sufficiently small $\varepsilon > 0$, then we can easily show that the viscosity criterion is violated for $t_\alpha < t < t_\alpha + \varepsilon$, so that a new way to build the solution is required (cf., Figure 6).

Let us consider the relation $H'(q) = \frac{H(p)-H(q)}{p-q}$ around (q_0, p_0) with $q_0 \neq p_0$, $H'(q_0) = \frac{H(p_0)-H(q_0)}{p_0-q_0}$ and $H''(q_0) \neq 0$. By the implicit function theorem, there exists a smooth function ψ around p_0 such that the above relation is equivalent to $q = \psi(p)$. We first construct *the contact discontinuity shock curve* as the solution of the following initial value problem

$$\begin{cases} \chi_c'(t) = H'(\psi(y_x^+(t, \chi_c(t)))), \\ \chi_c(t_\alpha) = \chi(t_\alpha). \end{cases}$$

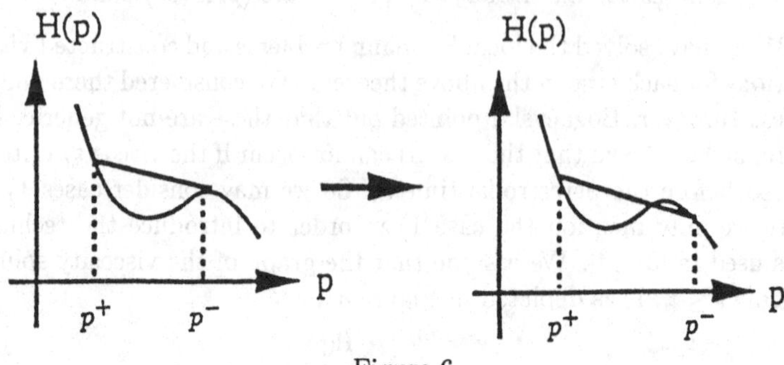

Figure 6.

The characteristic which is started at a point $(\tau, \chi_c(\tau))$ should satisfy the following:

$$\begin{cases} x'(t) = H'(p(t)), \\ p'(t) = 0, \\ y'(t) = -H(p(t)) + p(t)H'(p(t)), \end{cases}$$

with initial condition $x(\tau) = \chi_c(\tau)$, $y(\tau) = y^+(\tau, \chi_c(\tau))$ and $p(\tau) = \psi(y_x^+(\tau, \chi_c(\tau)))$. So the solution is exactly given as follows:

$$\begin{cases} \tilde{x}(t) = \chi_c(\tau) + (t-\tau)H'(\psi(y_x^+(\tau, \chi_c(\tau)))), \\ \tilde{p}(t) = \psi(y_x^+(\tau, \chi_c(\tau))), \\ \tilde{y}(t) = y^+(\tau, \chi_c(\tau)) + (t-\tau)\{-H(\psi(y_x^+(\tau, \chi_c(\tau)))) + \\ \qquad\qquad \psi(y_x^+(\tau, \chi_c(\tau)))H'(\psi(y_x^+(\tau, \chi_c(\tau))))\}. \end{cases}$$

We have shown that if the viscosity criterion is violated for $t > t_\alpha$, then the contact discontinuity curve χ is convex and the viscosity solution can be constructed. Then we can draw the picture of the graph of the viscosity solution for $t > t_\alpha$ and the shock curve around t_α, see Figure 7.

We have constructed the viscosity solutions for other cases. The perestroikas of the graphs of viscosity solutions are as depicted in Figure 8. In this case there appear, not only the contact discontinuity shock curves, but also the rarefaction wave type shock curve (see (3) of Figure 8). So we have two different kinds of shocks once the viscosity criterion is violated. The above arguments show that the graph of a viscosity solution is not always a subset of the graph of the geometric solution for (H). Until a couple of years ago very few people had believed this fact, which is now known to many mathematicians (cf. [45]).

To conclude this section, we present some problems.

Problem (3.a) For $n = 1$, the last argument describes the perestroikas of shocks around the first time t_α when the viscosity criterion is violated.

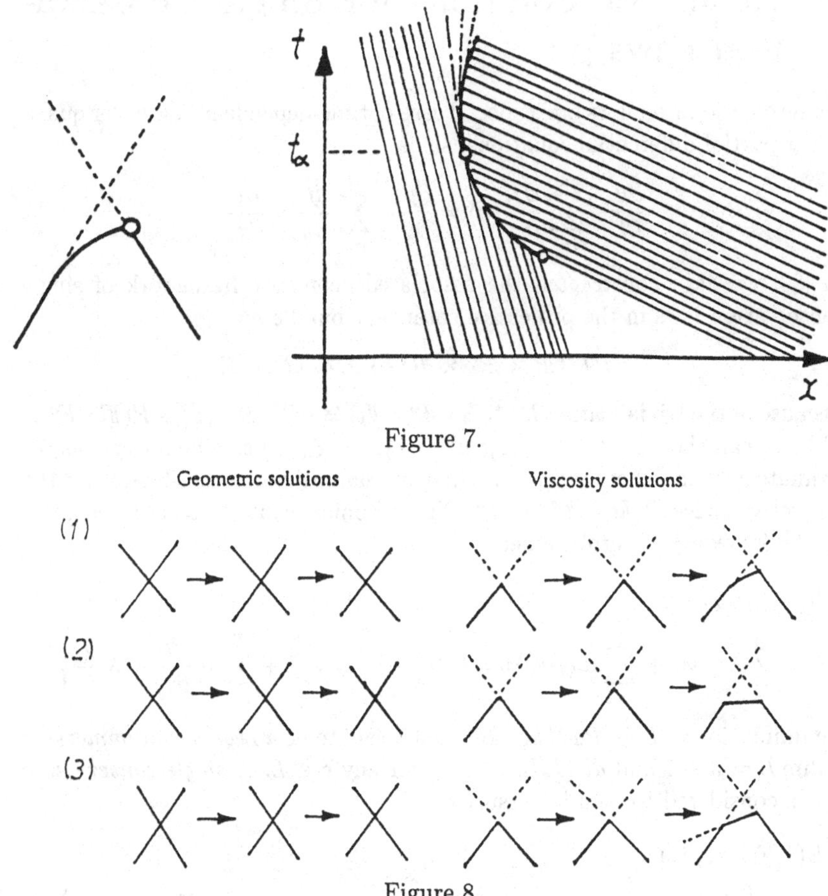

Figure 7.

Figure 8.

(1) What happens after that?

(2) Can we proceed with similar arguments in the case when the Hamiltonian depends also on (t, x)?

Problem (3.b) For the case $n \geq 2$, we have the following questions:

(1) How should the viscosity criterion be violated?

(2) How can we construct the viscosity solution after the viscosity criterion is violated? What is the contact discontinuity in the case $n \geq 2$?

(3) Find a crucial example which describes the situation for problems (1) and (2).

(4) (A problem posed by Bogaevskii). Are there any useful formulae to classify the perestroikas of shocks such as the Hopf-Lax type formula?

4 Geometric Solutions for Single Conservation Laws

Single conservation laws are a special case of time-dependent first order quasi-linear partial differential equations, since

$$\frac{\partial y}{\partial t}+\sum_{i=1}^{n}\frac{\partial f_i}{\partial x_i}(y)=\frac{\partial y}{\partial t}+\sum_{i=1}^{n}\frac{df_i}{dy}(y)\frac{\partial y}{\partial x_i}.$$

In [30] we have constructed the associated geometric framework of single conservation laws in the projective cotangent bundle

$$\Pi : PT^*(I\!R\times I\!R^n\times I\!R)\to I\!R\times I\!R^n\times I\!R.$$

Because of the trivialization $PT^*(I\!R\times I\!R^n\times I\!R)\cong(I\!R\times I\!R^n\times I\!R)\times P(I\!R\times I\!R^n\times I\!R)^*$, we can choose $((t,x_1,\dots,x_n,y),[\sigma;\xi_1;\dots;\xi_n;\eta])$ as a homogeneous coordinate system, where $[\sigma;\xi_1;\dots;\xi_n;\eta]$ are homogeneous coordinates for the projective space $P(I\!R\times I\!R^n\times I\!R)^*$. The canonical contact structure is given by the following hyperplane field:

$$K_{((t,x,y),[\sigma;\xi;\eta])}=\left\{X \mid \nu\sigma+\sum_{i=1}^{n}\mu_i\xi_i+\lambda\eta=0,\ d\Pi(X)=\nu\frac{\partial}{\partial t}+\sum_{i=1}^{n}\mu_i\frac{\partial}{\partial x_i}+\lambda\frac{\partial}{\partial y}\right\}.$$

An immersion $i: L\to PT^*(I\!R\times I\!R^n\times I\!R)$ is said to be *a Legendrian immersion* if $\dim L=n+1$ and $di_q(T_qL)\subset K_{i(q)}$ for any $q\in L$. *A single conservation law* is considered to be a hypersurface

$$E(1,f_1,\dots,f_n)=\left\{((t,x,y),[\sigma;\xi;\eta])\in PT^*(I\!R\times I\!R^n\times I\!R)\mid \sigma+\sum_{i=1}^{n}\frac{df_i}{dy}(y)\xi_i=0\right\}.$$

A geometric solution of $E(1,f_1,\dots,f_n)$ is a Legendrian submanifold L of $PT^*(I\!R\times I\!R^n\times I\!R)$ contained in $E(1,f_1,\dots,f_n)$ with $\Pi|L$ an embedding. We consider the meaning of the notion of a geometric solution. Let S be a smooth hypersurface in $I\!R\times I\!R^n\times I\!R$, then we have a unique Legendrian submanifold $\hat{S}$ in $PT^*(I\!R\times I\!R^n\times I\!R)$ such that $\Pi(\hat{S})=S$. $\hat{S}$ is given by:

$$\left\{((t,x,y),[\sigma;\xi;\eta])\mid \sigma\frac{\partial}{\partial t}+\sum_{i=1}^{n}\xi_i\frac{\partial}{\partial x_i}+\eta\frac{\partial}{\partial y}\in TS^\perp\subset T(I\!R\times I\!R^n\times I\!R)\right\},$$

where we adopt the canonical inner product on $I\!R\times I\!R^n\times I\!R$. It follows that if L is a geometric solution of $E(1,f_1,\dots,f_n)$, then we have $L=\widehat{\Pi(L)}$.

By the classical theory of first order partial differential equations [1], the characteristic vector field of $E(1,f_1,\dots,f_n)$ is defined to be $X(1,f_1,\dots,f_n)=\frac{\partial}{\partial t}+\sum_{i=1}^{n}\frac{df_i}{dy}(y)\frac{\partial}{\partial x_i}$. Then we can easily verify the following characterization theorem for geometric solutions.

Proposition 4.1 *Let S be a smooth hypersurface in $\mathbb{R} \times \mathbb{R}^n \times \mathbb{R}$. Then $\hat{S}$ is a geometric solution of $E(1, f_1, \ldots, f_n)$ if and only if the characteristic vector field $X(1, f_1, \ldots, f_n)$ is tangent to S.*

We can solve the Cauchy problem (C) by the method of characteristics. Set $a_i(y) = \frac{df_i}{dy}(y)$, $1 \le i \le n$. Then the characteristic equation, (i.e. the equation corresponding to the characteristic vector field) associated with (C) through $(0, x_0)$ is given as follows:

$$\begin{cases} \dfrac{dx_i}{dt}(t) = a_i(y(t, x(t)), & x_i(0) = x_{i,0} \\ \dfrac{dy}{dt}(t, x(t)) = 0, & y(0, x(0)) = \phi(x_0). \end{cases}$$

In this case the solution of the characteristic equation can be exactly solved as follows:

$$x(t) = x_0 + ta(\phi(x_0)) \quad \text{and} \quad y(t, x(t)) = y(0, x(0)) = \phi(x_0).$$

We define the corresponding embedding $\mathcal{I}[a, \phi] : \mathbb{R} \times \mathbb{R}^n \to \mathbb{R} \times \mathbb{R}^n \times \mathbb{R}$ by $\mathcal{I}[a, \phi](t, u) = (t, u + ta(\phi(u)), \phi(u))$, where $a = (a_1, \ldots, a_n)$. By definition the characteristic vector field is always tangent to the image of this embedding, so that it gives the geometric solution of (C). We also define mappings $F[a, \phi] : \mathbb{R} \times \mathbb{R}^n \to \mathbb{R}^n$ by $F[a, \phi](t, u) = u + ta(\phi(u))$ and $f[a, \phi; t_0] : \mathbb{R}^n \to \mathbb{R}^n$ by $f[a, \phi; t_0](u) = F[a, \phi](t_0, u)$.

By the classical theory of characteristics the classical solution of (C) is expressed by $y(t_0, x) = \phi(f[a, \phi; t_0]^{-1}(x))$ at any x which is not a critical value of $f[a, \phi; t_0]$. Then our problem is to study the singularities of $f[a, \phi; t_0]$. We can calculate the Jacobian matrix of $f[a, \phi; t_0]$ at u_0 as follows:

$$J(f[a, \phi; t_o]) = \begin{pmatrix} 1 + c_{11} & c_{12} & \cdots & c_{1n} \\ c_{21} & 1 + c_{22} & \cdots & c_{2n} \\ \vdots & \vdots & \ddots & \vdots \\ c_{n1} & c_{n2} & \cdots & 1 + c_{nn} \end{pmatrix}$$

where $c_{ij} = t_0 \frac{da_i}{dy}(\phi(u_0)) \frac{\partial \phi}{\partial u_j}(u_0)$.

In order to study the perestroikas of singularities of $f[a, \phi; t]$ along t, we introduce the notion of a one-parameter unfolding of an immersion. Let R be an $(n+1)$-dimensional smooth manifold and $\mu : (R, u_0) \to (\mathbb{R}, t_0)$ be a submersion germ. We call a map germ $\mathcal{I} : (R, u_0) \to \mathbb{R} \times \mathbb{R}^n \times \mathbb{R}$ of the form $\mathcal{I}(t, x) = (\mu(u), x(u), y(u))$ *an unfolding of an immersion* if $\mathcal{I}$ is an immersion germ and $\mathcal{I}|\mu^{-1}(t)$ is also an immersion for each $t \in (\mathbb{R}, t_0)$.

For the study of the perestroikas of singularities of unfoldings of immersions, we introduce the following equivalence relation. Let $\mathcal{I}_i : (R, u_i) \to (\mathbb{R} \times \mathbb{R}^n \times \mathbb{R}, (t_i, x_i, y_i))$ $(i = 0, 1)$ be unfoldings of immersions. We say that $\mathcal{I}_0$ and $\mathcal{I}_1$ are *t-P-$\mathcal{A}$-equivalent* if there exist a diffeomorphism germ

$\Phi : (\mathbb{R} \times (\mathbb{R}^n \times \mathbb{R}), (t_0, x_0, y_0)) \to (\mathbb{R} \times (\mathbb{R}^n \times \mathbb{R}), (t_1, x_1, y_1))$ of the form $\Phi(t, x, y) = (\phi_1(t), \phi_2(t, x), \phi_3(t, x, y))$ and a diffeomorphism germ $\Psi : (R, u_0) \to (R, u_1)$ such that $\hat{\Phi} \circ \mathcal{I}_0 = \mathcal{I}_1 \circ \Psi$.

Theorem 4.2 [30,47] *The generic unfolding of an immersion is t-P-$\mathcal{A}$-equivalent to one of the unfoldings of immersions in the following list:*

$$({}^0A_k) \qquad (t, u_n^{k+1} + \sum_{i=1}^{k-1} u_i u_n^i, u_1, \ldots, u_n) \quad (0 \leq k \leq n)$$

$$({}^1A_k) \qquad (t, u_n^{k+1} + u_n^{k-1}(t \pm u_{k-1}^2 \pm \cdots \pm u_{n-1}^2) + \sum_{i=1}^{k-2} u_i u_n^i, u_1, \ldots, u_n)$$
$$(2 \leq k \leq n+1).$$

We remark that the germ of type ${}^1A_2$ describes how a singularity of the geometric solution appears or vanishes. We can also define the notion of *stable unfoldings of an immersion* in exactly the same way as the usual definition of the notion of stable map germs in [19]. Then the above theorem gives a generic classification of the stable unfoldings of an immersion.

Theorem 4.3 [30] *There exists a dense subset $\mathcal{O} \subset C^\infty(\mathbb{R}^n, \mathbb{R})$ with the following properties: For any $\phi \in \mathcal{O}$ and $(t_0, u_0) \in \mathbb{R} \times \mathbb{R}^n$, the germ $I[a, \phi]$ at (t_0, u_0) is t-P-$\mathcal{A}$-equivalent to one of germs in the list of Theorem 4.2.*

In [30] we have shown that the ${}^0A_k$-type germs in Theorem 4.2 can always be realized as geometric solutions for a given single conservation law. However, if an unfolding of an immersion has a non-trivial bifurcation, the situation is different. It is known that unfoldings of immersions of type ${}^1A_2$ are not t-P-$\mathcal{A}$-equivalent to any germs of geometric solutions of $\frac{\partial y}{\partial t} + y^2 \sum \frac{\partial y}{\partial x_i} = 0$.

Hence, we need a kind of non-degeneracy condition on the single conservation law to realize nontrivial bifurcations. The single conservation law $E(1, f_1, \ldots, f_n)$ is *non-degenerate* at (t_0, x_0, y_0) if $\frac{d^2 f}{dy^2}(y) \neq 0$ at (t_0, x_0, y_0), where $\frac{d^2 f}{dy^2}(y) = (\frac{d^2 f_1}{dy^2}(y), \ldots, \frac{d^2 f_n}{dy^2}(y))$. In [30] we have established the following realization theorem.

Theorem 4.4 [30] *Let $\mathcal{I} : (R, u_0) \to (\mathbb{R} \times \mathbb{R}^n \times \mathbb{R}, (t_0, x_0, y_0))$ be a stable unfolding of an immersion. Then*
(1) If $\mathcal{I}$ has a trivial bifurcation, then there exists an unfolding of an immersion $\mathcal{I}'$ such that $\widehat{\text{Image}\,\mathcal{I}'}$ *is a local geometric solution of $E(1, f_1, \ldots, f_n)$ and $\mathcal{I}$, $\mathcal{I}'$ are t-P-$\mathcal{A}$-equivalent.*
(2) If the equation $E(1, f_1, \ldots, f_n)$ is non degenerate at (t_0, x_0, y_0), then there exists an unfolding of an immersion $\mathcal{I}'$ such that $\widehat{\text{Image}\,\mathcal{I}'}$ *is a local geometric solution of $E(1, f_1, \ldots, f_n)$ and $\mathcal{I}$, $\mathcal{I}'$ are t-P-$\mathcal{A}$-equivalent.*
(3) If an ${}^1A_k$ singularity appears at a point (t_0, x_0, y_0), then the equation $E(1, f_1, \ldots, f_n)$ is non-degenerate at (t_0, x_0, y_0).

We have also classified unfoldings of multi-germ immersions in [30], however, we have no space here to describe that work.

5 Entropy Solutions for Single Conservation Laws

An entropy solution for (C) is a function y_e satisfying the following: for any $\psi \in C^\infty(\mathbb{R}^{1+n})$

$$\iint_{\mathbb{R}^+\times\mathbb{R}^n} (y_e\frac{\partial\psi}{\partial t} + \sum_{i=1}^n f_i(y_e)\frac{\partial\psi}{\partial x_i})\,dt\,dx + \int_{\mathbb{R}^n}\psi(0,x)\,dx = 0 \qquad \text{(W)}$$

and for any $\psi \in C_0^\infty(\mathbb{R}^{1+n}), g \geq 0$, and any $k \in \mathbb{R}$,

$$\iint_{\mathbb{R}^+\times\mathbb{R}^n} \operatorname{sgn}(y_e - k)\{(y_e - k)\frac{\partial\psi}{\partial t} + \sum_{i=1}^n (f_i(y_e) - f_i(k))\frac{\partial\psi}{\partial x_i}\}\,dt\,dx \geq 0. \qquad \text{(E)}$$

We call the condition (E) *an entropy condition.* For the existence and uniqueness of entropy solutions, see, for example Kruzkov [36].

If we solve (C) by the characteristic method as in §4, before the characteristics cross, the classical solution $y(t,x) = \phi(f[a,\phi;t]^{-1}(x))$ is the entropy solution. After the critical time t_0 when the characteristic cross, that solution develops discontinuities (shocks).

Firstly we consider the case when $n = 1$. Under the assumption that $f'(y)$ is uniformly convex, Lax [37] has given an explicit formula for the entropy solution. There are some results on the shocks for the entropy solution by using this formula. For non-convex $f'(y)$, Ballow [3], Guckenheimer [21], Jennings [32], Chen [10] and Dafermos [14], etc., studied how shocks generate and propagate.

In this case the entropy condition can be replaced by a rather familiar condition appearing in the next theorem; first some notation. Assume that $\mathcal{O} \subset (0,\infty)\times\mathbb{R}$ is open and that there is a smooth curve $t \to x = \chi(t),\ t \in (t_1,t_2) \subset \mathbb{R}^+$, with negative slope, which divides $\mathcal{O}$ into two open sets $\mathcal{O}^+$ and $\mathcal{O}^-$, with $\mathcal{O} = \Gamma \cup \mathcal{O}^+ \cup \mathcal{O}^-$, $\Gamma = \{(t,x) :\ x = \chi(t)\}$, where $\mathcal{O}^+$ lies on the right of Γ. Then we have the following theorem.

Theorem 5.1 [32,42] *Let y_e be the solution satisfying* (W) *with $y_e = y_e^+$ in $\mathcal{O}^+ \cup \Gamma$, $y_e = y_e^-$ in $\mathcal{O}^- \cup \Gamma$ where $y_e^\pm \in C^1(\mathcal{O}^\pm \cup \Gamma)$. Then y_e satisfies the entropy condition* (E) *in $\mathcal{O}$ if and only if the following conditions hold:*
a) y_e^+ and y_e^- are classical solutions of (C) *in $\mathcal{O}^+$ and $\mathcal{O}^-$ respectively,*
b) if $y_e^- > y_e^+$ (resp. $y_e^- < y_e^+$) across Γ, then

$$f\left((1-\lambda)y_e^+ + \lambda y_e^-\right) - (1-\lambda)f(y_e^+) - \lambda f(y_e^-) \leq 0 \quad (\textit{resp.}, \geq 0),$$

where $\lambda \in [0,1]$. That is, the graph of f lies respectively below or above the line segment joining the points $(y_e^+, f(y_e^+))$ and $(y_e^-, f(y_e^-))$.

We recognize that the viscosity criterion is the analogue of the entropy condition (cf., Theorem 3.2). We also have the following Rankin-Hugoniot condition for the shock curve $\chi(t)$ of the entropy solution [32,42]:

$$\chi'(t) = \frac{f(y_e^+) - f(y_e^-)}{y_e^+ - y_e^-}.$$

On the other hand, there is a relation between (H) and (C) as follows. Differentiating the viscosity solution of $\frac{\partial y}{\partial t} + H(\frac{\partial y}{\partial x}) = 0$ with respect to x in the region of smoothness we have $\frac{\partial^2 y}{\partial t \partial x} + \frac{dH}{dp}(\frac{\partial y}{\partial x})\frac{\partial^2 y}{\partial x^2} = 0$. By Theorem 5.1, we can show that the derivative of the viscosity solution is the entropy solution of the scalar conservation law $\frac{\partial u}{\partial t} + \frac{dH}{dp}(u)\frac{\partial u}{\partial x} = 0$. So the classification of Bogaevskii [6] gives a complete list of the perestroikas of shocks for (C) in the case when $f'(y)$ is uniformly convex. Figure 8 also gives a classification of the perestroikas of shocks at the first time t_α when the entropy condition is violated.

In the higher dimensional case, there are very few results on this subject. Nakane [39] has constructed the entropy solution for (C) by selecting the proper discontinuous branch of the geometric solution past the first critical time. Here, we only refer to the articles of Guckenheimer [22] and Wagner [46]. A different geometric framework for the study of shock waves for single conservation laws and systems of conservation laws using the concept of generalized characteristics has been given by Dafermos [15].

We also state some problems concerning perestroikas of the shocks of the entropy solution of (C).

Problem (5.a) What is the notion of "convexity" of $(f_1(y), \ldots, f_n(y))$ for $n \geq 2$? One candidate is that each $f_i(y)$ is uniformly convex for $i = 1, 2, \ldots, n$. Under this condition, can we have a (useful) explicit formula for the entropy solution of (C) such as the Hopf-Lax type formula for the viscosity solution?

Problem (5.b) Solve the Riemann problem for each normal form of the multi-unfoldings of immersions given in [30].

Problem (5.c) Consider the system case. Geometric singularities of multi-valued solutions for first order systems have been studied by Rakhimov [40] and Caflisch et al [8], however, they have not classified the perestroikas of singularities for the Cauchy problem.

(1) Classify the perestroikas of geometric singularities.

(2) How can we solve the Riemann problem for each normal form?

(3) (Important!) What is the correct notion of weak solution?

References

[1] V. I. Arnol'd, *Geometric Methods in the Theory of Ordinary Differential Equations,* Springer-Verlag, 1983.

[2] V. I. Arnol'd, S. M. Gusein-Zade and A. N. Varchenko, *Singularities of Differentiable Maps,* Birkhauser, (1986).

[3] D. P. Ballou, Solutions to nonlinear hyperbolic Cauchy problems without convexity conditions, *Trans. Amer.Math. Soc.*, **152** (1970), 441–460.

[4] M. Bardi and L. C. Evans, On Hopf's formulas for solutions of hamilton-Jacobi equations, *Nonlinear Analysis,* **8** (1984), 1373–1389.

[5] P. Bernhard, *Singular surfaces in differential games, an introduction; in Differential Games and Applications, Lecture Notes in Control and Information Sciences ed. by P. Hagedorn et al.* **3** Springer Verlag (1977), 1–33.

[6] I. A. Bogaevskii, Modifications of singularities of minimum functions and bifurcations of shock waves at the Burgers equation with vanishing viscosity, *Leningrad Math. J.*, **1** (1990), 807-823.

[7] I. A. Bogaevskii, Perestroikas of fronts in evolutionary families, *Proceedings of the Steklov Institute of Math.*, **209** (1995), 57–72.

[8] R. E. Caflisch, N. Ercolani, T. Y. Hou and Y. Landis, Multi-valued solutions and branch point singularities for nonlinear hyperbolic systems, *Comm. Pure Appl. Math.*, **46** (1993), 453-499.

[9] M. Chaperon, Lois de conservation er géométrie symplectique, *C. R. Aca. Sc. Paris*, **312** (1991), 345–348.

[10] N. M. Chen, On types of singularities for solutions of nonlinear hyperbolic systems, *Bull. Inst. Math. Acad. Sinica,* **10** (1982), 405-416.

[11] M. G. Crandall and P.-L. Lions, Viscosity solutions of Hamilton-Jacobi equations, *Trans. Amer. Math. Soc.*, **277** (1983), 1–42.

[12] M. G. Crandall, L.C. Evans and P.-L. Lions, Some properties of viscosity solutions of Hamilton-Jacobi equations, *Trans.Amer. Math. Soc.*, **282** (1984), 487–502.

[13] M. G. Crandall, H. Ishii and P.-L. Lions, User's guide to viscosity solutions of second order partial differential equations, *Bull. Amer. Math. Soc.*, **27** (1992), 1–67.

[14] C. M. Dafermos, Regularity and large time behavior of solutions of a conservation law without convexity, *Proc. Roy. Soc. Edinburgh*, **99A** (1985), 201–239.

[15] C. M. Dafermos, Generalized characteristics in hyperbolic systems of conservation laws, *Arch. Rational Mech. Anal.*, **107** (1989), 127–155.

[16] L. C. Evans and P. E. Souganides, Differential games and representation formulas for solutions of Hamilton-Jacobi-Isaacs equations, *Indiana Univ. Math. J.*, **33** (1984), 773–797.

[17] W. H. Fleming and H. M. Soner, *Controlled Markov Processes and Viscosity Solutions,* Springer-Verlag (1993).

[18] V. A. Florin, Some simplest nonlinear problems of the consolidation of an aqueously saturated earthen medium, *Izv. Akad. Nauk SSSR Otdel. Tekhn. Nauk*, **9** (1948), 1389–1397.

[19] C. G. Gibson, *Singular points of Smooth Mappings,* Pitman, London 1979.

[20] J. Glim, D.Marchesin and O. Mcbryan, Unstable finger in two phase flows, *Comm. Pure Appl. Math.*, **34** (1981), 53–75.

[21] J. Guckenheimer, *Solving a single conservation law,* Lecture notes in Mathematics **468** Springer Verlag, New York, 1975, 108–134.

[22] J. Guckenheimer, Shocks and rarefaction in two space dimensions, *Arch. Rational Mech. Anal.*, **59** (1975), 281–291.

[23] E. Hopf, Generalized solution of non-linear equations of first order, *Jour. of Math. and Mech.*, **14** (1965), 951–973.

[24] R. Isaacs, *Differential Games,* John Wiley, New York (1965).

[25] S. Izumiya, The theory of Legendrian unfoldings and first order differential equations, *Proc. Royal Soc. Edinburgh*, **123A** (1993), 517–532.

[26] S. Izumiya, Perestroikas of optical wave fronts and graphlike Legendrian unfoldings, *J. of Differential Geometry*, **38** (1993), 485–500.

[27] S. Izumiya, Geometric singularities for Hamilton-Jacobi equation, *Advanced Studies in Pure Math.*, **22** (1993), 89–100.

[28] S. Izumiya and G. T. Kossioris, Semi-local classification of geometric singularities for Hamilton-Jacobi equations, *J. of Diff. Equations*, **118** (1995), 166–193.

[29] S. Izumiya and G. T. Kossioris, Realization theorem of geometric singularities for Hamilton-Jacobi equations, to appear in *Commun. Anal. and Geom.*.

[30] S. Izumiya and G. T. Kossioris, Geometric singularities for solutions of single conservation laws, to appear in *Arch. Rational Mech. Anal.*.

[31] S. Izumiya and G. T. Kossioris, Bifurcations of shock waves for viscosity solutions of Hamilton-Jacobi equations of one space variables, to appear in *Bull. Sciences Math.*.

[32] G. Jennings, Piecewise smooth solutions of a single conservation law exist, *Adv. in Math.*, **33** (1979), 192–205.

[33] T. Joukovskaia, *Singularités de minimax et solutions faibles d'équations aux dérivées partielles,* These, Université Denis Dierot (Paris 7), (1994).

[34] G. T. Kossioris, Propagation of singularities for viscosity solutions of Hamilton-Jacobi equations in one space variable, *Comm. P.D.E.*, **18** (1993), 747–770.

[35] G. T. Kossioris, Formation of singularities for viscosity solutions of Hamilton-Jacobi equations in higher dimensions, *Comm. P.D.E.*, **18** (1993), 1085–1108.

[36] S. N. Kruzkov, First order quasilinear equations in several independent variables, *Math. USSR Sb.*, **10** (1970), 217–243.

[37] P.D. Lax, Hyperbolic systems of conservation laws II, *Comm. Pure Appl.*, **10** (1957), 537–566.

[38] S. Nakane, Formation of shocks for a single conservation law, *SIAM J. Math. Anal.*, **19** (1988), 1391–1408.

[39] S. Nakane, Formation of singularities for Hamilton-Jacobi equations in several space variables, *J. Math. Soc. Japan*, **43** (1991), 89–100.

[40] A. K. Rakimov, Singularities of Riemannian Invariants, *Funct. Anal. Appl.*, **27** (1993), 39–50.

[41] H. Rund, *The Hamilton-Jacobi theory in the calculus of variations,* D. Van Nostrand, London (1966).

[42] J. Smoller, *Shock waves and Reaction-Diffusion Equations,* Grund. der Math. Wiss., **258** (1980), Springer.

[43] M. Tsuji, Solution globale et propagation des singularites pour l'equation de Hamilton-Jacobi, *C. R. Acad. Sc. Paris*, **289** (1979), 397–400.

[44] M. Tsuji, Formation of singularities for Hamilton-Jacobi equation II, *J. Math. Kyoto Univ.*, **26** (1986) 299–308.

[45] C. Viterbo, Generating functions, symplectic geometry and applications, *Proceedings of the Intern. Cong. Math.*, (1994) 537–547.

[46] D. Wagner, The Riemann problem in two space dimensions for a single conservation law, *SIAM J. Math. Ann.*, **14** (1983), 534–559.

[47] V. M. Zakalyukin, Reconstructions of fronts and caustics depending on a parameter and versality of mappings, *J. of Soviet Math.*, **27** (1984), 2713–2735.

Shyuichi IZUMIYA,
Department of Mathematics,
Hokkaido University,
Sapporo 060,
JAPAN

e-mail: izumiya@math.hokudai.ac.jp

Printed in the United States
By Bookmasters